Fundamentals
of
Polymerization

FUNDAMENTALS OF POLYMERIZATION

Broja M Mandal
Indian Association for the Cultivation of Science, India

NEW JERSEY · LONDON · SINGAPORE · BEIJING · SHANGHAI · HONG KONG · TAIPEI · CHENNAI

Published by

World Scientific Publishing Co. Pte. Ltd.
5 Toh Tuck Link, Singapore 596224
USA office: 27 Warren Street, Suite 401-402, Hackensack, NJ 07601
UK office: 57 Shelton Street, Covent Garden, London WC2H 9HE

British Library Cataloguing-in-Publication Data
A catalogue record for this book is available from the British Library.

FUNDAMENTALS OF POLYMERIZATION

Copyright © 2013 by World Scientific Publishing Co. Pte. Ltd.

All rights reserved. This book, or parts thereof, may not be reproduced in any form or by any means, electronic or mechanical, including photocopying, recording or any information storage and retrieval system now known or to be invented, without written permission from the Publisher.

For photocopying of material in this volume, please pay a copying fee through the Copyright Clearance Center, Inc., 222 Rosewood Drive, Danvers, MA 01923, USA. In this case permission to photocopy is not required from the publisher.

ISBN 978-981-4322-46-1

Typeset by Stallion Press
Email: enquiries@stallionpress.com

Printed in Singapore by B & Jo Enterprise Pte Ltd

To
the memory of
my parents
and
Prof. Santi Ranjan Palit

Preface

This book deals with the basic aspects of polymerization. It has nine chapters. An introduction to polymers and polymerizations has been given in Chapter 1. Then the various kinds of polymerization have been treated in Chapters 2 through 9.

The step polymerization and the various chain polymerizations have been treated each in a separate chapter. Living polymerization constitutes a large part of chain polymerization. Accordingly, it has been given the due space and treated in some depth and extent in the various chapters on chain polymerization. Heterophase polymerization, which offers many process advantages and a wide range of sizes of dispersed polymer particles, has also been dealt with in a separate chapter. It has been endeavored to make a straightforward presentation of the subjects and strike the right balance between the size of the book and the comprehensive treatment of the subjects.

Each chapter ends with a set of problems. The author may be contacted for a soft copy of the solutions to the problems. The answers to the numerical problems have been given at the end of the problems.

The book is intended to serve as a text for the advanced undergraduate and the graduate students. Professionals may use it to refresh their knowledge and as a source of information. Those who would like to be self-taught may also find it a suitable guide. What is required is only a good knowledge of the standard undergraduate courses in physical and organic chemistry.

The author gratefully acknowledges the support he received from the authorities of the Indian Association for the Cultivation of Science for writing the book while he was an honorary member of the teaching faculty from 2007–2010 following his retirement in mid 2006 from the post of Professor of Polymer Chemistry. He is indebted to many authors of books, monographs, reviews, articles, *etc.*, from which he received much information and materials for the book. He is also grateful to Dr. D. Baskaran, Dr. A. K. Chaudhuri, and Prof. A. K. Nandi for helping him in various ways, to Prof. D. C. Mukherjee for valuable advice, to many friends and students, notably, Prof. N. K. Das, Prof. Nira Misra, Dr. Suparna Guha, Dr. B. Ray, Dr. D. Chatterjee, Dr. S. Jewrajka, and Dr. Uma Chatterjee, who suggested him to write the book, and, above all, to his wife and family for understanding and encouragement all the way. He also thanks Mr. A. Chakraborty and Mr. G. Manna for assistance with typing and artwork respectively.

Broja M. Mandal
Kolkata
March 2012
psubm@iacs.res.in

Contents

Preface vii

List of Abbreviations xix

Chapter 1. Introduction 1
    1.1 Nomenclature 2
    1.2 Structural and Repeating (or Repeat) Units 3
    1.3 Classification 3
        1.3.1 Condensation and addition polymers and polymerizations . 3
        1.3.2 Step-growth and chain-growth polymers and polymerizations 7
    1.4 Functionality 9
    1.5 Designed Branched Polymers 12
        1.5.1 Star polymers 12
        1.5.2 Hyperbranched polymers 13
        1.5.3 Dendrimers 14
    1.6 Physical State 14
    1.7 Structure-Property Relationship 17
        1.7.1 Nonconjugated polymers 17
            1.7.1.1 Stereoregularity in vinyl polymers 23
            1.7.1.2 Regioregularity in vinyl polymers 24
        1.7.2 Conjugated polymers 25
    1.8 Thermodynamics of Polymerization 28
    1.9 Polymerizability of Internal Olefins 30
    1.10 Molecular Weights and Molecular Weight Distributions 31
    References 33
    Problems 35

Chapter 2. Step Polymerization 37
    2.1 Principle of Equal Reactivity of Functional Groups and Kinetics of Polymerization 37
    2.2 Ring vs. Chain Formation 41
    2.3 Intermolecular Interchange Reactions 46

2.4	Degree of Polymerization			47
	2.4.1	Carothers equation		47
		2.4.1.1	Bifunctional polycondensation	48
		2.4.1.2	Control of degree of polymerization	49
			2.4.1.2.1 Polymerization of nonequivalent monomer mixtures	49
			2.4.1.2.2 Departure from Carothers equation in interfacial polycondensation	51
2.5	Molecular Weight Distribution			52
	2.5.1	Molecular weight distribution in bifunctional polycondensation		52
	2.5.2	Average degree of polymerization for most probable distribution		55
2.6	Prediction of Gel Point in Polyfunctional Polycondensation			56
	2.6.1	Prediction based on Carothers equation		56
	2.6.2	Prediction based on statistical method of Flory		59
		2.6.2.1	Gel-forming range of monomer proportions	61
			2.6.2.1.1 Polycondensation of A_f and B–B monomers	62
			2.6.2.1.2 Polymerization of a ternary mixture of A–A, B–B, and A_3 monomers	62
2.7	Thermosetting Resins			62
	2.7.1	Phenolic resins		62
	2.7.2	Amino resins		65
	2.7.3	Unsaturated polyesters and alkyd resins		66
	2.7.4	Polyurethanes and polyureas		67
	2.7.5	Epoxy resins		69
	2.7.6	Polysiloxanes		70
	2.7.7	Aliphatic polysulfides		72
2.8	Engineering Plastics			72
	2.8.1	Polyamides		74
		2.8.1.1	Aliphatic polyamides	74
		2.8.1.2	Aromatic polyamides	75
	2.8.2	Polyesters		76
		2.8.2.1	Aliphatic-aromatic polyesters	77
		2.8.2.2	Aromatic polyesters	78
	2.8.3	Polycarbonates		79
	2.8.4	Poly(phenylene oxide)s		80
2.9	High Performance Polymers			81
	2.9.1	Liquid crystalline polymers (LCPs)		81
	2.9.2	Poly(p-phenylene sulfide)		83
	2.9.3	Poly(arylene ether)s		83

	2.9.4	Polyimides	84
		2.9.4.1 Thermoplastic polyimides	85
		2.9.4.2 Thermosetting polyimides	86
	2.9.5	Polybenzimidazoles	87
	2.9.6	Polybenzoxazoles and polybenzothiazoles	88
2.10	Nonconventional Step Polymerization	88
References	90
Problems	95

Chapter 3. Radical Polymerization — 97

3.1	General Features	97
	3.1.1 Monomer types	97
	3.1.2 Reactivity of monomers and radicals	98
3.2	Kinetics of Homogeneous Radical Polymerization	99
3.3	Reaction Orders in Initiator and Monomer	104
3.4	Initiators	105
	3.4.1 Thermal initiators	106
	3.4.2 Photoinitiators	107
	3.4.3 Redox initiators	108
3.5	Determination of Polymer End Groups	110
3.6	Initiator Efficiency	111
3.7	Thermal Polymerization and its Kinetics	112
3.8	Kinetic Chain Length, Degree of Polymerization, and Chain Transfer	113
	3.8.1 The general DP equation	114
	3.8.1.1 Solvent transfer	115
	3.8.1.2 Monomer transfer	116
	3.8.1.3 Initiator transfer	118
	3.8.1.4 Regulator transfer	120
	3.8.1.5 Addition-fragmentation chain transfer (AFCT)	120
	3.8.1.6 Polymer transfer	121
	3.8.1.7 Redox transfer	122
	3.8.1.8 Catalytic chain transfer (CCT)	123
	3.8.1.9 Instantaneous chain length distribution (CLD) and chain transfer	123
	3.8.2 Factors influencing chain transfer	124
	3.8.2.1 Quantifying the resonance and polar effects	126
	3.8.2.1.1 The Q, e Scheme	126
	3.8.2.1.2 The radical reactivity pattern scheme	127
	3.8.3 Telomerization	128
3.9	Inhibition and Retardation of Polymerization	129
	3.9.1 Oxidative and reductive termination	131

	3.9.2	Oxygen as an inhibitor	133
3.10		Rate Constants of Propagation and Termination	133
	3.10.1	The PLP-SEC method	136
	3.10.2	The single pulse-pulsed laser polymerization (SP-PLP) method for determining k_t	138
	3.10.3	Propagation rate constants	139
	3.10.4	Termination rate constants	140
	3.10.5	Variation of k_t with monomer conversion	141
	3.10.6	Chain length dependence of k_p and k_t	142
	3.10.7	Temperature dependence of rate of polymerization and molecular weight	142
3.11		The Course of Polymerization and Gel Effect	143
3.12		Popcorn Polymerization	144
3.13		Dead End Polymerization	144
3.14		Molecular Weight Distribution	145
3.15		Living Radical Polymerization (LRP)	148
	3.15.1	Persistent radical effect in reversible deactivation	151
	3.15.2	Kinetic equations for LRP with reversible deactivation	153
	3.15.2.1	Gel effect in living radical polymerization	154
	3.15.3	Molecular weight	155
	3.15.3.1	Molecular weight distribution and polydispersity index	156
	3.15.4	Nitroxide mediated polymerization	158
	3.15.4.1	Selection of initiator	161
	3.15.4.2	Polymerization of styrene	162
	3.15.4.3	Polymerization of (meth)acrylates and other monomers	164
	3.15.4.4	Rate constants and equilibrium constants	165
	3.15.4.5	Synthesis of block copolymers	166
	3.15.5	Metalloradical mediated polymerization	167
	3.15.6	Atom transfer radical polymerization	167
	3.15.6.1	Initiators	168
	3.15.6.1.1	Halide exchange effect	171
	3.15.6.2	Ligands	172
	3.15.6.3	Solvents	174
	3.15.6.4	Kinetics of ATRP	175
	3.15.6.5	Molecular weight and polydispersity	178
	3.15.6.6	Effect of temperature	178
	3.15.6.7	ATRP with activator generated in situ	179
	3.15.6.7.1	Reverse ATRP	179
	3.15.6.7.2	SR and NI ATRP	180
	3.15.6.7.3	AGET ATRP	180

			3.15.6.7.4	ARGET ATRP	181
			3.15.6.7.5	ICAR ATRP	182
		3.15.6.8	Synthesis of high molecular weight polymers		182
		3.15.6.9	Synthesis of block copolymers		184
	3.15.7	Reversible addition–fragmentation chain transfer (RAFT) polymerization			186
		3.15.7.1	Tuning the RAFT agent		188
			3.15.7.1.1	Selection of Z group	188
			3.15.7.1.2	Selection of R group	190
		3.15.7.2	Kinetics of RAFT polymerization		192
			3.15.7.2.1	Retardation	193
			3.15.7.2.2	Chain transfer constant	195
			3.15.7.2.3	Rate constant for exchange and activation	197
		3.15.7.3	Synthesis of block copolymers		198
		3.15.7.4	Star polymers		198
References					199
Problems					211

Chapter 4.	Anionic Polymerization				213
4.1	Living Anionic Polymerization				214
	4.1.1	General features			214
		4.1.1.1	Molecular weight and molecular weight distribution		216
		4.1.1.2	Long term stability of living polymers		218
		4.1.1.3	Initiators		219
			4.1.1.3.1	Initiation by nucleophile addition	219
			4.1.1.3.2	Initiation by electron transfer from alkali or alkaline earth metal	219
			4.1.1.3.3	Initiation by electron transfer from aromatic radical anions	220
			4.1.1.3.4	Electron transfer initiation by electrochemical means	222
	4.1.2	Living anionic polymerization of nonpolar monomers in hydrocarbon solvents			222
		4.1.2.1	Kinetics		223
		4.1.2.2	Effect of additives		225
		4.1.2.3	Living anionic polymerization at elevated temperatures — retarded anionic polymerization		226
		4.1.2.4	Stereospecificity		226
	4.1.3	Living Anionic Polymerization of Styrene in Ethereal Solvents			227

		4.1.3.1	Kinetics	227
		4.1.3.2	Temperature dependence of $k_p^{\pm}$	233
		4.1.3.3	Ion association beyond ion-pair	235
	4.1.4	Living anionic polymerization of polar monomers		236
		4.1.4.1	Side reactions	236
		4.1.4.2	Living anionic polymerization of methyl methacrylate in polar solvents in the absence of additives	237
			4.1.4.2.1 Nature of ion pairs and their aggregates	238
			4.1.4.2.2 Rate constants of propagation	239
		4.1.4.3	Ligated living anionic polymerization	240
	4.1.5	End functionalized polymers		242
	4.1.6	Synthesis of block copolymers		243
	4.1.7	Synthesis of star polymers		244
	4.1.8	Synthesis of comb polymers		246
	4.1.9	Group transfer polymerization		247
	4.1.10	Metal–free living anionic polymerization		250
References				251
Problems				257

Chapter 5.	Coordination Polymerization			259
5.1	Ziegler–Natta Catalysts			259
	5.1.1	Isotactic polypropylene		260
	5.1.2	Syndiotactic polypropylene		263
	5.1.3	Polyethylene and ethylene copolymers		264
	5.1.4	Polymerization of conjugated dienes		264
	5.1.5	Degree of stereoregularity		265
	5.1.6	Mechanism		266
		5.1.6.1	Isotactic propagation	267
		5.1.6.2	Syndiotactic propagation	271
	5.1.7	Polymerization of acetylene		272
	5.1.8	Chain transfer and regulation of molecular weight		272
	5.1.9	Branching in polyethylene		273
5.2	Metallocene Catalysts			274
	5.2.1	Mechanism of activation		275
5.3	Late Transition Metal Catalysts			276
5.4	Living Polymerization of Alkenes			279
References				279
Problems				283

Chapter 6.	Cationic Polymerization	285
6.1	The Nucleophilicity and Electrophilicity Scales	285

6.2	Bronsted Acids as Initiators		286
6.3	Lewis Acids as Coinitiators		289
	6.3.1	Controlled initiation	290
	6.3.2	Controlled termination	291
	6.3.3	Chain transfer	291
	6.3.4	Molecular weight dependence on temperature of polymerization	296
	6.3.5	Reversible termination	296
	6.3.6	Living carbocationic polymerization	298
		6.3.6.1 Selection of initiator	299
		6.3.6.2 Effect of added nucleophiles, proton traps, and common anion salts	301
		6.3.6.3 Test of livingness	303
		6.3.6.4 Kinetics	305
		6.3.6.5 Molecular weight and molecular weight distribution	306
		6.3.6.6 Block copolymerization	308
6.4	End Functionalized Polymers		308
	6.4.1	Living polymerization method	308
	6.4.2	Inifer method	309
6.5	Photoinitiated Cationic Polymerization		310
6.6	Propagation Rate Constants		311
	6.6.1	The diffusion clock method	311
		6.6.1.1 Competition experiment (method 1)	311
		6.6.1.2 Competition experiment (method 2)	313
	6.6.2	Values of rate constants of propagation	314
References			314
Problems			321

Chapter 7.	Ring-Opening Polymerization and Ring-Opening Metathesis Polymerization		323
7.1	General Features		324
	7.1.1	Polymerizability	324
	7.1.2	Ring-chain equilibrium	326
	7.1.3	The nature and reactivity of propagating species	327
	7.1.4	Backbiting and intermolecular interchange	328
	7.1.5	Bridged cyclic monomers	330
7.2	Cyclic Ethers		331
	7.2.1	Anionic polymerization	331
		7.2.1.1 Kinetics	333
	7.2.2	Cationic polymerization	334
		7.2.2.1 Kinetics	336

			7.2.2.1.1	Macroion-macroester interconversion	339

- 7.2.2.2 Backbiting and intermolecular interchange ... 339
- 7.2.3 Activated monomer mechanism of polymerization 340
- 7.3 Cyclic Acetals 342
- 7.4 Cyclic Esters 343
  - 7.4.1 Anionic polymerization 344
  - 7.4.2 Coordination polymerization 345
    - 7.4.2.1 Backbiting and intermolecular interchange ... 348
  - 7.4.3 Cationic polymerization 349
  - 7.4.4 Enzymatic polymerization 349
- 7.5 Lactams 350
  - 7.5.1 Anionic polymerization 350
  - 7.5.2 Cationic polymerization 351
  - 7.5.3 Hydrolytic polymerization 352
- 7.6 N-Carboxy-α-aminoacid Anhydrides 353
- 7.7 Oxazolines (Cyclic Imino Ethers) 355
- 7.8 Cyclic Amines 355
- 7.9 Cyclic Sulfides 356
- 7.10 Cyclosiloxanes 357
  - 7.10.1 Anionic polymerization 357
  - 7.10.2 Cationic polymerization 358
- 7.11 Cyclotriphosphazenes 359
- 7.12 Cyclic Olefins 360
  - 7.12.1 Initiators 361
  - 7.12.2 Livimg character 362
  - 7.12.3 Initiators with high functional group tolerance 363
  - 7.12.4 Synthesis of polyacetylene 363
  - 7.12.5 Backbiting and intermolecular interchange 364
  - 7.12.6 Alicyclic diene metathesis (ADMET) polymerization 364
- References 365
- Problems 371

Chapter 8. Chain Copolymerization 373

- 8.1 Terminal Model of Copolymerization 374
  - 8.1.1 Copolymer composition equation 374
    - 8.1.1.1 Types of Copolymerization 377
    - 8.1.1.2 Determination of monomer reactivity ratios ... 381
    - 8.1.1.3 Substituent effect on monomer reactivity 382
    - 8.1.1.4 Substituent effect on radical reactivity 383
    - 8.1.1.5 Resonance and polar effects in radical addition . 384
      - 8.1.1.5.1 Q, e Scheme 384
      - 8.1.1.5.2 The revised pattern scheme 386

		8.1.2	Statistical treatment of copolymerization	386
			8.1.2.1 Monomer sequence distribution	388
			8.1.2.1.1 Triad fractions and monomer reactivity ratios	389
		8.1.3	Kinetics of radical copolymerization based on terminal model	392
		8.1.4	Ionic copolymerization	393
		8.1.5	Validity of the terminal model	395
	8.2	Penultimate Model of Copolymerization		395
		8.2.1	Mean rate constant of propagation	397
		8.2.2	The copolymer composition equation	399
	8.3	Living Radical Copolymerization		400
	References			402
	Problems			405
Chapter 9.	Heterophase Polymerization			407
	9.1	Particle Stabilization Mechanisms		407
	9.2	Suspension Polymerization		409
	9.3	Emulsion Polymerization		410
		9.3.1	The three periods of polymerization	413
		9.3.2	Harkins qualitative theory	413
		9.3.3	Quantitative theory of Smith and Ewart	414
		9.3.4	Applicability of Smith-Ewart theory with oil soluble initiators	416
		9.3.5	Number density of particles	417
		9.3.6	Particle nucleation in the aqueous phase	417
	9.4	Inverse Emulsion Polymerization		419
	9.5	Miniemulsion Polymerization		419
	9.6	Microemulsion Polymerization		420
	9.7	Dispersion Polymerization		421
	9.8	Heterophase Living Radical Polymerization		422
	References			423
	Problems			427
Index				429

List of Abbreviations

AA = acrylamide
ACE = activated chain end
ADMET = alicyclic diene metathesis
AFCT = addition-fragmentation chain transfer
AGET = activator generated by electron transfer
AIBN = 2, 2'- azobisisobutyronitrile
AM = activated monomer
AN = acrylonitrile
ARGET = activator regenerated by electron transfer
ATRA = atom transfer radical addition
ATRP = atom transfer radical polymerization
BA = n-butyl acrylate
tBA = t-butyl acrylate
BD = butadiene
BDE = bond dissociation energy
tBHP = t-butyl hydroperoxide
bpy = 2,2'-bipyridine
Bz_2O_2 = benzoyl peroxide
CCT = catalytic chain transfer
CHP = cumene hydroperoxide
CLD = chain length distribution
CMC = critical micelle concentration
CMP = cobalt mediated polymerization
CRP = conventional radical polymerization
CT = chain transfer
CumCl = cumyl chloride
CumOAc = cumyl acetate
CumOMe = cumyl methyl ether
D_n = cyclodimethylsiloxane containing n numbers of –Si(Me)$_2$O- units (n ≥ 3)
DBN = di-t-butyl nitroxide
DtBP = 2,6-di-t-butylpyridine

DCP = dicumyl peroxide
DiCumCl = dicumyl chloride
DiCumOAc = dicumyl acetate
DiCumOMe = dicumyl methyl ether
DEPN = N-*tert*-butyl-N-[1-diethylphosphono(2,2-dimethylpropyl)]-N-oxy
dHbpy = 4,4'-di-n-hexyl-2,2'-bipyridine
dNbpy = 4,4'-di-(5-nonyl)-2,2'-bipyridine
DP = degree of polymerization
DPPH = 2,2-diphenyl-1-picrylhydrazyl
DT = degenerative transfer
EBriB = ethyl 2-bromoisobutyrate
EO = ethylene oxide
FRP = free radical polymerization
GPC = gel permeation chromatography
GTP = group transfer polymerization
HDPE = high density polyethylene
HLB = hydrophilic lipophilic balance
HMTETA = 1,1,4,7,10,10-hexamethyltriethylenetetramine
IB = isobutylene
ICAR = initiator for continuous activator generation
Inifer = initiator-transfer
Iniferter = initiator-transfer-terminator
i.p. = induction period
LAP = living anionic polymerization
LCP = liquid crystalline polymer
LDPE = low density polyethylene
LLDPE = linear low density polyethylene
LP = living polymerization
MA = methyl acrylate
MADIX = macromolecular design by interchange of xanthates
MAN = methacrylonitrile
Me$_6$TREN = tris[(2-dimethylamino)ethyl]amine
MMA = methyl methacrylate
MWD = molecular weight distribution
NMP = nitroxide mediated polymerization
NPPMI = N-(n-propyl)pyridylmethanimine
NOPMI = N-(n-octyl)pyridylmethanimine
o-phen = 1,10-phenanthroline
PA = polyamide
PAI = polyamideimide

List of Abbreviations

PAN = polyacrylonitrile
PBA = poly(n-butyl acrylate)
P*t*BA = poly(*t*-butyl acrylate)
PBD = polybutadiene
PC = polycarbonate
PDI = polydispersity index
PDMS = poly(dimethyl siloxane)
PE = polyethylene
PEEK = poly(ether ether ketone)
PEEKEK = poly(ether ether ketone ether ketone)
PEKEKK = poly(ether ketone ether ketone ketone)
PEK = poly (ether ketone)
PEI = polyetherimide
PEO = poly(ethylene oxide)
PES = poly(ether sulfone)
PET = poly(ethylene terephthalate)
PF = phenol formaldehyde
PI = polyimide
PIB = polyisobutylene
PMA = poly(methyl acrylate)
PMDETA = N,N′,N″,N‴- pentamethyldiethylenetriamine
PMMA = poly(methyl methacrylate)
PP = polypropylene
PPO = poly(*p*-phenylene oxide)
PPS = poly(*p*-phenylene sulphide)
PSt or PS = polystyrene
PTHF = polytetrahydrofuran
PVA = poly(vinyl acetate)
PVC = poly(vinyl chloride)
PVME = poly(vinyl methyl ether)
PVP = poly(N-vinyl pyrrolidone)
PU = polyurethane
RAFT = reversible addition-fragmentation chain transfer
ROP = ring-opening polymerization
ROMP = ring-opening metathesis polymerization
RP = radical polymerization
SBR = styrene butadiene rubber
SET = single electron transfer
SR & NI = simultaneous reverse and normal initiation
S-TEMPO = (1-phenylethyl)-TEMPO

TEMPO = 2,2,6,6-tetramethyl-1-piperidinyl-1-oxy
TIPNO = 2,2,5-trimethyl-4-phenyl-3-azahexane-3-oxy
TMPCl = 2-chloro-2,4,4-trimethylpentane
TMPOAc = 2,4,4-trimethyl-2-pentyl acetate
TPEN = N,N,N′,N′-tetrakis(2-pyridylmethyl)ethylenediamine
TPMA = tris[(2-pyridyl)methyl]amine
VBC = *p*-vinylbenzyl chloride

Chapter 1

Introduction

Prior to the early 1930s, the researchers were divided for a long period on their view of the nature of the molecules constituting naturally occurring substances, such as cellulose, starch, proteins, and natural rubber, which exhibit unique physical and mechanical properties.[1,2] These substances or their derivatives form colloidal solutions when dissolved in solvents. For example, in solutions they diffuse very slowly and do not pass through semipermeable membranes. There were two schools of thought. One school held the view that the substances are constituted of colloidal molecules, which are aggregates of ordinary molecules held together by secondary valence forces.[1,3–6] This primary small molecular constitution found support for some time even in X-ray crystallographic study, which incorrectly interpreted the unit cells of cellulose and of crystallized rubber to be representing the whole molecules. It was held that since the unit cells are similar in size with those of simple compounds the molecules must be small.[7]

However, negligible freezing point depressions of the colloidal solutions or some tens of thousands molecular weights determined in some cases, by cryoscopy or osmometry, strongly suggested the alternative view that the molecules are intrinsically big (macromolecules) being comprised of atoms bonded together by normal covalent bonds. Nevertheless, the proponents of the colloidal molecular hypothesis brushed such evidences aside on the unsubstantiated ground that the solution laws underlying the methods would be inapplicable to colloidal solutions.[1]

Against this backdrop, Staudinger relentlessly championed the macromolecular hypothesis presenting evidences in support. For example, he demonstrated that the soluble macromolecular substances give colloidal solutions, even when variously derivatized, in whichever solvents they are dissolved. This is very unlike the colloidal associates of simple molecules such as the soap micelles, which disintegrate to yield molecular solutions on making appropriate changes in solvents. In addition, the derivatives have the degrees of polymerization about the same as the original polymer had. He also made extensive investigations with synthetic polymers, polystyrene, and polyoxymethylene, in particular, and suggested chain structures for them as early as 1920. He was also the first to relate intrinsic viscosity of polymer with molecular weight.[2] That the relation was imperfect does not diminish its importance as a step in the right direction.[1]

X-ray crystallographic evidences in favor of the chain molecular structure also came along. Thus, Sponsler and Dore interpreted the crystallographic data on cellulose fibers in terms of unit cell, which represents a structural unit of the polymer chain rather than the whole molecule.[8] However, this interpretation based on rather limited data was not fully correct in being incompatible with the chemical evidence of cellobiose as the repeating unit.[10] The correction was subsequently done by Meyer and Mark who successfully applied the crystallographic method to establish the chain structures also of silk fibroin, chitin, and natural rubber.[9]

It must not be understood that these are the only work, which won the case for the macromolecular hypothesis. There were several other work, which also contributed.[1,10] The foremost of these are the work of Carothers, which we shall be acquainted with later in this and next chapter.[11] For the present, we shall be content with learning that Carothers was the first to formulate the principles for synthesizing condensation polymers and predicting their molecular weights. The successful application of these principles left no room for doubting the chain molecular nature of polymers.

However, while the term 'macromolecule' points to the big size of the molecule, the term 'polymer' not only connotes macromolecule (although Berzelius first used the term naming butene a polymer of ethene[10]), but also reflects the manner how it is built up. Derived from Greek words 'poly' meaning 'many' and 'mer' meaning 'part,' polymer (= many parts) means a molecule built up of many units (parts) joined. The small molecules, which are the precursors of the units, are called 'monomers' – 'mono' meaning 'one'. Sometimes the terms 'high polymer' and 'low polymer' are used to emphasize the big difference in molecular weights between them.

Before 1930, high polymeric materials used commercially were the naturally-occurring types or their derivatives, *e.g.*, cellulose, silk, wool, rubber, regenerated cellulose, cellulose acetate, nitrocellulose, *etc.*, the only exception being Bakelite, a phenolic resin, patented by Baekeland in 1909. However, no sooner had the macromolecular hypothesis gained ground than a rapid growth of synthetic polymer industry ensued. Thus, within ten years beginning 1930 many of the commodity polymers came into industrial production: poly(vinyl chloride) (1930), poly(methyl methacrylate) (1935), polystyrene (1935), low density polyethylene (1937), nylon 66 (1938), and nylon 6 (1940).

1.1 Nomenclature

The most widely used nomenclature system for polymers is the one based on the source of the polymer. When the polymer is made from a single worded monomer, it is named by putting the prefix "poly" before the name of the monomer without a space or a hyphen. Thus, the polymers from ethylene, styrene, and propylene are named polyethylene, polystyrene, and polypropylene respectively. However, when the name of the monomer is multiworded or abnormally long or preceded by a letter or a number or carries substituents it is placed under parenthesis, *e.g.*, poly(vinyl chloride), poly(ethylene oxide), poly(chlorotrifluoroethylene), poly(ϵ-caprolactam), *etc.* For some polymers, the source (monomer) may be a hypothetical one, the polymer being prepared by a modification of another polymer, *e.g.*, poly(vinyl alcohol). The vinyl alcohol monomer does not exist; the polymer is prepared by the hydrolysis of poly(vinyl acetate).

In the above examples, a single monomer forms the repeating unit of the polymer (*vide infra*). However, many polymers exist in which the repeating units are derived from two monomers. Such polymers are named by putting the prefix "poly" before the structural name of the repeating unit placed within parenthesis without a space or a hyphen. Thus, the polyamide with the repeating unit derived from hexamethylenediamine and adipic acid is named poly(hexamethylene adipamide),

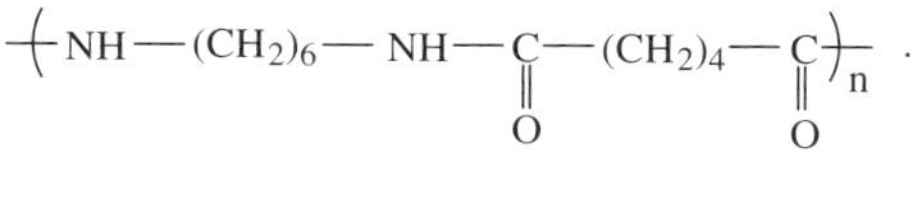

$$\left[NH-(CH_2)_6-NH-\underset{O}{\underset{\|}{C}}-(CH_2)_4-\underset{O}{\underset{\|}{C}} \right]_n .$$

Similarly, the polyester with the repeating unit derived from ethylene glycol and terephthalic acid is named poly(ethylene terephthalate).

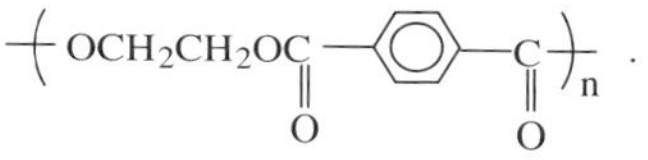

$$\left[OCH_2CH_2O\underset{O}{\underset{\|}{C}}-\!\!\bigcirc\!\!-\underset{O}{\underset{\|}{C}} \right]_n .$$

However, the International Union of Pure and Applied Chemistry have developed a detailed structure-based nomenclature.[12a,b] In this book, we shall follow the commonly used source-based nomenclature described above.

1.2 Structural and Repeating (or Repeat) Units

Polymers are identified by the structural and repeating units in their chain formulae. A structural unit (or "unit" in short) in a polymer chain represents a residue from a monomer used in the synthesis of the polymer, whereas a repeating unit, or a repeat unit, refers to a structural unit or a covalently bonded combination of two complementary structural units, which is repeated many times to make the whole chain.

The structural and repeating units are identical when a single monomer is used in the preparation of the polymer but not so when more than one monomer is used. Examples are given in Table 1.1. In each of the first five examples in the table, the repeating unit has only one structural unit, whereas in the last two examples the repeating unit in each case is composed of two covalently bonded unlike but complementary structural units.

1.3 Classification

1.3.1 *Condensation and addition polymers and polymerizations*

Carothers classified polymers into two groups, condensation, and addition. In condensation polymer, the molecular formula of the structural unit is devoid of some atoms from the ones present in the monomer(s) from which the polymer is formed or to which it may be degraded

Table 1.1 Examples of some polymers and their structural and repeating units.

Entry No.	Polymer	Monomer(s)	Structural unit(s)	Repeating unit
1	Polyethylene	$CH_2=CH_2$	$-CH_2-CH_2-$	$-CH_2-CH_2-$
2	Polystyrene	$CH_2=CH\ C_6H_5$	$-CH_2-CH(C_6H_5)-$	$-CH_2-CH(C_6H_5)-$
3	Poly(vinyl chloride)	$CH_2=CHCl$	$-CH_2-CH(Cl)-$	$-CH_2-CH(Cl)-$
4	Poly(chlorotrifluoroethylene)	$ClCF=CF_2$	$-ClCF-CF_2-$	$-ClCF-CF_2-$
5	Poly(ϵ-caprolactam)	$O=C-NH\ (CH_2)_5$ (ring)	$-C(=O)-(CH_2)_5-NH-$	$-C(=O)-(CH_2)_5-NH-$
6	Poly(hexamethylene adipamide)	$HOOC(CH_2)_4COOH$ and $H_2N(CH_2)_6NH_2$	$-C(=O)-(CH_2)_4-C(=O)-$ and $-HN-(CH_2)_6-NH-$	$-C(=O)-(CH_2)_4-C(=O)-NH(CH_2)_6NH-$
7	Poly(ethylene terephthalate)	$HOCH_2CH_2OH$ and $HOOC-C_6H_4-COOH$	$-OCH_2CH_2O-$ and $-C(=O)-C_6H_4-C(=O)-$	$-OCH_2CH_2OC(=O)-C_6H_4-C(=O)-$

by chemical means. In contrast, in addition polymer, the molecular formula of the structural unit is the same as that of the monomer from which the polymer is formed.[11,13]

By way of illustration, we may refer to Table 1.1. In entries 6 and 7, two typical examples of condensation polymers, a polyamide and a polyester respectively, are given. Clearly, the structural units lack certain atoms present in the corresponding monomers. This is due to the elimination of water taking place in the condensation of the functional groups.

$$n\ NH_2(CH_2)_6\ NH_2 + n\ HOOC(CH_2)_4\ COOH = H\text{-}[HN(CH_2)_6\ NHCO(CH_2)_4CO]_n\text{-}OH + (2n\text{-}1)\ H_2O.$$

$$n\ HOCH_2CH_2OH + n\ HOOC\text{-}\bigcirc\text{-}COOH = H\text{-}[OCH_2CH_2O\underset{O}{\overset{\|}{C}}\text{-}\bigcirc\text{-}\underset{O}{\overset{\|}{C}}]_n\text{-}OH + (2n\text{-}1)H_2O.$$

Like the polymer, the polymerization process is called condensation polymerization, it being defined as the one, which occurs by the condensation of mutually reactive (coreactive) functional groups in monomers with the elimination of small molecules (*e.g.*, water in the above examples). As we shall see later, in order that a linear polymer forms, each of the two monomers involved in polymerization must be bifunctional. However, the two coreactive functional groups may also be present in the same monomer molecule as in the two examples given below

$$n\ NH_2(CH_2)_r\ COOH \longrightarrow \text{-}[NH(CH_2)_rCO]_n\text{-}\ .$$

$$n\ HO(CH_2)_r\ COOH \longrightarrow \text{-}[O(CH_2)_r\ CO]_n\text{-}\ .$$

Cellulose and proteins are grouped under condensation polymers since on hydrolysis they are degraded to monomers, which have more atoms in their molecules than are present in the structural units of the polymers. Thus, cellulose yields glucose on hydrolysis, which has two hydrogen atoms and one oxygen atom more in the molecule than the numbers present in the anhydroglucose structural unit.

$$(C_6H_{10}O_5)_n + nH_2O = nC_6H_{12}O_6.$$

Similarly, proteins on hydrolysis yield α-amino acids, which differ from the corresponding peptide structural units having more atoms equivalent to a molecule of water in the molecular formulae.

The most prevalent addition polymers are the vinyl polymers prepared by chain polymerization of unsaturated monomers, which may be ordinarily ethylene, monosubstituted ethylenes, or 1,1-disubstituted ethylenes. For example, the polymerization of a monosubstituted ethylene may be represented as

$$n\ CH_2\text{=}CHX \longrightarrow \text{-}[CH_2\text{-}\underset{X}{CH}]_n\text{-}\ ,$$

where X = alkyl, aryl, F, Cl, Br, CN, COOR, OCOR, $CONH_2$, and so on. In the chemical structure of the polymer, the terminal units are not specified since these depend on the initiator used, the termination reaction, and the occurrence of chain transfer reaction in the polymerization (*vide infra*). However, no small molecule is eliminated unlike in the condensation polymerization.

However, not all polymers and polymerizations fit into the Carothers classification. Several anomalous cases appeared as more polymers become available ever since. For example, linear polyurethane is prepared by intermolecular reaction between a diol and a diisocyanate.

$$n\ HO-R-OH\ +\ n\ OCN-R'-NCO$$
$$= H{-}\!\!\left(\!O-R-O-\underset{\underset{O}{\|}}{C}-NH-R'-NH\underset{\underset{O}{\|}}{C}\!\right)_{\!n-1}\!\!O-R-O-\underset{\underset{O}{\|}}{C}-NH-R'-NCO\ .$$

In the reaction, no small molecule is eliminated. The polymerization, therefore, should be grouped under the addition class according to the Carothers classification. However, structurally, the polymer resembles a condensation polymer (*vide infra*).

Similarly, the ring-opening polymerization of ϵ-caprolactam yields the polyamide (nylon 6). No small molecule is eliminated during polymerization. Therefore, it should be placed in the addition class.

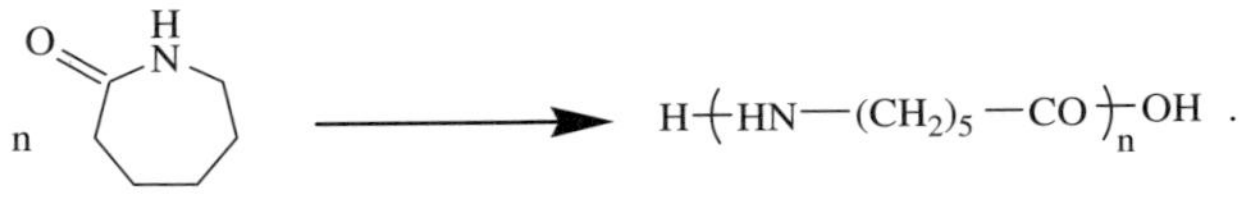

$$n\ \underset{\text{(caprolactam)}}{\bigcirc} \longrightarrow H{-}\!\!\left(HN-(CH_2)_5-CO\right)_{\!n}\!\!{-}OH\ .$$

It may be argued, however, that water elimination had already occurred during the synthesis of the lactam monomer justifying the inclusion of the polymerization in the condensation class. In addition, the same polymer may be prepared by the condensation polymerization of 6-aminocaproic acid, which strengthens the justification.

$$n\ NH_2(CH_2)_5COOH\ =\ H{-}\!\!\left(HN(CH_2)_5CO\right)_{\!n}\!\!{-}OH\ +\ (n\text{-}1)\ H_2O\ .$$

In order to get rid of the anomalies as the ones mentioned above, Carothers classification was supplemented by a structural criterion. The condensation polymers usually have polar groups, such as ether, amide, ester, carbonate, *etc.*, as interunit linkages at regularly placed intervals in the chain backbone. Although many addition polymers also have polar substituents in the structural units, these exist as lateral substituents and not as links in the chain such as

$$-CH_2-\underset{\underset{COOH}{|}}{CH}-\ ,\quad -CH_2-\underset{\underset{COOCH_3}{|}}{CH}-\ ,\quad -CH_2-\underset{\underset{OH}{|}}{CH}-\ ,\quad etc.$$

Thus, a polymer fulfilling Carothers classification and/or having polar groups as interunit linkages should be called a condensation polymer.

1.3.2 Step-growth and chain-growth polymers and polymerizations

A classification of polymers and polymerizations, which is substantially free from ambiguities and gives better insight into the polymerization, is based on the mechanism of the latter. Step (or step-growth) polymerization proceeds by stepwise intermolecular reaction of coreactive functional groups in monomers. To start with, a monomer reacts with another monomer forming a dimer. The latter, in turn, reacts with a monomer or one of its own kind to form a trimer or a tetramer respectively. The trimer reacts with a monomer, dimer, trimer, or tetramer to form a tetramer, pentamer, hexamer, or heptamer respectively. The polymerization progresses stepwise in this manner by random intermolecular reactions between the various species including the monomer. At every stage of polymerization, reactive species of various sizes starting from monomer coexist.

The average molecular weight increases slowly until near the end of the polymerization when it increases rapidly as shown in Fig. 1.1. Like the polymerization, the polymer is called a step-growth polymer.

In contrast, a chain-growth polymer forms by adding hundreds to thousands of monomer molecules in quick succession to each of certain active centers, which may be a radical, a cation, an anion, a coordinated ion, etc. The active centers exist in very small concentrations, ca., $10^{-6} - 10^{-8}$ mol/ L in respect of the monomer and have very short life times (milliseconds to seconds) in the conventional chain polymerization. They are continuously generated at one or the other end of the initiated new chains during the whole course of polymerization (Scheme 1.1). Thus, high polymers are obtained from the very beginning of

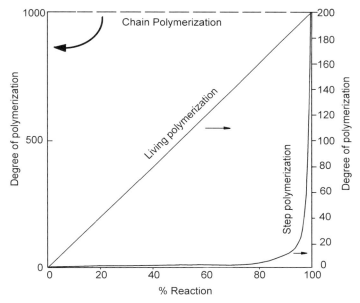

Figure 1.1 Evolution of molecular weight with progress of polymerization in three types of polymerization (for the dashed line at the top see text). Note the difference in ordinate scales between the chain polymerization and the other two. The difference in degree of polymerization is a typical one.

polymerization, although, the whole polymerization process may take hours to reach high conversion. With time, the number of polymer molecules formed increases as more and more chains are initiated, propagated, and terminated. Figure 1.1 shows the molecular weight evolution with conversion in a chain polymerization. At any instant, the polymerizing mass contains the unconverted monomer and high polymer but no intermediates, which is in sharp contrast to step polymerization. The average molecular weight remains nearly constant at low conversions (*ca.*, <10% conversion). It may then change to different extents with conversion depending on various factors, such as depletion of concentrations of monomer and initiator, dependence of termination rate on medium viscosity, *etc.* Accordingly, the molecular weight evolution above conversions larger than *ca.*, 10% has been represented by a dashed line in Fig. 1.1.

Like all chain reactions, the chain polymerization comprises at least three elementary steps: initiation, propagation, and termination as shown in Scheme 1.1 for vinyl polymerization.

An asterisk mark on a terminal carbon atom of the polymer chain represents an active center, which as mentioned above, may be a radical, a cation, an anion, a coordinated ion, *etc.* The nature of termination depends on the type of the active center, which has been discussed in Chapters 3 through 9.

In the initiation step, chains are initiated by the reaction of an initiator or its fragments with the monomer. Only very small amount of initiator is needed *ca.*, 0.01–0.1% of monomer since only one initiator molecule is used up in polymerizing some hundreds to thousands of monomer molecules. The initiating species adds to the monomer thus starting a new chain with the active center at one or the other end of the chain depending on the mechanism of initiation (*vide infra*).

Initiation is a slow reaction. Propagation follows by rapid addition of monomer molecules in quick succession to the active center, which is shifted to the end of the

Initiation

$$\text{Initiator} \longrightarrow R^*$$

$$R^* + CH_2=CHX \longrightarrow R-CH_2-\overset{*}{C}HX$$

Propagation

$$R-CH_2-\overset{*}{C}HX \xrightarrow{(n-1)CH_2=CHX} R\text{+}CH_2-CHX\text{+}_{n-1}CH_2-\overset{*}{C}HX$$

Termination

$$R\text{+}CH_2-CHX\text{+}_{n-1}CH_2-\overset{*}{C}HX \longrightarrow \text{Dead polymer}$$

Scheme 1.1 Three elementary steps in the chain polymerization of vinyl monomers.

added monomer unit following each monomer addition. The termination step refers to the destruction of the active center. However, an active center once formed may not remain attached to the same chain until its destruction. It may be transferred to other molecules making the former chain dead and starting a new chain. This reaction, called chain transfer, is not included in the scheme because it is not an essential element.

However, there are other chain polymerizations where initiation is very fast so that all active centers are generated almost simultaneously at the outset of polymerization. In addition, termination and transfer reactions are absent. Under these circumstances, if all chains have equal opportunity to grow, all the polymer molecules have nearly the same molecular weight. These are called living polymerizations. The molecular weight increases linearly with conversion as shown in Fig. 1.1. Similar results are also obtained when termination and transfer are reversible. These are discussed in details in Chapters 3, 4 and 6. Besides the vinyl monomers, the cyclic monomers (heterocycles or cyclic olefins) also undergo polymerization following the chain mechanism with the rate of initiation comparable with that of propagation. Many of these polymerizations are also living.

1.4 Functionality

The functionality (f) of a monomer is defined as the number of chemical bonds it can form with other reactants.[12C] Thus, in the examples given in Table 1.1, each monomer may link with two other monomers and hence each is bifunctional. Monomers with functionality greater than two are referred to as polyfunctional. However, a given functional group may not confer the same functionality to different monomers if the reaction types are different. Thus, in polyamidation, a diamine monomer links with two diacid monomers and, accordingly, is bifunctional. Thus, each amino group confers a single functionality to the diamine. In contrast, in the synthesis of amino resins, the triamine monomer melamine may join with six molecules of formaldehyde, which is a bifunctional monomer, and is, accordingly, hexafunctional (*vide* Chap. 2, Sec. 2.7.2). In this case, each amino group confers two functionalities to melamine.

A linear polymer forms when the monomers involved are strictly bifunctional each. The coreactive functional groups may be present either in the same monomer or in two different bifunctional monomers. The former type is customarily referred to as A – B monomer, A and B representing two unlike but coreactive functional groups.

$$n\, NH_2-R-COOH \longrightarrow H\!\!-\!\![HN-R-CO]_n\!\!-\!\!OH$$

$$n\, OH-R-COOH \longrightarrow H\!\!-\!\![O-R-CO]_n\!\!-\!\!OH$$

When the coreactive functional groups are present in different monomers, the bifunctional monomers are customarily referred to as A–A and B–B types.

$$n\ NH_2-R-NH_2\ +\ n\ HOOC-R'-COOH \longrightarrow H\left[HN-R-NH-\underset{O}{\underset{\|}{C}}-R'-\underset{O}{\underset{\|}{C}}\right]_n OH\ \cdot$$

$$n\ OH-R-OH\ +\ n\ HOOC-R'-COOH \longrightarrow H\left[O-R-O-\underset{O}{\underset{\|}{C}}-R'-\underset{O}{\underset{\|}{C}}\right]_n OH\ \cdot$$

However, some monomers exist in which the like functional groups undergo self-condensation. A bifunctional monomer of this kind (represented as A – A) is dimethylsilanediol, which is formed during the hydrolytic polymerization of dichlorodimethylsilane in the presence of water (*vide* Chap. 2, Sec. 2.7.6).

$$n\ HO-\underset{CH_3}{\underset{|}{\overset{CH_3}{\overset{|}{Si}}}}-OH \longrightarrow HO\left(\underset{CH_3}{\underset{|}{\overset{CH_3}{\overset{|}{Si}}}}-O\right)_n H\ .$$

Vinyl monomers are bifunctional inasmuch as each monomer joins with two other monomers in order to form a polymer.

$$n\ CH_2=CHX \longrightarrow \left(CH_2-\underset{X}{\underset{|}{CH}}\right)_n\ .$$

In contrast to the strictly linear polymers formed from bifunctional monomers, nonlinear polymers result from monomers, at least, one of which is polyfunctional. Reaction of glycerol, a trifunctional monomer of A_3 type, and phthalic anhydride, a bifunctional monomer of B–B type, may be cited as a typical example.

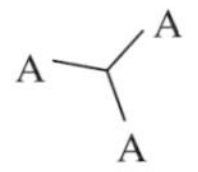

$$A-\overset{A}{\underset{A}{\diagup}}\qquad (A_3)$$

Each trifunctional unit incorporated in the polymer chain following the reaction of two of the three hydroxyl functions in a glycerol molecule becomes a potential branching site (Fig. 1.2). The branches introduce additional branching sites on their own chains. Larger the molecule becomes higher becomes its functionality. It only requires just one functional group amongst the many present to condense intermolecularly to link two molecules together. Bigger molecules, therefore, grow in size at faster rates and the growth assumes the character of an autoaccelerating reaction.

However, the reaction does not remain intermolecular all while. The number of unreacted coreactive functional groups in highly branched polymer molecules being large, the intramolecular reaction competes with the intermolecular one. Eventually, at some critical extent of reaction, the polymerizing mass transforms suddenly from a viscous liquid to an immobile gel with a three dimensionally crosslinked structure extending throughout the whole polymerizing mass. Such structure is referred to as infinite network and the polymer

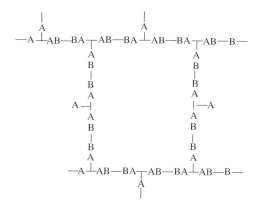

Figure 1.2 An arbitrary two-dimensional representation of a portion of crosslinked network formed in the reaction of a trifunctional monomer of A_3 type with a bifunctional one of B – B type. The positions of branches and crosslinks are also arbitrarily chosen. AB and BA represent interunit linkages such as an ester group in an alkyd resin.

as network polymer. An arbitrary two-dimensional representation of a portion of the network is shown in Fig. 1.2.

The point of sudden gelation is called the gel point. The gel makes its first appearance at this point. However, gelation does not necessarily mean that no molecules of finite size are present. This is substantiated by the fact that the polymerized mass at gel point contains a soluble fraction. A network polymer does not go into solution; at best, it swells in solvents in which a linear or a branched polymer of the same chemical type goes completely into solution. The soluble fraction (sol fraction) in the gelled mass comprises molecules of finite size, linear or branched, as well as unreacted monomers. Continuation of polymerization beyond gel point increasingly integrates finite sized molecules into the network. As a result, gel fraction increases and sol fraction decreases. The critical extent of reaction at gel point can be theoretically predicted. This has been discussed in the next chapter (Sec. 2.6).

However, gelation does not occur if the ratio of the coreactive functional groups is far from equivalent. For example, a mixture of three moles of phthalic anhydride and one of glycerol will not gel as will not a mixture of three quarters of a mol of phthalic anhydride and one of glycerol. However, for gelation to occur, a polyfunctional monomer need not necessarily be the sole reactant providing functional group of its kind. In fact, a major portion of it can be replaced with a bifunctional monomer still producing a network polymer provided the stoichiometric requirements discussed in the next chapter (Sec. 2.6) are met.

In chain copolymerization of a vinyl monomer, a network polymer is obtained by using a divinyl monomer as a comonomer. Thus, crosslinked polystyrene, shown in Fig. 1.3, is obtained by copolymerizing a mixture of styrene (f = 2) and divinylbenzene (f = 4). The density of the crosslinks and with it the property of the gel varies with the mol fraction of divinylbenzene used.

Apart from polyfunctional monomers, polyfunctional polymers may also be converted to network polymers by way of crosslinking preformed polymer molecules. Thus, the polyfunctionality of highly unsaturated polydiene rubbers is utilized to crosslink them

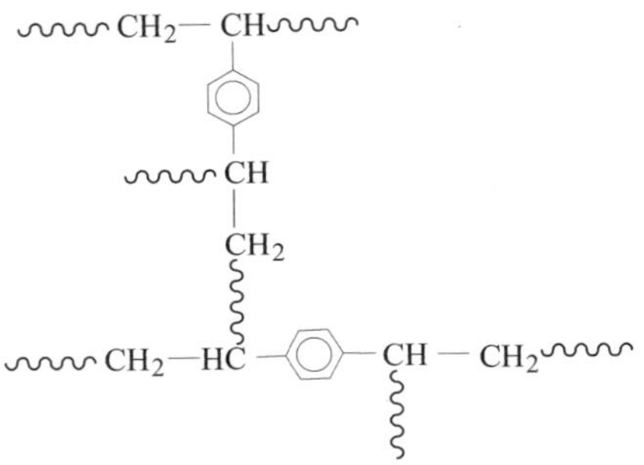

[Figure: crosslinked network structure of styrene and p-divinyl benzene copolymer]

Figure 1.3 An arbitrary two-dimensional representation of a portion of the crosslinked network formed in the copolymerization of styrene and a small quantity of *p*-divinyl benzene.

through vulcanization. Crosslinking of polymer chains occurs through mono- and/or polysulfide linkages.

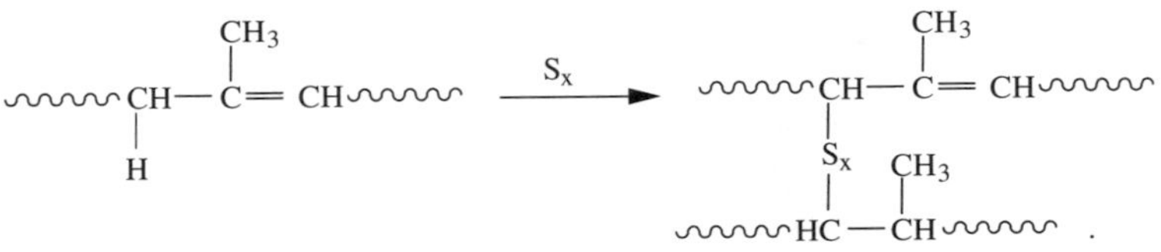

[Scheme: vulcanization crosslinking via S_x linkages]

Network formation, however, does not require high unsaturation content. Thus, low unsaturation (0.5 to 2% of chain bonds) rubbers, such as butyl and EPDM, are also crosslinked through vulcanization.

Similarly, prepolymers can be chain extended as well as simultaneously crosslinked to form network polymers. This practice is often followed in the synthesis of thermosetting polymers as would be found in many examples of their synthesis discussed in the next chapter (Sec. 2.7).

1.5 Designed Branched Polymers

Besides the randomly branched and crosslinked polymers, discussed above, branched polymers of various architectures may also be synthesized by both step and chain polymerizations. The synthesis strategies for some of the architectures using the step polymerization method are outlined schematically below. Strategies using living anionic chain polymerization are discussed in Chap. 4.

1.5.1 *Star polymers*

Star polymers can be prepared by self-condensing an A–B monomer in the presence of a small amount of an f-functional monomer containing either A or B functional groups, *e.g.*, A_f. Thus, when $f = 3$, a 3-arm star polymer is obtained (Fig. 1.4).

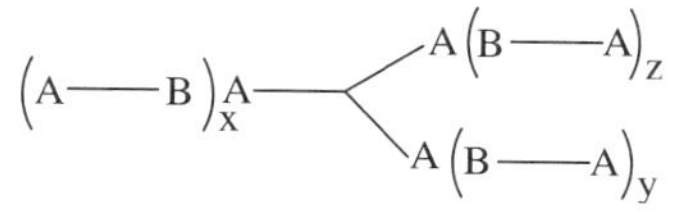

Figure 1.4   A three-arm star polymer.

1.5.2 *Hyperbranched polymers*

Hyperbranched polymers may be prepared by step polymerization of AB_x monomers where x is two or more provided A reacts only with B intermolecularly and *vice versa* (Scheme 1.2). In each n-meric species will be present only one unreacted A and $(x-1)$ n+1 unreacted B terminal groups,[14,15] as may be verified from structure (**1**) in which $x=2$ and $n=10$ and numbers of unreacted A and B groups are 1 and 11 respectively. Polymer growth can be limited using in small proportion a polyfunctional comonomer having a single kind of functions, which forms the core of the polymer molecules as shown in the scheme using a trifunctional (B_3) comonomer. The size of the molecule is determined by the mol ratio of AB_2 to B_3.

1. Hyperbranched polymer with potential for unlimited growth

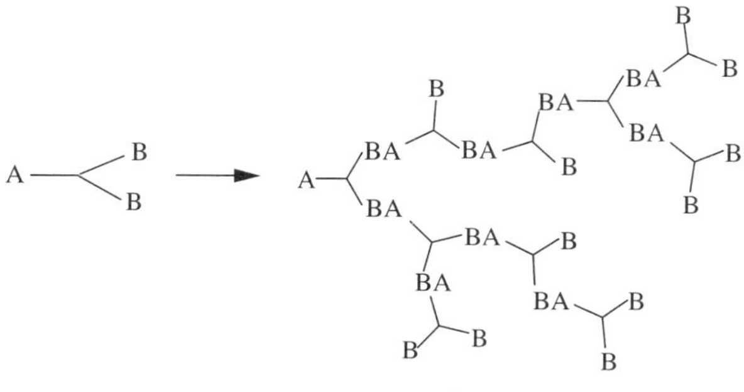

2. Hyperbranched polymer with limited growth

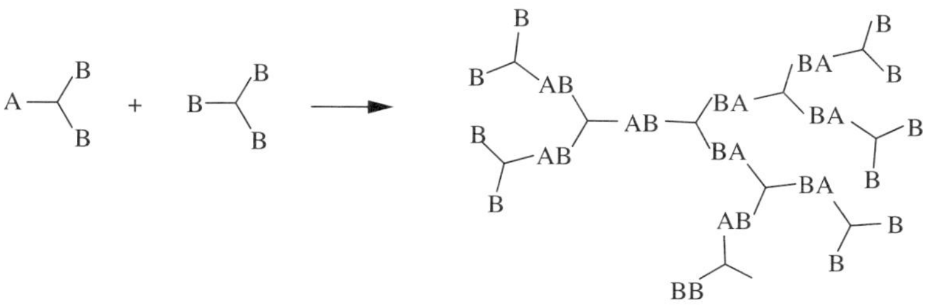

Scheme 1.2   Schemes of hyperbranched polymer synthesis.

Hyperbranched polymers have very low melt and solution viscosities due to their compact and branched structure with chains too short for chain entanglement.

1.5.3 *Dendrimers*

The term dendrimer coined by Tomalia *et al.* was derived from the Greek words *dendri* (meaning branch, tree-like) and *mer-* (meaning part). It refers to polymer with highly regular branched structure that follows a strict geometric pattern. Dendrimers are perfectly branched and can be prepared in highly monodisperse form.[15] Apart from their unique rheological properties, the dendrimers may act as very good hosts for various guest molecules. In this latter capacity, the three distinct regions of the molecule, *viz.*, (i) the core, (ii) the interior comprising generations of shells, and (iii) the outer periphery constituted of terminal functional groups may play specific roles. They are prepared using either divergent or convergent strategy. In the divergent strategy, the structure is built starting from the core,[16] whereas in the convergent strategy preformed dendrons are coupled around a core.[17]

However, the synthesis is laborious and expensive since it involves multiple steps requiring purification between steps. This is in contrast with the one step synthesis method used in the preparation of hyperbranched polymers, which are, of course, not perfectly branched. The synthesis of dendrimers by divergent and convergent routes using protection and deprotection chemistry and condensation reactions follows Scheme 1.3. The interunit AB or BA linkages are omitted in the dendrimer structures for the sake of simplicity.

1.6 Physical State

A polymer can be a viscous liquid (rubber), an amorphous solid (glass), or a semicrystalline solid containing both crystalline and amorphous regions depending on its molecular properties like chain flexibility, intermolecular attraction, structural regularity, chain packability into crystal lattice, and external factors – principally temperature (*vide* next section). Fully crystalline polymers are not ordinarily encountered. This is because perfect chain folding required for packing into the crystal lattice is hard to achieve even though the molecular properties are favorable for crystalline polymer formation. Besides, chain length polydispersity, presence of chain ends (which are excluded from the crystal lattice), and occluded impurities contribute to increase the amorphous content.

The physical state may be identified by studying thermal transitions. Two types of thermal transitions are observed, *viz.*, glass transition, and melting, although not all polymers exhibit both. Amorphous polymers exhibit only glass transition temperature at which the glassy regions change to liquid (rubbery) regions, whereas semicrystalline polymers exhibit both glass transition and melting.

Figure 1.5(a) shows the idealized changes in enthalpy of a semicrystalline polymer with temperature. The physical states in the various regions of the curves are also shown. As the

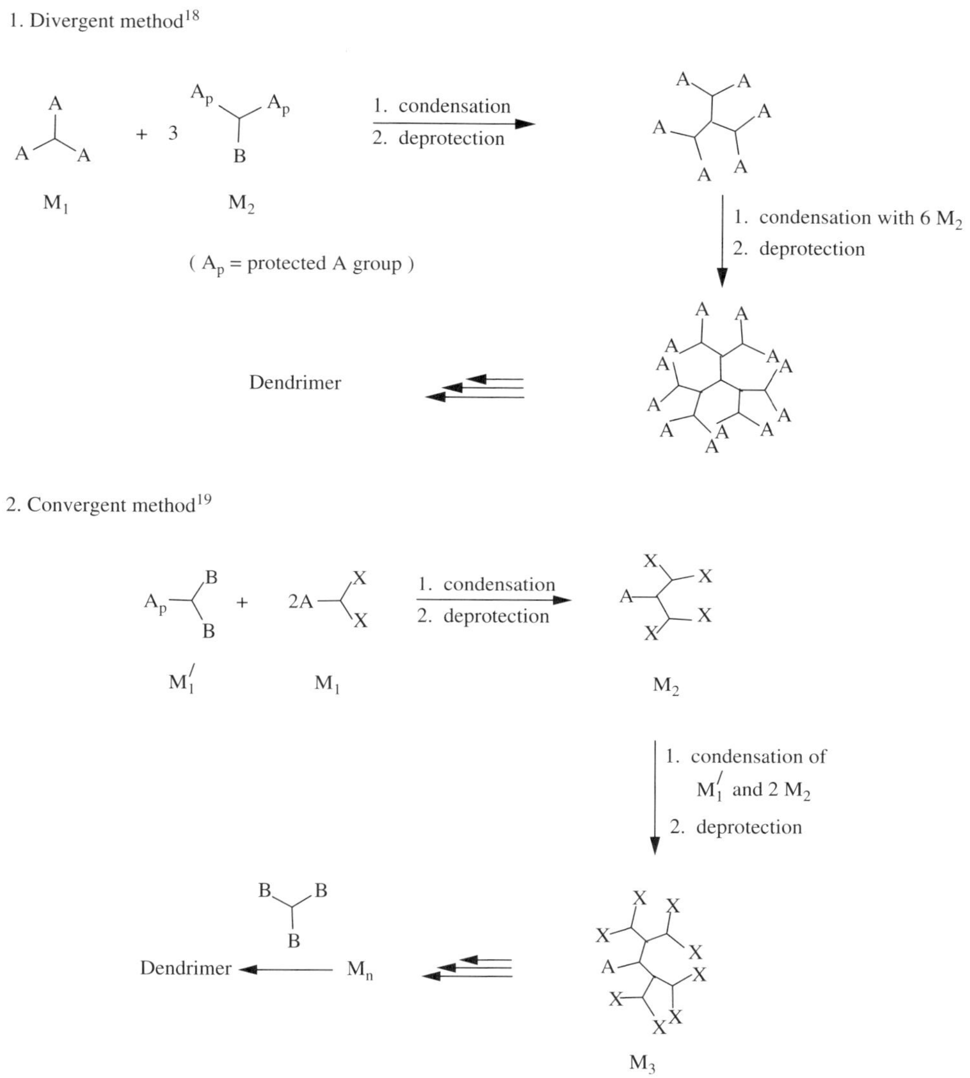

1. Divergent method[18]

2. Convergent method[19]

Scheme 1.3 Schemes of dendrimer synthesis (the interunit linkages are not shown).

material is heated, the enthalpy increases linearly until at the glass transition temperature (T_g) an increase in the slope of the line (but no discontinuity in transition) occurs due to the higher heat capacity of the rubbery polymer. The linear increase in enthalpy continues until the melting temperature (T_m) is reached where a discontinuous increase occurs due to the absorption of the latent heat of fusion. This is a phase transition and a first order one, which the glass transition is not.

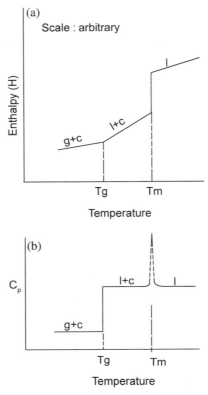

Figure 1.5 Idealized changes in enthalpy (upper figure) and heat capacity (lower figure) of a semicrystalline polymer with temperature. Letters c, l, and g on the lines represent crystal, liquid, and glass respectively. The dotted peak in the lower figure approximates the transition experimentally observed (see text).

The change in slope at T_g is better identified by plotting the first derivative, *i.e.*, heat capacity ($\delta H/\delta T = C_p$), against temperature as shown in Fig. 1.5(b). Now, the graph shows discontinuity at the T_g and, hence, T_g is referred to as the second order transition temperature. The graph is not defined at the melting temperature. However, in reality, it shows a peak (shown by dotted lines), which is although not necessarily symmetric as it is shown to be the one. This is indicative of the polymer melting over a range of temperatures, the presence of crystallites of different sizes being one principal reason. The smaller crystallites melt at lower temperatures.[20] Like the T_m, the T_g also would not be as sharp as it is shown due to the complexities of the rotational motion of polymer segments that sets in at glass to rubber transition.

Similar features would be observed in the plots of specific volume (in place of heat content) and volume expansion coefficient (in place of heat capacity) *vs.* temperature.

T_g and T_m are conveniently measured using a differential scanning calorimeter, which gives the heat capacity *vs.* temperature plot. The transition temperatures of several polymers are presented in Table 1.2.

Table 1.2 Glass transition and melting temperatures of some common polymers.[a,b]

Polymer	T_g, °C	T_m, °C
Polyethylene		
LDPE	−135 to −103	105–115
HDPE	−113 to −133	146
LLDPE		105–110, 121–125[c]
Polypropylene		
-*isotactic* (100%)	2.5	186
-syndiotactic (100%)		214
-atactic	−3	
Polystyrene	100	
Poly(methyl methacrylate)	110	
Poly(vinyl chloride)	75	
Poly(vinyl acetate)	30	
Polyacrylonitrile	97–125	320
Poly(vinylidene chloride)	−18	210
Poly(vinylidene fluoride)	−35	170–200
Polytetrafluoroethylene		314
Polyisobutylene	−70	2 to 44
cis-1,4-polyisoprene	−74 to −69	35
trans-1,4-polyisoprene	−70	60–67
cis-1,4-polybutadiene	−103	1
trans-1,4-polybutadiene	−103	97–145
Polydimethylsiloxane	−123 to −150	−40
Poly(ethylene terephthalate)	61	267
Poly(hexamethylene adipamide)	50	260
Poly(∈-caprolactam)	47–57	220
*Bis*phenol A polycarbonate	150	

[a]T_g depends on the time scale of the measurement technique used. The range of values when given is due to the different methods and/or different heating rates used in the measurements.

[b]The range of T_m shown for some polymers is due to different percentages of crystallinity in the samples arising out of differences in microstructure or due to polymorphs.

[c]Dual endotherms observed in differential scanning calorimetry.

1.7 Structure-Property Relationship

1.7.1 *Nonconjugated polymers*

Polymers are also classified according to their mechanical properties into three broad groups: rubbers, plastics, and fibers. The mechanical properties of one group are characteristically different from those of the other. Thus, rubber (scientifically called elastomer) has low initial tensile modulus (<1 MPa) but moderate tensile strength (<50 MPa), high elongation at break (*ca.*, 1000%), and also the ability to maintain the properties over a wide temperature range. The elongated material rapidly recovers to original dimensions after the stress is withdrawn even after long periods of the stress application. Fibers have very high tensile modulus (>1000 MPa), high tensile strength (>100 MPa), and low elongation at break (<50%). Plastics have moderate to very high tensile modulus (150–3000 MPa) and moderate tensile strength (*ca.*, 20–80 MPa). Both values depend upon chain rigidity and/or degree of crystallinity; higher are these molecular properties larger are the values. The elongation at break varies from as low as <3% for rigid plastics such as polystyrene to values close to those of rubbers for flexible plastics such as polyethylene. However, unlike rubbers, flexible plastics have low yield strain.

The characteristic physical and mechanical properties of fibers, plastics, and rubbers depend broadly on such molecular properties of the constituent polymers as chain flexibility, intermolecular attraction, and packability into crystal lattice.[21] Chain flexibility is related with the ease of rotation of chain bonds constituting the chain backbone. Intermolecular attraction increases with the introduction of polar or polarizable groups in the polymer chain, more so, when the groups are placed in the chain backbone at regular intervals. Packability into crystal lattice is high for chains, which are "slim", free from kinks, and high in structural and microstructural regularity. The last named criterion refers to the degree of stereo- and regioregularity, and to the extent, length, and distribution of branches (if any).

Rubbers

For a polymer to exhibit elastomeric (rubbery) property, chain flexibility should be high, intermolecular attraction low, chain packability low or absent, and the glass temperature lower than the use temperature.[21]

The elasticity of rubber arises from the ease of transformation from the extended (or helical) conformation of the molecules in the stretched state to the thermodynamically favored coiled conformation of highest entropy when the stretching force is withdrawn. This is possible when the chains are highly flexible with the material, of course, above its glass transition temperature. In addition, for the recovery to take place, the molecules should be crosslinked at several points along the chains, with molecular weights between crosslinks being ideally around 20 000 to 25 000.[21] Without crosslinks, the molecules would flow past each other as in a liquid on the application of stress and no elastic recovery would occur.

The energy barrier to rotation (E_b) for some bonds of interest in this discussion is given in Table 1.3. Higher is the barrier lower is the chain flexibility. The rotation frequency is given by the fraction $e^{-E_b/RT}$ of the torsional frequencies of single bonds, which lie in the range of 10^{12}–10^{13} s^{-1}.[21] Thus, chain flexibility increases with increase in temperature. From the E_b values it is not surprising that high unsaturation hydrocarbon elastomers, *e.g.*, *cis*-1,4-polybutadiene, have the bond sequence C–C=C in their structural units.

However, even though the intrinsic flexibility of each individually isolated polymer molecule may be high, the molecular mobility of chain segments in a polymeric mass may be sharply reduced due to strong intermolecular attraction.[21] In the polydienes, however, the intermolecular attraction is weak since, being nonpolar, they have only the weak dispersion forces as the source of intermolecular attraction. Their high segmental mobility is reflected in the low glass transition temperatures, which are −70°C and −115°C respectively for *cis*-1,4-polyisoprene and *cis*-1,4-polybutadiene.

The chain packability into crystal lattice is also low for the *cis*-polydienes (*vide infra*) so that they do not crystallize at ambient temperature. However, they undergo crystallization

Table 1.3 Approximate height of rotational energy barrier (E_b) for selected bonds.

Bond	E_b kJ/mol	Ref.
$H_3C \!\!-\!\! CH_3$	11.3–12.5	22
$H_3C \!\!-\!\! O \!\!-\!\! CH_3$	11.4	23
$H_3C \!\!-\!\! S \!\!-\!\! CH_3$	8.9	23
$H_3C \!\!-\!\! CH \!\!=\!\! CH_2$	8.2	22

when stretched, the crystallinity increasing with increasing elongation. The crystallites act as additional crosslinks and increase the retractive force as well as the tensile strength of the rubber. However, the intermolecular force is not strong enough to prevent melting at ambient temperature as the stress is withdrawn. This is, in fact, desirable since the stretched rubber would not otherwise have reverted to the original conformation and the elasticity would have been lost.

Unlike their *cis* counterparts, the *trans*-1,4-polyisoprene and the *trans*-1,4-polybutadiene are not elastomeric. The chains of the *trans* isomers are "slimmer" than those of the *cis* isomers, as may be judged from the extended chain conformation shown below for a part of the chain skeleton in the two isomers of polybutadiene.[21]

cis *trans*

This structural feature allows closer packing of the chain segments of the *trans* polymer into the crystallites. Accordingly, *trans*-1,4-polyisoprene and *trans*-1,4-polybutadiene have relatively high melting points, *viz.*, 65°C and 145°C respectively . Hence, they are not elastomeric at ambient temperature.

In contrast, neoprene rubber which is prepared by the emulsion polymerization of chloroprene has only about 90% *trans*-1,4-structure. This lower microstructural purity impedes its packing into crystallites so that the melting temperature is about 45°C compared to 105°C for pure *trans*-1,4-polychloroprene.[24]

Amongst the other high unsaturation rubbers are the random copolymers of styrene and butadiene (SBR) containing *ca.*, 25 mol% styrene and those of butadiene and acrylonitrile (nitrile rubber) containing *ca.*, 25 to 35 mol% acrylonitrile. The double bonds in these copolymers amount to about 15% of the chain bonds. Although this amount is much smaller than in the homopolydiene rubbers, where it is 25%, it is large enough to impart good chain flexibility.[24] However, the chain packability of these copolymers is extremely poor due to the random placement of the comonomer residues in the chain backbone. As a result, these elastomers are not crystallizable even under the stretched condition.[21]

In low unsaturation hydrocarbon elastomers based on polyisobutylene (PIB) or ethylene propylene (EP) copolymers (35 to 65% of either monomer) chain flexibility is markedly lower than in the above discussed high unsaturation ones. However, the intermolecular attraction is considerably weaker. Overall, segmental mobility is considerably high as manifested in glass transition temperatures, which are −50°C for EP copolymers and −70°C for PIB.

The structural irregularity in EP precludes its crystallization. On the other hand, although polyisobutylene has a highly regular and symmetric structure, the steric influence of the two methyl substituents on alternate carbon atoms prevents its spontaneous crystallization. It crystallizes, though, when stretched and the crystallites melt as soon as the stretching force is withdrawn, which is required of stress-crystallizable elastomers. Small amount of unsaturation to the extent of about 0.5 to 2% of chain bonds is purposefully introduced in PIB and EP copolymers in order to provide sites for vulcanization.

The presence of other flexible bonds, $viz.$, C–O–C or C–S–C, in structural units is responsible respectively for the elastomeric property of polyethers such as poly(propylene oxide), polytetrahydrofuran, and polysulfides.[21]

Compared to the C–O–C bonds, the Si–O–Si bonds are more flexible, which accounts for the elastomeric property of silicone rubber, $e.g.$, polydimethylsiloxane,

$$\left(\begin{array}{c} CH_3 \\ | \\ Si-O \\ | \\ CH_3 \end{array} \right)_n .$$

Fibers and plastics

Plastics are constituted of polymer molecules covering a wide range of chain packability so that they may be either amorphousxy or semicrystalline. Intermolecular attraction may be moderate or high, and chain flexibility low to moderate. Some examples of commodity plastics of the amorphous variety are polystyrene, poly(vinyl chloride), and poly(methyl methacrylate). Among the semicrystalline variety, notable examples are polyethylene, stereoregular polypropylene, nylons, and poly(ethylene terephthalate).

Fibers are made from polymer molecules, which have low chain flexibility, high intermolecular attraction, and high chain packability into crystal lattice. These molecular characteristics are opposite to those which give rubbery property.

The intermolecular attraction is generally high in step polymers due to the presence of polar groups at regular intervals in the chain skeleton. However, chain packing is enhanced when the crystallizable polymers are spun from melt or solution to form fibers. In addition, if the fibers are also drawn, crystallinity increases even more. On drawing, the extended or helical conformation of the chains is further enhanced, which facilitates crystallization. The scope of maximized hydrogen bonding in extended chain conformation of poly(hexamethylene adipamide) is evident from Fig. 1.6, which gives a schematic picture of the polar amide groups forming layers.[25]

However, when the structural regularity is lost as occurs, for example, when the monomers of nylon 6 and nylon 66 are copolymerized, the copolymer fails to form a fiber.

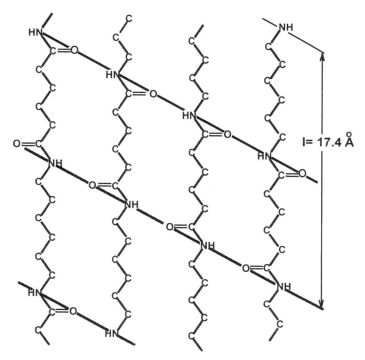

Figure 1.6 Schematic representation of polar layer structure formed by the association of amide groups in nylon 66. "Reprinted with permission from Ref. 25. Copyright ©1942 American Chemical Society."

Table 1.4 Repeating units and melting points in polyesters.

Polymer	Repeating unit	m.p. °C
Poly(ethylene suberate)	$-O(CH_2)_2 O\underset{O}{\overset{O}{\underset{\|}{C}}}(CH_2)_6\underset{O}{\overset{O}{\underset{\|}{C}}}-$	45
Poly(ethylene terephthalate)	$-O(CH_2)_2O-\underset{O}{\overset{O}{\underset{\|}{C}}}-\bigcirc-\underset{O}{\overset{O}{\underset{\|}{C}}}-$	256
Poly(hexamethylene adipamide)	$-NH(CH_2)_6 NH-\underset{O}{\overset{O}{\underset{\|}{C}}}(CH_2)_4\underset{O}{\overset{O}{\underset{\|}{C}}}-$	250
Poly(hexamethylene terephthalamide)	$-NH(CH_2)_6 NH-\underset{O}{\overset{O}{\underset{\|}{C}}}-\bigcirc-\underset{O}{\overset{O}{\underset{\|}{C}}}-$	350

Crystallinity and melting point increase with the introduction of rigid groups in structural units. This would be evident from the melting points given in Table 1.4 for some polyesters and polyamides.

The fully aliphatic polyester in the table has a much lower melting point than poly(ethylene terephthalate) (PET). This is attributable to the replacement of the flexible hexamethylene group in the repeating unit of the aliphatic polyester with the rigid aromatic ring in PET. The planar ring also effects better packing of the chain segments in the crystallites. This feature also raises the melting point.

The intermolecular attraction in aliphatic polyamides is much greater than in aliphatic polyesters due to the existence of hydrogen bonds between the amide groups, and, accordingly, the polyamides have higher melting points than the polyesters. Thus, the aliphatic polyester, poly(ethylene suberate), has a low m.p., 45°C, whereas the aliphatic polyamide, poly(hexamethylene adipamide), has a much higher m.p., 250°C (Table 1.4). It is of historical interest that Carothers was unsuccessful in his attempt to develop synthetic fibers based on aliphatic polyesters, which led him turn to aliphatic polyamides that resulted in the discovery of nylon 66. As in the case of polyesters, the introduction of rigid ring in the repeating unit of polyamide raises the melting point. This would be evident from the comparison of melting point data of the last two entries in Table 1.4.

Very long chains of condensation polymers are not needed to achieve good mechanical properties because of the strong intermolecular attraction, regularity in molecular structure, and chain rigidity due to the presence of aromatic or heterocyclic groups in chain skeleton. In fact, molecular weights in the range of 10 000–20 000 are good enough. This is in sharp contrast to the molecular weights of some hundred thousands required for addition polymers. Further discussion on the structure property relationship in condensation polymers containing rigid chains are given in Chap. 2 (Sec. 2.8).

Amongst vinyl polymers, polyethylene has the simplest structural unit, *viz.*, —CH_2—CH_2—. The intermolecular attraction in polyethylene is low due to the absence of polar or polarizable groups, but the high symmetry of the molecules and narrow chain width facilitate crystallization so that in the best crystallized sample the melting temperature is ca., 130°C.[21] However, polyethylene does not have the ideal linear structure, —$(CH_2$—$CH_2)_n$—. Various kinds of polyethylene exist, which differ in microstructures. They are differentiated according to density as low density polyethylene (LDPE, $\rho = 0.910 - 0.925$), high density polyethylene (HDPE, $\rho = 0.941 - 0.965$), and linear low density polyethylene (LLDPE, $\rho = 0.915 - 0.940$). The density difference is due to the difference in the degree of crystalline order, which is dependent on the microstructure resulting from the method of synthesis. LDPE is highly branched with most of the branches being short alkyl groups but some being long chains. LLDPE has only short alkyl groups as branches. HDPE is only lightly branched with short branches. The microstructure of each of these three classes of polyethylene is shown schematically in Fig. 1.7.

The crystallinity in polyethylene may range from *ca.*, 40% to 75% and the melting point from 105°C to 135°C. The crystalline regions are interspaced with amorphous regions,

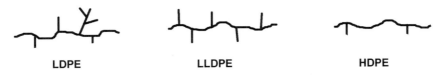

Figure 1.7 Schematic representation of chain microstructures in various polyethylenes.

1.7.1.1 Stereoregularity in vinyl polymers

Vinyl polymers with a single substituent in the structural unit, —CH_2—$CH(X)$—, has an irregular chemical structure due to the random variation of the configuration (d or l) of the monosubstituted alternate carbon atoms in the chain. This is evident from structure (**2**) in which part of the extended chain is shown. In this conformation, the carbon atoms constituting the chain backbone lie in a plane and the substituents bonded to the chain atoms in the backbone lie either above or below the plane. Polymers with such irregular stereochemical structures are called *atactic* polymers.

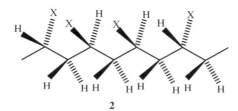

2

The structural irregularity makes these vinyl polymers amorphous or lowly crystalline at best. However, the substituent hinders rotation around the C–C bond. Greater is the degree of substitution or the size of the substituent greater is the hindrance to rotation and, accordingly, lower is the chain flexibility. The low chain flexibility may make the polymer behave as a plastic. Polar substituent increases intermolecular attraction. Accordingly, despite being structurally irregular, the polymer may exhibit some degree of crystallinity as in polyacrylonitrile or poly (vinyl alcohol). However, the crystalline order is not high.[26]

On the other hand, in stereoregular polymers where the substituted alternate carbon atoms assume either the same (d or l) configuration (*isotactic*) or alternating d and l configurations (*syndiotactic*) the polymers undergo crystallization, examples being the two stereoregular polypropylenes.

isotactic polypropylene   *syndiotactic polypropylene*

If the alternate carbon atoms have two different substituents each, as in poly(methyl methacrylate), the same problem of irregularity in stereochemical structure arises as occurs in vinyl polymers with monosubstituted alternate carbon atoms. Accordingly, the *atactic* polymer is amorphous.

In contrast, when the two substituents on alternate chain atoms are identical, the structural regularity becomes high so that the polymer may undergo crystallization, poly (vinylidine fluoride) and poly (vinylidine chloride) being examples.[26]

1.7.1.2 Regioregularity in vinyl polymers

Microstructural regularity not only depends on the regularity of configuration of the asymmetric carbons in vinyl polymers but also on the regiochemical regularity of the placement of the monomer units in the polymer. In Chap. 5, we shall see that the stereochemical regularity depends as well on the regiochemical regularity in the coordinated anionic polymerization of propylene.

In general, vinyl polymerization proceeds through head-to-tail addition of monomers, the more substituted carbon end of the double bond in the monomer is called the head, and the other end is called the tail. The head-to-tail and head-to-head additions are shown in Scheme 1.4.

Head-to-tail addition is favored due to the active center being more stabilized when located on the more substituted carbon through resonance interaction with the substituent groups. In radical addition, however, the role of radical stability is found to play a minor role when small reactants are involved, unless the radical is conjugated with C=C bond. Instead, steric and polar factors control the regiochemistry of addition (*vide* Chap. 3, Sec. 3.1).

In most vinyl polymers, the proportion of head-to-head addition is small. For example, in poly(vinyl acetate) which is prepared by radical polymerization, the proportion varies from 1 to 2% depending on the temperature of polymerization.[27] It increases with the increase in temperature of polymerization since the activation energy of this mode of addition is higher. An occasional head-to-head addition is immediately followed by a tail-to-tail addition (unless a termination or transfer event occurs) so that the radical center is restored to the head end of the terminal unit.

In asymmetrically substituted fluoroolefins, such as vinyl fluoride or vinylidene fluoride, the proportion of head-to-head, tail-to-tail addition is somewhat larger than in the other vinyl polymers, which is attributable to a lower steric factor, the steric bulk of fluorine being not much different from that of hydrogen (the van der Waals radii of hydrogen and fluorine atoms being 1.2 Å and 1.35 Å respectively[28]).

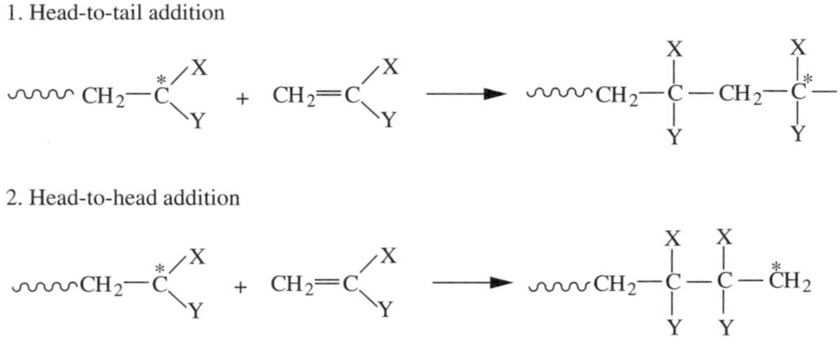

1. Head-to-tail addition

$$\sim\sim\sim CH_2-\overset{*}{C}\overset{X}{\underset{Y}{\diagup}} + CH_2=C\overset{X}{\underset{Y}{\diagup}} \longrightarrow \sim\sim\sim CH_2-\underset{Y}{\overset{X}{C}}-CH_2-\overset{X}{\underset{Y}{\overset{*}{C}}}-$$

2. Head-to-head addition

$$\sim\sim\sim CH_2-\overset{*}{C}\overset{X}{\underset{Y}{\diagup}} + CH_2=C\overset{X}{\underset{Y}{\diagup}} \longrightarrow \sim\sim\sim CH_2-\underset{Y}{\overset{X}{C}}-\underset{Y}{\overset{X}{C}}-\overset{*}{C}H_2$$

Scheme 1.4 Modes of addition of vinyl monomer.

1.7.2 Conjugated polymers

In the above discussion of structure property relationship, we focused mainly on the physical and mechanical properties of the polymers. However, conjugated polymers, which have alternate single and double bonds along the chain backbone, exhibit remarkable electrical and electronic properties under suitable conditions. Polyacetylene is the simplest member of this class of polymers. Both the *cis* and *trans* polyacetylenes prepared in film forms have shiny metal like appearance. The *trans* polymer is silvery, whereas the *cis* polymer is copper colored.[29] The synthesis is discussed in Chap. 5. However, their conductivities are nowhere metal like.

cis-polyacetylene   trans-polyacetylene

The conductivity of the *cis*-polyacetylene is 1.7×10^{-9} S cm^{-1}, whereas that of the *trans* polymer is 4.4×10^{-5} S cm^{-1}.[30,31] However, the conductivity increases by several orders of magnitude and becomes metal like when the polymer is partially oxidized or reduced.[32,33]

The partial oxidation or reduction of the polymer introduces charge carriers by removing some of the π-electrons or introducing some extra electrons into the polymer respectively. In analogy with doping of inorganic semiconductors, oxidation is called p-doping and reduction is called n-doping.

However, whereas in inorganic semiconductors doping is done at ppm level, in conjugated organic polymers it has to be done at a level *ca.*, 10 to 50% of repeating units.[34] This is due to the low mobility of charge carriers in doped organic polymers. Quantitatively speaking, conductivity depends on the number density (n) of charge carriers and their mobility (μ)[35]

$$\sigma = ne\mu,$$

where e is the electronic charge. The mobility of the charge carriers would have been high but for the high Coulombic energy of binding with the counterions, which are not very mobile. At high concentrations of counterions, the screening of the Coulombic charge of the counter ions allows higher mobility. However, by the application of an alternating electric field perpendicular to the film, the counter ions may be made to diffuse from or into the polymer resulting in undoping or doping respectively. Thus, a switching action can be realized.[36] Apart from the chemical method of doping shown in Scheme 1.5, an electrochemical method may also be used.

Besides polyacetylene, several other conjugated polymers exhibit conductivity on doping.[35] Some are listed in Table 1.5.[37] The conductivities of the doped polymers and the types of doping are also given. Unlike polyacetylene, which is environmentally unstable being easily oxidized by oxygen of the air and attacked by moisture, the conducting polyheterocycles, in particular, are environmentally stable.

Polyaniline is unique in that it can be doped with protonic acid.[31,38,39] It may be prepared in various average oxidation states, the base forms of which may be represented by the general structure (**3**) with y varying from 0 to 1 (Fig. 1.8). The half-oxidized base

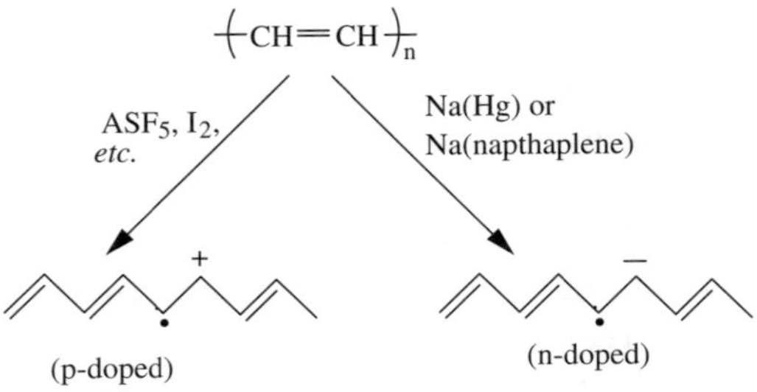

Scheme 1.5 Chemical doping of polyacetylene.

Table 1.5 Some examples of conducting polymers.[37]

Polymer	σ^a S cm^{-1}	Type of doping
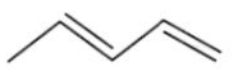 Polyacetylene	1.7×10^5	n, p
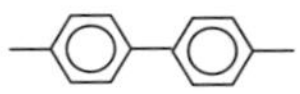 Poly(p-phenylene)	1×10^3	n, p
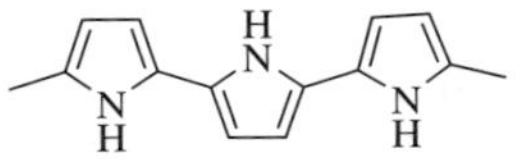 Polypyrrole	7.5×10^3	p
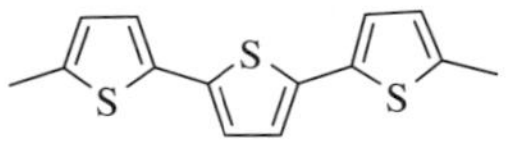 Polythiophene	1×10^3	p
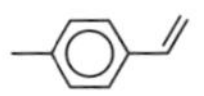 Poly(p-phenylene vinylene)	5×10^3	p
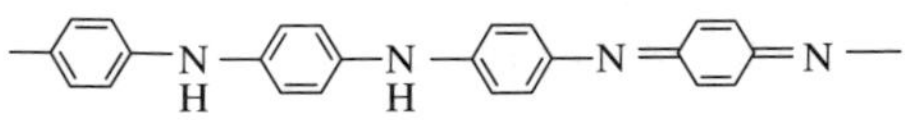 Polyaniline	1×10^2	p

^{a}Maximum conductivity reported for doped polymer.

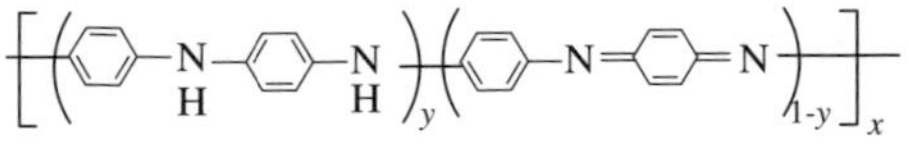

3

Figure 1.8 Base form of polyaniline in various oxidation states represented by y, which may vary from 0 to 1.

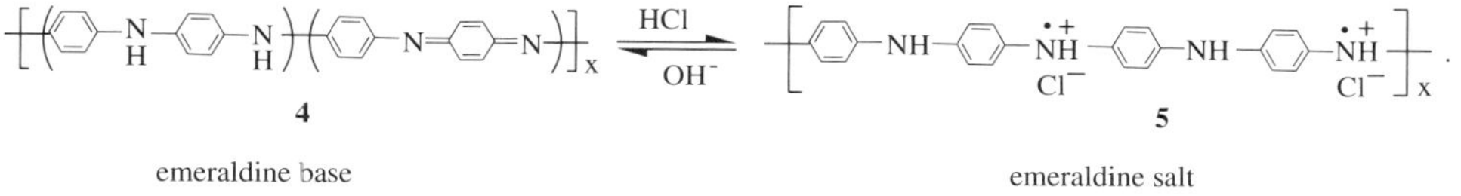

Scheme 1.6 Reversible acid doping of half oxidized polyaniline base.

form known as the emeraldine base (**4**) is made conducting by treatment with protonic acid, which protonates the imino nitrogens. Acid doping, unlike redox doping, does not change the number of electrons associated with the chain backbone. The degree of protonation depends on the pH of the aqueous acid. The conducting material may be reverted to the nonconducting state by treatment with a base (Scheme 1.6).

Complete protonation of the imino nitrogens results in the formation of the delocalized polysemiquinone radical cation (**5**) and the increase in conductivity by about ten orders of magnitude. The doped polymer may, however, be prepared also by chemical oxidation of the fully reduced form (structure **3** with y = 1).

Apart from its conducting property, polyaniline is electrochromic exhibiting a whole range of colors by virtue of its many protonation and oxidation forms.

Many uses of conducting polymers are reported or envisaged. The most commercially successful large application of conducting polymers is in anti-static coating of photographic films. In this application, the water soluble salt of a polythiophene derivative, *viz.*, poly(ethylenedioxythiophene) (PEDOT), doped with polystyrenesulfonic acid (**6**) is used. Among other applications are electromagnetic shielding (*e.g.*, shields for computer screens against electromagnetic radiation), printed circuit boards, antistatic clothing, light rechargeable batteries, electrochromics, supercapacitors, sensors, and corrosion inhibitors.

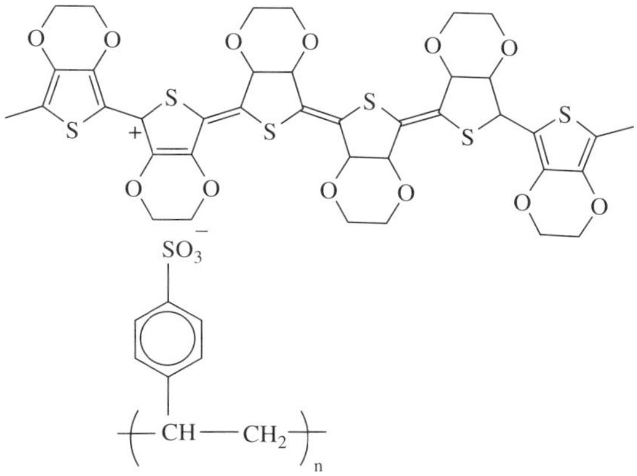

Besides being electrically conducting, when doped, high-purity conjugated polymers in the undoped state hold great promise as active components in semiconductor devices, *e.g.*, normal transistors, field-effect transistors, photodiodes, light-emitting diodes (LED), and solar cells.[40]

In these applications, the appeal of polymer stems from the combination of properties, *viz.*, film forming and semiconductivity. Solution processing of polymers makes device fabrication cheaper in cost. Besides, fabrication covering large area is possible as in large area flat panel displays.

Poly(*p*-phenylene vinylene) (Table 1.5) is the prototype of polymeric LEDs (PLED). It has a band gap of about 2.5 eV and it produces yellow green luminescence on electroexcitation, the spectrum being the same as that produced by photoexcitation. In the LED device, the light emissive polymer is asymmetrically surrounded by a transparent hole-injecting electrode, usually the indium-tin oxide (ITO) electrode, on one side, and a low work function, electron-injecting metal, such as aluminum, magnesium, or calcium, on the other side. With proper bias, electrons and holes are injected. These recombine in the polymer layer emitting light that comes out through the transparent ITO clectrode.[41–43] The color of the emitting light can be changed by varying the substituents in the phenylene ring[44] or by chemically varying the conjugation lengths in the chain backbone.[45,46]

Very large increase in efficiency of PLEDs based on PPVE occurs with the use of a layer of hole-injecting conducting polymer, *e.g.*, PEDOT doped with polystyrene sulfonic acid (**6**), in between the ITO and the emissive polymer layer.[40]

Several other conjugated polymers such as poly(arylene ethynylene)s, polyfluorenes, polythiophenes, and polycarbazoles also have proved to be efficient electroluminescent polymers.[47] Chemical modification of the various conjugated polymers has made possible to tune the emission color so that full-color displays can be fabricated.

The transition metal coupling polymerization has proved attractive for the synthesis of many conjugated polymers. Synthesis of polyarylenes, poly(arylene vinylene)s, poly(arylene ethynylene)s, and polythiophenes by this route has been briefly discussed in Chap. 2 (Sec. 2.1.8).

1.8 Thermodynamics of Polymerization

For polymerization to occur, the free energy of polymerization (ΔG_{poly}) must be less than zero.

$$\Delta G_{poly} = \Delta H_{poly} - T\Delta S_{poly}. \tag{1.1}$$

In most polymerization, ΔS_{poly} is negative. This is due to the loss of translational entropy of monomer. In exothermic chain polymerization, ΔH_{poly} is negative and ΔG_{poly} becomes zero at certain sharp temperature called ceiling temperature (T_c) above which no polymerization occurs.[48] However, when the polymerization is highly exothermic, the ceiling temperature may be too high to be experimentally accessible. The ceiling temperatures of some vinyl and cyclic monomers are given in Table 1.6.

Polymerizations of unsaturated monomers are exothermic since the π bond in the monomer is broken to form a stronger σ bond in the polymer

$$CH_2=CHX \longrightarrow -CH_2-CHX-.$$

Amongst the common vinyl monomers, α-methylstyrene has relatively low ceiling temperature.

Table 1.6 Ceiling temperatures and equilibrium monomer concentrations in the polymerizations of some vinyl and cyclic monomers.[50]

Monomer	Solvent	$[M]_e$ mol/L[a]	$T_c(°C)$[b]
Styrene	Benzene	1.2×10^{-4}	110
α-Methylstyrene	None	Pure monomer	61
MMA	Ethyl benzoate	0.61	135
5-membered rings			
Cyclopentene			
(*cis*Polymer)	Benzene	0.6	0
(*trans*Polymer)	Toluene	3.8	0
THF	None	Pure monomer	80
1,3-Dioxolane	Dichloromethane	1.0	1
6-membered ring			
Trioxane	Benzene	0.05	30
7-membered ring			
Oxepane	Dichloromethane	0.08	30
8-membered ring			
Sulfur	None	Pure monomer	159[c]

[a]Equilibrium monomer concentration for the temperature listed in column 4;
[b]ceiling temperature for the concentration of monomer listed in column 3;
[c]floor temperature. "Reprinted with permission from Ref. 50, Copyright © 1988 John Wiley & Sons, Inc."

Ring-opening polymerizations of small ring monomers (number of ring atoms 3 and 4) are highly exothermic (*vide* Chap. 7). Hence, they do not exhibit T_c at low to moderate polymerization temperatures unlike the common ring monomers (number of ring atoms 5 to 7), which do, their polymerizations being only mildly exothermic.[49]

The sharpness of ceiling temperature is a consequence of hundreds of monomer molecules adding one after another in rapid succession to the active center for the polymer formation. Reactions involving small molecules do not exhibit discontinuity since the reactions can proceed to some extent even when the equilibrium constant is somewhat unfavorable. In contrast, a somewhat unfavorable equilibrium constant in each propagation step makes the overall equilibrium constant of the monomer polymer equilibrium exceedingly small ($K_{poly} \approx 0$), since the former is many times multiplied yielding the latter.[50] The polymerization, therefore, exhibits discontinuity, K_{poly} suddenly dropping from a favorable value to an unfavorably small value at T_c. The phenomenon is analogous to the physical aggregation processes, such as the freezing of liquid, condensation of vapor, and micellization of amphiphiles, all of which are exothermic and exoentropic (ΔS–ve) exhibiting discontinuities at certain temperatures.[50,51]

An effect of temperature opposite to that giving T_c is observed in polymerizations in which both ΔH_{poly} and ΔS_{poly} are positive. In such cases, polymerization occurs only above a certain sharp temperature called the floor temperature at which $\Delta G_{poly} = 0$. Well known examples are ring-opening polymerizations of cyclic sulfur (S_8) and selenium (Se_8).[50]

At temperature much lower than T_c, propagation is irreversible. However, the rate constant of depropagation (k_{dp}) increases with temperature at a faster rate than the rate constant of propagation (k_p) since the former has a higher activation energy.[50,51] As ceiling temperature is approached, propagation becomes increasingly reversible.

$$P_n^* + M \underset{k_{dp}}{\overset{k_p}{\rightleftarrows}} P_{n+1}^* \quad . \tag{1.2}$$

The rate of polymerization before equilibrium is reached is given by

$$R_p = k_p \sum_{n=0}^{\infty} [P_n^*][M] - k_{dp} \sum_{n=0}^{\infty} [P_{n+1}^*]. \tag{1.3}$$

When n is not too small, the summation terms of the concentrations of the active centers of various chain lengths in Eq. (1.3) are equal. Representing them as $[P^*]$ one obtains from Eq. (1.3)

$$R_p = (k_p[M] - k_{dp})[P^*]. \tag{1.4}$$

At T_c, k_p [M] becomes equal to k_{dp}. As a result, R_p is reduced to zero and the system is brought to dynamic equilibrium. When there is no discernible polymerization, the monomer concentration at equilibrium, $[M]_e$, is equal to that of the monomer concentration used. Thus, T_c decreases with decrease in monomer concentration. At equilibrium temperatures lower than T_c, $[M]_e$ is less than [M].

The equilibrium monomer concentration, $[M]_e$, is related to the equilibrium constant in the following way

$$K = \frac{[P_{n+1}^*]}{[P_n^*][M]_e} = \frac{1}{[M]_e}. \tag{1.5}$$

(Since for n not too small, $[P_{n+1}^*] = [P_n^*]$.)
$[M]_e$ is also related to the standard free energy of propagation

$$\Delta G_p^0 = -RT_c \ln K = RT_c \ln[M] = RT\ln[M]_e, \tag{1.6}$$

which is independent of the nature and concentration of the active center and so is T_c.[48,51] T_c and $[M]_e$ values of some monomers are included in Table 1.6.

1.9 Polymerizability of Internal Olefins

Internal olefins either do not polymerize or do so with difficulty. Thus, maleic anhydride and maleimide undergo polymerization under rather harsh conditions of high temperature and high initiator concentrations, whereas 2-butene does not polymerize at all.[52,53] The nonpolymerizability of 2-butene and that of similarly substituted olefins is due to kinetic rather than thermodynamic factors.[50] In fact, *syndiotactic* poly(2-butene) is essentially strain-free, which means that polymerization of 2- butene is thermodynamically feasible.[54]

Polymerizations of 1,2-halosubstituted olefins have greater thermodynamic feasibility. The electron-withdrawing effect of halogens reduces σ-π overlap in the monomer, which weakens the π bond resulting in the increase in the polymerization exotherm. The effect increases with increase in the degree of halogen substitution as would be evident from the following data of ΔH^0_{gc} (for polymerization from gas to condensed amorphous phase), which are 108, 146 and 172 kJ/mol for ethylene, vinyl fluoride, and tetrafluoroethylene respectively.[50]

1.10 Molecular Weights and Molecular Weight Distributions

Synthetic polymers invariably do not have unique molecular weights unlike simple low molecular substances. Molecules of various sizes comprise a polymeric material. This is due to the random nature of the polymerization reactions. Accordingly, polymers are ascribed average molecular weights. Various kinds of averages are used, which can be determined experimentally. These are as follows.

The number average molecular weight

$$\overline{M}_n = \frac{\sum N_i M_i}{\sum N_i} = \sum n_i M_i,$$

where N_i and n_i represent respectively the number and number fraction of molecules having molecular weight M_i. The summation terms extend from $i = 1$ to infinity. Thus, this average is the sum of the number-fraction-weighted molecular weights of molecules of various sizes. It is determined by measuring colligative properties of dilute polymer solutions after due correction of non idealities. The usually used methods are membrane osmometry and vapor pressure osmometry. The upper limit of molecular weight determined by the former method is *ca.*, 1 million.[55] This limit is due to the osmotic pressure becoming too small to measure accurately. The lower limit depends on the molecular weight distribution (MWD) due to membranes being permeable to low polymers. For very narrow MWD polymers the lower limit may be *ca.*, 20 000. Vapor pressure osmometry is usually used to measure molecular weights below *ca.*, 20 000. However, VPO instrument was reported to be developed which made possible to measure M_n as high as 400 000.[56]

The weight average molecular weight

$$\overline{M}_w = \sum w_i M_i = \frac{\sum W_i M_i}{\sum W_i} = \frac{\sum N_i M_i^2}{\sum N_i M_i},$$

where W_i and w_i refer respectivly to the weight and weight fraction of molecules of molecular weight M_i. Thus, this average is the sum of the weight-fraction-weighted molecular weights of molecules of various sizes. It is principally determined by the light scattering method, which is based on the measurement of the intensity of light scattered from dilute polymer solutions after due correction of both intermolecular interaction effect and intramolecular interference.[57] The scattering intensity is proportional to the concentration in weight per unit volume and the size of the molecules (molecular weights). The light scattering method is capable of measuring molecular weights ranging from a few hundreds to several millions.[55]

The Z- and Z+1-average molecular weight

$$\overline{M}_z = \frac{\sum W_i M_i^2}{\sum W_i M_i} = \frac{\sum N_i M_i^3}{\sum N_i M_i^2},$$

$$\overline{M}_{z+1} = \frac{\sum W_i M_i^3}{\sum W_i M_i^2} = \frac{\sum N_i M_i^4}{\sum N_i M_i^3}.$$

The Z-, Z + 1-, as well as weight-average molecular weight may be determined by studying the sedimentation equilibrium in dilute polymer solution in a θ solvent in an ultracentrifuge.[57]

The viscosity average molecular weight

$$\overline{M}_v = \left(\frac{\sum W_i M_i^\alpha}{\sum W_i}\right)^{1/\alpha} = \left(\frac{\sum N_i M_i^{1+\alpha}}{\sum N_i M_i}\right)^{1/\alpha}.$$

The viscosity average molecular weight is determined by the viscometric method, which involves the measurement of intrinsic viscosity of the polymer in a solvent and uses the Mark-Houwink-Sakurada equation relating intrinsic viscosity with the viscosity average molecular weight

$$[\eta] = K\overline{M}_v^\alpha.$$

The parameters K and α depend on the thermodynamic interaction of the polymer with the solvent and hence vary from solvent to solvent and with temperature. As the solvent quality increases, the value of α increases from 0.5 in a θ solvent to the maximum 0.8 in a good solvent for randomly coiled linear polymers.[58]

Polydispersity index

The various average molecular weights will have the same value if the polymer is perfectly monodisperse, *i.e.*, if all the molecules are of the same molecular weight. Otherwise, the averages increase in the following order

$$\overline{M}_n < \overline{M}_v < \overline{M}_w < \overline{M}_z < \overline{M}_{z+1}.$$

This is because the molecular weights of the individual molecules are given increasingly greater weightage in the same order. However, a measure of the degree of dispersity in molecular weight is usually obtained from the ratio of weight average to number average molecular weights. This ratio is called the polydispersity index (PDI),

$$\text{PDI} = \frac{\overline{M}_w}{\overline{M}_n}.$$

A much better knowledge of the dispersity in molecular weight is obtained from the molecular weight distribution, which is conveniently and routinely measured by size exclusion chromatography, commonly called the gel permeation chromatography. This chromatographic method separates the molecules according to size. The larger is the molecule greater is the exclusion from various sized pores in the stationary phase material. Accordingly, higher is the molecular weight lower is the elution volume. Near monodisperse polymer standards are required to translate elution volumes to molecular weights. From the distribution curve, the various molecular weight averages may be calculated.

References

1. P.J. Flory, *Principles of Polymer Chemistry*, Cornell Univ. Press, Ithaca, New York, Chap. 1 (1953).
2. H. Staudinger, Nobel Lectures — Chemistry, 1942–1962, Elsevier Pub. Co., Amsterdam, p. 397 (1964).
3. P. Karrer, *Helv. Chim. Acta* (1920) 3, 620.
4. M. Bergmann, *Angew. Chem.* (1925) 38, 1141.
5. R. Pummerer, H. Nielsen, W. Gundel, *Ber.* (1927) 60, 2167.
6. K. Hess, *Chemie der Cellulose*, Leipzig, p. 590 (1928).
7. E. Ott, *Physik. Z.* (1926) 27, 174.
8. O.L. Sponsler, W.H. Dore, *Colloid Symp. Monograph* (Chemical Catalog Co., New York) IV, p. 174 (1926).
9. K.H. Meyer, H.F. Mark, *Ber.* (1928) 61, 593, 1932, 1936, 1939.
10. For a review see H. Morawetz, *Angew. Chem, Intl. Ed.* (1987) 26, 93.
11. W.H. Carothers, *Chem. Rev.* (1931) 8, 353.
12. (a) IUPAC, *Pure App. Chem.* (1994) 66, 2483; (2002) 74, 1921. (b) IUPAC "Compendium of Macromolecular Nomenclature", W.V. Metanomski Ed., Blackwell Scientific, Oxford (1991). (c) Glossary of terms related to kinetics, thermodynamics and mechanisms of polymerization (IUPAC Recommendation 2007).
13. W.H. Carothers, *J. Am. Chem. Soc.* (1929) 51, 2548.
14. P.J. Flory, *J. Am. Chem. Soc.* (1952) 74, 2718.
15. Dendrimers and Other Dendritic Polymers, J.A. Frechet, D.A. Tomalia Eds., Wiley Series in Polymer Science, John Wiley & Sons Ltd. (2001).
16. D. A. Tomalia *et al.*, *Polym. J.* (1985) 17(1), 117.
17. G.R Newcombe *et al.*, *J. Org. Chem.* (1985) 50, 2003.
18. M. Fischer, F. Vogtle, *Angew. Chem. Int. Ed. Eng.* (1999) 28, 884.
19. (a) C. J. Hawker, J.M.J. Frechet, *J. Am. Chem. Soc.* (1990) 112, 7638; Macromolecules (1990) 23, 4726. (b) T.M. Miller, T.X. Neenan, *Chem. Mater.* (1990) 2, 346.
20. B. Wunderlich, *Macromolecular Physics — vol. 3. Crystal Melting*, Academic Press, New York (1980).
21. H.F. Mark, *J. App. Polym. Sci. App. Polym. Symp.* (1984) 39, 1.
22. E.B. Wilson, *Proc. Natl. Acad. Sci.*, USA (1957) 43, 816.
23. E.L. Eliel, *Stereochemistry of Carbon Compounds*, Tata-McGrew-Hill Publishing Co, New Delhi, p. 124 (1962).
24. J.B. Campbell, *Science* (1963) 141, Issue 3578, 329.
25. W.O. Baker, C.S. Fuller, *J. Am. Chem. Soc.* (1942) 64, 2399.
26. P.J. Flory, *Principles of Polymer Chemistry*, Cornell Univ. Press, Ithaca, New York, Chaps. 2 and 6 (1953).
27. P.J. Flory, F.S. Leutner, *J. Polym. Sci.* (1948) 3, 880; (1950) 5, 267.
28. L. Pauling, *The Nature of the Chemical Bond*, 3rd ed., Cornell University Press, New York, p. 261 (1960).

29. T. Ito, H. Shirakawa, S. Ikeda, *J. Polym. Sci. Polym. Chem. Ed.* (1974) 12, 11.
30. H. Shirakawa, *Angew. Chem. Int. Ed.* (2001) 40, 2574.
31. A.G. MacDiarmid, *Angew. Chem. Int. Ed.* (2001) 40, 2581.
32. C.K. Chiang et al., *Phys. Rev. Lett.* (1977) 39, 1098.
33. H. Shirakawa, E.J. Louis, A.G. MacDiarmid, C.K. Chiang, A.J. Heeger, *J. Chem. Soc. Chem. Comm.* (1977) 579.
34. S. Parker, A. Janossy, Chemistry of Doping and Distribution of Dopants in Polyacetylene, in, *Handbook of Conducting Polymers*, T.K. Skotheim Ed., Marcel Dekker Inc., New York, Vol. 1, p. 45 (1986).
35. J.E. Frommer, R.R. Chance, Electrically Conducting Polymers, in, *Ency. Polym. Sci. & Eng.*, 2nd ed., H. Mark, N.M. Bikales, C.G. Overberger, G. Menges Eds., Wiley — Interscience, New York, p. 462 (1985).
36. B. Norden, E. Krutmeijer, The Nobel Prize in Chemistry, 2000: Conductive Polymers (Advanced Information), The Royal Swedish Academy of Sciences, Information Department, Stockholm, Sweden.
37. B.M. Mandal, *J. Ind. Chem. Soc.* (1998) 75, 121.
38. A.G. MacDiarmid, A.J. Epstein, *Faraday Discuss. Chem. Soc.* (1989) 88, 317.
39. A.G. MacDiarmid, J.-C. Chiang, A.F. Richter, A.J. Epstein, *Synth. Met.* (1987) 18, 285.
40. R.H. Friend, *Pure Appl. Chem.* (2001) 73, 425.
41. J.H. Burroughes et al., *Nature* (1990) 347, 539.
42. R.H. Friend et al., *Nature* (1999) 397, 121.
43. L.B. Groenendaal et al., *Adv. Mater.* (2000) 12(7), 481.
44. P.L. Burn et al., *J. Chem. Soc. Perkin Trans.* 1 (1992) 3225.
45. R. Gowri, D. Mandal, B. Shivkumar, S. Ramakrishnan, *Macromolecules* (1998) 31, 1819.
46. G. Padmanabhan, S. Ramakrishnan, *J. Am. Chem. Soc.* (2000) 122, 2244.
47. I.F. Perepichka, D.F. Perepichka, H. Meng, F. Wudl. *Adv. Mater.* (2005) 17, 2281.
48. F.S. Dainton, K.J. Ivin, *Proc. Roy. Soc.* (1952) A212, 207.
49. F.S. Dainton, K.J. Ivin, *Quart Rev.* (1958) 12, 61.
50. K.J. Ivin, W.K. Busfield, in, *Ency. Polym. Sci. & Eng.*, H.F. Mark, N.M. Bikales, C.G. Overberger, G. Menges Eds., Wiley-Interscience, New York, Vol. 12, p. 555 (1988).
51. K.J. Ivin, *J. Polym. Sci. Polym. Chem. Ed.* (2000) 38, 2137.
52. R.M. Joshi, *Makroml. Chem.* (1962) 55, 35.
53. R.M. Joshi, *Makromol. Chem.* (1963) 62, 140.
54. R.M. Joshi, B.J. Zwolinski in G.E. Ham Ed., Vinyl Polymerization, Edward Arnold, London, p. 485 (1967).
55. A.R. Cooper, *Molecular Weight Determination*, in, *Ency. Polym. Sci. & Tech.*, 3rd ed. H. Mark Ed., Wiley-Interscience, Vol. 10, p. 445 (2004).
56. A.A. Wachter, W. Simon, *Anal. Chem.* (1969) 41, 90.
57. P.J. Flory, *Principles of Polymer Chemistry*, Cornell University Press, Ithaca, New York, Chap. 8 (1953).
58. P.J. Flory, *Principles of Polymer Chemistry*, Cornell University Press, Ithaca, New York, Chap. 14 (1953).

Problems

1.1 Explain the following facts:

   (i) Polyesters have lower melting points than polyamides.
   (ii) Amorphous polymer is obtained in the polycondensation of hexamethylenediamine, adipic acid, and 6-aminocaproic acid.
   (iii) Nylon 66 may be used both as fiber and plastic.
   (iv) Poly(vinyl chloride) prepared at ambient to moderate temperatures is poorly crystalline, whereas similarly prepared poly(vinylidene chloride) is highly crystalline.

1.2 Living polymerization is a chain polymerization even though the pattern of molecular weight evolution with conversion is drastically different from that observed in conventional chain polymerization. Explain why it is so.

1.3 How can you explain the increase in the flexibility of polymers with temperature?

1.4 A semicrystalline polymer has T_m at 130°C and T_g at −50°C. What type of material it is at ambient temperature?

1.5 Show graphically how would the specific volume and the volume expansion coefficient of a semicrystalline polymer change with temperature.

1.6 Explain why the next addition following a head-to-head addition in radical polymerization should be invariably a tail-to-tail one. What is likely to be the second next addition?

1.7 Explain why the proportion of the head-to-head addition of monomers in radical polymerization increases with temperature.

Chapter 2

Step Polymerization

The scientifically controlled synthesis of the first step polymer Bakelite, a phenol formaldehyde resin, was reported by Baekeland in 1909. However, it was not until the early 1930s that the principles of step polymerization were unraveled through the pioneering works of Carothers[1] and Flory.[2] We have already discussed some of these principles in the previous chapter and shall discuss the others in this one. It may be recalled from the discussions in the previous chapter that the step polymers are formed by the stepwise reaction (usually condensation) of functional groups. As a result, molecular weight increases slowly with the progress of reaction until the latter has highly advanced, say, >95% (Chap. 1, Fig. 1.1). Linear polymers form when the monomers are strictly bifunctional, whereas nonlinear ones form when at least one of the monomers has functionality more than two. Nonlinear polymerization may also lead to gelation provided the proportion of the coreactive functional groups lies within certain range and the polymerization is carried to high enough conversion. In this chapter, we shall treat these features quantitatively. Besides, we shall discuss the underlying principles in the synthesis of various step polymers.

2.1 Principle of Equal Reactivity of Functional Groups and Kinetics of Polymerization

Due to the random nature of the polymerization, not only not all the polymer molecules have the same chain lengths but also the number fractions of the different sized molecules are different. Any one of these molecules bearing a functional group may react with any other bearing a coreactive functional group. Thus, there will be innumerable reactions involving molecules of different chain lengths. Under these circumstances, if the reactivity of functional groups attached to the ends of the molecules of different chain lengths were different, one would have to treat the various reactions as separate ones with unequal rate constants, although the reactions are chemically the same. As a result, the treatment of the kinetics would be highly complicated. On the other hand, if the functional groups of a given kind were of equal reactivity, the individual reactions would have the same rate constant. It would then be perfectly legitimate to treat the reactions as those between the functional groups rather than between the molecules and the rate expressed in terms of the global functional group concentration using a single rate constant.[2-4]

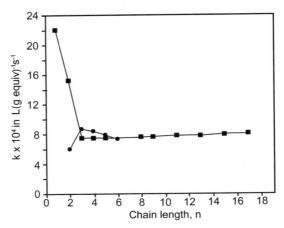

Figure 2.1 Variation of rate constant of esterification of carboxylic acids with ethyl alcohol catalyzed by HCl at 25°C with chain lengths of the acids (■) H(CH$_2$)$_n$COOH, (●) (CH$_2$)$_n$(COOH)$_2$. k [H$^+$]refers to the pseudo first order rate constant with the esterification carried out in large excess of ethanol containing HCl.[5]

Indeed, evidence of equal reactivity of functional groups is available from reaction rate studies in a series of homologous compounds under comparable conditions.[5–7] The rate constants of functional group reactions of individual members of a homologous series and a given coreactant are virtually equal for all members except for the smallest ones with too short chain lengths. For instance, Fig. 2.1 shows the variation of the rate constants of the acid catalyzed esterification of carboxylic acids, belonging to two homologous series, with ethyl alcohol.[5] One series is constituted of monocarboxylic acids, H(CH$_2$)$_n$COOH, whereas the other of dicarboxylic acids, (CH$_2$)$_n$(COOH)$_2$. In the monocarboxylic acid series, the rate constant decreases sharply as n increases from one to three and remains nearly the same with further increase in n. In the dicarboxylic acid series, the data are not as extensive as in the other series. Nevertheless, the data reveal that the rate constant reaches the limiting value observed with the monocarboxylic acids when n = 6. Similar results are reported for the saponification of esters or the etherification of alkyl iodides.[6,7]

Flory used the principle of equal reactivity of functional groups of the same kind to treat the kinetics of polycondensation.[2,3] As pointed out above, the assumption of equal reactivity immensely simplifies the kinetics and allows one to write the rate equation in terms of the functional group concentration using a single rate constant.

Thus, one may write for the rate of self-catalyzed polyesterification of a dicarboxylic acid with a diol as [3]

$$\frac{-d[COOH]}{dt} = k[COOH]^2[OH], \qquad (2.1)$$

with the concentration terms expressed in equivalent/volume unit.

The third order reaction is a consequence of the carboxylic acid group acting as both a catalyst and a reactant. Integration of Eq. (2.1) for equal concentration (c) of COOH and

OH groups gives

$$\frac{1}{c^2} = 2kt + \text{constant}. \qquad (2.2)$$

Representing initial concentrations of both COOH and OH groups as c_o and the fraction of c_o reacted at time t as p, called the extent of reaction, Eq. (2.2) transforms to

$$\frac{1}{(1-p)^2} = 2c_0^2 kt + \text{constant}. \qquad (2.3)$$

We shall see later that in the step polymerization of bifunctional monomers with the coreactive functional groups present in equivalent amounts the number average degree of polymerization ($\overline{DP_n}$) at a given extent of reaction is given by $1/(1-p)$. Substituting $\overline{DP_n}$ for $1/(1-p)$ in Eq. (2.3) one obtains Eq. (2.4), which quantitatively describes $\overline{DP_n}$ evolution with time.

$$\overline{DP_n}^2 = 2c_0^2 kt + \text{constant}. \qquad (2.4)$$

Figure 2.2 shows the kinetic plots according to Eq. (2.3) for two systems: (i) the polyesterification of diethylene glycol with the dibasic adipic acid and (ii) the esterification of diethylene glycol with the monobasic caproic acid at 166°C.[3] It transpires that both the monoesterification and the polyesterification follow the same course. The two curves in the figure superimpose by merely multiplying all time values in one of them by a suitable factor. The lower rate for the monoesterification arises largely from the use of lower initial concentrations of the reactants.[3] The results provide one of the strongest proofs of the validity of the assumption of equal reactivity of functional groups.

The third order rate law, however, does not apply to the major part of the reaction, viz., the region of $p = 0$ to about 0.8. This has been attributed to the large changes in the nature of the medium caused by the consumption of the relatively abundant polar functional groups in this region.[2,3] Esterification being proton catalyzed, the rate constant is affected by the

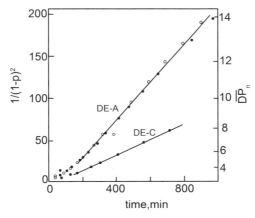

Figure 2.2 Kinetic plots of esterification of (1) diethylene glycol and adipic acid (DE-A) and (2) diethylene glycol and caproic acid (DE-C) at 166°C. The right-hand ordinate gives the degree of polymerization for the former. "Reprinted in part with permission from Ref. 3. Copyright © 1939 American Chemical Society."

large changes that occur in dielectric constant, polarity, and proton-solvating power of the medium. At large extents of reaction, the concentrations of functional groups are far too low to affect significantly the nature of the medium by their consumption.

With the use of a strong acid catalyst, one may express the polyesterification as a second order reaction[3,4]

$$-\frac{dc}{dt} = k'c^2, \qquad (2.5)$$

where c is the concentration each of COOH and OH functional groups, and k' is proportional to the catalyst (H^+) concentration. Proceeding as above, one gets

$$\frac{1}{1-p} = c_0 k't + \text{constant}, \qquad (2.6)$$

where p is the extent of reaction and c_0 is the initial concentration each of COOH and OH groups. Substituting $\overline{DP}_n$ for $1/1 - p$ in Eq. (2.6) gives

$$\overline{DP}_n = c_0 k't + \text{constant}. \qquad (2.7)$$

Figure 2.3 shows that the acid catalyzed polyesterification of diethylene glycol with adipic acid follows second order kinetics, over the extent of reaction ranging from $p = 0.9$ to at least 0.989. In this range of p, $\overline{DP}_n$ increases from 10 to 90 and the viscosity of the polymerizing mass increases more than 2000 fold. The results not only prove the independence of the reactivity of the functional end groups on the lengths of the polymer chains, but also that the high viscosity of the polymerizing mass does not affect the rate constant.[3]

Ordinarily, one would expect that the increase in the medium viscosity should decrease the translational diffusion rate of the polymer molecules and, hence, decrease the collision frequency and the rate constant of the reaction between them. However, collision between

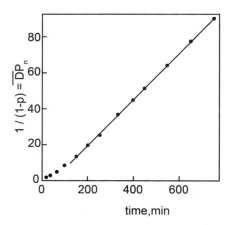

Figure 2.3 Kinetic plot for the esterification of diethylene glycol and adipic acid catalyzed by 0.4 mol % p-toluenesulfonic acid at 109°C. "Reprinted with permission from Ref. 3. Copyright © 1939 American Chemical Society."

molecules occurs in sets in the liquid phase. In each set, a pair of molecules undergoes multiple collisions before seperating.[8,9] However, the molecules diffuse away from the colliding state at the same slower rate as they diffused in. Accordingly, even if the average inter set time increases due to the lower mobility of the molecules, the duration of collision in a set is proportionately increased. However, it is the rate of collision between the coreactive functional end groups present in different polymer molecules that should be of ultimate concern. If the actual collision rate of the molecules themselves does not decrease that of the coreactive end groups also should not. This is because the mobility of the end groups occurs through the diffusion of segments in the polymer coil, which is almost as rapid as the diffusion of molecules in simple liquids is.[2]

Nevertheless, if the activation energy of the reaction is small ($<\sim 10$ kJ/mol), the reaction rate may be too fast for the translational diffusion rate to keep pace with it in maintaining the equilibrium concentration of pairs of reactants within colliding distance of each other.[3] As a result, the rate constant falls and the reaction becomes translational diffusion-controlled. However, to reach this state the viscosity must exceed certain threshold; the bimolecular termination of polymer radicals in chain polymerization is a typical example (*vide* Chap. 3, Sec. 3.10.5). In contrast, the activation energy of step polymerization is large enough so that the above situation does not arise. The rate constant remains unaffected by the increase in viscosity. However, it may not remain so when the viscosity becomes far too high.

2.2 Ring vs. Chain Formation

In the previous section, we have considered only the intermolecular reaction between coreactive functional groups. However, intramolecular reaction may also occur. This gives rise to cyclic products (Scheme 2.1).[2] Both reactions are reversible, at least to some extent, in almost all step polymerization.[10–12]

In principle, the rings may form from all chains with coreactive terminal groups starting from the chain with one repeating unit. Scheme 2.1 shows the reactions with an A–B type monomer. Similar reactions also occur with A-A type monomers as well, dimethylsilanediol being an example. When the monomers are mixtures of A–A and B–B types, rings can form only if the chains contain equal numbers of structural units of the two types.

Ring formation is a unimolecular reaction. The rate constant for ring formation (k_i $i = 1, 2, 3$ for $n = 1, 2, 3$ respectively) depends on the ring strain as well as on the probability that the chain assumes ring conformation. Higher is the ring strain higher is the

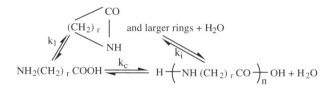

Scheme 2.1 Intramolecular and intermolecular condensations.

Table 2.1 Ring strain in cycloalkanes.[13]

Ring size[a]	Strain, kJ/mol	Ring size[a]	Strain, kJ/mol
3	115.5	11	46.0
4	109.6	12	15.0
5	27.2	13	21.8
6	0	14	0
7	26.4	15	6.2
8	40.2	16	6.7
9	52.7	17	−14.2
10	50.2		

[a] Number of carbon atoms in ring.

activation energy and higher is the probability of the ring conformation more favorable is the entropy of activation.[13] Hence, a low ring strain and a high probability of ring conformation facilitate ring formation. Whereas ring strain varies with the ring size in a complex way, as will be discussed shortly, the probability of a chain assuming ring conformation decreases monotonically with increase in the length of the chain.

Ring strain may be conveniently discussed using cycloalkanes as models. Table 2.1 records the magnitude of ring strain for rings varying in size from 3 to 17 ring atoms.[13] The strain is calculated from the difference between the heat of combustion per methylene group of a cycloalkane and that of a long chain molecule, multiplying the difference with the number of ring atoms. A scale model helps to understand the strain pattern.

The small rings (3 and 4 ring atoms) have high strain, which is due to the distortion in the valance angle. Among the common rings (5 to 7 ring atoms), the 5-membered ring has some angle strain, the valence angle being $108°$ compared to the normal tetrahedral angle of $109°28'$. Six-membered ring is free from strain in its puckered chair conformation. Seven- to eleven-membered rings are also puckered but not free from strain, part of which is due to the presence of eclipsing hydrogen atoms. However, 8- to 11-membered rings (the medium rings) are more strained than the 7-membered one is. This is due to the additional strain caused by the crowding hydrogen atoms in the ring interior (transannular strain).[13,14] This type of strain decreases sharply as the ring size increases to 12-membered and becomes lowered still with further increase in ring size. The total strain becomes vanishingly small in the rings containing 14 or more ring atoms. The rings containing 12 atoms or larger are referred to as the large rings.

Considering together the effects of both the ring strain and the probability of the ring conformation the experimentally observed ease of ring formation may be understood. The ease of ring formation is relatively high for the 3-membered ring due to the very favorable probability factor. It drops sharply for the 4-membered ring due to the large deterioration in the probability factor.[13] It then rises sharply for the 5-membered ring due to the large reduction in strain, then decreases for the 6-membered ring due to the deteriorated probability factor overcoming the effect of the slightly improved strain factor. It decreases further for the 7-membered ring due to both the strain factor and the probability factor deteriorating. The decrease continues for the medium rings due to the continued deterioration of both factors. Then the ease increases for the large rings up to about 18 ring atoms due to the decrease in strain.[10–14] With further increase in the ring size, it decreases monotonically as the probability factor deteriorates even as the ring strain vanishes.[14] However, the trend may not be exactly the same for rings of all chemical types, *e.g.*, cyclic esters, anhydrides, amides, carbonates, and so on.[12] This is due to the changes in bond lengths, valence angles, and steric hindrance to ring conformation. However, the ease of ring formation discussed above is based on the kinetically controlled yield of rings. A different result may follow in the thermodynamically equilibrated systems.[10–12]

In fact, cyclic monomers smaller than 5-membered and larger than 7-membered are thermodynamically unstable *vis-à-vis* their chain polymers. On the other hand, the stability of 5- to 7-membered rings (the common ring monomers) depends critically on their chemical structure.[15] Thus, whereas the 5-membered cyclic γ-butyrolactone is stable relative to its chain polymer at temperatures ranging from low to the melting temperature, γ-butyrolactam becomes so only at elevated temperatures, *e.g.*, at the melt polymerization temperature of γ-aminobutyric acid. This is because the favorable enthalpy factor for the transformation of the 5-membered ring to the linear polymer is outweighed by the unfavorable entropy factor at elevated temperatures (*vide* Chap. 7, sec. 7.1.1). This is true also for the 6-membered δ-valerolactam. Thus, the ring $\rightleftharpoons$ chain equilibrium is unfavorable at the melt polymerization temperatures of the above-mentioned amino acids resulting in ring formation. Even, β-aminopropionic acid gives the 4-membered cyclic β-propiolactam, which is not expected in view of the high strain in the 4-membered ring. However, the melting temperature of the polymer is so high that the polymer becomes unstable[16]

$$NH_2(CH_2)_r COOH \longrightarrow \overline{NH\,(CH_2)_r\,CO} \quad \text{for } (r = 2 \text{ to } 4).$$

Thus, from the studies of the attempted melt polymerizations of A–B monomers the following generalizations are made. Linear polymers are obtained if the rings to be formed are 3- or 4-membered except when the melting temperatures are too high to give stable polymers (the β-amino acid case discussed above being an example). The products are exclusively rings if the latter are 5-membered. They are either rings or linear polymers or both if 6- or 7-membered ring is to be formed and invariably linear polymers if the rings to be formed are 8-membered or larger.[2]

However, rings not only form through intramolecular condensation of end functional groups in n-mers (n $\geq$ 1) but also through intramolecular exchange reaction, which is known as backbiting. The latter involves intramolecular reaction between a terminal functional group and an interunit linkage in the polymer chain as shown below in the case of polyamide. Although the nucleophilic attack of the amino group on the penultimate carbonyl is shown; in principle, it can occur on any of the interunit carbonyls leading to rings of various sizes

$$HO+\overset{O}{\overset{\|}{C}}-R-\overset{O}{\overset{\|}{C}}-NH\overset{}{\underset{x}{\rightarrow}}\overset{O}{\overset{\|}{C}}-R-\overset{O}{\overset{\|}{C}}-NH_2 \rightleftharpoons HO+\overset{O}{\overset{\|}{C}}-R-\overset{O}{\overset{\|}{C}}-NH\overset{}{\underset{x}{\rightarrow}}H + R\overset{\overset{O}{\overset{\|}{C}}}{\underset{\underset{O}{\overset{\|}{C}}}{\diagdown}}NH.$$

The proportion of large rings may be theoretically determined using Jacobson–Stockmayer treatment of ring ⇌ chain equilibrium[17]

$$C_{x+y} \rightleftharpoons C_y + R_x, \qquad (2.8)$$

where C and R represent respectively chain and ring and the subscripts refer to the number of structural units. Since there is no change in the number of bonds or the bond types, the reaction enthalpy should be negligibly small provided the rings are strain-free. Thus, the entropy change solely determines the equilibrium constant. It equals the sum of the entropy changes of the following reactions.

$$C_{x+y} \rightleftharpoons C_x + C_y. \qquad (2.9a)$$

$$C_x \rightleftharpoons R_x. \qquad (2.9b)$$

Using statistical thermodynamic principles the entropy change of reaction (2.9a) may be related to the probability that a skeletal chain atom present at one or the other of the chain ends is situated away from a bonding partner present at an end of another chain at distances greater than that of the bonding. This probability increases with the system dilution. The entropy change of reaction (2.9b) may be related to the probability of the ring conformation, which may be estimated using the statistical theory of chain conformation with the freely rotating chain as a simplified model. Utilizing the sum of these entropy changes the equilibrium constant K of the ring ⇌ chain equilibrium (2.8) may be evaluated and then the concentration of R_x.

The treatment provides Eq. (2.10) for the concentration of large rings of a given size (15 or more ring atoms) present in equilibrium with linear polymer chains prepared from an A–B monomer as[17]

$$[R_x] = 2B(p')^x x^{-5/2}, \qquad (2.10)$$

where

x = number of structural units,
R_x = number of x-mer rings per cm^3,
p' is the extent of reaction in the chain polymer portion,
B = $\frac{[3/(2\pi\nu)]^{3/2}}{2b^3}$, ν being the number of chain atoms per structural unit, and b the effective bond length in cm.

The concentration (in g/cm^3) and the weight fraction of the combined large rings follow respectably as

$$c_R = \frac{M_0}{N_A} \sum x[R_x] = \frac{2BM_0}{N_A} \sum_{x=1}^{\infty} (p')^x x^{-3/2}, \qquad (2.11)$$

and

$$w_R = \left(\frac{2BM_0}{cN_A}\right) \sum_{x=1}^{\infty} (p')^x x^{-3/2}, \qquad (2.12)$$

where M_0 is the molecular weight per structural unit, c is the total concentration of structural units (present in both rings and chains) in g/cm³, and N_A is the Avogadro number. All other symbols have been defined above.

Inasmuch as rings do not have unreacted functional groups, p' may be related to the overall extent of reaction, p, in the total polymerization system as[19]

$$1 - p' = (1-p)/(1-w_R). \quad (2.13)$$

For $p' = 1$, i.e., when the chains are infinitely long, the values of the summation functions in the above equations are[17,18]

$$\sum_{x=1}^{\infty}(p')^x x^{-5/2} = 1.341. \quad (2.14a)$$

$$\sum_{x=1}^{\infty}(p')^x x^{-3/2} = 2.612. \quad (2.14b)$$

Equation (2.11) predicts a constant total concentration of rings, which is independent of the amount of monomer polymerized provided the rings are large and the chains present in equilibrium are long, i.e., p' is near unity. It may be noted that the summation functions also include the contribution of the ring containing a single repeating unit, which may not have enough number of chain atoms for the statistical theory to be applicable.[2] However, Eq. (2.12) predicts that the weight fraction of rings decreases with the increase in the amount of monomer polymerized. The theory also predicts a critical concentration of structural units below which the products will be all rings, i.e., $w_R = 1$. This may be derived from Eq. (2.12) for $w_R = 1$ using the value of the summation term for $p' = 1$. The procedure gives $BM_0/c_{crit}N_A = 0.19$.[17]

When the monomer is A–A type in which the A groups can self-condense, e.g., a silanediol, the numerical factor 2 is to be omitted from the above equations.

For step polymers prepared from equivalent amounts of A–A and B–B monomers v in the B parameter refers to the number of chain atoms per repeating unit, and M_0 refers to the mean molecular weight per structural unit. Besides, since the rings contain only even number (x) of structural units, necessary changes in the summation term must be made. Thus, the weight fraction of the total amount of rings is given by

$$w_R = \left(\frac{2BM_0}{cN_A}\right) \sum_{\frac{x}{2}=1}^{\infty} (p')^x \left(\frac{x}{2}\right)^{-3/2}. \quad (2.15)$$

It has been calculated with b estimated from dilute solution viscosity measurements of polymers that the ring content in high molecular weight polyester or polyamide of A–AB–B type with 10–20 skeletal atoms in the repeating unit would be only about 2.5 percent. For the same values of v, and b, the ring content in A–B type polymer would be about twice as much. However, these values of the ring content are valid only in ring $\rightleftharpoons$ chain equilibrated systems, but not in kinetically controlled ones.[19]

The ease of ring formation discussed so far concerns rings, which have carbon as the predominant member. In rings constituted of dimethylsiloxane units, the 6-membered ring is moderately strained, whereas the rings containing 8 ring atoms or more are strain-free.[20] Accordingly, the 6-membered ring exists only in trace amount in the ring ⇌ chain equilibrated system and the 8-membered ring is the most abundant in the ring fraction.[21–23]

$$HO{-}(Si(CH_3)_2{-}O)_n{-}H \rightleftharpoons (Si(CH_3)_2{-}O)_n + H_2O \cdot \quad (2.16)$$

The yield of large rings monotonically decreases in accordance with Jacobson-Stockmayer equation.[21]

2.3 Intermolecular Interchange Reactions

Besides the intramolecular interchange reaction discussed above, intermolecular interchange may also occur during polymerization, processing, or blending.[24] In this reaction, sections of polymer chains are exchanged between the chains themselves, which brings about a change in the molecular weight distribution. However, the number average molecular weight does not change because the number of molecules remains unchanged. The reaction may either involve an end functional group in one molecule and an interunit linkage in another or two interunit linkages belonging to two different molecules. As an example, Scheme 2.2 shows the interchange reactions in polyamide that are of both kinds. Reactions analogous to those shown in the scheme take place in many other polymers such as polyesters, polyanhydrides, polycarbonates, polysulfides, and silicones.[2]

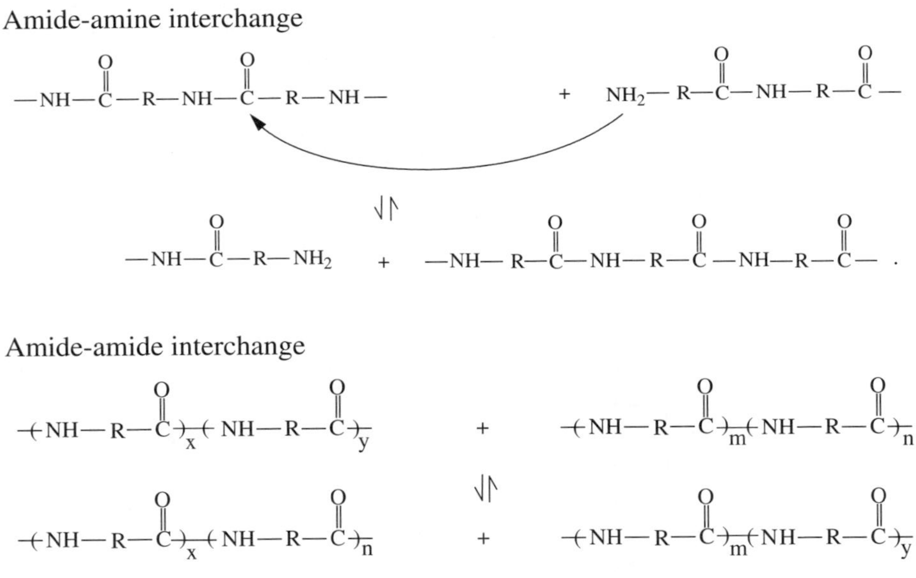

Amide-amine interchange

$$-NH-\underset{O}{\overset{\|}{C}}-R-NH-\underset{O}{\overset{\|}{C}}-R-NH- \quad + \quad NH_2-R-\underset{O}{\overset{\|}{C}}-NH-R-\underset{O}{\overset{\|}{C}}-$$

$$\updownarrow$$

$$-NH-\underset{O}{\overset{\|}{C}}-R-NH_2 \quad + \quad -NH-R-\underset{O}{\overset{\|}{C}}-NH-R-\underset{O}{\overset{\|}{C}}-NH-R-\underset{O}{\overset{\|}{C}}- \; .$$

Amide-amide interchange

$$+NH-R-\underset{O}{\overset{\|}{C}}\!\!\!\!+_x\!\!+NH-R-\underset{O}{\overset{\|}{C}}\!\!\!\!+_y \quad + \quad +NH-R-\underset{O}{\overset{\|}{C}}\!\!\!\!+_m\!\!+NH-R-\underset{O}{\overset{\|}{C}}\!\!\!\!+_n$$

$$\updownarrow$$

$$+NH-R-\underset{O}{\overset{\|}{C}}\!\!\!\!+_x\!\!+NH-R-\underset{O}{\overset{\|}{C}}\!\!\!\!+_n \quad + \quad +NH-R-\underset{O}{\overset{\|}{C}}\!\!\!\!+_m\!\!+NH-R-\underset{O}{\overset{\|}{C}}\!\!\!\!+_y \; .$$

Scheme 2.2 Intermolecular interchange in polyamide.

As will be discussed later in this chapter, using a statistical method Flory deduced the equation for the distribution of polymer molecular weights in equilibrated interchange and provided experimental evidence in support.[26] The same distribution was predicted also in linear step polymerization.[25,27] The distribution is not ordinarily altered even when interchange occurs during polymerization.

Interchange provides an attractive method of preparing random copolymers by melt blending of two polymers prepared from different monomers. However, it may prove damaging also. For example, block copolymers may transform to random copolymers during melt processing and, accordingly, lose their molecular architecture and the associated attractive properties.[28]

2.4 Degree of Polymerization

2.4.1 Carothers equation

Carothers deduced an equation relating the degree of polymerization with the average degree of functionality of monomers and the extent of reaction in step polymerization with the coreactive functional groups present in equivalent amounts, as presented below.[29,30]

Let

f = degree of functionality, *i.e.*, functionality per monomer molecule present in the reaction system. When two monomers with different functionalities are used in equivalent amounts, f is given by the average functionality per monomer molecule.

N_0 = number of molecules (monomer) initially present.

N = number of molecules (monomer + r-mers) present at an extent of reaction p.

At each step of polymerization, for each two coreactive functional groups reacted intermolecularly one molecule decreases from the system. Therefore, assuming that the reactions are all intermolecular and there are no side reactions, $2(N_0 - N)$ = number of functional groups lost.

The extent of reaction may be expressed as

$$p = \frac{2(N_0 - N)}{N_0 f}, \quad (2.17)$$

$N_0 f$ being the number of functional groups initially present.

Now, the number average degree of polymerization, $\overline{DP}_n$, refers to the average number of structural units per molecule. Thus, it may be related to N_0 and N as

$$\overline{DP}_n = \frac{N_0}{N}. \quad (2.18)$$

Substituting $\overline{DP}_n$ for $\frac{N_0}{N}$ in Eq. (2.17) and rearranging one obtains the Carothers equation

$$\overline{DP}_n = \frac{2}{2 - fp}. \quad (2.19)$$

The number average molecular weight follows as

$$\overline{M}_n = M_0 \overline{DP}_n,$$

M_0 being the mean molecular weight per structural unit,

Application of Carothers equation to various polymerization systems are discussed below.

2.4.1.1 Bifunctional polycondensation

Consider the polycondensation of a hydroxy acid or of a diol and a diacid.

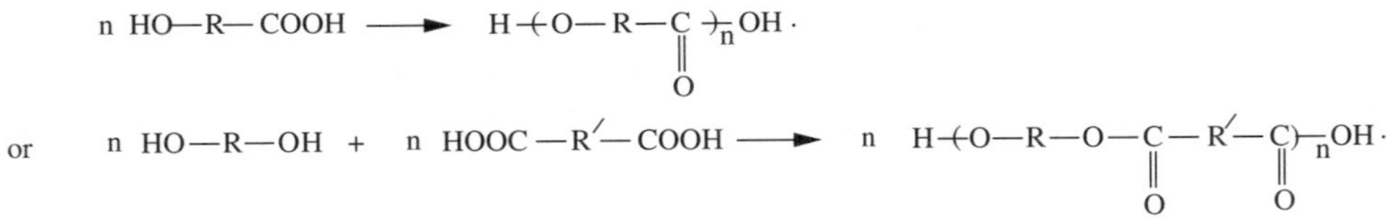

$$n\ HO-R-COOH \longrightarrow H{\leftarrow}O-R-\underset{\underset{O}{\|}}{C}{\rightarrow}_n OH$$

or

$$n\ HO-R-OH + n\ HOOC-R'-COOH \longrightarrow n\ H{\leftarrow}O-R-O-\underset{\underset{O}{\|}}{C}-R'-\underset{\underset{O}{\|}}{C}{\rightarrow}_n OH$$

In both reactions, all the molecules would have combined into a single linear molecule of virtually infinite molecular weight if it were possible to condense intermolecularly all the hydroxyl and the carboxyl groups excepting, of course, the pair constituting the two end groups in the ultimate molecule. However, since a chemical reaction cannot ever be complete, the polymers produced will always have finite molecular weights, which, of course, will increase with the increase in the extent of reaction.

In this case, the degree of functionality is two so that the Carothers equation gives

$$\overline{DP}_n = \frac{1}{1-p}. \tag{2.20}$$

The $\overline{DP}_n$ evolution with the extent of reaction, already shown in Fig. 1.1 in Chapter 1, follows from Eq. (2.20).

The numerical values of $\overline{DP}_n$ at different extents of reaction as calculated from the equation are given in Table 2.2. It is obvious that the reaction is to be carried to beyond 99% completion to get useful polymer ($\overline{DP}_n > \sim 100$). This requires long time. For instance, the acid catalyzed polyesterification of adipic acid with diethylene glycol follows second order kinetics over the extent of reaction 0.9 to at least 0.989, as already discussed in Sec. 2.1. With the use of 0.4 percent p-toluenesulfonic acid catalyst, the reaction half-life following 95% completion is 2.5 h at 109°C.[3] Since the half-life varies inversely as the initial concentration (the reaction being second order and the two reactants being used at equimolar concentration)

Table 2.2 Degree of polymerization at selected extents of reaction according to the Carothers equation.

% reaction	$\overline{DP}_n$
50	2
90	10
95	20
99	100
99.5	200
99.9	1000
100	∞

each successive half-life is doubled. Thus, if it takes the reaction 2.5 h to reach 97.5% from 95%, it will take 5 h more to reach 98.75% and a further 10 h to reach 99.375%. However, the degree of polymerization reaches only 160 from the value of 25 at 95% reaction.

In order to cut down the inordinately long time required for the reaction to reach 99% and above, it is the usual practice to carry out the last few percent of reaction at elevated temperature and under vacuum (0.1 to 0.5 mm Hg) or dry nitrogen purge. The increase of temperature increases the reaction rate, while the application of vacuum or nitrogen purging efficiently removes the volatile by-product, which would otherwise have set up an equilibrium resulting in the cessation of reaction.[3]

2.4.1.2 Control of degree of polymerization

It follows from Eq. (2.20) that the $\overline{DP}_n$ can be controlled by stopping the reaction at suitable extents of reaction, for example, by quenching the reaction to ambient temperature. However, the method suffers from the problem that the polymer, being essentially an A–B type macromonomer, retains its potential for growth. The degree of polymerization would increase whenever the polymer be subjected to heat treatment, e.g., during fabrication of the end-use material.

The best way to obtain a stable polymer is to have like functional groups at its both ends. This can be achieved in the condensation of A–A and B–B type monomers by using one monomer in stoichiometric excess over the other. As a result, the functional end groups of the polymer will be the same as those of the excess monomer. This method of stabilization however succeeds provided the like functional groups do not condense with each other. An additional way would be to add a monofunctional reactant into a stoichiometrically balanced monomer mixture (A–A + B–B). The monofunctional reactant should be able to condense with one or the other end-functional group, A or B, and thus defunctionalize that end of the polymer. Thus, when a –B type monofunctional reactant is added into an equivalent mixture of A–A and B–B monomers, the polymer formed will have the B functional group at one end only at the completion of reaction, the other end having no functional group.

$$A-A \ + \ B-B \ \longrightarrow \ A-\!\!\!-\!\!\!-\!\!\!-\!\!\!- B.$$

$$A-\!\!\!-\!\!\!-\!\!\!-\!\!\!- B \ + \ -B \ \longrightarrow \ -\!\!\!-\!\!\!-\!\!\!-\!\!\!- B.$$

In A–B type monomers, the equivalence between the amounts of the coreactive functional groups is built-in. However, the addition of a monofunctional reactant of –A or –B type into the polymerization system sets up nonequivalence and controls the molecular weight. The same effect occurs with the addition of a difunctional monomer of A–A or B–B type.

2.4.1.2.1 Polymerization of nonequivalent monomer mixtures

In order to apply Carothers equation (2.19) to the bifunctional polycondensation of nonequivalent monomer mixtures, f is to be defined as the average number of reactable functional groups per combined molecule of monomer and any other reactant, if used, since the excess functional groups remain unreacted.[31]

Thus,

$$f = \frac{2\times \text{number of functional groups which are not in excess}}{\text{total number of molecules}}. \quad (2.21)$$

The factor 2 in the equation comes in because the number of reactable functional groups of the excess kind is equal to the number of functional groups of the other kind all of which are reactable.

Case 1. Nonstoichiometric mixtures of A–A and B–B monomers with the latter in excess

Let

N_A = total number of A functional groups.
N_B = total number of B functional groups.

$$r = \frac{N_A}{N_B} < 1. \quad (2.22)$$

Since A functional groups are not in excess,

$$f = \frac{2N_A}{1/2(N_A + N_B)} = \frac{4N_A}{N_A + N_B}. \quad (2.23)$$

Substituting Eq. (2.22) for N_A/N_B in Eq. (2.23) gives

$$f = \frac{4r}{1+r}. \quad (2.24)$$

Using this value of f in the Carothers equation one obtains

$$\overline{DP}_n = \frac{1+r}{1+r-2pr}. \quad (2.25)$$

However, p now refers to the extent of reaction of the functional groups, which are stoichiometrically deficient.

At the completion of polymerization (p = 1), Eq. (2.25) reduces to

$$\overline{DP}_n = \frac{1+r}{1-r}. \quad (2.26)$$

As discussed above, at p = 1 all chain ends are capped with functional groups which are in excess providing stability to molecular weight. Thus, the effect of stoichiometric imbalance (r < 1) on $\overline{DP}_n$ may be determined quantitatively using Eq. (2.25).

Table 2.3 presents the calculated values of $\overline{DP}_n$ for selected values of r and p. It is evident that larger is the stoichiometric imbalance in the monomer mixtures larger is the reduction in the molecular weight. For instance, 0.5 mol percent excess of one monomer reduces the degree of polymerization by 20% at 99% reaction, whereas 1 mol percent excess effects 33% reduction.

Case 2. The effect of addition of a monofunctional reactant

The effect of the addition of a monofunctional reactant on the $\overline{DP}_n$ may also be predicted quantitatively using the Carothers equation. As in the preceding case, f is calculated using

Table 2.3 Effect of stoichiometric imbalance between monomers on degree of polymerization in bifunctional polycondensation.

Excess of one monomer (mol%)	r	p	$\overline{DP}_n^a$
0	1	1	∞
0	1	0.99	100
0.5	0.995	0.99	80
1	0.9901	0.99	67
1	0.9901	1	201

[a]Calculated using Eq. (2.25).

Eq. (2.21). Thus, for a binary mixture comprising 1 mol of A–A monomer, 1 mol of B–B monomer, and 0.01 mol of a monofunctional reactant containing B functional group, f is 4/2.01. Use of this value in Carothers Eq. (2.19) yields $\overline{DP}_n = 201$ at p = 1. This degree of polymerization is the same as that obtained with the use of 1 mol % excess of one monomer (Table 2.3, last row). Thus, excess of one monomer or the monofunctional reactant has the same effect on $\overline{DP}_n$, mol for mol. This is because only one functional group of the excess monomer, on average, takes part in the reaction. Hence, Eq. (2.25) or (2.26) also may be used to predict $\overline{DP}_n$ with r defined as if the monofunctional reactant is a bifunctional one, *i.e.*,

$$r = \frac{N_A}{N_B + 2N'_B},$$

where N'_B is the number of monofunctional reactant molecules, and N_A and N_B have been defined earlier. Thus, for the reactant amounts given in the above example, $r = 2/2.02 = 0.9901$. Use of this value of r in Eq. (2.26) gives $\overline{DP}_n = 201$ at p = 1, which is the same as that calculated above. An equivalent case would be to use 0.5 mol % of a monofunctional A or B reactant in the polymerization of an A–B monomer. Thus, using 1 mol of an A–B monomer and 0.005 mol of the monofunctional –B reactant, $r = 1/1.01 = 0.9901$; accordingly, $\overline{DP}_n = 201$ at p = 1.

Nonequivalence between the functional groups may also be inadvertently set up resulting in the loss of control on $\overline{DP}_n$. This may occur due to error in weighing, presence of impurities such as monofunctional reactants in monomers, side reactions, and volatilization loss of monomers. Appropriate preventive measures may, of course, greatly reduce or even eliminate the problem. The side reactions, for instance, may be minimized by carrying out most of the polymerization but the last few percent at lower temperatures. The rest may need to be carried out at higher temperatures to reduce the reaction time. The volatilization loss of monomers may be prevented using the appropriate reaction temperature and pressure or the lost amounts have to be replaced.

2.4.1.2.2 Departure from Carothers equation in interfacial polycondensation

Interfacial bifunctional polycondensation involves contacting two highly reactive monomers of A–A and B–B types separately dissolved in two immiscible liquids at ambient or near ambient temperatures. For the polymerization to be successful, the condensation reaction must be very fast such as the one involving an acid chloride and a compound containing an active hydrogen atom, especially amines, alcohols, and thiols.[32] For instance, polycondensation of a diamine and a diacid chloride may be carried out by reacting the diacid chloride dissolved in an organic solvent, usually an aromatic hydrocarbon or a chlorinated hydrocarbon, and the diamine in water, with or without stirring. An inorganic base is also used in the aqueous phase to neutralize the hydrochloric acid generated in the reaction.

Due to the much higher solubility of the amine in the organic solvent than of the acid chloride in water, the diffusing monomers meet near the interface at the organic solvent side. The rate is controlled by the mass transfer of the diamine.[33] The condensation products are located at or near the interface due to their somewhat amphiphilic character. As a result, the diffusing monomers have the first opportunity to react with the already formed products before coreacting with their own selves. Since the reaction is very fast, growth of already formed chains occurs in preference to starting of new chains. Thus, the mechanism of growth is different from that of a regular step-growth polymer. Hence, Carothers equation is not applicable. High molecular weight polymer rapidly forms at conversions, which are far less than complete. Polymers with $\overline{M}_w$ as high as 500 000 have been prepared with especially pure reactants under optimum condition.[33,34]

The stoichiometric balance between monomers is not critical to obtain polymer of useful molecular weight. In addition, the very fast rate of polymerization makes side reactions unimportant, which is more so due to the low temperature used in polymerization. Accordingly, the polymerization is appreciably tolerant to impurities in monomers, monofunctional reactants excepted. In fact, relatively impure monomers have reportedly yielded polymers with molecular weights of 10 000 to 15 000 in some cases, which is good enough by condensation polymer standard.

2.5 Molecular Weight Distribution

2.5.1 *Molecular weight distribution in bifunctional polycondensation*

The molecular weight distribution in bifunctional polycondensation was deduced theoretically by Flory using a statistical method.[25,27] The deduction makes use of the principle of equal reactivity of like functional end-groups irrespective of the molecular size as established in a previous section (Sec. 2.1). This principle allows equating the probability of a functional group having reacted with the extent of reaction p. Consider the bifunctional polycondensation of an A–B type monomer forming a linear polymer. The probability (P_x) that a molecule selected at random will have exactly x structural units may be determined in the following way. The molecule will have x − 1 interunit linkages and an A group at one end and a B group at the other end. Starting from one of the ends, say A, the probability that the B group of the first unit has reacted is p. The probability that the B group of the second unit has reacted also equals p, this being independent of whether or not the preceding interunit linkage has been formed. In this way, the probability that a sequence of x − 1 B groups have reacted to form x − 1 interunit linkages is given by the product of these separate probabilities, i.e., p^{x-1}. Finally, the probability that the B group of the xth unit is unreacted is $1 - p$. The product of p^{x-1} and $1 - p$ gives the probability P_x that we proceeded to determine. Thus,

$$P_x = p^{x-1}(1-p). \tag{2.27}$$

However, P_x equals the mol fraction of x-mers if intramolecular condensation and other side reactions are absent. Thus, defining N_x and N as the number of x-mers and total molecules respectively we have

$$P_x = \frac{N_x}{N}. \tag{2.28}$$

Equating (2.27) with (2.28) gives

$$N_x = Np^{x-1}(1-p), \tag{2.29}$$

$$\therefore \quad n_x = N_x/N = p^{x-1}(1-p). \tag{2.30}$$

Equation (2.30) is the mol fraction or number distribution of the degree of polymerization. N relates to the total number of units N_0 as

$$N = N_0(1-p). \tag{2.31}$$

Substituting Eq. (2.31) for N in Eq. (2.29) gives

$$N_x = N_0(1-p)^2 p^{x-1}. \tag{2.32}$$

The weight distribution relation follows by substituting w_x/x for N_x/N_0 in Eq. (2.32)

$$w_x = x(1-p)^2 p^{x-1}, \tag{2.33}$$

w_x being the weight fraction of x-mers with the contribution of end groups neglected.

Calculated number and weight distribution curves of degree of polymerization in linear step polymers are shown in Figs. 2.4 and 2.5 respectively. The number distribution curves monotonically decline with the increase in the number of units at every stage of polymerization, whereas the weight distribution curves exhibit maxima very near the number average value of units, which equals $1/1 - p$.[25] Monomer is present in the highest number but not in weight at all stages of polymerization.

The same number and weight distribution relations were derived for polymers prepared by polycondensation of precisely equivalent amounts of A–A and B–B monomers with x now representing the combined number of structural units of both kinds in the polymer chain.

For nonequivalent systems yielding stabilized polymer at complete reaction ($p = 1$), the weight fraction distribution (Eq. 2.34) is similar to Eq. (2.33) in form

$$w_x = xr^{(x-1)/2} \frac{(1-r)^2}{(1+r)}, \tag{2.34}$$

$r < 1$ representing the stoichiometric imbalance and x assuming only odd integral values due to the stabilized polymer possessing only odd number of units.[25]

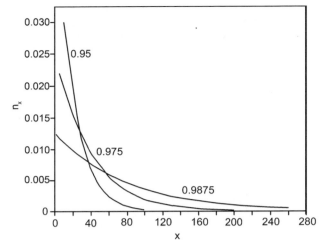

Figure 2.4 Mol fraction (n_x) distributions of degree of polymerization in linear step polymers at several extents of reaction (p) shown on the curves. "Adapted with permission from Ref. 25. Copyright © 1936 American Chemical Society."

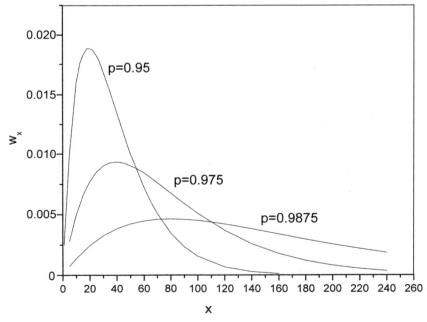

Figure 2.5 Weight fraction distributions of degree of polymerization in linear step polymers at different extents of reaction shown on curves. "Adapted with permission from Ref. 25. Copyright © 1936 American Chemical Society."

The above distribution relations are based on the assumption that an interunit linkage is permanently formed. However, this may not be so. Intermolecular interchange of interunit linkages may occur with varied facility resulting in the exchange of sections of chains between the chains themselves (*vide* Sec. 2.3). This may affect the size distribution relations if the equilibrium is not reached.

Flory considered the case in which the interchange has reached dynamic equilibrium. This may occur when a polymer undergoes intermolecular interchange for sufficiently long time but no further polymerization. Since the number of molecules does not change due either to the interchange or lack of further polymerization, dynamic equilibrium is established eventually in which the rate of formation of each molecular species is equal to the rate of its disappearance. The principle of equal reactivity as applied to the polymerization kinetics may apply to the interchange reaction as well so that the thermodynamic stability of an interunit linkage is independent of its position in the chain, and of the chain length. Based on this premise, Flory deduced the same distribution relations (2.30) and (2.33), which apply to the random linear step polymerization.[26,35] He also provided experimental confirmation of this theoretical result. Two stabilized (hydroxyl terminated) polyesters of widely different molecular weights were subjected to heat treatments in the presence of *p*-toluenesulfonic acid catalyst and the viscosity variation with time was followed. At equilibrium, viscosity reaches a constant value. The weight average degree of polymerization calculated from the viscosity at equilibrium agreed well with that derived from the distribution of the equilibrated polymer (Eq. 2.34).[26]

Distribution relations (2.30) and (2.33) also apply to polymers formed by random chain scission of infinitely long chain polymers.[36]

All the above examples seem to suggest that the distribution described by Eqs. (2.30) and (2.33) is quite common. Hence, it is referred to as the "most probable distribution".[27]

The first direct proof of the correctness of the predicted weight distribution was provided by Taylor who fractionated a poly(hexamethylene adipamide) sample into forty-six fractions and determined the number average degree of polymerization by end group titration. The experimental curve was in reasonable agreement with the theoretical one.[37]

The most probable distribution even applies to linear polymers formed in chain polymerization in which the chain breaking occurs exclusively by chain transfer and/or unimolecular termination. Besides, it also holds in radical chain polymerization with termination occurring bimolecularly by the disproportionation mode. Termination by the combination mode leads to narrower distribution (Chap. 3, Sec. 3.14).

2.5.2 *Average degree of polymerization for most probable distribution*

The distribution relations (2.30) and (2.33) may also be used to derive relations between number or weight average degree of polymerization and the extent of reaction. The number average degree of polymerization is given by[19]

$$\overline{DP}_n = \frac{\sum xN_x}{\sum N_x} = \frac{\sum xN_x}{N}. \quad (2.35)$$

Substituting from Eq. (2.29)

$$\overline{DP}_n = \sum_{x=1}^{\infty} x(1-p)p^{x-1}$$
$$= (1-p)\sum xp^{x-1}. \qquad (2.36)$$

Remembering that $p < 1$ and working out the summation, Eq. (2.36) gives

$$\overline{DP}_n = \frac{1}{1-p}$$

This is the same as the Carothers equation for bifunctional polycondensation (Eq. 2.20).
The weight average degree of polymerization is given by

$$\overline{DP}_w = \sum_{x=1}^{\infty} xw_x. \qquad (2.37)$$

Substituting Eq. (2.33) for w_x in Eq. (2.37) gives

$$\overline{DP}_w = \sum_{x=1}^{\infty} x^2 p^{x-1}(1-p)^2$$
$$= (1-p)^2 \sum x^2 p^{x-1}. \qquad (2.38)$$

Working out the summation gives

$$\overline{DP}_w = \frac{1+p}{1-p}. \qquad (2.39)$$

The polydispersity index follows as

$$PDI = \frac{\overline{DP}_w}{\overline{DP}_n} = 1 + p. \qquad (2.40)$$

Larger is the value of PDI broader is the distribution. The value approaches 2 as the reaction nears completion ($p \to 1$), *i.e.*, when the molecular weight is large.

2.6 Prediction of Gel Point in Polyfunctional Polycondensation

2.6.1 *Prediction based on Carothers equation*

Carothers equation may be used to treat gelation quantitatively. Several cases may be considered.

Case 1: Equivalent mixtures of a bifunctional and a polyfunctional monomer

For instance, for the polycondensation of 1 mol glycerol ($f = 3$) and 1.5 mol phthalic acid ($f = 2$)

$$f_{average} = 6/2.5 = 2.4.$$

Use of this value of f in the Carothers equation gives

$$1/\overline{DP}_n = 1 - 1.2p. \qquad (2.41)$$

Since gels have three dimensional infinite network structures, they may be considered to have infinite molecular weight. From Eq. (2.41), $\overline{DP_n} \to \infty$ as $p \to 0.83$.[29] Carothers pointed out that the p_c so predicted actually represents the maximum extent of reaction that may be possible before gelation occurs. In fact, the autoaccelerated nature of the increase of molecular size discussed in the previous chapter (Sec. 1.4) makes the gel point appear before all molecules become part of the network. Accordingly, the p_c predicted by Carothers equation exceeds the real value. For example, the experimental value of gel point in the reaction between glycerol and phthalic acid is $p_c = 0.77 \pm 0.015$,[39,40] as against the predicted value 0.83 derived above.[29]

Case 2: Nonequivalent binary mixtures of a bifunctional and a polyfunctional monomer

In the polymerization of a nonequivalent binary mixture of a bifunctional monomer and a polyfunctional monomer, gelation does not take place if the mixture composition is far from equivalent. Consider the polymerization of a bifunctional and a trifunctional monomer B–B (B_2) and A_3 respectively with the mol ratio of the two monomers $B_2/A_3 = x$, where $x < 1.5$ (*i.e.*, A functional groups are in excess).

$$A-\!\!<^{A}_{A}$$

$$(A_3)$$

For the monomer mixtures under consideration

$$f = 4x/1 + x \quad \text{(from Eq. 2.21)}.$$

Use of this value of f in the Carothers equation gives

$$\frac{1}{\overline{DP_n}} = 1 - \frac{2px}{1+x}.$$

It follows that $\overline{DP_n} \to \infty$ as $p \to 1$ at $x = 1$. Since p cannot exceed unity, this value of x sets the lower limit of B_2/A_3 below which gelation is not possible.

Similarly, with $x > 1.5$, B functional groups are in excess. Accordingly, from Eq. (2.21)

$$f = 6/(1+x).$$

Proceeding as above, it follows from the Carothers equation that $\overline{DP_n} \to \infty$ as $p \to 1$ at $x = 2$. This value of x sets the upper limit of B_2/A_3 above which gelation is predicted to be not possible. Thus, the predicted range of gel forming compositions is

$$1 \leq B_2/A_3 \leq 2. \tag{2.42}$$

However, since the Carothers equation predicts the gel point to occur at a somewhat higher p the actual range will be somewhat wider than the above, extended from both ends.

Case 3: Ternary mixtures

Stoichiometric requirements for gelation in ternary mixtures comprising three monomers of which two are bifunctional, A–A (A_2) and B–B (B_2) respectively, and the third is polyfunctional,A_f, may be predicted as follows.

Consider the specific case in which the functionality of the polyfunctional monomer is three and let ρ be the ratio of A functional groups belonging to A_3 monomers to total A groups and r be the ratio of total A to B groups. Then

$$\rho = N_{A3}/(N_{A2} + N_{A3}) = N_{A3}/N_A, \qquad (2.43)$$

and

$$r = N_A/N_B, \qquad (2.44)$$

where N_{A2} and N_{A3} represent the numbers of functional groups from A_2 and A_3 monomers respectively, and N_A and N_B are the total numbers of functional groups of A and B kinds respectively.

Three situations may arise:

(a) $r = 1$, for which

$$f = (N_A + N_B)/(N_{A2}/2 + N_B/2 + N_{A3}/3). \qquad (2.45)$$

Substituting from Eqs. (2.43) and (2.44)

$$f = \frac{6(1+r)}{3+3r-\rho r}. \qquad (2.46)$$

For gelation to occur, f must be greater than 2.

$$\therefore \quad 6 + 6r > 6 + 6r - 2\rho r, \qquad (2.47)$$

$$\text{or } 2\rho r > 0. \qquad (2.48)$$

Since $r = 1$, and ρ is always greater than zero, gelation is possible with all compositions of the A_2 and A_3 monomer mixtures.

(b) $r < 1$, for which f follows from Eq. (2.21) as

$$f = \frac{2N_A}{N_{A2}/2 + N_{A3}/3 + N_B/2}. \qquad (2.49)$$

Substituting from Eqs. (2.43) and (2.44)

$$f = \frac{12r}{3 + r(3-\rho)}. \qquad (2.50)$$

For gelation to take place, r and ρ should be such that $f > 2$.

$$\therefore \quad \frac{12r}{3 + 3r - \rho r} > 2. \qquad (2.51)$$

$$\text{or } r > 3/(3+\rho). \qquad (2.52)$$

(c) $r > 1$, for which f follows from Eq. (2.21) as

$$f = \frac{2N_B}{N_{A2}/2 + N_{A3}/3 + N_B/2}. \qquad (2.53)$$

Substituting from Eqs. (2.43) and (2.44)

$$f = \frac{12}{3 + 3r - \rho r}. \tag{2.54}$$

For gelation to take place, f must be greater than 2.

$$\therefore \quad 6 + 6r - 2\rho r < 12, \tag{2.55}$$

$$\text{or } r < 3/(3 - \rho). \tag{2.56}$$

2.6.2 Prediction based on statistical method of Flory

A statistical treatment of gelation was provided by Flory, which makes prediction of gel point possible.[27,38] Consider a polymerization system involving three monomers of which two are bifunctional, A–A and B–B, and another trifunctional, A_3, and the A and B functional groups are mutually reactive. Polymerization leads to the formation of highly branched molecules, which continue to increase in size with the progress of polymerization leading to indefinitely large three dimensional polymer structure at gel point, provided the polymerization is carried to high enough extent. A two dimensional representation of a portion of the highly branched molecule is shown in Fig. 2.6. A structural element linking one branch unit to the next (e.g., the portion ab in the figure) or extending from one branch unit and ending in an unbranched end (e.g., the portion cd in the figure) is defined as a chain.[27]

Defining branching coefficient α as the probability that a chain ends in a branch unit producing two new chains, the number of such chains formed from n chains is given by $2n\alpha$. The indefinitely large structure referred to above forms only when the number of new

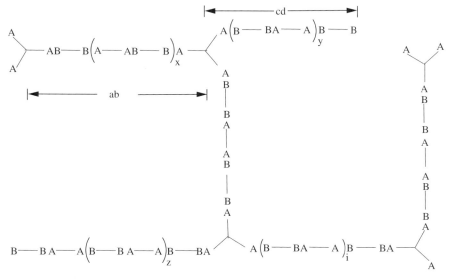

Figure 2.6 A schematic figure representing a portion of a highly branched molecule. "Adapted with permission from Ref. 27. Copyright © 1946 American Chemical Society."

chains formed through branching in a generation exceeds the number formed in the previous generation, i.e., when $2n\alpha > n$.[27,38]

Thus the critical value of α_c is

$$\alpha_c = \frac{1}{2}. \tag{2.57}$$

Similarly, for polymerizations in which the branch unit is f-functional

$$\alpha_c = \frac{1}{f-1}. \tag{2.58}$$

In order to relate α with the experimentally measurable quantities, we turn to the specific case in which $f = 3$ and determine the probability that a chain starting from a branch unit ends in a branch unit.[27,38] For instance, we consider the chain ab at the left of the portion of the structure shown in Fig. 2.6. It represents a chain composed of a sequence of $2x + 1$ numbers of bivalent units and bounded by two trifunctional units (x having any value from 0 to ∞, AB and BA representing interunit linkages). The assumption is made that the principle of equal reactivity applies to functional groups of the same kind and that the reaction is exclusively intermolecular.

The probability that the A group in the first trifunctional unit on the left of the chain has reacted is given by p_A, the extent of reaction of the A groups. The probability that the B group on the right of the first reacted B–B unit is connected to a bifunctional A–A unit is given by $p_B(1-\rho)$ where p_B is the extent of reaction of the B groups and ρ is the proportion of the A groups belonging to the trifunctional units in the total A groups. Since B can react with an A group of either A–A or A_3, the probability that it has reacted with an A group of A_3 is $p_B\rho$ and of A–A is $p_B(1-\rho)$.

Thus, the probability that the above chain bounded by two branch units has formed is given by[27,38]

$$p_A(p_B(1-\rho)p_A)^x \rho p_B. \tag{2.59}$$

The probability, α, that the chain ends in a branch unit irrespective of its length is given by

$$\alpha = \sum_{x=0}^{\infty} (p_A p_B(1-\rho))^x \rho p_A p_B. \tag{2.60}$$

Working out the summation gives

$$\alpha = p_A p_B \rho / [1 - p_A p_B (1-\rho)]. \tag{2.61}$$

If the ratio of the number of A to B groups, N_A/N_B, is r

$$p_B = r p_A. \tag{2.62}$$

Using Eq. (2.62) to eliminate either p_B or p_A from Eq. (2.61) gives

$$\alpha = r p_A^2 \rho / [1 - r p_A^2 (1-\rho)] \tag{2.63}$$

$$= p_B^2 \rho / [r - p_B^2 (1-\rho)]. \tag{2.64}$$

Equation (2.64) reduces to simpler forms in some especial cases.[27,37]

Case 1

(a) When the bifunctional A–A monomers are not used, and only B–B and A_3 monomers are used, $\rho = 1$ and α follows as

$$\alpha = rp_A^2 = p_B^2/r. \tag{2.65}$$

(b) Additionally, for equivalent amounts of A and B groups (r = 1)

$$\alpha = p^2, \quad \text{where } p = p_A = p_B. \tag{2.66}$$

Case 2

When the bifunctional A–A monomers are used along with A_3 and the coreactive A and B groups are present in equivalent amounts, r = 1 and $p_A = p_B = p$; α is given by

$$\alpha = p^2 \rho/[1 - p^2(1 - \rho)]. \tag{2.67}$$

Gel formation is possible with all compositions of A–A and A_3 mixtures, as was found earlier using Carothers equation (*vide* p. 58).

Case 3

When bifunctional and f-functional monomers containing only self-condensable groups, *e.g.*, A–A and R–A_f monomers are used, α follows from Eq. (2.61) putting $p_A = p$ and omitting p_B

$$\alpha = p\rho/[1 - p(1 - \rho)]. \tag{2.68}$$

Case 4

When the system contains only a polyfunctional monomer with self-condensable groups, *e.g.*, R–A_f, α is obtained by putting $\rho = 1$ in Eq. (2.68), which gives

$$\alpha = p. \tag{2.69}$$

The theoretically predicted point may be compared with the experimental value to check the efficacy of the method. For instance, polymerizations using equivalent amounts of glycerol and dibasic acids belong in case 1b discussed above. Using Eq. (2.66) and $\alpha = 0.5$, it is predicted to occur at p = 0.707. Experimentally, it is found at p = 0.77 ± 0.015.[39,40] Flory attributed the higher experimental value to some intramolecular reaction occurring, which makes lesser number of functional groups available for intermolecular reaction.[27,38] However, although the possibility of side reactions, intraesterification and anhydride formation, was indicated by experimental studies using some diacids, no evidence of such reactions exists using some others.[39,40]

2.6.2.1 Gel-forming range of monomer proportions

Besides predicting extent of reaction at gel point, the statistical theory can also predict the range of proportion of monomers outside of which no gelation can occur in a polyfunctional polycondensation. We consider here two of the four cases discussed above.

2.6.2.1.1 Polycondensation of A_f and B–B monomers

In binary polycondensation of A_3 and B_2 monomers, gelation is predicted in the range of r, viz., $2 \geq r \geq 0.5$, since, from Eq. (2.65), p_A exceeds unity for $r < 0.5$ at $\alpha_c = 0.5$, whereas p_B does so for $r > 2$. The range of monomer mol ratios outside of which gelation cannot take place follows from that of r, this being

$$0.75 \leq [B-B]/[A_3] \leq 3. \qquad (2.70)$$

However, the extent of reaction at gel point for each composition differs. For example, all B groups must react with A groups for gelation to occur when the monomer mol ratio is at the lower boundary of the above range. Similarly, all A groups must react with B groups when the monomer mol ratio is at the upper boundary of the above range. The assumption behind these predictions is that the reaction is exclusively intermolecular. However, since intramolecular reaction is unavoidable, the range would be narrower with upper and lower boundaries moved inward. As discussed earlier, the gel forming range (Eq. 2.42) predicted by Carothers equation is narrower than the actual. Thus, the actual range should lie in between the two ranges predicted by the two methods.

Similarly, proceeding as above, it may be shown using Eq. (2.65) with $\alpha_c = 1/3$ that in the polymerization of a tetrafunctional monomer (A_4) and a bifunctional monomer (B_2), geletion can occur only when the ratio (r) of A to B groups lies in the range $1/3 \leq r \leq 3$. This range corresponds to the range of monomer mol ratios

$$0.67 \leq B_2/A_4 \leq 6. \qquad (2.71)$$

At the lower end of the range all B groups and at the upper end all A groups must react for gelation to take place.

2.6.2.1.2 Polymerization of a ternary mixture of A–A, B–B, and A_3 monomers

The statistical theory predicts gelation to occur in the range of r,[38]

$$1/1+\rho \leq r \leq 1+\rho. \qquad (2.72)$$

The upper limit follows from solving Eq. (2.64) for p_B at $\alpha_c = 0.5$. The procedure gives $p_B = r^{1/2}/(1+\rho)^{1/2}$, which assumes the absurd value greater than unity, when r exceeds $1+\rho$. Similarly, the lower limit follows proceeding with Eq. (2.63) and solving for p_A at $\alpha_c = 0.5$. This gives $p_A = 1/[r(1+\rho)]^{1/2}$. Hence, p_A exceeds unity for values of r smaller than $1/(1+\rho)$.

2.7 Thermosetting Resins

The early-developed step polymers are mostly of the thermosetting type. Notable examples are phenolic, alkyd, amino, urethane, urea, epoxy, and unsaturated polyester resins.

2.7.1 *Phenolic resins*

Phenolic resins are prepared by the reaction of phenols and formaldehyde. However, the most important resin of this class is made from phenol (P) and formaldehyde (F). These resins find major application in plywood industries as adhesives. Other important application areas are matrices for composites, insulating parts in electrical and thermal gadgets, protective coatings, adhesives and binders of abrasives and of wood.

Phenol is trifunctional having three reactive sites, the two *ortho* positions and one *para* position, which can be involved in the reaction, whereas formaldehyde existing in water in equilibrium with methylene glycol is bifunctional[41,42]

$$CH_2O + H_2O \rightleftharpoons HOCH_2OH.$$

Phenol being a monomer of A_3 kind and formaldehyde being one of B–B kind, Eq. (2.70) may be used to predict the gel point based on the statistical method of Flory. On this basis, polymerization leads to gelation when the starting mol ratio of the two monomers lies in the range $0.75 \leq F/P \leq 3$, provided the reaction be carried to high enough extent of reaction. For example, it may be calculated using Eq. (2.65) that with $F/P = 0.75$ (r = 2) all the methylol groups in formaldehyde, and with $F/P = 3$ (r = 0.5) all the reaction sites in phenol have to react to reach the gel point in the respective systems. Prediction based on Carothers equation (Eq. 2.42), however, gives the gel-forming range of monomer mol ratios as $1 \leq F/P \leq 2$. As discussed earlier, the actual range should lie in between the two ranges predicted by the two methods.

Both acids and bases catalyze the reaction. However, in the manufacture of PF resins two procedures are followed. In the two-stage procedure, a soluble and fusible polymer, called novolac, is prepared in the first stage using an acid catalyst and excess phenol, the F/P ratio used being in the range of 0.7 to 0.85. From the discussion made in the previous paragraph, these systems may not reach gel point even at very high extents of reaction.

The conversion of the oligomer novolac to the thermoset is carried out in the second stage, where the oligomer is reacted with hexamethylenetetramine (HMTA) in molds under heat (130–180°C) and pressure (14–35 MPA) in compression molding.[43] HMTA acts as a basic catalyst and a source of formaldehyde. However, formaldehyde is not actually released from HMTA, which thermally dissociates only at temperatures above 250°C.[42] Of course, hydrolysis of some HMTA is effected by the residual water in novolac making some formaldehyde directly available for curing

$$C_6H_{12}N_4 + 6H_2O \rightleftharpoons 6CH_2O + 4NH_3.$$

The mechanism of novolac formation involves electrophilic aromatic substitution. In the first step, methylene glycol reacts with proton generating the hydroxymethylene cation (reation 2.73a), which attacks phenol at *o*- or *p*-position forming the methylolphenol (**A**). Protonation of the latter produces the benzylic carbenium ion (**B**), which reacts with another molecule of phenol producing the methylene-bridged bisphenol (**C**). Some dimethylene ether linkages also form between phenols through the condensation of two methylol groups (reaction 2.73d). Step polymerization follows through the repetition of similar reactions. Since both *o–p* and *p–p* bridging occur, the polymer formed is structurally irregular.[42]

$$HOCH_2OH + H^+ \rightleftharpoons {}^+CH_2OH + H_2O. \qquad (2.73a)$$

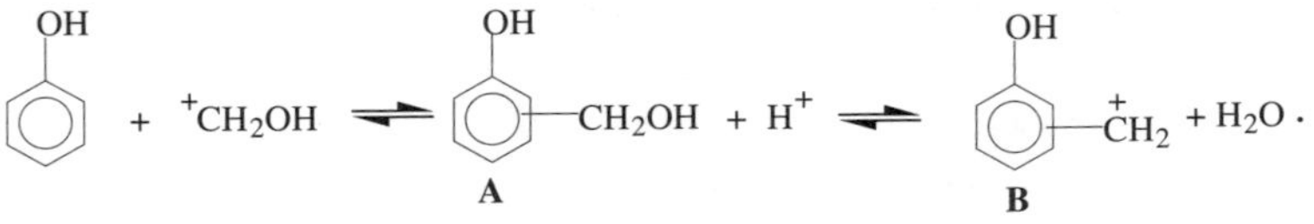

$$\text{C}_6\text{H}_5\text{OH} + {}^+\text{CH}_2\text{OH} \rightleftharpoons \underset{\textbf{A}}{\text{HOC}_6\text{H}_4\text{-CH}_2\text{OH}} + \text{H}^+ \rightleftharpoons \underset{\textbf{B}}{\text{HOC}_6\text{H}_4\text{-}\overset{+}{\text{C}}\text{H}_2} + \text{H}_2\text{O}. \qquad (2.73\text{b})$$

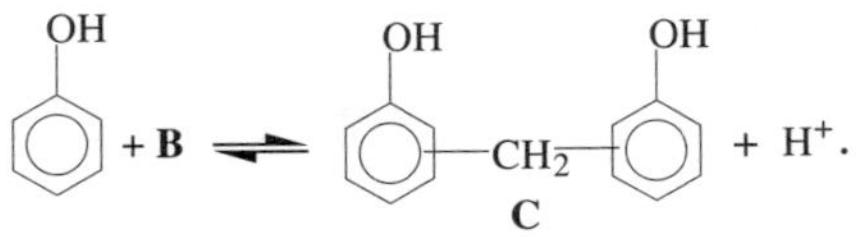

$$\text{C}_6\text{H}_5\text{OH} + \textbf{B} \rightleftharpoons \underset{\textbf{C}}{\text{HOC}_6\text{H}_4\text{-CH}_2\text{-C}_6\text{H}_4\text{OH}} + \text{H}^+. \qquad (2.73\text{c})$$

$$\textbf{A} + \textbf{B} \rightleftharpoons \underset{\textbf{D}}{\text{HOC}_6\text{H}_4\text{-CH}_2\text{-O-CH}_2\text{-C}_6\text{H}_4\text{OH}} + \text{H}^+. \qquad (2.73\text{d})$$

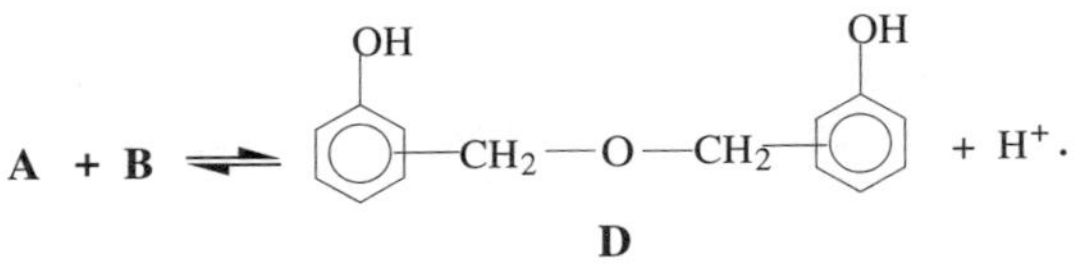

In the one-stage procedure, polymerization is carried out at 60–80°C to a stage short of gel point using a basic catalyst and a F/P mol ratio ranging from 1 to 3 producing a liquid resin known as resole, which is neutralized or made slightly acidic and then cured. The curing temperature varies from the ambient under acidic condition to 130–200°C under neutral condition.[44] Typically, resole is a mixture mainly of mono-, di-, and tri-methylolphenols and only small amounts of higher oligomers containing multiple methylol groups.[42]

The mechanism involves C-alkylation of a phenolate by formaldehyde resulting in the formation of o- and p-methylolphenol as shown in reaction (2.74) for the former. O-alkylation is suppressed due to the strong hydrogen bonding of the phenoxide ion with water and phenol, and the consequent large reduction in the reactivity of the oxide.[45,46]

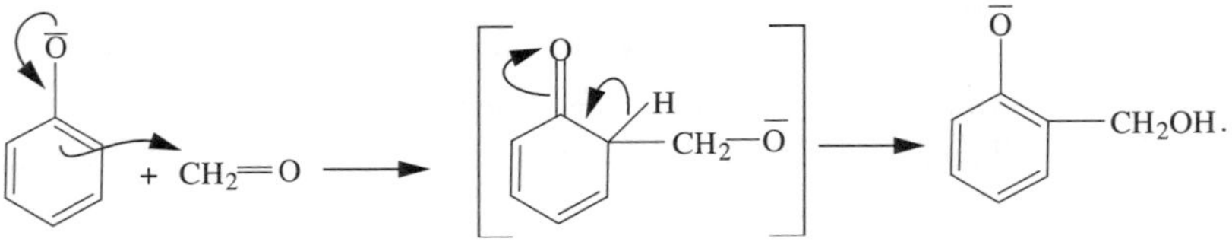

$$\text{C}_6\text{H}_5\text{O}^- + \text{CH}_2\text{=O} \longrightarrow [\text{intermediate}] \longrightarrow \text{o-HOC}_6\text{H}_4\text{-CH}_2\text{OH}. \qquad (2.74)$$

Subsequent substitutions yield dimethylol- and trimethylolphenols.

[Structures of 2,6-bis(hydroxymethyl)phenol, 2,4-bis(hydroxymethyl)phenol, and 2,4,6-tris(hydroxymethyl)phenol]

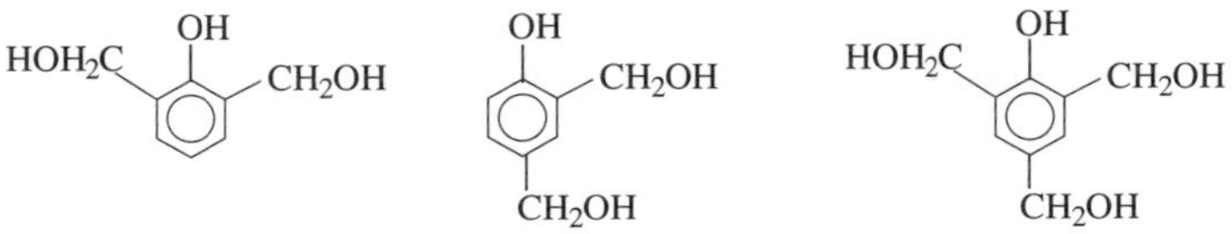

Further condensation occurs linking the phenolic rings with methylene bridges. Representative reactions involving the monomethylol derivatives are shown in (2.75) and (2.76).

In the neutral or slightly acidic condition of the curing stage, the self-condensation of methylol groups forming ether linkages becomes important (reaction 2.77). However, at temperatures above 130°C formaldehyde is also eliminated forming methylene bridges.[42]

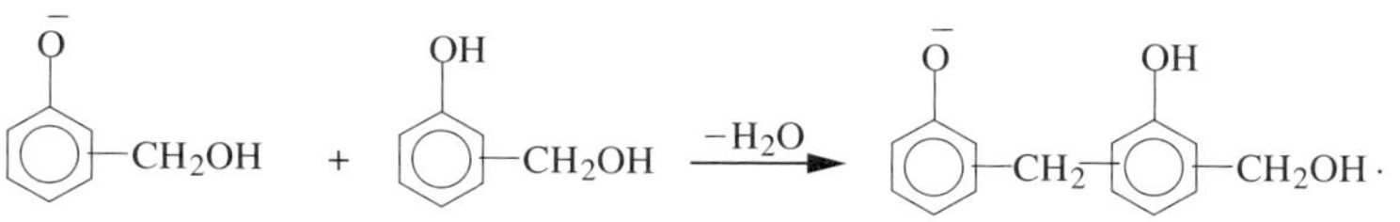

$$\text{(phenol-CH}_2\text{OH)} + \text{(phenol-CH}_2\text{OH)} \xrightarrow{-H_2O} \text{(phenol-CH}_2\text{-O-CH}_2\text{-phenol-CH}_2\text{OH)} \quad (2.75)$$

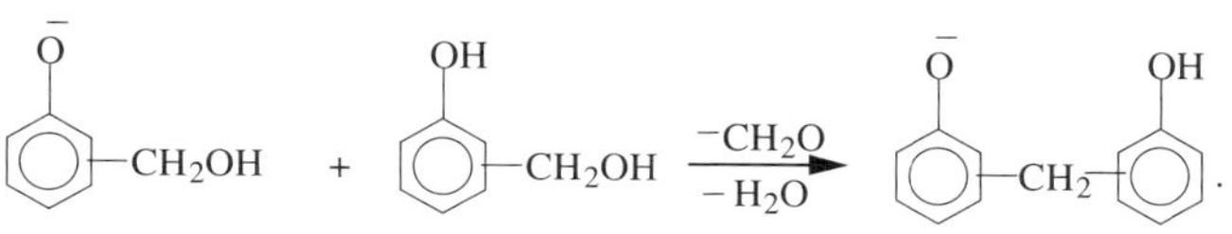

$$\text{(phenol-CH}_2\text{OH)} + \text{(phenol-CH}_2\text{OH)} \xrightarrow[-H_2O]{-CH_2O} \text{(phenol-CH}_2\text{-phenol)} \quad (2.76)$$

$$2\,\text{(phenol-CH}_2\text{OH)} \xrightarrow[-H_2O]{H^+} \text{(phenol-CH}_2\text{-O-CH}_2\text{-phenol)} \xrightarrow[-CH_2O]{>130°C} \text{(phenol-CH}_2\text{-phenol)} \quad (2.77)$$

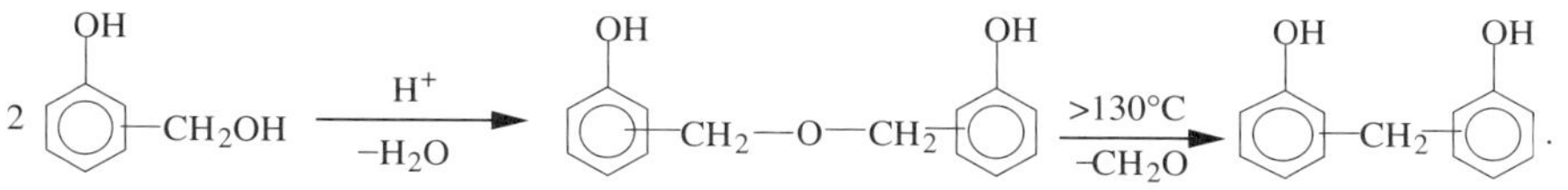

The self-condensation reactions of methylol groups (reactions 2.76 and 2.77) are particularly important in systems using high F/P mol ratios. It may be recalled that at F/P = 3, all the phenolic reactive sites must react for gelation to occur, according to the statistical theory of Flory (Eq. 2.70). Carothers equation predicts that monomer mixtures with F/P ratios greater than even 2 will not gel (Eq. 2.42). With the self-condensation of methylol groups, the F/P ratio is effectively lowered allowing crosslinked network formation even when the initial F/P = 3 at an extent of reaction of phenol lesser than unity, which is easier to accomplish.

2.7.2 Amino resins

Amino resins are prepared by the condensation of formaldehyde and amines or amides. Urea-formaldehyde (UF) and melamine-formaldehyde (MF) resins are the two principal members of the family. Urea is tetrafunctional inasmuch as each amido group can react with two molecules of formaldehyde, which is bifunctional. Equation (2.71) predicts the mol ratio (F/U) of formaldehyde to urea should range between 0.67 and 6 for gelation to occur. In fact, in the commercial preparation of the resin, formaldehyde is used in molar excess, which may be as high as five times that of urea.[47] The reaction is carried out in two stages. The first stage starts with hydroxymethylation of urea generating mono- and dimethylolureas.[48]

$$NH_2CONH_2 + CH_2O \rightleftharpoons NH_2CONHCH_2OH.$$

$$NH_2CONHCH_2OH + CH_2O \rightleftharpoons HOCH_2NHCONHCH_2OH.$$

These compounds and the low molar mass polymers obtained in this stage are soluble in water. Both bases and acids catalyze the reaction in this stage, although in commercial production a base is used.

In the second stage, acid catalyzed condensation of methylol derivatives is effected forming mainly methylene and some dimethylene ether linkages between urea units leading ultimately to the polymer network.[49]

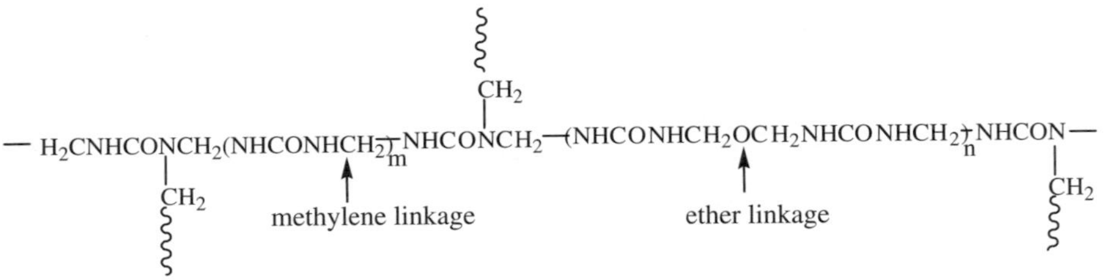

$$-\text{H}_2\text{CNHCONCH}_2(\text{NHCONHCH}_2)_m\text{NHCONCH}_2-(\text{NHCONHCH}_2\text{OCH}_2\text{NHCONHCH}_2)_n\text{NHCON}-$$

(with CH₂ branches labeled as methylene linkage and ether linkage)

Figure 2.7 Part of a chain in the network of UF resin with the branches leading to crosslinks arbitrarily positioned.

Like PF resins, UF resins also find major applications as plywood adhesives and particleboard binders. Among other applications are as additives to impart mechanical strength to paper and as textile finishes imparting crease resistance to fibers.

Synthesis of melamine formaldehyde resins also uses formaldehyde in molar excess over melamine. The initial products are methylol derivatives of melamine (reaction 2.78), which undergo condensation forming mainly methylene and some dimethylene ether linkages between melamine units in the curing stage. MF resins and melamine urea-formaldehyde resins find use as high performance wood binders. Besides, MF resins find application as textile finishes imparting wash-and-wear properties to cellulosic fabrics. In addition, MF molding powders are used in making dinner wares.

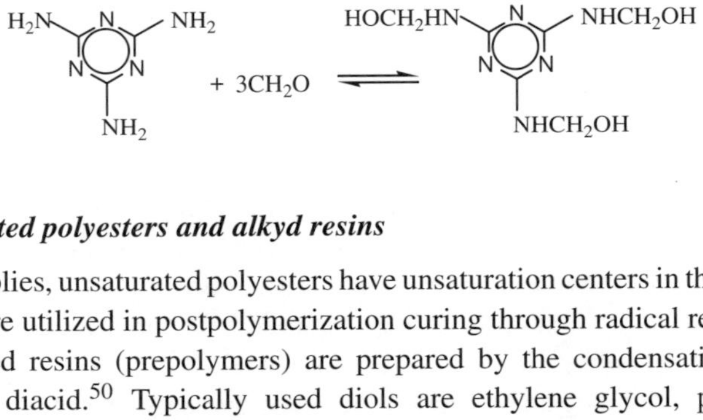

$$\text{melamine} + 3\text{CH}_2\text{O} \rightleftharpoons \text{tris(methylol)melamine} \quad (2.78)$$

2.7.3 Unsaturated polyesters and alkyd resins

As the name implies, unsaturated polyesters have unsaturation centers in the polyester chain. These centers are utilized in postpolymerization curing through radical reactions.

The uncured resins (prepolymers) are prepared by the condensation of a diol and an unsaturated diacid.[50] Typically used diols are ethylene glycol, propylene glycol, 1,3-propanediol, 1,4-butanediol, and diethylene glycol. Maleic, fumeric, and itaconic acid (anhydride) are the typical unsaturated diacids used. Replacement of some unsaturated acids by saturated ones helps to regulate the unsaturation content in the prepolymer chain. Besides, the saturated acids also regulate the physical properties of the thermosets. Aromatic diacids, such as phthalic or isophthalic anhydride, are used to improve hardness and heat resistance. The uncured resin prepared, for example, from 1,3-propanediol, maleic anhydride, and phthalic anhydride is essentially a random copolymer having the chemical structure

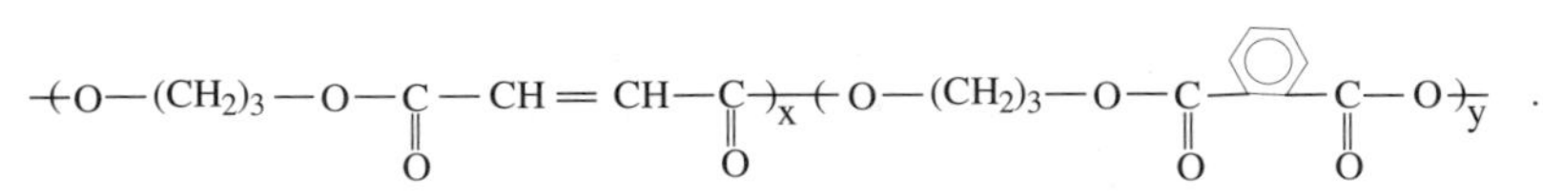

$$+\!\text{O}-(\text{CH}_2)_3-\text{O}-\underset{\text{O}}{\overset{\|}{\text{C}}}-\text{CH}=\text{CH}-\underset{\text{O}}{\overset{\|}{\text{C}}}\!\!\!-\!\!\!\!\!\!\!\!\!\!)_x(\text{O}-(\text{CH}_2)_3-\text{O}-\underset{\text{O}}{\overset{\|}{\text{C}}}-\!\!\text{C}_6\text{H}_4\!-\!\underset{\text{O}}{\overset{\|}{\text{C}}}-\text{O})_y \; .$$

The resin is blended with a vinyl monomer such as styrene and cured by polymerizing the mixture at ambient temperature using a thermally decomposable radical initiator, e.g., benzoyl peroxide or a redox initiator, e.g., the combination of methyl ethyl ketone hydroperoxide and cobalt napthenate.[47]

Almost all unsaturated polyester thermosets are composites containing reinforcing fibers and fillers. Glass fibers are the usual reinforcing fibrous material used. Replacement of some of the diols by polyols (f > 2) in the reaction formulations for polyesters leads to alkyd resins (name derived from alcohol and acid), which are used in surface coating formulations. Glycerol is the most widely used polyol and pentaerythritol is the second most widely used. The alkyds are modified by different fatty acids or oils of both drying and nondrying types. They may be modified further by styrene as is done in the curing of unsaturated polyesters discussed above. Modification by drying oils or their fatty acids gives oxidizing alkyds, which crosslink by autoxidation. When oil is used, it is to be first reacted with sufficient glycerol using a transesterification catalyst to form monoglycerides.

2.7.4 *Polyurethanes and polyureas*

The major application of polyurethanes is in the area of foams, which vary in flexibility. Among other application areas are coatings, adhesives, sealants, encapsulants, cast elastomers, thermoplastics, and elastomeric fibers, *etc.*[51]

Polyurethanes are formed by the reaction of di- and/or polyisocyanates and di- and/or polyols. The repeating units have the urethane linkage as shown in reaction (2.79). The closely related polymers, the polyureas, are formed by the reaction of di- and/or polyisocyanates and di- and/or polyamines (reaction 2.80). In fact, many polyurethanes are polyurethanes/ureas, being produced using mixtures of both alcohol and amine monomers instead of solely alcohols for the nonisocyanate monomers.

$$OCN-R-NCO + HO-R'-OH \longrightarrow \left[\begin{matrix} C-HN-R-NH-C-O-R'-O \\ \| \\ O \end{matrix} \begin{matrix} \\ \| \\ O \end{matrix} \right]_n$$

(2.79)

$$OCN-R-NCO + H_2N-R'-NH_2 \longrightarrow \left[\begin{matrix} C-HN-R-NH-C-NH-R'-NH \\ \| \\ O \end{matrix} \begin{matrix} \\ \| \\ O \end{matrix} \right]_n$$

(2.80)

An aromatic diisocyanate, *e.g.*, 2,4- or 2,6-toluene diisocyanate (**1**), is more reactive towards nucleophiles, *e.g.*, those of concern here (alcohols and amines), than an aliphatic one such as hexamethylene diisocyanate (**2**) or an alicyclic one such as 4,4'-methylene dicyclohexyl diisocyanate (**3**).

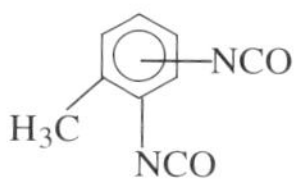

(**1**) $OCN-(CH_2)_6-NCO$ $OCN-\langle\rangle-CH_2-\langle\rangle-NCO$

(**1**)   (**2**)   (**3**)

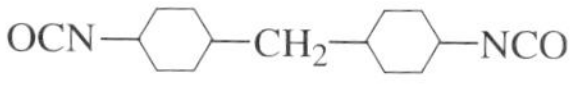

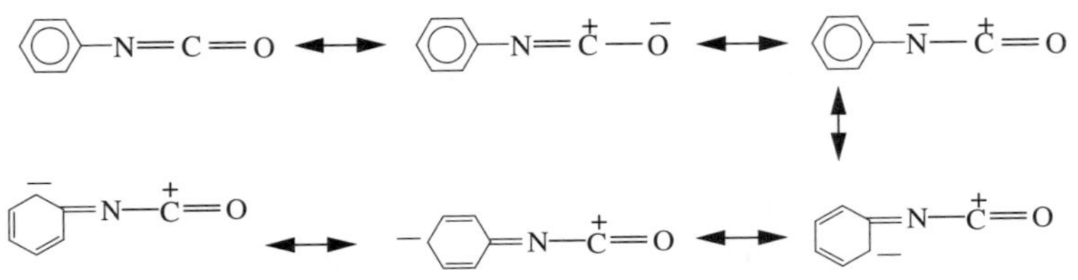

Figure 2.8 Resonating structures of aromatic isocyanate.

Conjugation of aromatic ring with isocyanate group in aromatic isocyanates contributes additional resonance structures. This makes the carbon in NCO more electrophilic than in aliphatic or alicyclic isocyanates (Fig. 2.8).[51]

The di- or polyols used may be small molecular (e.g., 1,4-butanediol (f = 2), trimethylolpropane (f = 3), triethanolamine (f = 3), etc., or oligomeric (e.g., polyether polyols, polyester polyols, polycarbonate polyols, polybutadiene polyols, etc.).

Apart from the main interunit linkages of urea and urethane groups, some minor linkages also form during thermal curing. For example, reaction of amido nitrogen in urea or urethane linkages with isocyanate leads respectively to allophanate or biuret crosslinks.[51]

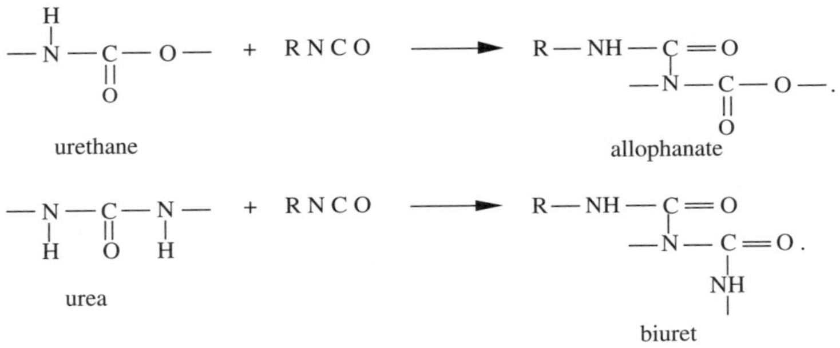

Adventitious water present in monomers, diluents, and/or atmosphere reacts with isocyanate forming amine and carbon dioxide

$$RNCO + H_2O \rightarrow RNH_2 + CO_2. \quad (2.81)$$

The generated amine reacts with the isocyanate forming urea linkage; the carbon dioxide formed remains as bubbles in the product impairing product quality in many cases. However, atmospheric moisture helps to cure polyurethane, used as sealants, adhesives, coatings, binders, and encapsulants in moisture-cure formulations. On the other hand, the use of water in larger quantities provides an environment-friendly method for the synthesis of flexible foams, the large amount of liberated carbon dioxide providing the gas bubbles required for making the foam.

Segmented polyurethanes containing multiple blocks of alternating soft (flexible) and hard (rigid) segments provide materials for fibers, cast elastomers, and thermoplastics. These materials are prepared either by copolymerizing a low T_g polyether diol or a polyester diol

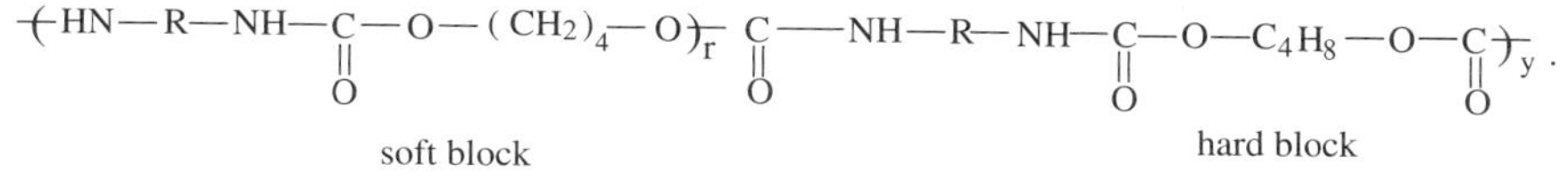

$$\{HN-R-NH-\underset{\underset{O}{\|}}{C}-O-(CH_2)_4-O\}_{\overline{r}}\underset{\underset{O}{\|}}{C}-NH-R-NH-\underset{\underset{O}{\|}}{C}-O-C_4H_8-O-\underset{\underset{O}{\|}}{C}\}_y.$$

soft block   hard block

Figure 2.9 Chemical structure of a polyether-based segmented polyurethane.

and 1,4-butane diol with a diisocyanate, or by chain extending an isocyanate-capped low T_g polyether or a polyester with a small molecular diol, *e.g.*, 1,4-butanediol. The flexible polyether or the polyester segments of the multiblock copolymer constitute the soft blocks, whereas the rigid segments formed from the diisocyanate and 1,4-butane diol constitute the hard blocks (Fig. 2.9).

These segmented copolymers have micro phase-separated morphology, the soft and hard blocks being incompatible with each other due to the absence of any significant specific interaction between them. The soft phase, being present in a much larger proportion by weight or volume, provides the matrix of the system. The hard phase is dispersed in the continuous matrix of the soft phase forming micro domains, which act as physical crosslinks and reinforcing fillers, for the matrix polymer.

2.7.5 *Epoxy resins*

An important representative of the epoxy resins is diglycidyl ether of bisphenol A (DGEBA) (**7**) which is prepared by the condensation of bisphenol A (**4**) with excess epichlorohydrin in the presence of alkali.[52]

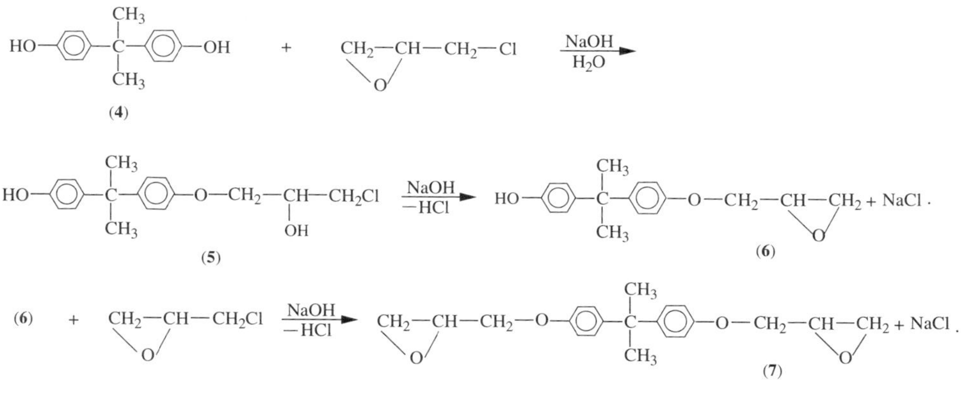

The mechanism involves nucleophilic attack of the bisphenolate anion on the primary carbon of the epoxy group in epichlorohydrin causing the epoxy ring opening and bonding of the phenolate with the *primary* carbon. Subsequent dehydrohalogenation of the chlorohydrin (**5**) formed produces the monoglycidyl ether of bisphenol A (**6**). Identical reaction on the phenolate end of the latter produces DGEBA (**7**).

However, even with the use of excess epichlorohydrin the condensation of bisphenol A with some DGEBA cannot be avoided. As a result, DGEBA in the epoxy resin technology is represented by (**8**) with $n \simeq 0.2$.[53]

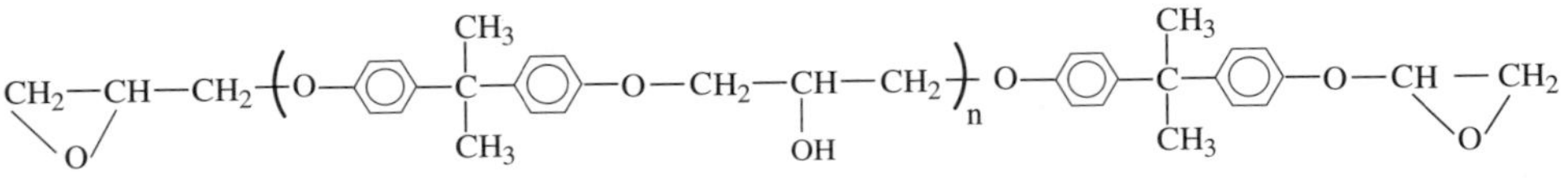

(structure **8**: DGEBA oligomer)

(**8**)

DGEBA is liquid at ambient temperature. The resin may be cured in a variety of ways. These involve: (1) chain extension of DGEBA associated with branching and cross-linking and/or (2) linking of *secondary* hydroxyl groups of the chain extended resin by anhydrides. The curing agents include among others multifunctional *primary* or *secondary* amines, *tertiary* amines, Lewis acids, dicarboxylic acids or their anhydrides, Lewis acids complexed with amines, and cationic photoinitiators.[47,52]

Primary and *secondary* polyamines extend the chains by way of addition to epoxy groups, each N–H bond being potentially reactive. Reaction (2.82) shows chain extension by *primary* amine. Inasmuch as the amine is polyfunctional, R also contains amino groups.

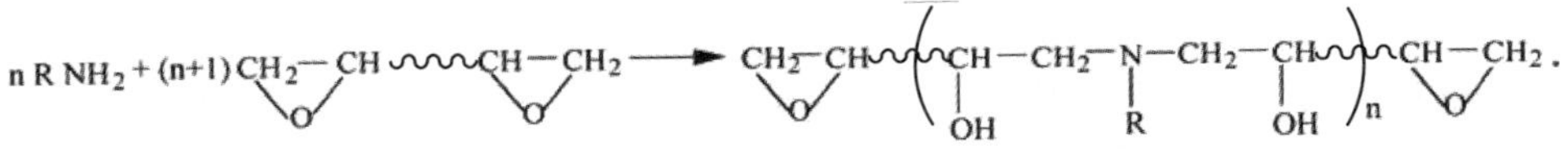

(reaction scheme 2.82)

(2.82)

As a result, a network polymer easily results. With the use of *t*-amines or cationic photoinitiators chain extension occurs through ring-opening polymerization of the chain end epoxides (*vide* Chap. 7). Curing by complexes of Lewis acids such as $BF_3 \cdot RNH_2$ takes place at elevated temperatures. One of the suggested mechanisms is the dissociation of the complex forming a proton, which initiates ring-opening polymerization of the chain end epoxides.[54]

$$BF_3 \cdot NH_2R \xrightarrow{\Delta} [BF_3NHR]^- + H^+. \qquad (2.83)$$

Epoxy resins find important applications as adhesives, coatings, and in structural composites, among others.

2.7.6 Polysiloxanes

Linear polysiloxanes (silicone polymers) are produced by the hydrolysis and condensation of dichlorosilanes or dialkoxysilanes (reaction 2.84) followed by the equilibration of the products in the presence of acids or bases.

The proportion of the three end-functionalized polysiloxanes formed in the hydrolytic polymerization depends on the ratio of the silane to water used. In the commercial silicone polymers, R is usually methyl. However, in some products, other groups, such as higher alkyl, fluoroalkyl, phenyl, and vinyl, among others, replace methyl to various extents for specific purposes.

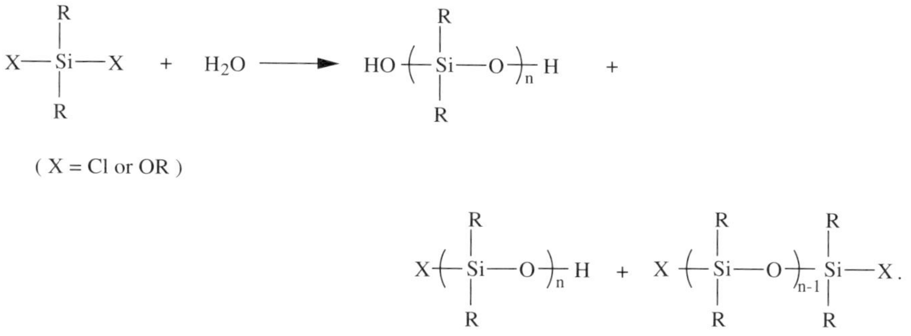

$$X-\underset{R}{\overset{R}{\underset{|}{Si}}}-X \ + \ H_2O \ \longrightarrow \ HO{\left(\underset{R}{\overset{R}{\underset{|}{Si}}}-O\right)}_n H \ +$$

(X = Cl or OR) \hfill (2.84)

$$X{\left(\underset{R}{\overset{R}{\underset{|}{Si}}}-O\right)}_n H \ + \ X{\left(\underset{R}{\overset{R}{\underset{|}{Si}}}-O\right)}_{n-1}\underset{R}{\overset{R}{\underset{|}{Si}}}-X.$$

The hydrolyzate obtained in reaction (2.84) is an oil containing approximately equal amounts of cyclic siloxanes [$(CH_3)_2SiO]_n$, represented usually as D_n, and linear polysiloxanes. The products, however, do not represent an equilibrium mixture, the cyclic fraction being much larger than the equilibrium value. In the preparation of the silicone fluids, the hydrolyzate is equilibrated in the presence of acids or bases and a monofunctional siloxane. For instance, the monofunctional hexamethyldisiloxane, $(CH_3)_3SiOSi(CH_3)_3$, is used to stabilize polysiloxane with unreactive timethylsiloxy groups at chain ends. Reactive end groups such as silanol, alkoxy, vinyl, and hydride may be introduced using water, alcohol, divinyltetramethyldisiloxane, and tetramethyldisiloxane respectively as monofunctional reactants in the place of hexamethyldisiloxane.[23] The proportion of the monofunctional reactant decides the average molecular weight of the linear polymer formed. Some cyclics, expectedly, remain in equilibrium with linear polymers (*vide* Sec. 2.2). The lower cyclics are separated from the hydrolyzate and subjected to ring-opening polymerization (*vide* Chap. 7, Sec. 7.11) to give the high molecular weight products.

Silicones with appropriate reactive end groups provide room temperature vulcanizing silicone elastomers supplied as one-component or two-component products. The one-component product contains silanol-terminated polysiloxane, the trifunctional curing agent methyltriacetoxysilane, and a tin soap catalyst. The curing agent caps the polysiloxane chain ends by the Si $(OAc)_2$Me groups.

$$HO(SiMe_2O)_nH \ + \ 2MeSi(OAc)_3 \ \longrightarrow \ Me(AcO)_2SiO(SiMe_2O)_n Si(OAc)_2Me \ + \ 2\ AcOH.$$

Curing is triggered by exposure to atmospheric moisture, which cleaves the acetoxy groups forming the silanol groups. The latter react with other acetoxy groups leading eventually to a crosslinked product. Volatilization of acetic acid helps to drive the reactions toward completion.[23]

The two-component formulation does not require the triggering by atmospheric moisture. The formulation supplies the catalyst and the polysiloxane in two different packages and the curing agent in either or both of the two packages. The curing agents can be of different types. Reaction (2.85) represents the curing effected by methyltriethoxysilane.[23]

$$3\ HO{-}(Si\ Me_2O)_n{-}H + Me\ Si(OEt)_3 \xrightarrow{tin\ soap} \left[HO{-}(Si\ Me_2O)_n\right]_3 Si\ Me + 3\ Et\ OH \quad (2.85)$$

$$\downarrow$$

Cross linked product

2.7.7 Aliphatic polysulfides

Poly(alkylene sulfide) chains are more flexible than polymethylene chains due to the substitution of the more flexible C–S–S–C bonds for some C–C bonds in the chain skeleton. This makes polysulfide chains with sufficiently large C–S bond concentrations elastomeric. Some of the C–C bonds may also be replaced by another more flexible C–O bonds. In fact, in the commercial elastomeric polysulfide both C–S and C–O bonds (*vide* Chap. 1, Table 1.3) substitute for some C–C bonds at regular intervals in the chain skeleton. The polymer as marketed is a liquid rubber, which is a low molecular weight nonlinear polysulfide with the main chain as well as the side chains terminated with SH groups. It is prepared by the polycondensation of bis(2-chloroethyl)formal in the presence of small proportion (<2 mol %) of the trifunctional monomer, 1, 2, 3-trichloropropane, and sodium polysulfide (reaction 2.86).[55,56] The high molecular weight branched polymer formed is converted to liquid polymer with SH end groups by treatment with sodium hydrosulfide (NaSH) and sodium sulfite (Na_2SO_3). The former breaks some of the disulfide linkages forming terminal SH and SSH groups, whereas the latter converts the SSH groups to SH groups.[56]

$$Cl{-}CH_2{-}CH_2{-}O{-}CH_2{-}O{-}CH_2{-}CH_2{-}Cl + \underset{\underset{Cl}{|}}{CH_2}{-}\underset{\underset{Cl}{|}}{CH}{-}\underset{\underset{Cl}{|}}{CH_2} + Na_2S_x\ (x = 2 - 2.5)$$

$$\downarrow$$

$$-(CH_2{-}CH_2{-}O{-}CH_2{-}O{-}CH_2{-}CH_2{-}S_x)_n(CH_2{-}\underset{\S}{CH}{-}CH_2{-}S_x)_m{-}.$$

(2.86)

The liquid polymer with the SH groups present at the ends of both the main chain and the side chains can be transformed to a solid elastomer by oxidation of the SH groups back to the disulfides at low temperature in applications such as adhesives, coatings, and sealants.

2.8 Engineering Plastics

The step polymers discussed so far owe their dimensional stability and mechanical properties to the three dimensionally cross-linked network structures they possess. However, linear polymers with a good proportion of chain bonds replaced by rigid rings, aromatic or heterocyclic, or, to some extent, cycloaliphatic, have thermal stability, dimensional stability, chemical resistance, and mechanical properties improved to become high performing.[57,58] Among these, the ones with lower rigidity and lower but good enough performance for engineering

Table 2.4 Examples of step polymers classified as engineering plastics and high performance polymers.

Engineering plastics	High performance polymers
1. Aliphatic polyamides	1. Liquid crystalline polymers
2. Aromatic-aliphatic polyesters	2. Poly(arylene ether)s
3. Polyarylates	3. Poly(p-phenylene sulfide)
4. Polycarbonates	4. Polyimides
5. Polysulfones	5. Polyamide imides
6. Poly(phenylene oxide)	6. Polyetherimides
	7. Polybenzimidazoles
	8. Polybenzoxazoles
	9. Polybenzothiazoles

applications, are referred to as engineering plastics. Others having highly rigid molecular structure with very good above-mentioned properties, which remain good even at high temperatures, are referred to as high performance polymers.[59] The various families of step polymers belonging to these two classes are listed in Table 2.4, which is by no means complete.

In the examples given in the table, all the polymers except the aliphatic polyamides have rigid rings present to various extents in their chain skeletons. The rings periodically placed in the chain skeleton not only decrease the chain flexibility but also increase the intermolecular attraction due to the polarizability of the rings particularly when they are *para* connected.[60] These molecular characteristics endow the polymer with high glass transition temperature, high crystallinity, and, accordingly, high melting temperature. In aliphatic polyamides, the intermolecular attraction is high enough due to the strong interchain hydrogen bonding between the amide groups (*vide* Chap. 1, Sec. 1.7.1) that the polymer exhibits good mechanical properties even in the absence of rigid rings.

Too much abundance of rigid rings may make a polymer intractable. Thus, poly(p-phenylene) having solely *para* linked phenylene groups in the chain skeleton is insoluble and infusible. Tractable polymers are obtained by inserting flexible chain bonds between rings. Atoms or groups of atoms, such as O, S, CH_2, $(CH_2)_{2-4}$, COO, CO, SO_2, $C(CH_3)_3$, $C(CF_3)_3$, and CONH, among others, are used to link the phenylenes for this purpose.[28,57]

Engineering plastics have the most flexible of these linkages, namely, O, $(CH_2)_{2-4}$, or COO. High performance polymers have the lesser flexible linkages. For example, replacement of hydrogens in CH_2 links by alkyl or phenyl groups decreases chain flexibility and increases T_g accordingly. When the linking group is CO, SO_2, or amide, the relatively low flexibility is due to the resonating structures of the repeating units, which confer some double bond character to the chain bonds. The resonance structures are shown in some small model molecules in Fig. 2.10. Although, some double bond character is imparted to the chain bonds even when the ether linkage is used, it is relatively small.

The aromatic and heterocyclic rings impart high resonance stabilization to the polymer molecule, resulting in high heat resistance, which reaches its extreme when the polymer is wholly aromatic or heterocyclic, *e.g.*, polyimides, poly (1,4-benzamide), polybenzimidazoles, polybenzoxazoles, polybenzothiazoles, *etc.*

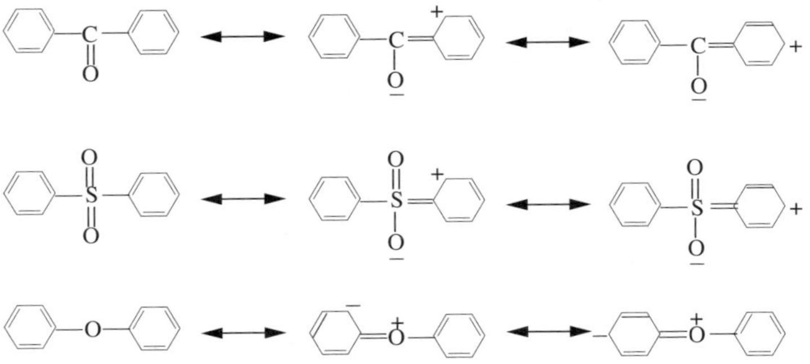

Figure 2.10 Resonating structures in diphenyl ketone, diphenyl sulfone, and diphenyl ether. Structures involving the other ring in all three compounds and the other oxygen atom in the sulfone are not shown.

2.8.1 *Polyamides*

Polyamides (trade name nylons) can be of two types; one type is prepared from an A–B monomer, *i.e.*, an amino acid, and the other type is prepared from the combination of an A–A and a B–B monomer, which represent a diamine and a diacid respectively. They are identified by the structural units, which are referred to by the number of carbon atoms they contain. Thus, the A–B type polyamide is referred to as nylon- or PA-5, -6, and -7 according as the structural unit contains 5, 6, and 7 carbon atoms respectively

$$-[\mathrm{NH-(CH_2)_x-CO}]_n-.$$

When two types of structural units are present as in the A–AB–B type polyamide, the diamine is referred to first, *e.g.*, nylon- or PA- 4,6, -6,6, and -6,10, according as the diamine contains 4, 6, and 6 carbon atoms and the diacid contains 6, 6, and 10 carbon atoms respectively

$$-[\mathrm{NH-(CH_2)_x-NH-CO-(CH_2)_y-CO}]_n-.$$

When the diacid is aromatic, it is referred to by the first letter of its name in the capital, *e.g.*, T for terephthalic acid and I for isophthalic acid. Thus, the nylon prepared from hexamethylene diamine and terephthalic acid is called nylon-6, T.

2.8.1.1 *Aliphatic polyamides*

As has been discussed in Sec. 2.2, the ring ⇌ chain equilibrium in the melt polymerizations of the lower polyamides, nylon -3 through nylon -5, is in the favor of rings. Hence, these polymers are prepared at lower temperatures (well below the melting temperatures of the polyamides) by the ring-opening polymerizations of the respective lactams.[16] However, Nylon-6 can be prepared both by the melt polymerization of ε-aminocaproic acid and the low temperature ring-opening polymerization of ε-caprolactam. Industrially, the latter is favored (*vide* Chap. 7). Some amount of the monomeric lactam exists in equilibrium with the polymer. Besides, small amounts of large ring lactams also form through the cyclization as well as backbiting reaction (Sec. 2.2). Polyamides of the A–AB–B type, *e.g.*, nylon-4,6, nylon-6,6, nylon-6,10, and nylon-6,12, are prepared by the melt polycondensation of the corresponding diamines and the diacids. As discussed in Sec. 2.2, according to the Jacobson–Stockmayer theory of ring ⇌ chain equilibrium, the proportion of rings in the A–AB–B type polyamides was predicted to be about 2.5%. In the A–B type polyamides, with the same number of chain atoms in the repeating unit and the same effective bond length as in the other type the proportion would be about twice as much.[19]

The preparation of the A–AB–B type polyamides is typified by that of nylon-6,6, which was discovered by Carothers and is the most important polymer of this class. The polymer is industrially prepared by the reaction of hexamethylenediamine and adipic acid. The diamine and the diacid have to be present in equivalent amounts in order that polymer of high molecular weight (by step polymer standard) is produced. The reaction mixture at the start comprises 50% solution of the diamine and the diacid in water.[16] The presence of the two monomers in equivalent amounts is easily verified by measuring the pH of the diluted reaction mixture.[61,62]

The reaction is carried out in three stages of which the first stage is conducted at high pressure in an autoclave at 210°C. The procedure prevents the loss of the more volatile diamine maintaining thereby the stoichiometric balance for the monomers. This stage marks the formation of the prepolymer. The second stage polymerization is carried out at a higher temperature ca., 290°C and at atmospheric pressure.[16] The volatilization of water in this stage prevents the reaction from reaching equilibrium making high conversion possible for obtaining high molecular weight polymer. However, the rising melt viscosity sets an operationally practicable limit to conversion. However, this problem may be avoided by carrying out the subsequent polymerization in the solid state at 10–40°C below the melting temperature of the polymer.[62,63] The higher chain end concentration in the amorphous region of the semicrystalline polymer allows further polymerization to occur at temperatures lower than the melt.[16,62,64]

2.8.1.2 Aromatic polyamides

The low reactivity of the aromatic acids towards the aromatic amines precludes the synthesis of the wholly aromatic polyamides by the direct bifunctional polycondensation.[65,66] Morgan pioneered the low temperature synthesis of these polymers using acid chlorides in place of acids. Both solution and interfacial polymerization processes were developed (Sec. 2.4.1.2.2).[65–68] The reaction yields hydrogen chloride as a by-product, which is to be removed, or else, it would react with the aromatic amine causing the reaction to stop. The acid is removed either by using an aliphatic tertiary amine or an alkali.

$$ClOC{-}\bigcirc{-}COCl + H_2N{-}\bigcirc{-}NH_2 + R_3N \longrightarrow {\left(\!{-}\underset{O}{\overset{\|}{C}}{-}\bigcirc{-}\underset{O}{\overset{\|}{C}}{-}NH{-}\bigcirc{-}NH\!\right)}_n$$

$$+ R_3\overset{+}{N}H\; \overset{-}{Cl}.$$

Aromatic polyamides, which are soluble in water immiscible solvents, may be prepared in both solution and interfacial polymerization, an example being poly(m-phenylene isophthalamide). The polymer is soluble in chlorinated solvents such as chloroform and methylene chloride.[16,65,66] However, liquid crystalline polyamides are soluble in admixtures of polar aprotic solvents, such as N-methylpyrrolidone and N,N-dimethylacetamide, and salts like LiCl and $CaCl_2$. These solvents offer suitable media for their preparation.[66–68] The aramide fibers can be directly spun from the solution following polymerization.

2.8.2 Polyesters

Synthesis of thermosetting polyesters has been discussed in Sec. 2.7.3. In this section, we shall discuss the synthesis of linear polyesters and that too of the engineering plastics and the high performance types. In principle, the synthesis may be effected using any of the following methods.

Method 1: Direct esterification

$$n\ HOOC\text{-}R\text{-}COOH + n\ HO\text{-}R'\text{-}OH \rightleftharpoons HO{-}\!\!\left[\!\!\begin{array}{c}C\text{—}R\text{—}C\text{—}O\text{—}R'\text{—}O\end{array}\!\!\right]_{\!n}\!\!{-}H + (2n-1)H_2O.$$

$$n\ HOOC\text{-}R\text{-}OH \rightleftharpoons HO{-}\!\!\left[\!\!\begin{array}{c}C\text{—}R\text{—}O\end{array}\!\!\right]_{\!n}\!\!{-}H + (n-1)H_2O.$$

Method 2: Ester interchange reactions

(a) ester-alcohol interchange

$$n\ R'OOC\text{—}R\text{—}COOR' + n\ HO\text{—}R''\text{—}OH \rightleftharpoons R'O{-}\!\!\left[\!\!\begin{array}{c}C\text{—}R\text{—}C\text{—}O\text{—}R''\text{—}O\end{array}\!\!\right]_{\!n}\!\!{-}H + (2n-1)R'OH.$$

(b) ester-acid interchange

$$n\ R'COO\text{—}R'\text{—}OOCR'' + n\ HOOC\text{—}R\text{—}COOH \rightleftharpoons HO{-}\!\!\left[\!\!\begin{array}{c}C\text{—}R\text{—}C\text{—}O\text{—}R'\text{—}O\end{array}\!\!\right]_{\!n}\!\!{-}OCR'' + (2n-1)\ R'COOH.$$

(c) ester–ester interchange

$$n\ R''OOC\text{—}R\text{—}COOR'' + n\ R'''COOR'OOCR''' \rightleftharpoons R''O{-}\!\!\left(\!\!\begin{array}{c}C\text{—}R\text{—}C\text{—}O\text{—}R'\text{—}O\end{array}\!\!\right)_{\!n}\!\!{-}CR''' + (2n-1)R'''COOR''.$$

Method 3: Acid chloride–alcohol reaction

$$n\ ClOC\text{—}R\text{—}COCl + n\ HO\text{—}R'\text{—}OH \xrightarrow{R_3N} {-}\!\!\left[\!\!\begin{array}{c}C\text{—}R\text{—}C\text{—}O\text{—}R'\text{—}O\end{array}\!\!\right]_{\!n}\!\! + R_3\overset{+}{N}HCl^{-}.$$

The reactions involved in methods 1 and 2 are slow at the ambient temperature and, therefore, required to be carried out in the melt at high temperatures in the presence of suitable catalysts as well (particularly method 2), which are usually metal salts, metal oxides, or metal alkoxides.

Wholly aromatic polyesters cannot be prepared by direct esterification or alkyl arylate-phenol interchange reaction since phenols are weaker nucleophiles and better leaving groups than aliphatic alcohols. They may, however, be prepared by aromatic carboxylic acid–aryl acetate or phenyl arylate–phenol interchange reactions.[28]

$$HOOC\ Ar\ COOH\ +\ CH_3COO\ Ar'\ OOC\ CH_3\ \longrightarrow\ (OC\ Ar\ COO\ Ar'\ O)_n\ +\ CH_3COOH. \quad (2.87)$$

$$HO\ Ar\ OH\ +\ PhOOC\ Ar'\ COOPh\ \longrightarrow\ (OC\ Ar'\ COO\ Ar\ O)_n\ +\ Ph\ OH. \quad (2.88)$$

They may also be prepared conveniently by method 3 using diacid chloride and diphenol. The reaction is facile even at low temperature. A *tertiary* amine base is used, which acts as both a catalyst and a hydrogen chloride acceptor. Neutralization of hydrogen chloride not only prevents the reverse reaction but also the H^+ ion catalyzed side reactions from occurring.[69] Both solution and interfacial polymerization processes may be used.

Among the ester interchange reactions, the ester-ester interchange reaction (method 2c) is too slow to be industrially practiced. The ester-alcohol exchange reaction is the most preferred.[2] However, the ester-acid exchange reaction is also used particularly in the synthesis of the wholly aromatic polyesters including the liquid crystalline ones.[28]

2.8.2.1 Aliphatic-aromatic polyesters

The most important polymer of this family is poly(ethylene terephthalate). Earlier, the synthesis method exclusively used the ester exchange reaction of dimethyl terephthalate and ethylene glycol. However, due to the subsequent commercial availability of purified terephthalic acid (PTA) the direct esterification route gained ground, which is discussed here. The reaction is carried out in two stages.[28,70] In the first stage, the reaction is restricted only to the oligomer level using excess ethylene glycol (EG), *e.g.*, PTA : EG = 1 : 1.2, and removing water as it is formed. The stabilized oligomer thus formed is end-capped with hydroxyl groups.

Stage 1: Direct esterification

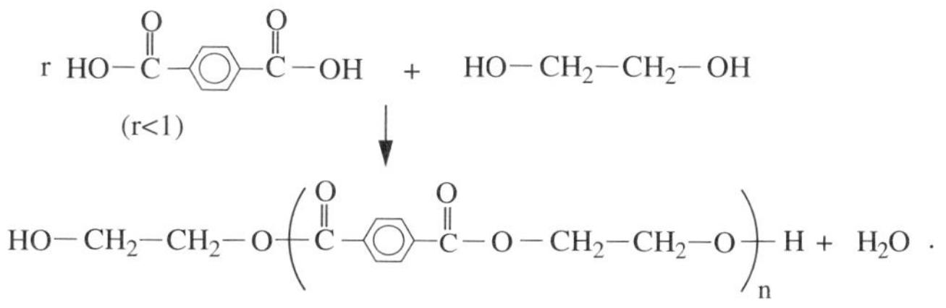

$$r\ HO-\underset{\underset{(r<1)}{}}{C}(=O)-\bigcirc-C(=O)-OH\ +\ HO-CH_2-CH_2-OH$$

$$\downarrow$$

$$HO-CH_2-CH_2-O\left(C(=O)-\bigcirc-C(=O)-O-CH_2-CH_2-O\right)_n H\ +\ H_2O\ . \quad (2.89)$$

(9)

With the use of 20 mol % excess glycol (r = 0.833) n in (**9**) is calculated to be 4.98 using Eq. (2.26) for complete esterification of all the carboxyl groups.

Stage 2: Ester interchange

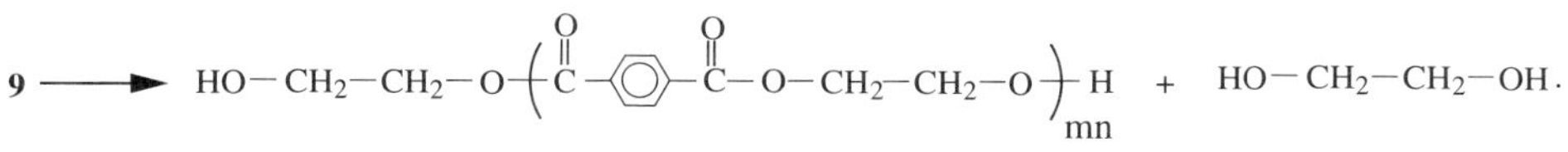

$$9\ \longrightarrow\ HO-CH_2-CH_2-O\left(C(=O)-\bigcirc-C(=O)-O-CH_2-CH_2-O\right)_{mn} H\ +\ HO-CH_2-CH_2-OH. \quad (2.90)$$

In the second stage, the hydroxyl end-capped oligomer is subjected to hydroxyl-ester interchange reaction in the presence of Sb_2O_3 catalyst while removing the liberated glycol under reduced pressure.

Reactions (2.89) and (2.90) are reversible. They are carried to high conversions by removing the eliminated water and ethylene glycol respectively. The molecular weight of the polymer formed in the second stage may be increased further by solid-state post polymerization as in the case of polyamides (Sec. 2.8.1.1). The water eliminated is removed continually under either nitrogen flow or vacuum.

The direct esterification method is also suitable for the synthesis of poly (trimethylene terephthalate) but not of poly (butylene terephthalate). The inapplicability in the latter case is due to the dehydration of 1,4-butanediol to tetrahydrofuran taking place in the presence of acid. Accordingly, alcohol-ester interchange reaction between dimethyl terephthalate and excess 1,4-butanediol is used in the first stage to synthesize hydroxy end-capped oligo (butylene terephthalate) (reaction 2.91) removing the eliminated methanol to drive the reaction forward[28]

$$r\ MeOOC-\phi-COOMe + HO-(CH_2)_4-OH \longrightarrow HO-(CH_2)_4-O\left[\overset{O}{\underset{\|}{C}}-\phi-\overset{O}{\underset{\|}{C}}-O-(CH_2)_4-O\right]_n H + MeOH,$$

(r < 1)

(**10**)

(2.91)

where n is related to r and may be calculated using Eq. (2.26).

In the second stage, the oligomer is polymerized further by alcohol-ester interchange reaction eliminating excess 1, 4-butanediol that is removed

$$\mathbf{10} \longrightarrow HO-(CH_2)_4-O\left[\overset{O}{\underset{\|}{C}}-\phi-\overset{O}{\underset{\|}{C}}-O-(CH_2)_4-O\right]_{nm} H + HO-(CH_2)_4-OH.$$

(2.92)

Replacement of some of the 1,4-butanediol by dihydroxy-poly (oxytetramethylene) polyether, $HO-(CH_2-CH_2-CH_2-CH_2-O)_n-H$, gives multiblock copolymers containing randomly distributed hard poly(butylene terephthalate) and soft polyether blocks. Depending on the ratio of the two diols, material property varies from elastomeric to rigid engineering thermoplasitics.

2.8.2.2 *Aromatic polyesters*

As mentioned earlier, aromatic polyesters are prepared either by the reaction of a diacid chloride and a diol or by the aromatic carboxylic acid–aryl acetate or the phenyl arylate-phenol interchange reaction. Below is given one example each of the latter two reaction types.

Synthesis of poly(*p*-hydroxybenzoic acid), an LC polyester, is carried out by the exchange reaction between an aromatic carboxylic acid and an aryl acetate using *p*-acetoxybenzoic acid as the monomer (reaction 2.93) and magnesium turnings as the catalyst at elevated temperatures.[71] The liberated acetic acid is removed in order to drive the reaction toward completion.

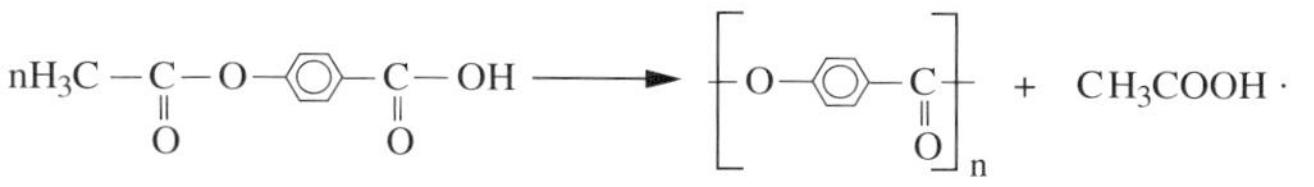

$$nH_3C-\underset{O}{\underset{\|}{C}}-O-\bigcirc-\underset{O}{\underset{\|}{C}}-OH \longrightarrow \left[O-\bigcirc-\underset{O}{\underset{\|}{C}}\right]_n + CH_3COOH \cdot \qquad (2.93)$$

Bisphenol A polyarylate (**11**) may be synthesized by the phenyl arylate–phenol exchange reaction (2.94) in which Sb_2O_3 is used as the catalyst.[28]

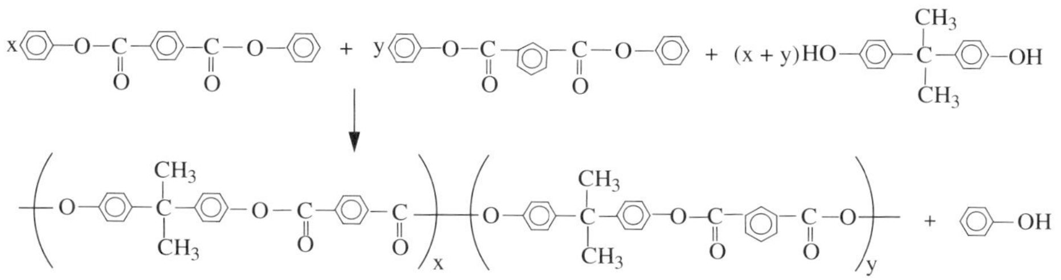

(**11**)

(2.94)

2.8.3 Polycarbonates

The most important representative of this class of engineering thermoplastics is bisphenol A polycarbonate. The polymer is produced industrially by an amine catalyzed interfacial polycondensation of bisphenol A and phosgene or by a base catalyzed melt transesterification of bisphenol A with diphenyl carbonate.

The interfacial polymerization process is represented by reaction (2.95). Alkali (NaOH) is used to convert some bisphenol A to phenolate and neutralize the hydrochloric acid liberated in the reaction. Accordingly, the pH of the aqueous phase is maintained at 10-12. Methylene chloride is the commonly used water-immiscible organic solvent. A phenolic monofunctional reactant such as phenol, p-t-butylphenol or p-cumylphenol is used in small amounts (1–5% of bisphenol A) to control the molecular weight and stabilize the polymer. Thus, the polycarbonate is end-capped with the corresponding phenyl or substituted phenyl groups.[72]

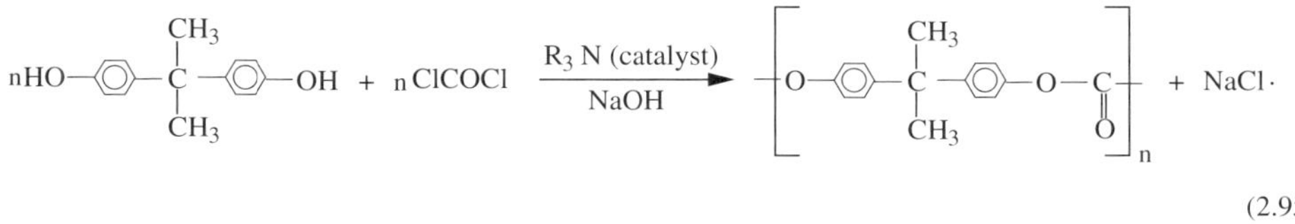

(2.95)

The melt transesterification process represents an environment-friendly method, which avoids the use of the toxic phosgene and the volatile organic solvent methylene chloride. It involves hydroxyl-ester interchange reaction (2.96). Phenol is eliminated as a by-product, which is removed by applying vacuum in order to drive the reaction toward completion. A basic catalyst such as lithium, potassium, tetraalkylammonium, tetraalkylphosphonium hydroxide or carbonate is used.[73,74] The polymer produced is not fully end-capped unlike the interfacially polymerized one. Usually, some of the ends are capped with phenolic functional group. Besides the basic catalysts mentioned above, several other catalysts such as the organotin compound, dibutyltin oxide,[75] and the rare earth salt, lanthanum acetate,[76] prove promising. The last one, in particular, has the attractive feature that it does not promote degradation of polycarbonate.

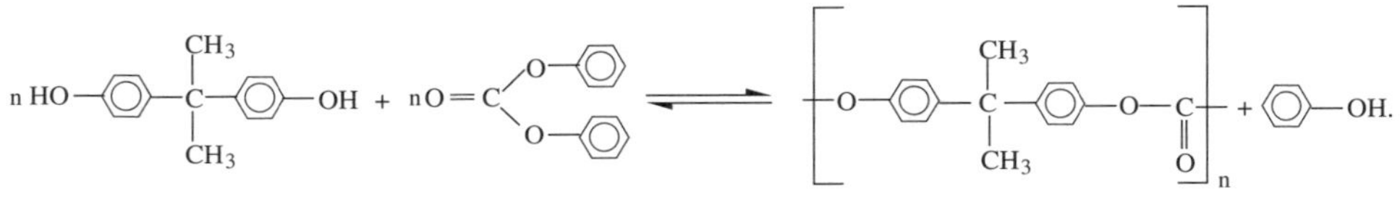

$$n\ HO-\underset{CH_3}{\underset{|}{\overset{CH_3}{\overset{|}{C}}}}-OH + nO=C\overset{O-\phi}{\underset{O-\phi}{\diagup}} \rightleftharpoons \left[O-\phi-\underset{CH_3}{\underset{|}{\overset{CH_3}{\overset{|}{C}}}}-\phi-O-\underset{O}{\underset{\|}{C}}\right]_n + \phi-OH. \qquad (2.96)$$

2.8.4 Poly(phenylene oxide)s

The most important polymer of this class is poly(2,6-dimethyl-1,4-phenylene ether), referred to as poly(phenylene oxide) (PPO). The polymer has high glass transition temperature (210°C) and melting temperature (307°C). Commercially, however, miscible blends of this polymer with polystyrene are marketed (trade name noryl) due to their easier processability (lower T_g) and cost-effectiveness. The polymer is prepared by the oxidative polymerization of 2,6-dimethylphenol in toluene using a copper(I) amine catalyst. Methyl substitution at the two *ortho* positions of the phenol blocks these reactive sites preventing thereby crosslinking of the polymer.[77]

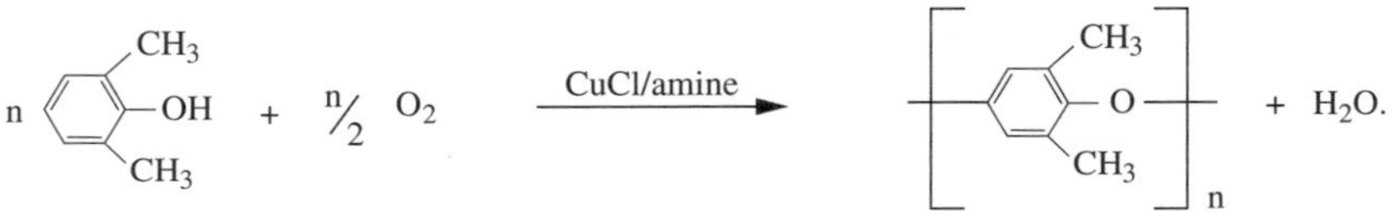

$$n \underset{CH_3}{\underset{CH_3}{\phi}}-OH + \tfrac{n}{2}\ O_2 \xrightarrow{CuCl/amine} \left[\underset{CH_3}{\underset{CH_3}{\phi}}-O\right]_n + H_2O. \qquad (2.97)$$

The mechanism of stepwise growth to the trimer stage is shown below.[77] Further growth proceeds in the similar way.

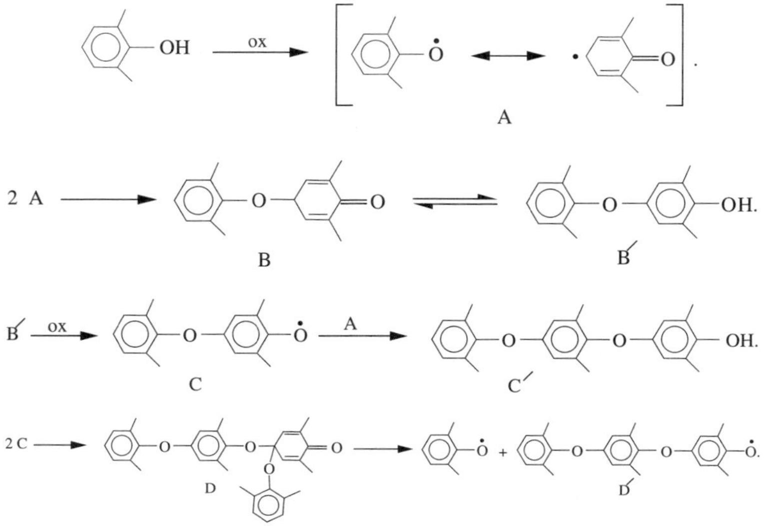

In the first step, the monomeric phenoxy radical A is generated from the monomer. Two such radicals couple together to form the dimer B, which tautomerizs to B'. The latter, in turn, is oxidized and the resultant radical couples with the monomeric phenoxy radical to form the trimer, and so on. The coupling can also take place between other combinations of radicals present in the system as shown for the dimer radical C, which on self-coupling gives the ketal D. The latter is unstable and dissociates either back to the original radicals or to a monomer radical plus a trimer radical (D'). Thus, the polymer grows by a step mechanism.

2.9 High Performance Polymers

An introduction to high performance polymers has been given in Sec. 2.8 and the families of polymers belonging to this class have been listed in Table 2.4. These polymer families are discussed separately in the following subsections.

2.9.1 *Liquid crystalline polymers (LCPs)*

Rod shaped polymer molecules form when the rigid rings in the structural units are linked so as to form the extended[65,66] or the helical chain.[71] Thus, benzene, biphenyl, and naphthalene rings are to be linked as 1,4-phenylene, 4,4'-biphenylene, and 2,6-, 1,4-, or 1,5-naphtylenes respectably. In rod like polyester or polyamide molecules, such rigid moieties are interlinked through ester or amide groups respectively to form the chain skeleton.[65]

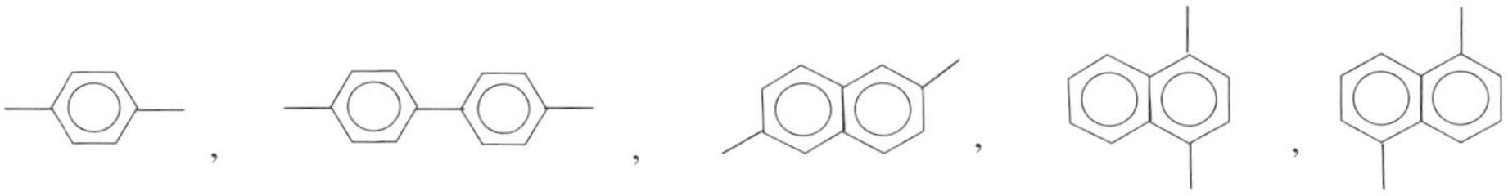

Such molecules tend to align parallel to each other imparting liquid crystalline character to polymer melts (thermotropic) or solutions (lyotropic). Structural units giving rise to liquid crystalline property are referred to as mesogens. Appropriate processing technique yields polymer with exceptional mechanical properties. For example, liquid crystalline solutions of poly(1,4-benzamide) (**12**) or poly(*p*-phenylene terephthalamide) (**13**) in N,N-dialkylamide–salt (LiCl, CaCl$_2$) solvents can be spun into fibers (trade name Kevlar for both), which have tensile strength higher than steel on a weight basis and initial modulus several times that of glass.[67,68]

(12) 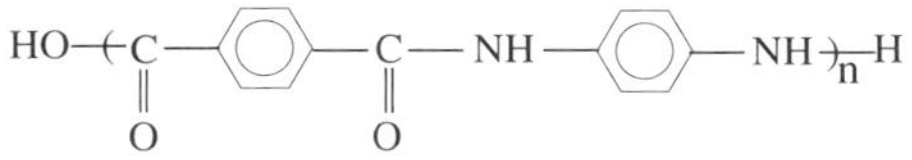   (13)

The domains of the rod like molecules aligned in parallel orient in the direction of flow during spinning as the solution is forced through the holes of the spinneret giving molecularly oriented fibers. Among other lyotropic main chain LCPs are polybenzimidazoles, polybenzoxazoles,[78] poly(oxadiazole)s,[78] and polybenzothiazoles.[79]

Thermotropic LCPs provide self-reinforced thermoplastics, which exhibit high tensile modulus and strength in flow direction in molded items. For instance, polyester LCPs have mechanical properties comparable with glass filled PET.[28,58] Poly(4-hydroxybenzoic acid) (**14**) and poly(*p*-phenylene terephthalate) (**15**), which are polyesters with structures analogous to those of Kevlar discussed above, are insoluble and infusible.

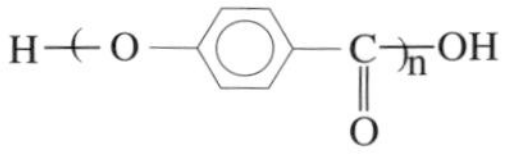

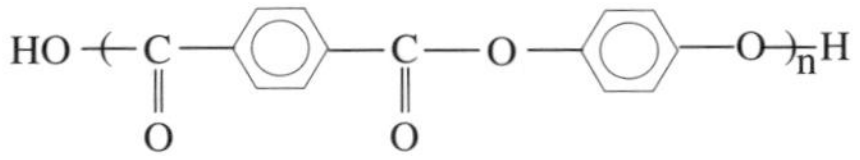

However, several random copolyesters exist, which are liquid crystalline as well as melt processable.[80–82] Some examples are given in Table 2.5. In these copolyesters, the loss of chain symmetry lowers the intermolecular attraction and melting point. Even a liquid crystalline homopolyester containing mesogenic groups may be melt procssable when bulky substituents are present in the structural unit, *e.g.*, **19**.[28] The bulky substituents increase the inter chain distance resulting in melting point lowering. Liquid crystalline polymer solutions are also obtained when the polymer chains contain mesogenic groups as side groups. However, it is the main chain LCPs, which exhibit exceptionally high mechanical properties.[28,80,83]

Table 2.5 Some examples of melt processable liquid crystalline polyesters.

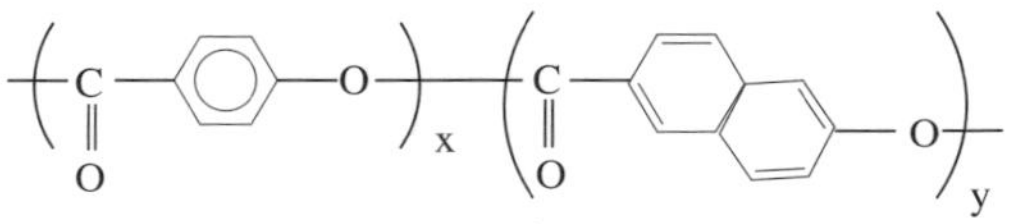

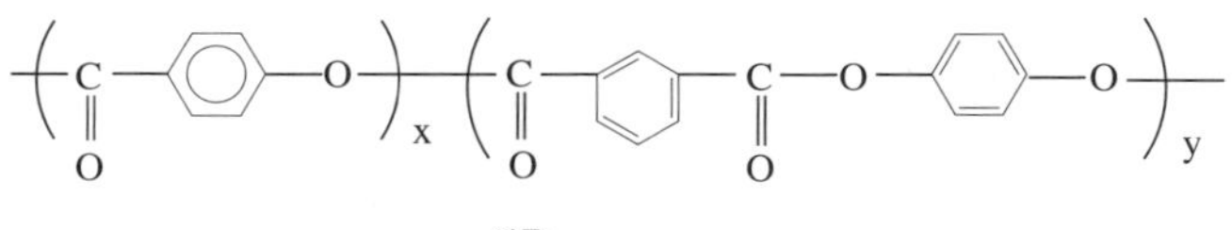

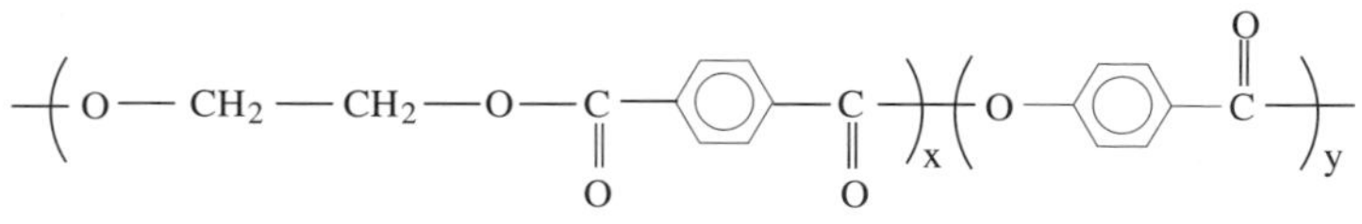

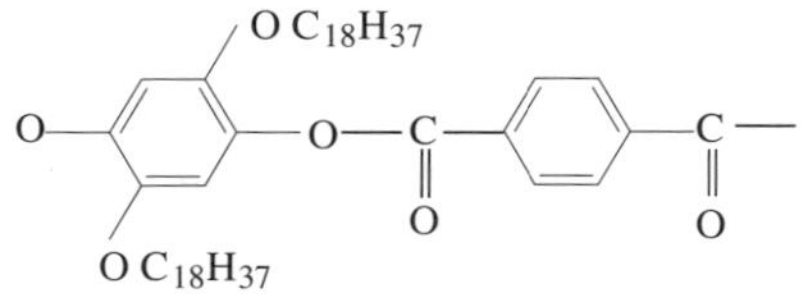

The synthesis of aromatic polyamides and aromatic polyesters has been discussed in Secs. 2.8.1.2 and 2.8.2.2 respectively. We shall now discuss that of the other high performance polymers.

2.9.2 Poly(p-phenylene sulfide)

The mechanical properties of linear PPS are not that high to justify the inclusion of the polymer in High Performance Polymer category. However, the properties are significantly improved by curing the polymer in air at elevated temperatures *ca.*, 260°C for precure molding and 370°C for coating applications.[84] PPS is commercially produced by the reaction of *p*-dichlorobenzene and sodium sulfide in a polar solvent, *e.g.*, N-methylpyrrolidone (reaction 2.98).[85]

$$n\,Cl-\!\!\!\bigcirc\!\!\!-Cl \;+\; n\,Na_2S \longrightarrow \left[\!\!\bigcirc\!\!-S\right]_n \;+\; NaCl \cdot \tag{2.98}$$

2.9.3 Poly(arylene ether)s

Poly(arylene ether)s have in their repeating units phenylene groups with ether and less flexible carbonyl, sulfonyl, or isopropylidene linkages in between, as shown in Table 2.6. PEEK is the most prevalent of these polymers. Synthesis may be carried out using electrophilic aromatic substitution, nucleophilic aromatic substitution, or transition metal catalyzed coupling reactions.[86] Of these, the nucleophilic aromatic substitution is the most practiced. Two typical examples are reactions (2.99)[87,88] and (2.100).[89,90]

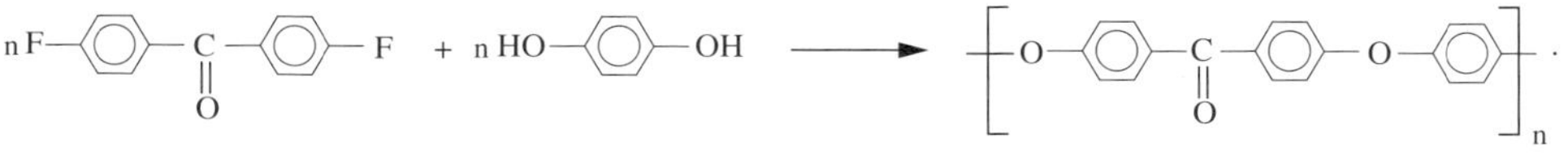

$$n\,F-\!\bigcirc\!-\overset{O}{\underset{\|}{C}}-\!\bigcirc\!-F \;+\; n\,HO-\!\bigcirc\!-OH \longrightarrow \left[O-\!\bigcirc\!-\overset{O}{\underset{\|}{C}}-\!\bigcirc\!-O-\!\bigcirc\!\right]_n \cdot$$

PEEK

(2.99)

$$n\,Cl-\!\bigcirc\!-\overset{O}{\underset{\underset{O}{\|}}{S}}-\!\bigcirc\!-Cl \;+\; n\,HO-\!\bigcirc\!-\underset{CH_3}{\overset{CH_3}{C}}-\!\bigcirc\!-OH \longrightarrow \left[O-\!\bigcirc\!-\overset{O}{\underset{\underset{O}{\|}}{S}}-\!\bigcirc\!-O-\!\bigcirc\!-\underset{CH_3}{\overset{CH_3}{C}}-\!\bigcirc\!\right]_n \cdot$$

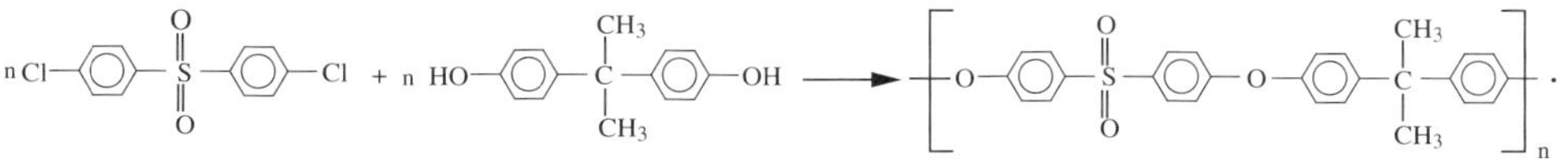

(2.100)

The reaction proceeds faster when the halogen is fluorine than when it is chlorine.[89] However, fluorides are costlier than chlorides, and accordingly, the latter are used wherever possible. In the two examples shown above, the weaker electron-withdrawing C=O group necessitates the use of a fluoride reactant, whereas the stronger electron-withdrawing SO_2 group makes it possible to work with a chloride reactant.

A polar aprotic solvent such as N,N-dimethylformamide (DMF), N,N-dimethyl-acetamide, (DMAc), N-methylpyrrolidone (NMP), diphenylsulfone, or sulfolane facilitates the reaction. The aprotic character of the solvent is important since it would otherwise have hydrogen-bonded the phenolate causing a reduction in the reactivity of the latter

Table 2.6 Repeating units of various poly(arylene ether)s.

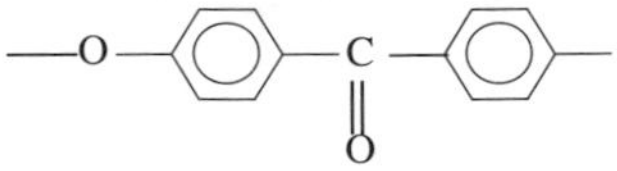

Poly(ether ketone) (PEK)

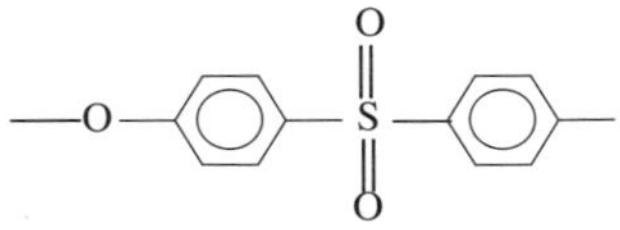

Poly(ether sulfone) (PES)

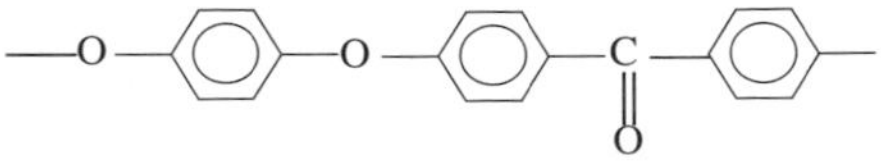

Poly(ether ether ketone) (PEEK)

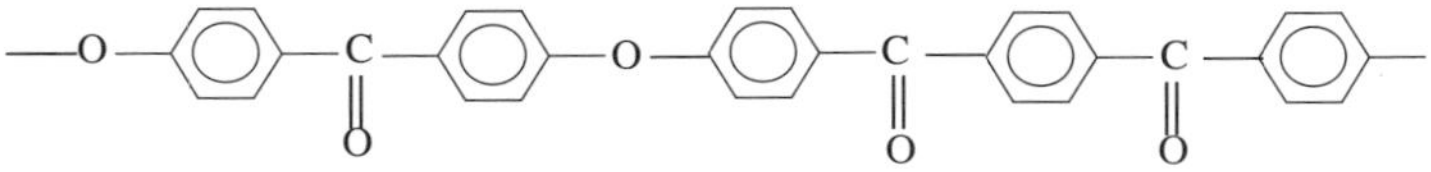

Poly(ether ketone ether ketone ketone) (PEKEKK)

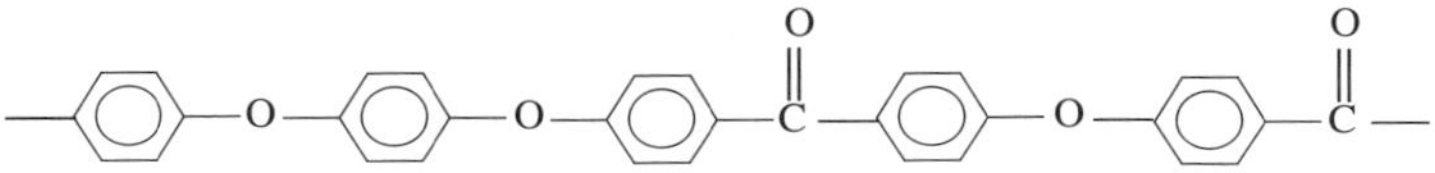

Poly(ether ether ketone ether ketone) (PEEKEK)

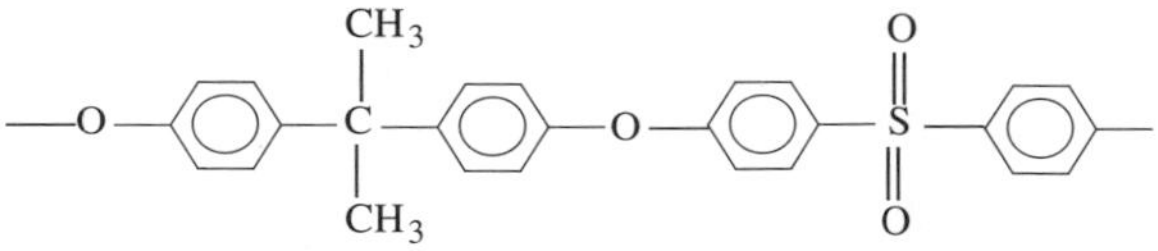

Bisphenol A polysulfone

(*cf.*, C-alkylation of phenolate in reaction with formaldehyde in protic media discussed in Sec. 2.7.1).

A high reaction temperature close to the melting point of PEEK is to be used in order to dissolve this semicrystalline polymer, whereas a much lower reaction temperature is good enough for the dissolution of polysulfone, which is amorphous. In addition, the reaction must be conducted under anhydrous condition lest hydrolysis of the diphenolate reactant occurs to generate hydroxyl ion, which in turn hydrolyzes the dihalide reactant causing stoichiometric imbalance. Anhydrous potassium carbonate is preferred to sodium hydroxide for this reason.[92–94]

2.9.4 Polyimides

Synthesis of polyimides by the reaction of A–A and B–B monomers typically uses a dianhydride and a diamine. Polyimide with a low ratio of flexible to rigid units in the chain backbone is intractable, being insoluble and infusible. The synthesis, therefore, is carried out in two steps. A typical example is represented by reaction (2.101). The first step involves polyamidation, which yields the soluble and fusible polyamic acid. The second step involves polyheterocyclization,[96] which converts the polyamic acid into the polyimide (**20**) on heating at temperatures above 150°C.

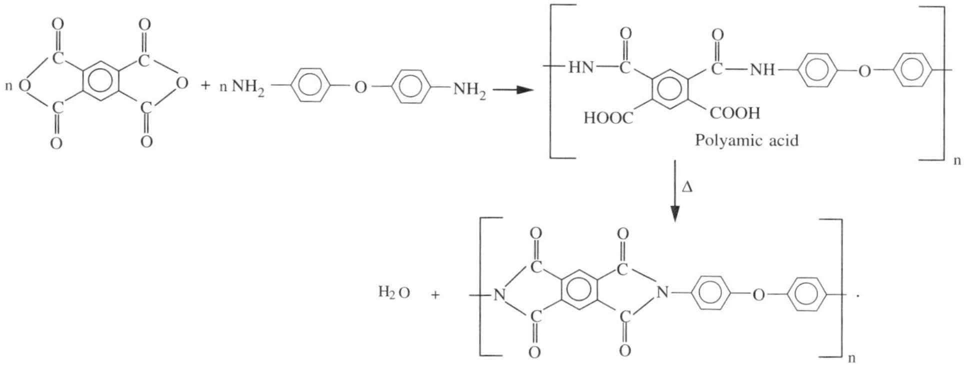

$$ \text{(2.101)} $$

Shelf life of the polyamic acid is, however, poor due to the slow imidation occurring during storage. In addition, water liberated during the imidation may cause degradation of the polymer, which is prevented by adding acetic anhydride and pyridine to the polyamic acid solution in NMP, the liberated water being removed by reaction with the acetic anhydride. The reaction is catalyzed by pyridine.[97] Interestingly, an eco-friendly processing of polyamic acid has been reported, which uses an aqueous or methanolic solution of the acid containing a *tertiary* amine.[98] The *tertiary* ammonium salt of the polyamic acid is stable toward hydrolysis and its imidation is faster.[99,100]

2.9.4.1 Thermoplastic polyimides

Polyetherimides and Polyamide imides

The proportion of flexible linking groups in the chain backbone of polyimide may be so increased as to make it processable using the usual plastics processing machineries. For instance, in polyetherimide (**22**) the ether groups provide the principal flexible links. Presence of the flexible links as well as the absence of full *p*-catenation reduces greatly T_g and T_m. The synthesis can be carried out by the method discussed above for polyimide using the appropriate diamine and the dianhydride. However, a different but convenient method uses the displacement reaction (2.102) between a bisphenolate and a dinitrobisimide. It involves nucleophilic aromatic substitution of the phenolate for nitro group in the dinitrobisimide, which is similar to the synthesis of poly(arylene ether)s discussed earlier (Sec. 2.9.3) with the difference that the displacement of halogen atoms rather than the nitro groups were involved in the former case.[101] The electron-withdrawing heterocyclic imide group activates the NO_2

group for displacement.

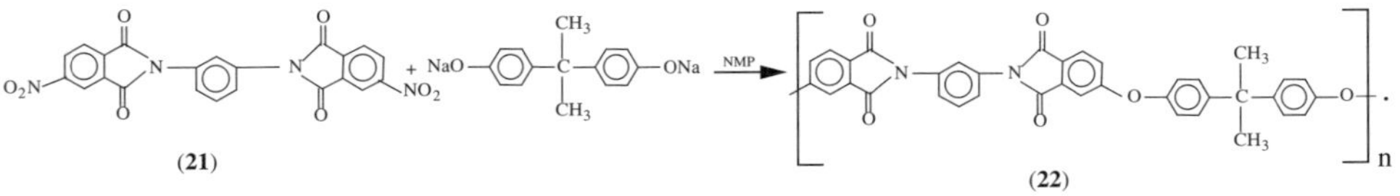

[Structural reaction scheme showing compound (21) + NaO—C(CH₃)₂—ONa in NMP yielding polymer (22)]

(2.102)

Other heterocyclic groups such as quinoxaline, benzoxazole, benzothiazole, oxadiazole, and benzimidazole can also act as activating groups to produce soluble and processable poly(arylene ether)s containing the respective groups in the chain backbone.[102–108]

Another melt processable polyimide is the polyamide imide **23**. The polymer contains both imide and amide groups in the chain backbone. It is synthesized by the reaction of trimellitic anhydride and 4,4′-diaminodiphenyl ether

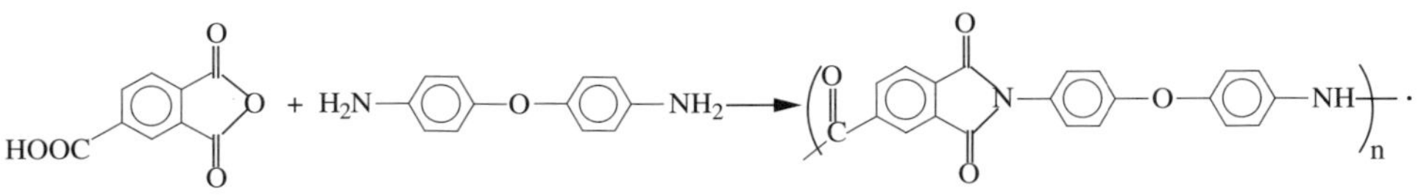

[Structural reaction scheme showing trimellitic anhydride + H₂N—C₆H₄—O—C₆H₄—NH₂ yielding polymer (23)]

(2.103)

2.9.4.2 Thermosetting polyimides

Polyimides find important applications as thermostable adhesives and matrices for structural composites in aerospace industries. Resins with good flow and surface-wetting properties are needed for adhesive application. Similarly, good wetting of the resin with reinforcing carbon fibers is essential for developing good mechanical properties in the structural composites. These requirements are met with liquid resins, which are ultimately converted to the thermosets for developing the desired mechanical properties.[100]

Oligomers with $\overline{M}_n$ in the range of 1000–5000 and end-capped with reactive groups, such as acetylene,[108,109] phenylacetylene,[110,111] biphenylene,[112] nadimide,[113–115] and maleimide,[115] among others, are used as the liquid resins. Making thermosetting polyamides through chain extension and crosslinking of these reactive oligomers is commonly referred to as PMR (polymerization of monomer reactants).[116] The end-capping is conveniently done by first preparing the oligomer using one monomer, say, the diamine, in appropriate excess. The amino end-capped oligoimide is reacted with maleic anhydride or nadic anhydride to have the maleimide (**24**) or the nadimide (**25**) end group respectively.[113–115] Similarly, using the dianhydride monomer in appropriate excess, the anhydride end-capped oligoimide may be prepared. This may be subsequently reacted with 4-ethynylaniline, 3-(phenylethynyl)aniline or 2-aminobiphenylene to obtain acetylene (**26**), phenylacetylene (**27**), or biphenylene (**28**) end-capped oligoimide respectively.[117–128]

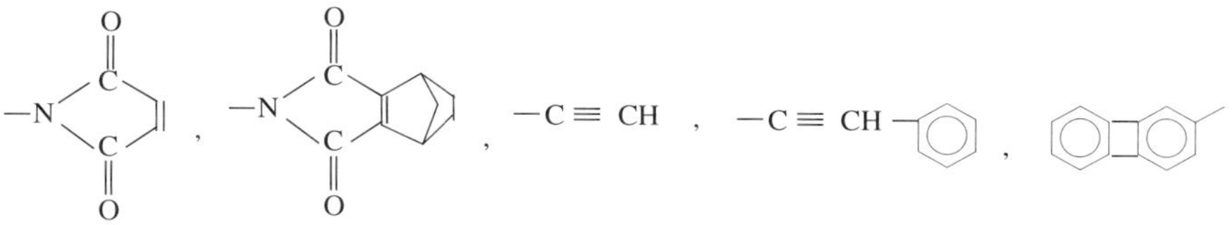

(24) (25) (26) (27) (28)

The resins are thermally cured at temperatures ranging from 200 to 370°C depending on the reactive end group present giving complex addition products. For example, the thermal curing of the maleimide[124] and the nadimide[125] end-capped oligomers involve reactions (2.104) and (2.105) respectively.

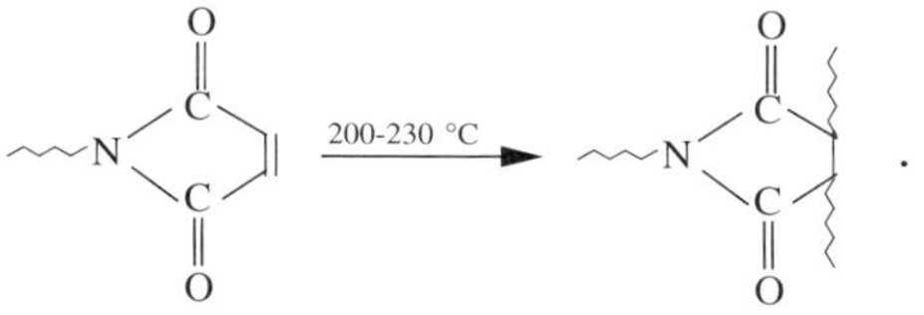

$$\text{maleimide} \xrightarrow{200-230\ °C} \text{product} \qquad (2.104)$$

$$\text{nadimide} \xrightarrow{300-350\ °C} \text{maleimide} + \text{cyclopentadiene} \longrightarrow \text{Chain copolymerized product of A, B and C.} \qquad (2.105)$$

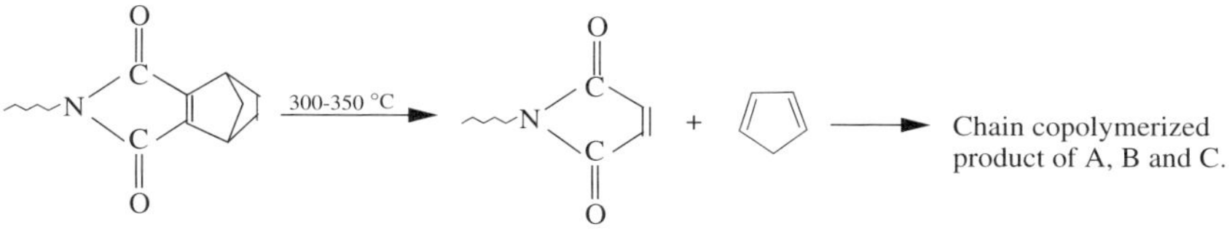

A   B   C

The maleimide end-capped oligomer (bismaleimide) may also be cured by a dinucleophile such as a diamine, which can add to the maleic double bond (Michael type addition)[127]

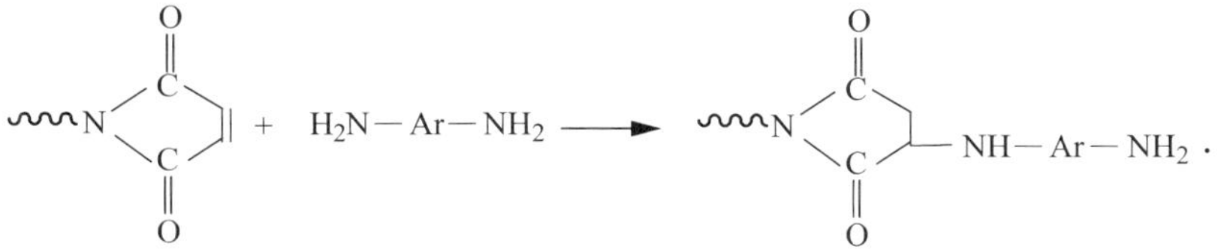

$$\text{maleimide} + H_2N-Ar-NH_2 \longrightarrow \text{adduct}-NH-Ar-NH_2 . \qquad (2.106)$$

Diels-Alder [4 + 2] cycloaddition of the maleimide end group with a diene such as dibenzocyclobutene, and substituted dicyclopentadiene may also be used in curing.[128]

The PMR approach has been used with other polymers as well such as polyphenylene,[108] poly(phenylquinoxaline),[108] polyquinolines,[129] and poly(ether ketosulfone)s.[108]

2.9.5 Polybenzimidazoles

Marvel pioneered the heterocyclization of the chain backbone during or following polymerization as a method of preparing polymers with heterocyclic groups in repeating units.[96] Thus, polybenzimidazoles containing both heterocyclic and aromatic rings in the structural units may be prepared by reacting diphenyl esters of aromatic diacids with aromatic tetramines as shown in reaction (2.107). The reaction is carried out in two stages.[96,130] In the first stage, low molecular weight polymers are prepared by heating the reactants at 260°C. In the second stage, further polymerization in the solid state at higher temperatures under high vacuum yields the high molecular weight polymer (*cf.*, synthesis of aliphatic polyamides, Sec. 2.8.1.1).

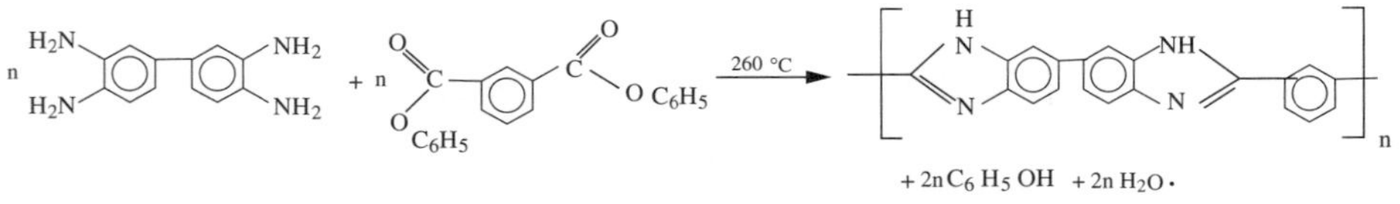

$$+ 2n\, C_6H_5OH + 2n\, H_2O.$$

(2.107)

However, the synthesis may be performed at much lower temperatures in polyphosphoric acid using a diacid in place of the diester.[131]

2.9.6 Polybenzoxazoles and polybenzothiazoles

Polybenzoxazoles and polybenzothiazoles may be prepared by reactions analogous to those used in the synthesis of polybenzimidazoles described above by reacting a bis-*o*-aminophenol or bis-*o*-aminothiophenol with a diphenyl ester of a diacid as shown in the example (reaction 2.108).[132]

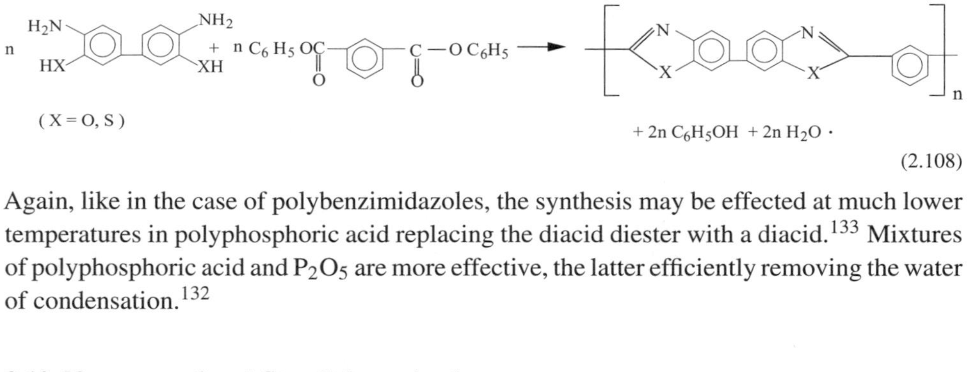

(X = O, S)

$$+ 2n\, C_6H_5OH + 2n\, H_2O.$$

(2.108)

Again, like in the case of polybenzimidazoles, the synthesis may be effected at much lower temperatures in polyphosphoric acid replacing the diacid diester with a diacid.[133] Mixtures of polyphosphoric acid and P_2O_5 are more effective, the latter efficiently removing the water of condensation.[132]

2.10 Nonconventional Step Polymerization

Besides the conventional functional group reactions discussed in the preceding sections 2.7 through 2.9, several nonconventional reactions have been used in step polymerization. These are classified into two types: acyclic diene metathesis (ADMET) and transition metal coupling.

The ADMET polymerization, which uses the Schrock and Grubbs ring-opening metathesis polymerization (ROMP) catalysts, has been discussed in Chap. 7 (Sec. 7.12.2).

Transition metal coupling polymerizations have been very useful in the synthesis of conjugated polymers, optically active polymers as well as hyperbranched and dendritic polymers.[134] The synthesis involves the coupling reaction between aryl/alkenyl halides with aryl/alkenyl organo metals or with free alkene/alkynes by Pd based catalysts. It uses mild conditions and exhibits high regio- and chemoselectivities as well as varying degree of functional group tolerance depending upon the nucleophilicity of the organometal used. Representative synthesis of some conjugated polymers, *e.g.*, poly(*p*-phenylene)s, polythiophenes, poly(phenylenevinylene)s, and poly(phenyleneethynylene)s are presented in the following.

The poly(*p*-phenylene)s may be synthesized using the coupling reaction between aryl halide and organometallic reagents. In the Kumuda coupling (reaction 2.109), the latter are the Grignard reagents, which are high in nucleophilicty and, accordingly, low in the functional group tolerance. Nickel(II) complexes are used as the catalysts.[135,136]

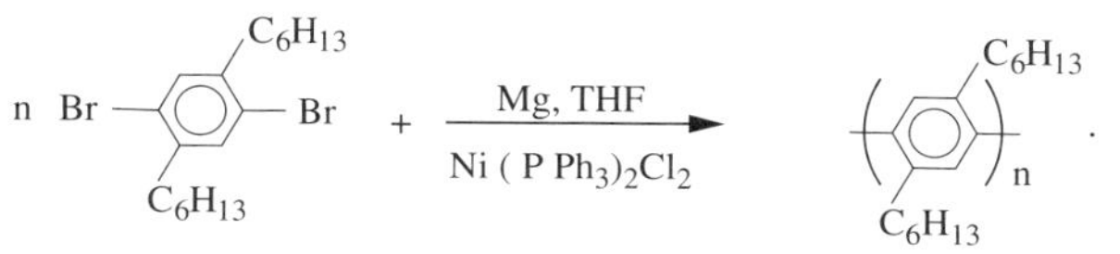

$$\text{n Br-C}_6\text{H}_3(\text{C}_6\text{H}_{13})_2\text{-Br} \xrightarrow{\text{Mg, THF}; \text{Ni(PPh}_3)_2\text{Cl}_2} \text{[-C}_6\text{H}_2(\text{C}_6\text{H}_{13})_2\text{-]}_n \quad (2.109)$$

Organozincs have lower nucleophilicity. Coupling reaction involving organozinc and aryl halide in the presence of a palladium complex catalyst (reaction 2.110), has provided an effective method for the synthesis of the soluble regioregular poly(alkylthiophene)s.[137–139]

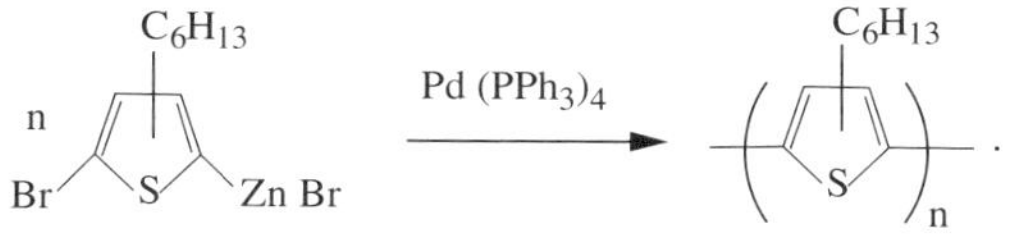

$$(2.110)$$

The low nucleophilicity of organotins also makes possible to synthesize functionalized conjugated polymers using functionalized monomers. This coupling reaction, known as the Stille coupling, is effected in the presence of Pd catalysts.[140–142]

Organoborons are very low in nucleophilicity and stable even in the presence of water. The cross coupling of the aryl/alkenyl boronic acids with the aryl/alkenyl halides in the presence of a base and a palladium catalyst, known as the Suzuki coupling, may be conducted even in aqueous organic mixtures.[143] For example, a water soluble polyphenylene is prepared as shown in reaction (2.111).[144]

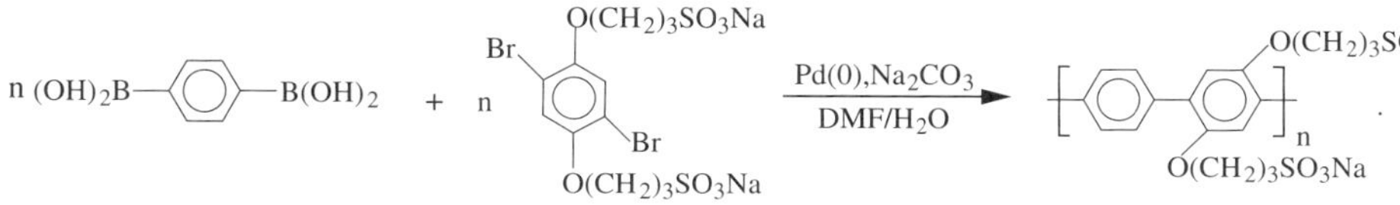

$$(2.111)$$

A facile route for the synthesis of poly(phenylenevinylene)s uses the Heck coupling.[145] This involves the coupling of aryl bromides with aryl-alkenes in the presence of a palladium catalyst as shown in reaction (2.112).[146]

$$n\,Br{-}Ar(R,R){-}Br + n\,CH{=}CH{-}Ar(R',R') \xrightarrow{Pd(0),\ base} {-}[Ar(R,R){-}CH{=}CH{-}Ar(R',R')]_n{-} \quad (2.112)$$

Poly(phenyleneethynylene)s may be prepared by coupling aryl halides with alkynes in the presence of a base using both palladium and cuprous iodide as catalysts.[147–150] In fact, the alkyne is converted to alkynylcopper species by reacting with CuI in the presence of the base. The organocopper so formed undergoes cross coupling with the aryl halide to form the polymer.

$$n\,HC{\equiv}C{-}Ar(OC_{12}H_{25}, OC_{12}H_{25}){-}C{\equiv}CH + n\,Br{-}Ar(OAc, OAc){-}Br \xrightarrow{Base,\ Pd(PPh_3)_4/CuI} {-}[Ar(OAc, OAc){-}C{\equiv}C{-}Ar(OC_{12}H_{25}, OC_{12}H_{25}){-}C{\equiv}C]_n{-} \quad (2.113)$$

References

1. Collected Papers of Wallace Hume Carothers on High Polymeric Substances, H. Mark, G,S. Whitby Eds. Interscience, N. Y. (1940).
2. P.J. Flory, Principles of Polymer Chemistry, Cornell University, Ithaca, New York Chap. 3 (1953).
3. P.J. Flory, *J. Am. Chem. Soc.* (1939) 61, 3334.
4. P.J. Flory, *J. Am. Chem. Soc.* (1940) 62, 2261.
5. B.V. Bhide, J.J. Sudborough, *J. Indian Inst. Sci.* (1925) 8A, 89.
6. D.P. Evans, J.J. Gordon, H.B. Watson, *J. Chem. Soc. (London)* (1938) 1439.
7. P.C. Haywood, *J. Chem. Soc. (London)* (1922) 121, 1904.
8. E. Rabinowitch, W.C. Wood, *Trans. Faraday Soc.* (1936) 32, 1381.
9. E. Rabinowitch, *Trans. Faraday Soc.* (1937) 33, 1225.
10. E.W. Spanagel, W.H. Carothers, *J. Am. Chem. Soc.* (1935) 57, 929; (1936) 58, 654.
11. J.W. Hill, W.H. Carothers, *J. Am. Chem. Soc.* (1933) 55, 5043.
12. J.W. Hill, W.H. Carothers, *J. Am. Chem. Soc.* (1933) 55, 5031.
13. E.L. Eliel, Stereochemistry of Carbon Compounds, Tata-McGrew-Hill Publishing Co., New Delhi, Chap. 7 (1962).
14. M. Stoll, A. Rouve, *Helv. Chim. Acta* (1935) 18, 1087.
15. F.S. Dainton, K.J. Ivin, *Quart. Rev. (London)* (1958) 12, 61.
16. R.J. Gaymans, Polyamides, in, *Synthetic Methods in Step-growth Polymers*, M.E. Rogers, T.E. Long Eds., Wiley-Interscience, Hoboken, New Jersey, Chap. 3 (2003).
17. H. Jacobson, W.H. Stockmayer, *J. Chem. Phys.* (1950) 18, 1600.
18. C.A. Truesdell, *Ann. Math.* (1945) 46, 144.
19. P.J. Flory, Ref. 2, p. 317.
20. (a) H.A. Liebhafsky, *Silicones under the Monogram*, John Wiley & Sons, Inc., New York (1978).
    (b) A.J. Barry, H.N. Beck, in, *Inorganic Polymers*, F.G.A. Stone, W.A.G. Graham Eds., Academic Press Inc., New York (1962).
21. J.F. Brown, G.M.J. Slusarczuk, *J. Am. Chem. Soc.* (1965) 87, 931.

22. J.B. Carmichael, R. Winger, *J. Polym. Sci.* (1965) A3, 971.
23. B. Hardman, A. Torkelson, Silicones, in, *Ency. Polym. Sci. & Eng.*, 2nd ed., H. Mark, N.M. Bikales, C.G. Overberger, G. Menges Eds., Wiley-Interscience, New York (1989), Vol. 15, p. 204.
24. S.R. Porter, L.M. Wang, *Polym. Rev.* (1992) 33, 2019.
25. P.J. Flory, *J. Am. Chem. Soc.* (1936) 58, 1877.
26. P.J. Flory, *J. Am. Chem. Soc.* (1942) 64, 2205.
27. P.J. Flory, *Chem. Rev.* (1946) 39, 137.
28. A. Fradet, M. Tessier, *Polyesters*, in, Synthetic Methods in Step-growth Polymers, M.E. Rogers, T.E. Long Eds. Wiley-Interscience, Chap. 2 (2003), Hoboken.
29. W.H. Carothers, *Trans. Faraday Soc.* (1936) 32, 39.
30. W.H. Carothers, *Chem. Rev.* (1931) 8, 353.
31. S.H. Pinner, *J. Polym. Sci.* (1956) 21, 153.
32. R.G. Beaman *et al.*, *J. Polym. Sci.* (1959) 40, 329.
33. E.L. Wittbecker, P.W. Morgan, *J. Polym. Sci.* (1959) 40, 289.
34. P.W. Morgan, S.L. Kowlek, *J. Polym. Sci.* (1959) 40, 299.
35. P.J. Flory, *J. Chem. Phys.* (1944) 12, 425.
36. W.B. Kuhn, *Ber.* (1930) 63, 1503.
37. G. B. Taylor, *J. Am. Chem. Soc.* (1947) 69, 638.
38. P.J. Flory, *J. Am. Chem. Soc.* (1941) 63, 3083.
39. R.H. Kienle, F.E. Petke, *J. Am. Chem. Soc.* (1941) 63, 481; (1940) 62, 1053.
40. R.H. Kienle, P.A. van der Meulen, F.E. Petke, *J. Am. Chem. Soc.* (1939) 61, 2258.
41. A. Knop, L.A. Pilato, *Phenolic Resins — Chemistry, Applications and Performance*, Springer–Verlag, Berlin (1985).
42. S. Lin–Gibson, J.S. Riffle, *Chemistry and Properties of Phenolic Resins and Networks in Synthetic Methods*, in, Step-growth Polymers, M.E. Rogers, T.E. Long Eds., John Wiley & Sons, Hoboken, New Jersey, Chap. 7 (2003).
43. N. Schreve, The Chemical Process Industries McGrew-Hill Book Co., Inc. New York, Chap. 35 (1956).
44. A. Knop, W. Scheib, *Chemistry and Applications of Phenolic Resins*, Springer–Verlag, New York (1979).
45. N. Kornblum, P. J. Berrigan, W.J, Lenoble, *J. Am. Chem. Soc.* (1963) 85, 1141.
46. F.A. Carey, R.J. Sundberg, *Advanced Organic Chemistry, Part B: Reactions and Synthesis*, 4th ed., Kluer Academic/Plenum Pub. New York, Chap. 35 (2001).
47. W.P. Sorenson, W. Sweeney, T.W. Campbell, Synthetic Resins and Composites, in, *Preparative Methods of Polymer Chemistry*, 3rd ed., Wiley-Interscience, New York, Chap. 11 (2001).
48. E.M. Yorkgitis, Adhesive Compounds, in, *Ency. Polym. Sci. & Tech.*, 3rd ed., H. Mark Ed. (2003), p. 256.
49. I.H. Updegraff, Amino Resins, in, *Ency. Polym. Sci. & Eng.*, 2nd Ed., H. Mark, N.M. Bikales, C.G. Overberger, G. Menges Eds., Wiley-Interscience, New York (1985) Vol. 1, p. 752.
50. Z.W. Wicks, Jr., Alkyd Resins, in, *Ency. Polym. Sci. & Tech.*, 3rd. ed., H.F. Mark Ed. (2003) Vol. 1, p. 318.
51. J. Dodge, Polyurethanes and Polyureas, in, *Synthetic Methods in Step-growth Polymers*, M.E. Rogers, T.E. Long Eds., Wiley Interscience, Hoboken, New Jersey, Chap. 4 (2003).
52. H. Lee, K. Neville, Epoxy Resins: Their Applications and Technology, McGraw Hill, New York (1957).
53. L.V. McAdams, J.A. Gannon, Epoxy Resins, in, *Ency. Polym. Sci. & Eng.*, 2nd Ed., H. Mark, N.M. Bikales, C.G. Overberger, G. Menges, Eds. (1986) Vol. 6, p. 322
54. R.J. Arnold, *Mod. Plast.* (1964) 41(4), 149.
55. M.B. Berenbaum, Polymer Containing Sulfur, Polysulfide, in, *Kirk-Othmer Ency. Chem. Tech.* 2nd ed., A. Standen Ed., John Wiley & Sons, Inc., New York (1968), p. 253.

56. S.M. Ellerstein, E.R. Bertozzi, *ibid*, 3rd ed., M. Grayson Ed. (1982) Vol. 18, pp. 814–831.
57. H.Lee, D. Stoffey, K. Neville, New Linear Polymers, McGraw Hill, New York (1967).
58. H.F. Mark, *Macromolecules* (1977) 10, 881.
59. F. Garbassi, R. Po, *Engineering Thermoplastics, Overview*, in, *Ency. Polym. Sci. & Tech.*, H. Mark Ed., 3rd ed. (2003) Vol. 2, p. 307.
60. H. F. Mark, *J. App. Polym. Sci. App. Polym. Symp.* (1984) 39, 1.
61. P.E. Beck, E.E. Margat, *Macromolecular Syntheses*, Vol. 6, C.G. Overberger Ed., Wiley, New York (1977), p. 57.
62. D.B. Jacobs, J. Zimmermann in *High Polymers* Vol. 29, C.E. Schildnecht, I. Skeist Eds., Wiley-Interscience, New York (1977), p. 424.
63. J. Zimmerman, Polyamides, in, *Ency. Polym. Sci. & Eng.*, 2nd ed., H. Mark, N.M. Bikales, C.G. Overberger, G. Menges, Eds., Vol. 11, Wiley-Interscience (1988), p. 315.
64. R.J. Gaymans, J. Amirtharay, H. Kamp, *J. App. Polym. Sci.* (1982) 27, 2513.
65. P.W. Morgan, *Macromolecules* (1977) 10, 1381.
66. P.W. Morgan, Condensation Polymers: By Interfacial and Solution Methods, Interscience, New York (1965).
67. R.W. Kwolek *et al.*, *Macromolecules* (1977) 10, 1390.
68. T.I. Blair, P.W. Morgan, F.I. Killian, *Macromolecules* (1977) 10, 1396.
69. V.A. Vasnev, S.B. Vinogradova, *Russ. Chem. Rev.* (1979) 48, 16.
70. D.E. James , L.G. Packer, *Ind. Eng. Chem. Res.* (1995) 34, 4049.
71. J. Economy *et al. J. Polym. Sci. Polym. Chem. Ed.* (1976) 14, 2207.
72. D.J. Brunelle, Polycarbonates, in *Ency. Polym. Sci. & Tech.* 3rd ed., H.F. Mark Ed. (2003) Vol. 7, 397.
73. S.N. Hersh, Y.K. Choi, *J. App. Polym. Sci.* (1990) 41, 1033.
74. Y. Kim, K.Y. Choi, *J. App. Polym. Sci.* (1993) 49, 747.
75. M. Yokoyama, H. Yoshitoka, H. Masam, JP Pat 06, 239, 989 (1944).
76. V.N. Ignatov *et al.*, *Macromol. Chem. Phys.* (2001) 202, 1941.
77. Ref. 57, p. 61.
78. L.S. Tan, *et al.*, *Macromol. Rapid Com.* (1999) 20, 16.
79. M. Hasegawa, *Polyhydrazides and Polyoxadiazoles*, in, Ency. Polym. Sci. & Tech., H. Mark, N.M. Bikales, N.G. Gaylord, Eds. (1969) Vol. 11, p. 169.
80. J.-I. Lin, C.-S. Kang, *Prog. Polym. Sci.* (1997) 22, 937.
81. H. Han, P.K. Bhowmick, *Prog. Polym. Sci.* (1997) 22, 1431.
82. J. Economy, K. Goranov, *Adv. Polym. Sci.* (1994) 117, 221.
83. J. Preston, Polyamides, Aromatic, in, *Ency. Polym. Sci. & Eng.*, 2nd ed., H. Mark, N.M. Bikales, C.G. Overberger, G. Menges Eds., Wiley-Interscience, New York (1988) Vol. 11, p. 381.
84. R.T. Hawkins, *Macromolecules* (1976) 9, 189.
85. A. Kultys, Sulfur — Containing Polymers, in, *Ency. Polym. Sci. & Tech.*, 3rd ed., Vol. 4, H.F. Mark Ed., John Wiley & Sons, Hoboken, New Jersey (2003), p. 336.
86. S. Wang, J.E. McGrath, Synthesis of Poly(arylene ether)s, in, *Synthetic Methods in Step — Growth Polymers*, M.E. Rogers, T.E. Long Eds., John Wiley & Sons, Inc. Hoboken, New Jersey, Chap. 6 (2003).
87. T.E. Attwood *et al.* Polymer (1981) 22, 1096.
88. J.B. Rose, R.A. Staniland, U.S. Pat 4, 320, 224. (1982).
89. R.N. Johnson *et al.*, *J. Polym. Sci., Polym. Chem. Ed.* (1967) 5, 2375.
90. A.C. Farnham, L.M. Robeson, J.E. McGrath, *J. Appl. Polym. Sci., Appl. Polym. Symp.* (1975) No. 26, 373.
91. A.J. Parker, *Quart. Rev.* (1962) 16, 163.
92. R.A. Clendining, A.G. Farnham, N.N. Zutly, D.C. Priest, Canad. Pat. 847, 963, Union Carbide Corpn. (1970).

93. R. Viswanathan, B.C. Johnson, J.E. McGrath, *Polymer* (1984) 25, 1927.
94. J.L. Hedrick et al., *J. Polym. Sci. Polym. Chem. Ed.* (1986) 24, 287.
95. T. Takekoshi, *Polyimides — Fundamentals and Applications*, H.K. Ghosh, K.L. Mittal Eds., Marcel Dekker, New York (1996), p. 7.
96. H. Vogel, C.S. Marvel, *J. Polym. Sci.* (1961) 50, 511.
97. C.E. Sroog,. *Polyimides, IUPAC Symp.*, Prague, Sept. (1965).
98. Y. Imai, T. Fueki, T. Inoue, M. Kakimoto, *J. Polym. Sci. Polym. Chem. Ed.* (1998) 36, 2663.
99. J.A. Kruez, A.L. Endrey, F.P. Gay, C.E. Sroog, *J. Polym. Sci.*, A-1 (1966) 4, 2607.
100. B. Sillion, R. Mercier, D. Picq, Polyimides and Other High — Temperature Polymers, in, *Synthetic Methods in Step — Growth Polymers*, M.E. Rogers, T.E. Long Eds., John Wiley & Sons, Inc. Hoboken, New Jersey, Chap. 5 (2003).
101. D.M. White et al., *J. Polym. Sci. Polym. Chem. Ed.* (1981) 19, 1635.
102. Ref. 95, p. 39.
103. J.N. Labadie, J.L. Hedrick, *Macromol. Symp.* (1992) 54/55, 313.
104. M. Lucas, P. Brock, J.L. Hedrick, *J. Polym. Sci. Polym. Chem. Ed.* (1993) 31, 2179.
105. J.L. Hedrick, R. Twieg, *Macromolecules* (1981) 25, 2021.
106. J.L. Hedrick, J.W. Labadie, *Macromolecules* (1990) 23, 1561.
107. J.W. Connell, P.M. Hergenrother, *Polym. Mater. Sci. Eng. Proc.* (1989) 60, 527.
108. P.M. Hergenrother, *Rev. Macromol. Chem.* (1980) C19 (1), 1.
109. A.L. Landes et al., *Polym. Prep. Div. Polym. Chem. ACS* (1974) 15(2), 533, 537.
110. F.W. Harris et al., *J. Macromol. Chem.* (1984) 421, 1117.
111. X. Fang, D.F. Rogers, D.A, Scola, M.P. Steven, *J. Polym. Sci. Polym. Chem. Ed.* (1998) 36, 461.
112. T. Takeichi, J.K. Stille, *Macromolecules* (1986) 19, 2093.
113. T.T. Serafin, P. Delvigs, G.R. Lightsey, *J. App. Polym. Sci.* (1972) 16, 905.
114. H.R. Lubowitz, *Polym. Prep. (ACS Div. Polym. Chem.)* (1971) 12(1), 329.
115. T.L. St. Clair, R.A. Jewell, *Solventless LARC-160, :Polyimide Matrix Resin*, in, Proc. 23rd Nat'l SAMPE Symp., 1 May (1978).
116. T.T. Serafini, P. Delvigs, G.R. Lightsay, *J. App. Polym. Sci.* (1972) 16, 905.
117. D.F. Lindow, L. Friedman, *J. Am. Chem. Soc.* (1967) 89, 1271.
118. M.D. Sevik et al., *Macromolecules* (1979) 12, 423.
119. S.A. Swanson, W.W. Fleming, D.C. Hofer, *Macromolecules* (1992) 25, 582.
120. J.J. Ratto, P.J. Dynes, C.L. Hammermesh, *J. Polym. Sci. Polym. Chem. Ed.* (1980) 18, 1035.
121. P.M. Hergenrother, J.G. Smith, *Polymer* (1994) 35, 4857.
122. L. Friedman, D.F. Lindow, *J. Am. Chem. Soc.* (1968) 90, 2324.
123. J.P. Droske, J.K. Stille, *Macromolecules* (1984) 17, 10.
124. P. Grundschober, J. Sambeth, U.S. Pat. 3, 380, 964 (30 Apr 1968).
125. A.C. Wang, M.M. Ritchey, *Macromolecules* (1981) 14, 825.
126. T.T. Serafini, P. Delvigs, *App. Polym. Symp.* (1973) No. 22, 89.
127. J.V. Crivello, *J. Polym. Sci. Polym. Chem. Ed.* (1973) 11, 1185.
128. R.G. Bryant, *Polyimides*, in, Ency. Polym. Sci.& Tech. 3rd ed., H.F. Mark Ed., Wiley-Interscience, Hoboken, New Jersey (2003) Vol. 7, p. 529.
129. J.P. Droske, U.M. Gaik, J.K. Stille, *Macromolecules* (1984) 17, 10.
130. T.S. Chung, *Rev. Macromol. Chem. Phys.* (1997) C37, 277.
131. E.W. Choe, *J. Appl. Polym. Sci.* (1994) 53, 497.
132. J.F. Wolfe, Polybenzothiazoles and Polybenzoxazoles, in, *Ency. Polym. Sci. & Eng.*, 2nd ed., H. Mark, N.M. Bikales, C.G. Overberger, G. Menges, Eds. (1988) Wiley-Interscience, New York, Vol. 11, p. 601.
133. (a). J.F. Wolfe, B.H. Loo, F.E. Arnold, *Macromolecules* (1981) 14, 915.
     (b). J.F. Wolfe, F.E. Arnold, *Macromolecules* (1981) 14, 909.

134. Q.-S. Hu, Nontraditional Step-Growth Polymerization: Transition Metal Coupling, in, *Synthetic Methods in Step-Growth polymers*, M.E. Rogers, T.E. Long Eds., Wiley-Interscience, Hoboken, New Jersey, Chap. 9 (2003).
135. M. Kumuda, *Pure App. Chem.* (1980) 52, 669.
136. T. Yamamoto, A. Yamamoto, *Chem. Lett.* (1977) 353.
137. T. Yamamoto, *Bull. Chem Soc. Jpn.* (1999) 72, 621.
138. T.A. Chen, R.D. Rieke, *J. Am. Chem. Soc.* (1992) 114, 10087.
139. M.J. Marsella, P.J. Carroll, T.M. Swager, *J. Am. Chem. Soc.* (1994) 116, 9347.
140. J.K. Stille, *Pure App. Chem.* (1985) 57, 1771.
141. M. Bochmann, K. Kelly, *Chem. Commun.* (1989) 532.
142. Q.T. Zhang, J.M. Tour, *J. Am. Chem. Soc.* (1998) 120, 5355.
143. M. Mitaura, T. Yanagi, A. Suzuki, *Synth. Commun.* (1981) 11, 513; *Chem. Rev.* (1995) 95, 2457.
144. A.D. Child, J.R. Reynolds, *Macromolecules* (1994) 27, 1975.
145. R.F. Heck, In Comprehensive Organic Synthesis B.M. Trost Ed., Pergamon, New York (1991), p. 833.
146. A. Greiner, W. Heitz, *Macromol. Chem. Rapid Commun.* (1988) 9, 58.
147. S. Takahash, Y. Kuroyama, K. Sonogashira, N. Hagihara, *Synthesis* (1980) 627.
148. K. Sanechika, A. Yamamoto, T. Yamamoto, *Bull. Chem. Soc. Japan* (1984) 57, 752.
149. Q. Zhou, T. M. Swager, *J. Am. Chem., Soc.* (1998) 120, 5321.
150. T. Yamamoto, T. Kimura, K. Shiraiashi, *Macromolecules* (1999) 32, 8886.

Problems

2.1 If cyclization of an A–B monomer is to give a 4-membered ring, the chain polymer will be the product in melt polymerization provided the melting temperature of the polymer is not too high. Explain.

2.2 What will be the degree of polymerization when an A–B monomer is polymerized in the presence of one mol percent of the following compounds: (i) a monofunctional reactant, (ii) an A–A monomer, or (iii) a B–B monomer?
(Ans: $\overline{DP}_n = 101$ in each case.)

2.3 Will the result be the same if in the Problem 2.2 the A–B monomer is replaced with an equivalent mixture of a pair of A–A and B–B monomers?
(Ans: Yes, if one mol of A-B monomer is replaced by 0.5 mol each of A-A and B-B monomers, but if replaced by one mol each of A-A and B-B monomers, the value would be 201.)

2.4 Is there any limiting value of the polydispersity index in bifunctional step polymerization? Explain your answer.
(Ans: Yes, 2.)

2.5 Using the statistical method of Flory find out the range of monomer composition, which would lead to gelation at high enough extents of reaction in each of the following combinations of monomers (1 and 2) with different functionalities:
(i) $f_1 = 2$, $f_2 = 3$ (ii) $f_1 = 2$, $f_2 = 4$ (iii) $f_1 = 2$, $f_2 = 6$.
(Ans: (i) $0.75 \leq m_1/m_2 \leq 3$, (ii) $0.66 \leq m_1/m_2 \leq 6$, (iii) $0.6 \leq m_1/m_2 \leq 15$, m_1 and m_2 being the number of mols of monomer 1 and 2 respectively.)

2.6 Work out the solution to the problem 2.5 using the Carothers equation.
(Ans: (i) $1 \leq m_1/m_2 \leq 2$, (ii) $1 \leq m_1/m_2 \leq 3$, (iii) $1 \leq m_1/m_2 \leq 5$, m_1 and m_2 being the mols of monomer 1 and 2 respectively.)

2.7 What changes in physical and mechanical properties take place if block copolyester is heated in the melt in the presence of an acid catalyst for long time?

2.8 In as-synthesized nylon 6 the proportion of cyclics would be more than that in as-synthesized nylon 66. Explain.

2.9 Does the prediction of ring fraction by Jacobson–Stockmayer theory using freely rotating chain model agree with the experimental result? Explain your answer.

2.10 In the synthesis of polyimide or polybenzimidazole one of the monomers has the functionality 4, the other has it 2, and yet the polymer formed is linear. Do you think the result is in conflict with the Carothers equation? If not, how would you apply it to predict the molecular weight?

Chapter 3

Radical Polymerization

Radical polymerization is a chain polymerization in which the chain carrier is a carbon centered radical. It is the most widely practiced of all chain polymerizations, which is mainly due to its applicability to a wide range of monomers, ease of operation, low operation cost, wide choice of operating temperature, and inertness toward the most environment–friendly solvent (water). Oxygen in atmospheric air ordinarily inhibits polymerization. However, it may be excluded easily by purging the reaction mixture and vessel with an inert gas, or else, an induction period is accepted during which a small fraction of the initiator is sacrificed. In contrast, oxygen in small amount may act as initiator of polymerization as well (*vide* Sec. 3.9.2).

The field acquired a new dimension in the mid 1990s when the living radical polymerization was discovered. The discovery made possible the synthesis of polymer with defined chain ends, targeted molecular weight, narrow MWD, and a wide range of molecular architecture, and, all these, in the easiest way and at the cheapest cost.

3.1 General Features

3.1.1 *Monomer types*

Ordinarily, both vinyl and vinylidene monomers undergo radical polymerization. However, when the monomer is an alkene, only the simplest member, *i.e.*, ethylene, is polymerizable; others are not homopolymerizable presumably due to autoinhibition caused by chain transfer to monomer (*vide* Sec. 3.8.1.2). However, they may copolymerize with electrophilic monomers. 1,2-Disubstituted ethylenes (internal olefins) either do not homopolymerize (stilbene being an example) or do so under drastic conditions, *e.g.*, maleic anhydride, due to steric hindrance to propagation. On the other hand, many of them can copolymerize with various vinyl monomers. However, when the substituents are fluorine, which has similar steric bulk as hydrogen atom,[1] even the tetrasubstituted ethylenes, *viz.*, tetrafluoroethylene and trifluorochloroethylene, are polymerizable (*vide* p. 24). Apart from vinyl monomers, conjugated dienes, *e.g.*, butadiene, isoprene, and chloroprene, readily undergo polymerization.

3.1.2 Reactivity of monomers and radicals

Carbon centered radicals with which we are concerned here are uncharged species unlike ions. However, they may be electrophilic or nucleophilic according as their α-substituents are electron withdrawing or electron releasing respectively. Similarly, monomers with electron-withdrawing substituents are electron acceptors, whereas those with electron-releasing substituents are electron donors. These polar properties influence the rate of radical addition to monomer and impart some selectivity in the preference of monomer by the radical. An electrophilic radical shows some specificity for addition to electron-donor monomers, whereas a nucleophilic radical does the opposite; it prefers to add to electron-acceptor monomers. This occurs due to the contribution of charge-transfer resonance structures to the transition state. Such structures for the addition of an electrophilic radical to an electron-donor monomer may be represented as

$$R\overset{\bullet}{-}\underset{Y}{C}H + CH_2\!=\!CH\!\!-\!\!\!\!<\!\!X \longrightarrow R-\underset{Y}{\overset{-}{C}H}\ \overset{+}{C}H_2-\overset{\bullet}{C}HX \longleftrightarrow R-\underset{Y}{\overset{-}{C}H}\ \underset{X}{\overset{+}{C}H}-\overset{\bullet}{C}H_2.$$

These structures lower the energy of the transition state and enhance the rate constant.

Besides the polarity, the stability of the radical also plays important role, the stability of the monomer playing lesser role due to the relatively low resonance energy of stabilization.[2-4]

The relative reactivities of monomers and radicals have been determined from the studies of radical copolymerization (Ch. 8).[4] The monomers have been arranged in the order of decreasing average reactivity, e.g., butadiene > α-methylstyrene > styrene > methyl methacrylate > methyl vinyl ketone > methacrylonitrile > acrylonitrile > methyl acrylate > vinyl chloride > vinyl acetate > allyl acetate > vinyl ethyl ether. The monomer reactivity may be correlated with the corresponding radical stability. The estimated resonance-stabilization energies of radicals corresponding to seven of the twelve monomers in the above series are reproduced from the literature in Table 3.1.[5] It follows that the monomer

Table 3.1 Substituent effect on radical stabilization energy (RSE) using RSE of CH_3 radical = 0.0 kJ/mol as reference. "Adapted with permission from Ref. 5. Copyright © 2005 American Chemical society."

Radical $CH_3-\overset{\bullet}{C}XY$		
X	Y	Radical stabilization energy kJ/mol^{-1}
$CH=CH_2$	H	−66.9
C_6H_5	CH_3	−53.1
C_6H_5	H	−49.3
CN	CH_3	−32.2
COOH	CH_3	−30.3
$COOCH_3$	CH_3	−29.7
CN	H	−28.4
COOH	H	−26.5
$COOCH_3$	H	−25.9

reactivity increases as the radical stability increases excepting a small deviation from this trend observed in the relative behavior of methacrylonitrile and methyl methacrylate. On the other hand, it was observed that radical reactivity decreases as the radical stability increases. For example, the following radicals may be arranged in decreasing order of reactivity as: vinyl acetate > methyl acrylate > methyl methacrylate > styrene > butadiene.[4]

The influence of product radical stability on monomer reactivity is suggestive of a late transition state in the addition of radical to monomer, *i.e.*, the transition state lies to the product side of the reaction coordinate largely assuming the product radical character. However, this view has been opposed on the ground that radical addition to monomer is exothermic due to the breakage of a weaker π bond and the consequent formation of a stronger σ bond

$$\dot{R} + H_2C=CHX \longrightarrow RCH_2-\dot{C}HX,$$

and that an exothermic reaction has an early transition state being located in the reactant side of the reaction coordinate. Accordingly, the transition state resembles the reactants rather than the product radical. Thus, the stability of the latter should have only minor, if at all any, influence on the reaction rate.[6–8]

Indeed, studies with small radicals have revealed minor influence of radical stability except for radicals conjugated with C=C π bonds such as the styrene and butadiene radicals, which have relatively high stability (Table 3.1). It is the steric and polar effects, which governs the reaction rate and the regiochemistry of addition.[6–8] These conclusions were based on the relative yield of products in suitably chosen competing reactions.

However, subsequent studies involving determination of absolute reaction rates and quantum mechanical calculations of the transition state structures and the various energy quantities that determine the reaction barriers revealed that the reaction enthalpy has a large influence on the rate of radical addition.[9] Generally, activation energy increases with decreasing reaction exothermicity. Thus, contrary to the earlier view, the radical stability, which influences bond strengths and, thus, the reaction enthalpy, plays important role. Polar effect decreases the activation energy. The results are well explained on the basis of the state correlation diagram.[10] Steric effect may increase the activation energy by lowering the reaction enthalpy due to any strain induced in the newly formed bond. It, of course, reduces the reaction rate also by lowering the frequency factor.[9]

3.2 Kinetics of Homogeneous Radical Polymerization

Rate equation for the homogeneous radical polymerization using a thermally dissociable initiator may be derived using the kinetic Scheme 3.1.

In the scheme, I, M, and T-X represent respectively a homolytically dissociable initiator, a monomer, and a transfer agent. $R^\bullet$ is a primary radical and $P_n^\bullet$ is a chain radical of chain length n (the number of propagation steps the chain undergoes), *i.e.*, a polymer radical with the degree of polymerization n. The rate constants of the various component reactions are written over the arrows showing the directions of the respective reactions. Termination of

Scheme 3.1 Kinetic scheme for radical polymerization.

$$I \xrightarrow{k_d} (2R^\bullet) \qquad \text{Initiator dissociation} \qquad (3.1)$$

$$(2R^\bullet) \longrightarrow \text{Stable products} \qquad \text{Cage reaction} \qquad (3.2)$$

$$(R^\bullet) \rightarrow R^\bullet \qquad \text{Escape from the cage} \qquad (3.3)$$

$$R^\bullet + M \xrightarrow{k_i} P_1^\bullet \qquad \text{Initiation} \qquad (3.4)$$

$$P_n^\bullet + M \xrightarrow{k_p} P_{n+1}^\bullet \qquad \text{Propagation} \qquad (3.5)$$

$$P_n^\bullet + P_m^\bullet \xrightarrow{k_t} \text{Dead polymer} \qquad \text{Termination} \qquad (3.6)$$

$$P_n^\bullet + R^\bullet \xrightarrow{k_t'} \qquad \text{Primary radical termination} \qquad (3.7)$$

$$R^\bullet + R^\bullet \longrightarrow \text{Stable products} \qquad \text{Primary radical self-termination} \qquad (3.8)$$

$$P_n^\bullet + T-X \xrightarrow{k_{tr,T}} P_n - X + T^\bullet \qquad \text{Chain transfer} \qquad (3.9)$$

$$T^\bullet + M \xrightarrow{k_i'} P_1^\bullet \qquad \text{Reinitiation} \qquad (3.10)$$

radicals takes place by combination and/or disproportionation, involving two of them (*vide infra*).

Derivation of an equation for the rate of polymerization is simplified with the following assumptions, which are valid.[11,12]

1. The radical reactivity is independent of the chain length. This assumption allows the use of a single rate constant each for k_p and k_t irrespective of the chain length. Indeed, the experimental determination of k_p has revealed that it does not change much except for the first few monomer additions (Sec. 3.10.6). In contrast, k_t decreases with increase in chain length. However, since small radicals are injected continuously into the system by the initiator throughout the course of polymerization effecting broad chain length distribution, a single k_t may apply in view of the random selection of two chains for termination.
2. The radical concentration becomes stationary very quickly. This means that the total concentration of chain radicals of all lengths reaches a level, which remains constant during polymerization. Under this condition, the rate of initiation equals the rate of termination.
3. The chains are long. This allows equating the rate of polymerization with the rate of propagation.

The initiation process is constituted of four component reactions. It starts with the generation of a pair of radicals from an initiator molecule. However, not all radicals so generated become available for initiation. A part is lost under all conditions.[13] In order to explain this phenomenon Noyes postulated that the radical pair is formed in a cage formed by the molecules of monomer and solvent, if used, (reaction 3.1, the parenthesis representing the cage). Before the partners of the pair undergo separation by as much as a molecular diameter, they may recombine. This is referred to as primary recombination. Radicals, which survive primary recombination, escape the cage and undergo random diffusion with a rate that decreases with the increase in the viscosity of the medium. During this process, some of the original partners of a radical pair may reencounter each other and undergo recombination, which is called secondary recombination. The combined primary and secondary recombination of the original partners is referred to as geminate or cage recombination (reaction 3.2) and the phenomenon is called a cage effect. Only those radicals that escape cage recombination (reaction 3.3) are captured by the monomer. These are referred to as primary radicals. If the cage recombination regenerates the original initiator (cage return) and disproportionation of the caged radicals (cage disproportionation) does not occur, no cage reaction products are formed. In contrast, if the cage recombination gives a stable molecule, some of the initiator molecules are wasted apart from those wasted through cage disproportionation, if any.

An excellent example of cage reaction is found in the decomposition of the azo initiator 2, 2'-azobisisobutyronitrile (AIBN).[14] It dissociates with the evolution of nitrogen forming a pair of 2-cyano-2-propyl radicals (**1**) in the cage

$$R-N=N-R \longrightarrow (\dot{R} + \dot{R} + N_2) \longrightarrow \dot{R} + \text{cage products} + N_2, \quad (3.11)$$

where $R^\bullet = C^\bullet(CH_3)_2CN$ (**1**). Some of the radical pairs react in the cage to give products of combination (**2** and **3**) as well as of disproportionation (**4** and **5**).

$NC(CH_3)_2C-C(CH_3)_2CN$ $NC(CH_3)_2C-N=C=C(CH_3)_2$ $CH_2=C(CH_3)CN$ $CH(CH_3)_2CN$

(**2**)   (**3**)   (**4**)   (**5**)

Almost all the radicals that escape cage reaction, which amounts to about 60% of the radicals generated, are captured by the monomer unless the concentration of the latter is low ca., < 1 mol/L.[15-17] The rate of initiation, therefore, may be written as

$$R_i = 2fk_d[I], \quad (3.12)$$

where f is the initiator efficiency defined as the fraction of the initiator-derived radicals that become available for initiation, the other symbols have been defined earlier. The factor 2 has been used to indicate explicitly that two radicals are formed in each event of initiator dissociation. Insofar as f is independent of monomer concentration within the above-mentioned limit, both self-termination (reaction 3.8) and cross-termination (reaction 3.7) of primary radicals are negligible. However, f usually decreases at high conversions due to the enhanced cage effect engendered by the increased viscosity of the medium as well as the decreased rate of primary radical capture at the lowered monomer concentrations.

The two modes of termination, referred to above, are illustrated in reactions (3.13) and (3.14) for poly(methyl methacrylate) radical

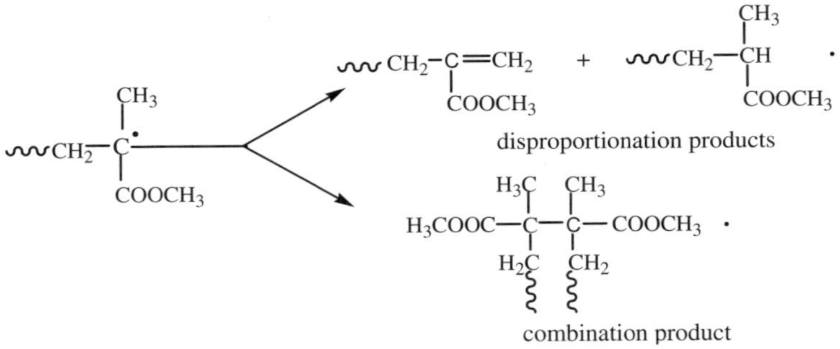

$$\text{\textasciitilde}CH_2-\underset{\underset{COOCH_3}{|}}{\overset{\overset{CH_3}{|}}{C}}\cdot \longrightarrow \begin{cases} \text{\textasciitilde}CH_2-\underset{\underset{COOCH_3}{|}}{C}=CH_2 \; + \; \text{\textasciitilde}CH_2-\underset{\underset{COOCH_3}{|}}{\overset{\overset{CH_3}{|}}{CH}} \cdot \quad (3.13) \\ \text{disproportionation products} \\ H_3COOC-\underset{\underset{H_2C}{|}}{\overset{\overset{H_3C}{|}}{C}}-\underset{\underset{CH_2}{|}}{\overset{\overset{CH_3}{|}}{C}}-COOCH_3 \cdot \quad (3.14) \\ \text{combination product} \end{cases}$$

In disproportionation, a hydrogen atom attached to the α carbon atoms (preferably from the α methyl group[18,19]) of one radical is transferred to another producing two dead polymer molecules, one terminally unsaturated and the other saturated. In general, radicals with α methyl groups terminate predominantly by disproportionation. The mode of termination has an influence over the degree of polymerization but not over the rate of polymerization (*vide infra*).

In Table 3.2, the values of the fraction of disproportionation to total termination for several polymer radicals are given. These are determined either directly from the estimation of the initiator-derived end groups[20–27] or using other indirect methods such as gel formation capability through condensation of the end groups with multifunctional monomers/oligomers[28–30] or analysis of molecular weight distribution (*vide* Sec. 3.14).[31] The proportion of the unfavored termination mode decreases with decrease in temperature due to higher activation energy.

The rate of termination is given by

$$R_t = 2k_t[P^\bullet]^2, \tag{3.15}$$

where $[P^\bullet]$ is the total concentration of all chain radicals irrespective of chain length. Factor 2 has been used to indicate that two radicals are terminated in each termination event. The chain transfer reaction (3.9) does not affect kinetics provided the transferred radical ($T^\bullet$) is at least as reactive as the chain radical ($P^\bullet$).

Application of the stationary state condition (assumption 2) gives

$$R_i = R_t, \tag{3.16}$$

Table 3.2 Fraction of termination by disproportionation of some polymer radicals.

Monomer	Temperature (°C)	$k_{td}/(k_{tc}+k_{td})$	Reference
Styrene	60	0.25	21,15,25
Acrylonitrile	60	0.08	27,28
Methyl acrylate	25	small	28
Vinyl acetate	25	small	28
Methacrylonitrile	25	0.65	28
Methyl methacrylate	60	0.9	24

or

$$2fk_d[I] = 2k_t[P^\bullet]^2. \tag{3.17}$$

$$\therefore [P^\bullet] = \left(\frac{fk_d[I]}{k_t}\right)^{1/2} = \left(\frac{R_i}{2k_t}\right)^{1/2}. \tag{3.18}$$

Typically, $[P^\bullet] = 10^{-7} - 10^{-8}$ mol/L, $k_d \approx 10^{-5} s^{-1}$, $f \approx 0.5 - 1$, $[I] = 10^{-3} - 10^{-1}$ mol/L and $k_t = 10^7 - 10^8$ L mol^{-1}s^{-1}.

For long chains, the consumption of monomer in the initiation step is negligible compared to that in the propagation. Rate of polymerization, therefore, may be considered equal to that of propagation

$$-\frac{d[M]}{dt} = R_p = k_p[P^\bullet][M]. \tag{3.19}$$

Substituting Eq. (3.18) for $[\overset{\bullet}{P}]$ in Eq. (3.19) gives

$$R_p = \frac{-d[M]}{dt} = \frac{k_p R_i^{1/2}[M]}{(2k_t)^{1/2}} = \frac{k_p(fk_d[I])^{1/2}[M]}{(k_t)^{1/2}}. \tag{3.20}$$

With initiators other than the thermally dissociable type, the appropriate kinetic equation should be substituted for R_i in Eq. (3.20). However, the polymerization kinetics is first order in monomer unless the concentration is relatively low, ca., < 1 mol/L, when f becomes dependent on it.[16] In the extreme case, when f becomes directly proportional to [M], does the polymerization become 1.5 order in monomer.[11]

The compound parameter $k_p/(2k_t)^{1/2}$ determines the rate of polymerization for a given R_i and monomer concentration. The value of the parameter can be obtained using Eq. (3.20) from the knowledge of R_p, R_i, and [M]. The methods used for the determination of R_i are discussed in a later section (Sec. 3.6). An alternative method of evaluating the parameter makes use of the Mayo equation (Sec. 3.8.1.2) and is known as the molecular weight method, which yields the value $(1 + x)k_t/k_p^2$ where x is the fraction of termination by disproportionation to total termination.

Table 3.3 gives the value of $k_p/(2k_t)^{1/2}$ at 60°C for several monomers. The value spans over a 220-fold range from 0.03 for styrene to 6.80 for acrylamide. The relative rate of

Table 3.3 The value of $k_p/(2k_t)^{1/2}$ for some monomers at 60°C.

Monomer	$\frac{k_p}{(2k_t)^{1/2}}$ (L$^{1/2}$mol$^{-1/2}$s$^{-1/2}$)	Method	Ref.
Acrylamide[a] (AA)	6.80	Mol.wt.	36,37
Methyl acrylate (MA)	0.68[c]	R_i	34
Vinyl acetate (VA)	0.30	R_i	32
Methyl methacrylate (MMA)	0.12	Mol. wt.	38
Acrylonitrile (AN)[b]	0.08	R_i	35
Styrene (St)	0.03	Mol.wt.	38,41

[a]In water; [b]in DMF; [c]about 2 from k_p (PLP-SEC method) and k_t (SP-PLP method)

polymerization of the monomers at 60°C follows approximately as

Monomer :	St	<	AN	<	MMA	<	VA	<	MA	<	AA.
Relative R_p :	1		3		4		10		75		220

3.3 Reaction Orders in Initiator and Monomer

The reaction order of 0.5 in initiator predicted by Eq. (3.20) agrees with the experimentally determined value for many kinds of initiators. Figure 3.1 shows it for some thermally dissociable initiators (also called thermal initiators). The figure shows that the initial rate of bulk polymerization of methyl methacrylate effected by AIBN, benzoyl peroxide, t-butyl hydroperoxide (t-BHP), or cumene hydroperoxide (CHP) increases in proportion to the square root of the initiator concentration as predicted by Eq. (3.20).[38] Similar results were obtained by other workers[39,40] and with other monomers as well.[41–45] It is evident from the figure that the two hydroperoxides, CHP and t-BHP, are much weaker initiators than benzoyl peroxide or AIBN. This is due to their slower rates of decomposition. In addition, they also partly decompose by a nonradical pathway.[46,47]

Apart from undergoing primary decomposition, peroxide and hydroperoxide initiators also undergo induced decomposition. For example, the self-induced decomposition of benzoyl peroxide (reaction c) shown in Scheme 3.2 is effected by primary benzoyloxy radical produced from its primary decomposition (reaction a). Induced decomposition does not change the radical concentration but increases the consumption of the initiator. It is a chain reaction, which causes an increase in the reaction order of decomposition. However,

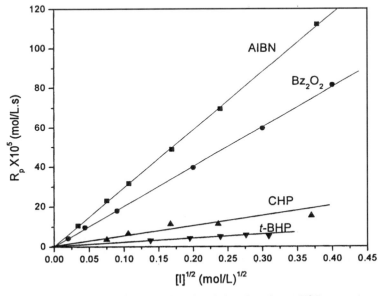

Figure 3.1 Plots of initial rate of bulk polymerization of methyl methacrylate at 60°C vs. square root of initiator concentration using azobisisobutyronitrile (AIBN), benzoyl peroxide (Bz_2O_2), cumene hydroperoxide (CHP), and t-butyl hydroperoxide (t-BHP) as initiators. "Reprinted with permission from Ref. 38. Copyright ©1962 John Wiley & Sons Inc."

$$C_6H_5-\underset{\underset{O}{\|}}{C}-O-O-\underset{\underset{O}{\|}}{C}-C_6H_5 \longrightarrow 2\,C_6H_5-\underset{\underset{O}{\|}}{C}-\overset{\bullet}{O}\,. \qquad (a)$$

$$2\,C_6H_5-\underset{\underset{O}{\|}}{C}-\overset{\bullet}{O} \longrightarrow C_6H_5-\underset{\underset{O}{\|}}{C}-O-C_6H_5 + CO_2\,. \qquad (b)$$

$$C_6H_5-\underset{\underset{O}{\|}}{C}-\overset{\bullet}{O} + C_6H_5-\underset{\underset{O}{\|}}{C}-O-O-\underset{\underset{O}{\|}}{C}-C_6H_5 \longrightarrow C_6H_5-\underset{\underset{O}{\|}}{C}-O-C_6H_5 + CO_2 + C_6H_5-\underset{\underset{O}{\|}}{C}-\overset{\bullet}{O}\,. \qquad (c)$$

Scheme 3.2 Self-induced decomposition of benzoyl peroxide

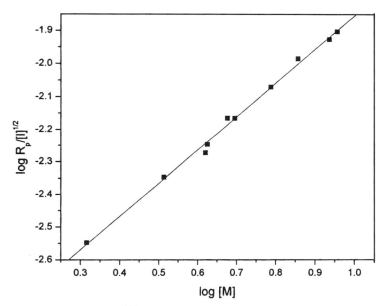

Figure 3.2 Logarithm of initial $R_p/[I]^{1/2}$ plotted against that of monomer concentration in the polymerization of methyl methacrylate in benzene using AIBN initiator at 77°C. The points in the graph are calculated from data given in Ref. 39.

it is nearly eliminated in the presence of monomer. The benzoyloxy radical is scavenged by the monomer; the resulting carbon centered radical or the subsequently grown chain radical is inefficient in inducing decomposition.[48–52]

The polymerization is first order in monomer. Figure 3.2 shows a plot of log $R_p/[I]^{0.5}$ vs. log [M] using initial values of rates and concentrations in the polymerization of methyl methacrylate.[39] The plot yields a straight line with a slope of 1.02 as against 1 predicted (Eq. 3.20).

3.4 Initiators

There are different classes of initiators, *viz.*, thermal, photochemical, redox, electrochemical, high energy radiation, *etc*. However, the first three are used most commonly and discussed here in some detail.

3.4.1 Thermal initiators

The thermal initiators used at concentrations of ca., 0.01–1% of monomer dissociate thermally to generate radicals at practically useful rates ca., 10^{-8}–10^{-6} mol L^{-1} s^{-1} at moderate temperatures ca., 40–80°C. The principal initiators belonging to this class are the azo and the peroxo compounds.

The azo initiators are represented by the chemical structure, R−N = N−R, where R is an α-substituted *tertiary* alkyl group with substituents like cyano, carbomethoxy, amidinium, *etc.*,

$$R = (CH_3)_2(CN)C, (H_3C)_2(CO_2CH_3)C, (H_3C)_2(C = \overset{+}{N}H_2(NH_2)C, \ etc.$$

The dissociation of the initiator (reaction 3.11) takes place in a concerted manner at the C–N bonds. The latter, as such, are too strong to undergo dissociation even at moderately high temperatures.[14] However, the large stabilization energy of the triple bonded N$_2$ molecule formed lowers the bond dissociation energy. The resonance stabilization of the alkyl radical by α-substituents lowers it still further.

Similarly, resonance stabilization of the C$_6$H$_5$COO$^\bullet$ radical reduces the activation energy of dissociation (primary decomposition) of the peroxide bond in benzoyl peroxide such that radicals are produced at useful rates at moderate temperatures. Other peroxides, *e.g.*, peroxydicarbonates, peroxyesters, dialkyl peroxides, and alkyl hydroperoxides as well as persalts, *viz.*, persulfates, also act efficiently at appropriate temperatures.

The initiator efficiency of benzoyl peroxide is close to unity.[53,54] However, it decreases to a great extent at high monomer conversions due to the large increase in the viscosity of the polymerization medium (Sec. 3.6).[55] On the other hand, aliphatic diacyl peroxides exhibit initiator efficiencies less than unity even at low conversions.[56]

Alkali or ammonium persulfates are efficient initiators of polymerization in aqueous medium. They decompose at much faster rates in acidic than in alkaline medium but the proton-assisted decomposition proceeds through a nonradical pathway.[61] Therefore, it is preferable to use them in buffered aqueous medium (pH $\geq$ 7). In Table 3.4, rate constants and activation energies of primary decomposition of some azo and peroxo initiators are given. It is evident that k$_d$ covering a wide range is obtainable by changing the structure of the initiators.

Table 3.4 First order rate constants and activation energies of decomposition of some thermal initiators.

Initiator	Solvent	T(°C)	$k_d(s^{-1})$	E_a(kJ/mol)	Ref.
2,2'-Azobisisobutyronitrile	Benzene	60	8.45×10^{-6}	123.4	57
	Xylene	77	9.5×10^{-5}	134.3	39
2,2'-Azobis-(2,4-dimethylvaleronitrile)	Xylene	77	5.8×10^{-4}	—	39
1,1-Azobis(1-cyclohexanenitrile)	Xylene	77	5.3×10^{-6}	—	39
Benzoyl peroxide	Benzene	60	2×10^{-6}	124.3	58
Lauroyl peroxide	Benzene	60	8.9×10^{-6}	125.3	59
Methyl ethyl ketone peroxide	Benzene	100	1.3×10^{-5}	101	60
Potassium persulfate	0.1M NaOH (water)	60	3.16×10^{-6}	140.2	61
4,4'-Azobis-4-cyanopentanoic acid	Water	60	4.9×10^{-6}	142	62

3.4.2 Photoinitiators

Photoinitiators act by way of photoexcitation followed by fragmentation or reaction with other molecules to generate radicals, which are capable of initiating polymerization. In industry, they have occupied a unique position in solvent-less ultrafast UV curing of multifunctional monomers, which have found many applications, *e.g.*, in printing inks, protective coatings, adhesives, and in making printing plates and microelectronics circuitry.[63] There are several classes of these initiators depending on the mechanism of photoinitiation.[63,64]

1. Photo fragmentation

Molecules containing keto group, such as benzoin, benzoin ethers and esters, benzil ketals, acetophenone derivatives, α-aminoalkyl phenones, acylphosphine oxides, *etc.*, act by this mechanism. They absorb in the ultraviolet. Their triplet state energy is higher than the energy required for α cleavage.

Fragmentation of a commercially used α-aminoalkyl phenone[63] (**6**) and an acylphosphine oxide[63] (**7**) is shown in reactions 3.21a and b respectively.

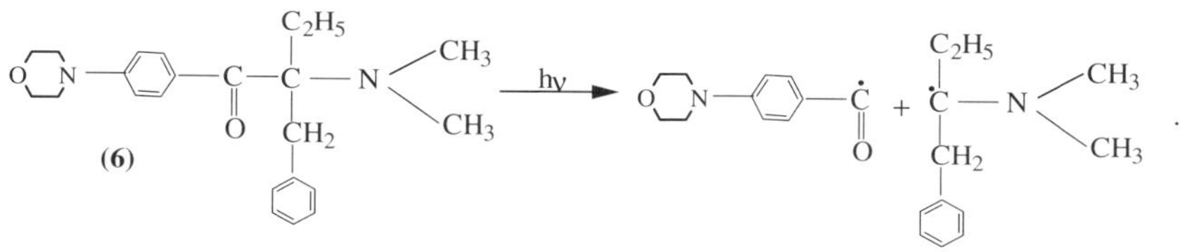

(3.21a)

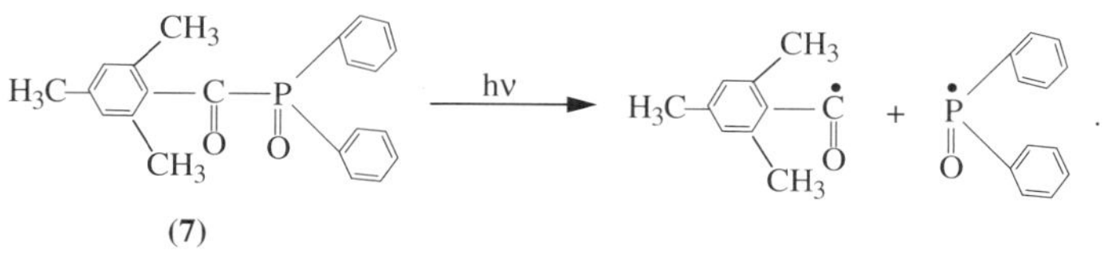

(3.21b)

2. Hydrogen atom abstraction by triplet excited state

Aromatic keto compounds belonging to this class are benzil, benzophenone, anthraquinone, thioxanthone derivatives, Michler's ketone, *etc*. These compounds in their triplet excited state abstract hydrogen atoms from suitable donors such as thiols, alcohols, amines, and alkyl ethers. An example is shown in reaction (3.22) with the commercially used 1-chloro-4-isopropoxy-thioxanthone (**8**) as the photoinitiator and a *tertiary* amine as the hydrogen donor.[63]

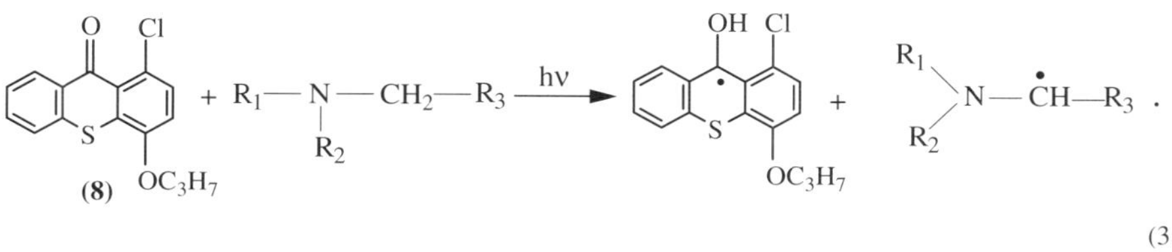

(3.22)

3. Triplet energy transfer to acceptor followed by fragmentation

An important commercial photoinitiator system belonging to this class uses a thioxanthone, e.g., **8**, as photosensitizer and quinone sulfonyl chloride as acceptor. Following energy transfer from the triplet state of **8** to the acceptor, the sulfonyl chloride group of the latter undergoes homolytic cleavage.

$$^3(\mathbf{8})^* + \text{[quinoline-SO}_2\text{Cl]} \longrightarrow \text{[quinoline-SO}_2^\bullet\text{]} + \text{Cl} + (\mathbf{8}) \cdot \quad (3.23)$$

4. Formation and fragmentation of an exciplex

Aromatic ketones in their excited triplet states may form complexes (exciplexes) with *tertiary* amines. The exciplex may then fragment by transferring a hydrogen atom from the α-carbon atom of the amine to the excited carbonyl group resulting in the formation of a semipinacol and an aminoalkyl radical as shown in reaction (3.24).[64]

$$\underbrace{\left[\text{Ar}_2\text{CO} \longleftarrow \text{NR}_3 \right]^*}_{\text{Exciplex}}{}^3 \longrightarrow \text{Ar}_2\overset{\bullet}{\text{C}}\text{OH} + \text{R}_2\text{N}-\overset{\bullet}{\text{C}}\text{HR}' \cdot \quad (3.24)$$

The photoinitiators discussed above use UV light for photoexcitation. However, for many applications visible light photoinitiators are required. Several photoreducible organic dyes such as eosin,[65] methylene blue,[66] and riboflavin[67] on excitation with visible light abstract hydrogen atoms from *tertiary* amines generating α-aminoalkyl radicals. Similarly, camphorquinone in combination with *tertiary* amines produces radicals on exposure to 450–500 nm radiation. This system finds applications for visible light curing of methyl methacrylate in bisphenol A-bis(glycidyl methacrylate) based dental and orthopedic filling materials.[68–70]

The photoinitiation efficiency is determined by the quantum yield of initiation (ϕ_i), which represents the number of polymer chains initiated per photon absorbed. The rate of initiation is given by Eq. (3.25) for the absorption of a very small fraction of incident light

$$R_i = (I_0 - I)\phi_i = I_0(1 - e^{-\varepsilon ct})\phi_i \approx I_0\phi_i \in ct \quad (\text{for } \epsilon\, ct \ll 1), \quad (3.25)$$

where I_0 is the intensity of incident light in mols of photon $L^{-1}\, s^{-1}$, I is the intensity of transmitted light, ϵ is the molar absorbancy index of the initiator at the excitation wave length, c is the photoinitiator concentration, and t is the thickness of the irradiated layer.

3.4.3 Redox initiators

The redox initiators, as the name indicates, involve redox reactions, which generate radicals as intermediates. The advantage with these initiators is that unlike thermal initiators they are useful at ambient to low temperatures due to much lower activation energy of redox reactions ca., 20–40 kJ/mol compared to 100–140 kJ/mol for dissociation of thermal initiators. A well-known initiator system of this class comprises cumene hydroperoxide and Fe(II) salt, which was first used commercially in the late 1940 s in the production of the so-called "cold rubber" by emulsion copolymerization of butadiene and styrene at a low temperature *ca.*, 5°C in an alkaline phosphate buffer medium. The radical generation involves electron transfer from Fe(II) to the hydroperoxide.[52] Dextrose is used to regenerate Fe(II) from Fe(III) produced in the process so that Fe(II) can be used at a concentration lower than the equivalent of the hydroperoxide. The reactions involved are shown in (3.26) and (3.27).

$$\text{C}_6\text{H}_5\text{C}(\text{CH}_3)_2\text{-OOH} + \text{Fe(II)} \longrightarrow \text{C}_6\text{H}_5\text{C}(\text{CH}_3)_2\text{-}\dot{\text{O}} + \text{OH}^- + \text{Fe(III)}. \quad (3.26)$$

$$\text{Fe(III)} + \text{dextrose} \longrightarrow \text{Fe(II)}. \quad (3.27)$$

Reaction (3.26) is analogous to the first step of the Haber-Weiss mechanism for the decomposition of H_2O_2 catalyzed by Fe(II)[71] (reaction 3.28)

$$\text{Fe(II)} + H_2O_2 \longrightarrow \dot{\text{O}}\text{H} + \text{OH}^- + \text{Fe(III)}. \quad (3.28)$$

The Fe(III) produced may act as chain terminator (*vide infra*) when this redox system is used as initiator.[72] For this reason Ti(III) is preferred to Fe(II) as the reducing agent.[73]

Similar to peroxides, persulfate ion also makes redox initiating systems with various metal ions, *e.g.*, Ag^+, Fe^{2+}, Ti^{3+}, *etc*.[74–76]

$$S_2O_8^{2-} + M^{n+} \rightarrow \dot{\text{O}}SO_3^- + SO_4^{2-} + M^{(n+1)+}. \quad (3.29)$$

Besides metal ions, various other reducing agents may also be used to form redox initiators with persulfate ion. Some examples are dithionite, thiosulfate, bisulfite ions, and amines such as tetramethylethylenediamine, triethanolamine, and 3-(dimethylamino) propionitrile.

A host of other redox initiators has been studied in aqueous polymerization[77,78] and radicals involved in initiation identified from end group analysis of polymers by dye partition method[79] (*vide infra*).

Redox initiators suitable for use in organic media also exist. A well-known example is the combination of benzoyl peroxide and N,N-dimethylaniline or better N,N-dimethyl-*p*-toluidine.[80–86] The reaction produces benzoyloxy and aminoalkyl radicals.[82–86] Both take part in initiation.

$$CH_3 C_6H_5 NMe_2 + C_6H_5 COO-OOCC_6H_5 \longrightarrow C_6H_5 COO^- + \left[C_6H_5 \overset{+}{N}Me_2 OOCC_6H_5\right]$$
$$\downarrow$$
$$C_6H_5 COOH + C_6H_5 N(CH_3)\dot{C}H_2 + C_6H_5 CO\dot{O}.$$

Another important class of initiators comprises oil soluble transition metal soaps in their lower oxidation states, *e.g.*, Co(II), Mn(II), Pb(II) or V(III) soaps, in combination with hydroperoxides, such as *t*-butyl hydroperoxide and cumene hydroperoxide. Radicals are produced in reactions analogous to (3.26) at ambient temperature.[87] A well-known application is that of cobalt(II) napthenate as a dryer in paint industry. This soap reacts with hydroperoxides generated *in situ* in the auto-oxidized paint films producing radicals, which accelerate crosslinking of the polyunsaturated chains present in paint films. Co(II) is regenerated by way of reduction of Co(III) soap by hydroperoxides

$$Co(III) + ROOH \rightarrow Co(II) + RO\dot{O} + H^+.$$

Thus, a catalytic amount of Co(II) is only required.[87]

3.5 Determination of Polymer End Groups

Knowledge of the nature of polymer end groups and their number per polymer chain provides insight into the chain end forming events such as initiation, termination, chain transfer, and deactivation (in living radical polymerization). Both chemical and spectroscopic methods have been used for the purpose. However, reliable results can be obtained only with low polymers. Thus, NMR and MALDI-TOF mass spectrometry are used almost routinely in estimating end groups in dormant polymers of low molecular weights ca., $\overline{M}_n < 3000$. However, high resolution 600 MHz ^{1}H NMR spectroscopy allows precise analysis of chain ends in polymers of somewhat higher $\overline{M}_n$, ca., 10 000.[88,89] With high polymers, more sensitive methods need to be used or response to methods magnified. The following methods meet one or the other of these requirements.

The dye partition (extraction, to be exact) method introduced by Palit[90,91] is of adequate sensitivity since it attaches ionic dyes to oppositely charged end groups and dyes have very high molar absorbancy indices ca., 10^5 L mol cm^{-1}. However, the method suffers from the limitation that it is applicable only to those end groups which are charged or which can be transformed to charged ones.[92,93]

The method is based on the extraction of an ionic dye from aqueous solution into chloroform by a polymer having an ionic end group of a charge opposite to that of the dye. For example, cationic dyes, viz., methylene blue (**9**) and pinacyanol chloride, are used respectively in the estimation of strong (SO$_3$H and OSO$_3$H) and weak acids (COOH) end groups and the anionic dye disulfine blue (**10**) is used in the estimation of positively charged end groups (NR$_3^+$, R = H, alkyl or substituted alkyl). Equilibria (3.30a) and (3.30b) represent extractions of cationic and anionic dyes respectively

$$\sim\!\!\sim\!\!SO_3^- M^+_{(o)} + Dye^+_{(w)} \rightleftharpoons \sim\!\!\sim\!\!SO_3^- Dye^+_{(o)} + M^+_{(w)}, \quad (3.30a)$$

$$\sim\!\!\sim\!\!NR_3^+ X^-_{(o)} + Dye^-_{(w)} \rightleftharpoons \sim\!\!\sim\!\!NR_3^+ Dye^-_{(o)} + X^-_{(w)}, \quad (3.30b)$$

where the subscripts o refers to the organic phase and w to the aqueous phase; M$^+$ and X$^-$ represent an alkali metal ion (or proton) and halide ion respectively.[25,94–96]

Radiochemical tracer technique is another sensitive method, which has been used very extensively. The method requires radio labeling of initiators, e.g., AIBN-α-^{14}C, benzoyl peroxide-*carbonyl*-^{14}C, K$_2^{35}$S$_2$O$_8$, etc.[13,20,21,97,98] It measures the specific activity of the purified polymer and compares the value with that of the initiator.

In the application of NMR spectroscopy, the resonance intensity is enhanced by appropriately enriching the population of the resonating atom in the initiator.[99,100] For example, in ^{13}C NMR this has been achieved with ^{13}C enriched initiators, such as, AIBN β, β-^{13}C, AIBN-α-^{13}C, benzoyl peroxide-*carbonyl*-^{13}C, etc. Besides ^{13}C, various other stable isotopes-labeled initiators have also been used.[101,102]

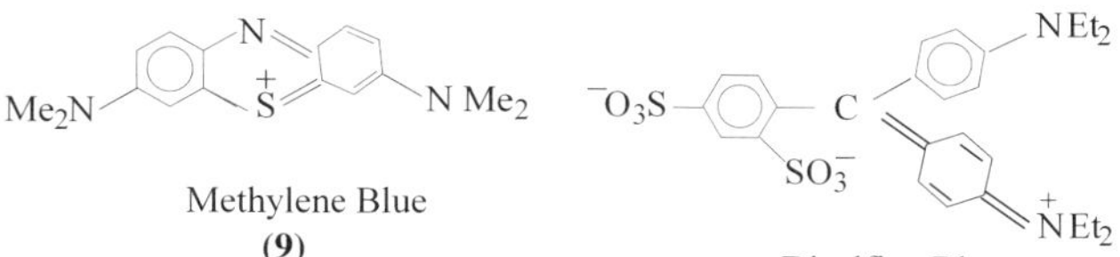

Methylene Blue
(9)

Disulfine Blue
(10)

3.6 Initiator Efficiency

Initiator efficiency, f, is defined by

$$f = \frac{\text{rate of initiation}}{n \times \text{rate of primary decomposition of initiator}} = \frac{R_i}{2nk_d[I]}, \quad (3.31)$$

where n is the number of radicals generated from the primary decomposition of a molecule of initiator. The rate constant k_d is determined by studying the kinetics of decomposition.

R_i may be determined by various methods but none of them is free from uncertainties. However, the end group method has been considered to be probably the best for many polymerization systems.[98] The method involves determination of the number average kinetic chain length (ν) which is given by Eq. (3.32).

$$\nu = R_p/R_i. \quad (3.32)$$

ν may be determined from the analysis of initiator-derived end group.

$$\nu = \frac{\text{amount of polymer}}{\text{mol initiator fragment} \times \text{molecular wt of monomer}}. \quad (3.33)$$

Knowing R_p and ν, R_i is calculated using Eq. (3.32).

Using this method, Bevington *et al.* found f ≈ 0.6 for AIBN, which is nearly independent of monomer concentration for [M] >∼ 1 mol/L.[15,20,103] However, this value applies to low conversion polymerization. As conversion increases, viscosity of the medium of polymerization increases. This facilitates cage reaction and f therefore decreases.[104] Furthermore, at high conversions, the monomer concentration may be too low to be able to capture all radicals that escape cage reaction. On this account also, f should decrease at high conversions. Thus, in the solution polymerization of styrene by AIBN at 70°C, f varies from 0.76 at low conversions to less than 0.2 at 90–95% conversions.[105] With benzoyl peroxide initiator, f is close to 1 at low conversions[53,54] and only 0.1–0.2 at high conversions ca., 80%.[55]

An alternative method to measure R_i would be to use a powerful scavenger for chain radicals. The scavenging should be a zero order reaction and the reaction stoichiometry should be known accurately. The stable radical 2,2-diphenyl-1-picryl-hydrazyl has been used for this purpose (Sec. 3.9).[32–34,106] Besides, transition metal halides (FeCl$_3$ and CuCl$_2$), which oxidatively terminate chain radicals by Cl atom transfer and retard polymerizations may also be used to measure R_i (Sec. 3.9.1).

3.7 Thermal Polymerization and its Kinetics

Polymerization may also occur thermally, albeit slowly, without the aid of an initiator. Kinetic studies of thermal polymerization, however, are beset with difficulties since traces of impurities, which may accelerate or retard the polymerization, bring about irreproducible results. However, scrupulous removal of oxygen, which is an inhibitor of polymerization (*vide infra*), and use of rigorously purified fresh monomer, which has not been exposed to air, give reproducible polymerization.[11]

With the above measures taken, styrene polymerizes thermally in bulk at 100°C at a rate of *ca.*, 20% in 10 h.[107–109] Methyl methacrylate does so at a slower speed, which is about one fifth that of styrene,[109] whereas vinyl acetate[110] and vinyl chloride[111] do not give detectable polymer.

Early researchers found the initial rate of thermal polymerization of styrene to be approximately proportional to the square of its concentration in homogeneous solution.[107,108] Bimolecular self-initiation generating a diradical was proposed to explain the second order kinetics (reaction 3.34).[11,111]

$$2CH_2 = CHC_6H_5 \xrightarrow{k_i} C_6H_5\overset{\bullet}{C}HCH_2CH_2\overset{\bullet}{C}HC_6H_5. \tag{3.34}$$

Using the stationary state assumption one gets

$$R_i = 2k_i[M]^2 = R_t = 2k_t[P^\bullet]^2. \tag{3.35}$$

Substituting from Eqs. (3.35) for $[P^\bullet]$ in Eq. (3.19) one gets

$$R_p = k_p[P^\bullet][M] = k_p \left(\frac{k_i}{k_t}\right)^{1/2} [M]^2. \tag{3.36}$$

However, the diradical mechanism was criticized on the ground that the diradical will undergo cyclization at some stages of its growth, predominantly at the trimer stage.[11,112] This view is substantiated by statistical calculations.[113] In addition, a consequence of cyclization is a change in kinetics. This is because cyclization entails a first order termination, unlike the second order termination assumed in Eq. (3.35), which would raise the monomer order to three.

However, according to Mayo, thermal polymerization of styrene in refluxing bromobenzene is explained best by third order initiation. The over all reaction order is 2.5.[114] Besides, thermal polymerization gives both high polymers and a mixture of dimers.[114,115] The third order initiation was strongly supported by Hiatt and Bartlett from the kinetic analysis of the results of thermal polymerization of styrene in the presence of the regulator ethyl thioglycolate.[116]

The initiation mechanism proposed by Mayo is shown in Scheme 3.3[115] which also explains the formation of the dimeric nonradical products. The first stage of initiation is postulated to be the formation of a Diels-Alder adduct (RH) involving two molecules of styrene, which is followed by transfer of a hydrogen atom from the adduct to styrene.

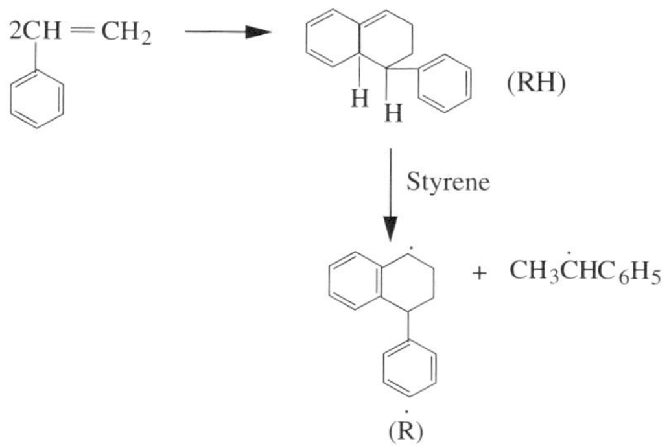

Scheme 3.3 Mechanism of 3rd order initiation in the thermal polymerization of styrene.[115]

Direct support for the mechanism came from the identification of the adduct radical R as an end group in polystyrene by ^{1}H NMR spectroscopy of styrene oligomers, which were prepared by thermal polymerization in the presence of FeCl$_3$ used as a retarder.[117] There are various indirect supports as well. For example, retardation of thermal polymerization of styrene in the presence of camphorsulfonic acid is explicable by the loss of the Diels-Alder adduct (RH) through acid catalyzed aromatization.[118,119] Also, Hui and Hamielec showed that the kinetic model based on 3rd order initiation gives a satisfactory fit of monomer conversion as well as of polymer molecular weight.[120]

3.8 Kinetic Chain Length, Degree of Polymerization, and Chain Transfer

In a chain polymerization, the number average kinetic chain length (v) is defined as the number of monomer units added to an active center from its inception to death. Therefore, v is given by Eq. (3.32a), which has been already introduced in a previous section (Sec. 3.6).

$$v = R_p/R_i \tag{3.32a}$$
$$= R_p/R_t \text{ (under stationary state condition).} \tag{3.32b}$$

Substituting Eq. (3.20) for R_p in Eqs. (3.32a) gives

$$v = k_p[M]/(2k_t R_i)^{0.5}. \tag{3.37}$$

Chain transfer does not affect kinetic chain length since the active center is only transferred to another molecule and not destroyed in the process. On the other hand, the number average degree of polymerization ($\overline{DP}_n$) decreases in the presence of chain transfer.

$\overline{DP}_n^v$ (the superscript v indicating the absence of chain transfer) is equal to v, if termination occurs solely by disproportionation, and $2v$, if it occurs solely by combination. For the general case of termination by both disproportionation and combination

$$\overline{DP}_n^v = \frac{2v}{1+x}, \tag{3.38}$$

where x is the fraction of termination by disproportionation to total termination.

Substituting R_p/R_i for ν in Eq. (3.38) and subsequently substituting from Eq. (3.20) for R_i one obtains

$$\overline{DP}_n^\nu = \frac{k_p^2[M]^2}{(1+x)k_t R_p}. \tag{3.39}$$

3.8.1 *The general DP equation*

Since a linear polymer molecule has a pair of chain ends, $\overline{DP}_n$ can be expressed in terms of the concerned reaction rates by Eq. (3.40)

$$\overline{DP}_n = \frac{\text{Rate of propagation}}{\text{Rate of formation of pairs of chain ends}}. \tag{3.40}$$

Various reactions in the radical chain polymerization contribute to chain end formation as follows: (i) an initiation event forms half a pair of chain ends; (ii) a transfer event forms a pair of chain ends; (iii) an act of termination by disproportionation forms a pair of chain ends, whereas that by combination does not produce any. Accordingly, Eq. (3.40) may be expressed as

$$\overline{DP}_n = \frac{R_p}{\frac{1}{2}R_i + \sum R_{tr,T} + \frac{1}{2}R_{td}}, \tag{3.41}$$

where the summation term in the denominator represents the sum of the contributions of chain transfer by various transfer agents that may be present in a polymerizing system; the factor $\frac{1}{2}$ with R_{td} is due to two radicals producing a pair of chain ends. All the rate terms except $R_{tr,T}$ have been defined earlier. The latter is given by

$$R_{tr,T} = k_{tr,T}[P^\bullet][T] \text{ (from reaction (3.9) and writing T for TX).} \tag{3.42}$$

Replacing R_i with R_t (stationary state approximation) and R_{td} with xR_t, one obtains from Eq. (3.41)

$$\overline{DP}_n = \frac{R_p}{\sum R_{tr,T} + \left(\frac{1+x}{2}\right)R_t}. \tag{3.43}$$

Substituting Eqs. (3.19), (3.42) and (3.15) for R_p, $R_{tr,T}$, and R_t respectively in Eq. (3.43) and subsequently inverting the equation one obtains

$$\frac{1}{\overline{DP}_n} = \frac{\sum k_{tr,T}}{k_p}\frac{[T]}{[M]} + \frac{(1+x)}{2k_p}\frac{(2k_t)}{[M]}[P^\bullet]. \tag{3.44}$$

The ratio $k_{tr,T}/k_p$ is referred to as the chain transfer constant, C_T, for the transfer agent T. Substituting Eq. (3.19) for $[P^\bullet]$ in Eq. (3.44) gives

$$\frac{1}{\overline{DP}_n} = \sum C_T \frac{[T]}{[M]} + \frac{(1+x)k_t}{k_p^2}\frac{R_p}{[M]^2}. \tag{3.45}$$

In initiator-aided solution polymerization, one gets from Eq. (3.45) after expanding the summation term[38,41,42]

$$\frac{1}{\overline{DP}_n} = C_M + C_I \frac{[I]}{[M]} + C_S \frac{[S]}{[M]} + C_P \frac{[P]}{[M]} + \frac{(1+x)k_t}{k_p^2}\frac{R_p}{[M]^2}, \tag{3.46}$$

where C_M, C_I, C_S, and C_P are transfer constants for chain transfer to monomer, initiator, solvent, and polymer respectively. Substitution of Eq. (3.20) for R_p in Eq. (3.46) gives the

DP equation (3.47) for solution polymerization using a thermal initiator

$$\frac{1}{\overline{DP}_n} = C_M + C_I \frac{[I]}{[M]} + C_S \frac{[S]}{[M]} + C_P \frac{[P]}{[M]} + \frac{(1+x)(fk_d[I])^{1/2}k_t^{1/2}}{k_p[M]}. \qquad (3.47)$$

Each term but the last, in the right side of Eq. (3.46) or (3.47) represents the number of polymer molecule formed per molecule of monomer polymerized by the corresponding transfer reaction. The last term is equivalent to the inverse of $\overline{DP}_n^v$ and accounts for the contribution of solely initiation and termination to chain end formation. C_I is negligibly small for azo initiators but not for peroxide and hydroperoxide initiators. As for polymer transfer, it is not important at low conversions of monomer to polymer for intermolecular transfer (since the concentration of polymer available for transfer would be small) but not for intramolecular transfer (*vide infra*).

However, if any of the terms in Eq. (3.46) or (3.47) is overwhelmingly large compared to all others combined, $\overline{DP}_n$ is controlled by it. For example, $\overline{DP}_n$ is nearly equal to the inverse of $C_S[S]/[M]$ when regulators such as thiols, which have high chain transfer constants, are used. Similarly, for the allylic monomers C_M is large enough to control $\overline{DP}_n$, which approximately equals $1/C_M$. The values of the various transfer constants may be determined using Eq. (3.46) or (3.47) from appropriately designed experiments measuring $\overline{DP}_n$ with the change in concentration of the transfer agent.

3.8.1.1 Solvent transfer

Chain transfer constants of solvents (C_S) may be determined using Eq. (3.46) from the slope of a plot of the reciprocal degrees of polymerization *vs.* [S]/[M] from several polymerization, which are so designed that the terms other than $C_S[S]/[M]$ in the equation remains either constant or negligible. The last term is held constant by using the same $[I]^{1/2}/[M]$ in all polymerization. An initiator with $C_I \approx 0$ (AIBN being an example) may be used to eliminate the initiator transfer term. Restricting polymerization to low conversion allows approximating [S]/[M] to be equal to its initial value. Besides, this makes the polymer transfer term negligible.[121]

Mayo used a simplified equation (3.48) for determining transfer constants of various solvents in the thermal polymerization of styrene[122]

$$\frac{1}{\overline{DP}_n} = \frac{1}{\overline{DP}_n^0} + C_S \frac{[S]}{[M]}, \qquad (3.48)$$

where $\overline{DP}_n^0$ refers to the degree of polymerization of the polymer formed in the absence of solvents.

The equation follows from Eq. (3.46) in which the last term becomes constant for the thermal polymerization of styrene following second order kinetics (Eq. 3.36). Straight lines were obtained by plotting reciprocal degrees of polymerization against [S]/[M] in a number of solvents.[123] From the slopes of such lines (Fig. 3.3), C_S values of various solvents were evaluated.[122-125] Transfer constants of some selected solvents in the polymerization of some selected monomers are given in Table 3.5. A discussion on the transfer constants is deferred to a later section (Section 3.8.2).

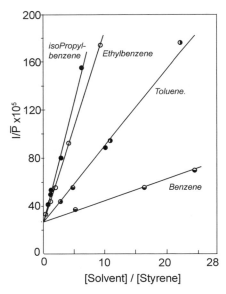

Figure 3.3 Mayo plots for chain transfer to solvents in the thermal polymerization of styrene at 100°C. "Reproduced from Ref. 123 by permission of the Royal Society of Chemistry."

Table 3.5 Chain transfer constants of selected transfer agents in the polymerizations of four selected monomers at 60°C (unless indicated otherwise).

Transfer agent		$C_T \times 10^4$			
	Monomer:	Styrene	Methyl methacrylate	Vinyl acetate	Acrylonitrile
Toluene		0.125[a]	0.52[b]	20.9[c]	5.8[d]
Ethylbenzene		0.67[a]	1.35[b]	55.2[c]	35.7[d]
Propylbenzene		0.82[a]	1.90[b]	89.9[c]	—
Butylbenzene		0.06[a]	0.26[b]	3.6[c]	1.9[d]
Methanol		0.74[j]	0.2[j]	2.26[o]	0.5[d]
Ethanol		1.32[j]	0.4[j]	25[f]	—
i-Propanol		3.05[j]	0.58[m]	44.6[n]	—
t-Butanol		0.22[k]	0.08[m]	1.3[f]	0.44[d]
Acetone		0.32[l]	0.195[m]	1.5[f]	—
Chloroform		0.5[e]	—	15[f]	5.6[d]
Carbon tetrachloride		92[a]	2.39	9600[f]	0.85[d]
Carbon tetrabromide		22 000[g]	2700[g]	≥390 000[g]	1900[i]
1-Butanethiol		220 000[h]	6700[h]	480 000[h]	—
Triethylamine		7.1[i]	8.3[i]	370[i]	5900[i]

[a]Refs. 123 and 124; [b]80°C, Ref. 126; [c]Ref. 128; [d]Ref. 129 (precipitation polymerization); [e]Ref. 125; [f]Ref. 130; [g]Ref. 144; [h]Ref. 143, 145; [i]Ref. 127; [j]Ref. 131; [k]Ref. 132; [l]Ref. 133; [m]Ref. 134; [n]80°C, Ref. 135; [o]Ref. 136.

3.8.1.2 Monomer transfer

C_M may be determined using Eq. (3.46) in bulk polymerization in which an initiator with $C_I \approx 0$ (e.g., AIBN) is used at several different concentrations. Since, in addition, $[S] = 0$ and $[P]$ is negligibly small at low conversions, a plot of $1/\overline{DP}_n$ vs. R_p should give a straight

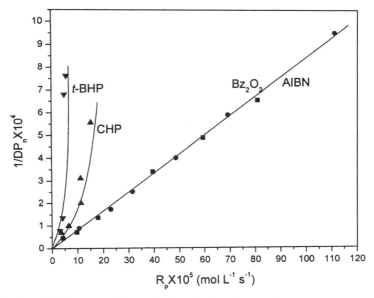

Figure 3.4 Reciprocal of the average degree of polymerization plotted against the rate of bulk polymerization of methyl methacrylate at 60°C using azobisisobutyronitrile (AIBN), benzoyl peroxide (Bz_2O_2), cumene hydroperoxide (CHP), and t-butyl hydroperoxide (t-BHP) as initiators. "Reprinted with permission from Ref. 38. Copyright ©1962, John Wiley & Sons Inc."

line. The intercept gives the value of C_M and the slope gives the value of $(1+x)k_t/k_p^2[M]^2$ from which the important kinetic parameter $k_p/(2k_t)^{1/2}$ is evaluated with the knowledge of x (Table 3.2) and [M]. Figure 3.4 includes such plots in low conversion polymerizations of methyl methacrylate with AIBN and Bz_2O_2 as initiators indicating C_I to be zero for these initiators and C_M nearly zero.[38]

The kinetic parameter of polymerization $k_p/(2k_t)^{1/2}$ for several monomers has already been presented in Table 3.3 and its significance has been discussed. For $C_I \neq 0$, a nonlinear plot is obtained, for example, with CHP or t-BHP initiator, as shown in Fig. 3.4.[38] The nonlinearity is easily explained by substituting from Eq. (3.20) for [I] in Eq. (3.46). This gives the quadratic Eq. (3.49) for bulk polymerization at low conversion ($[P] \approx 0$).

$$\frac{1}{\overline{DP_n}} = C_M + \frac{(1+x)k_tR_p}{k_p^2[M]^2} + C_I\left(\frac{k_t}{fk_p^2k_d}\right)\frac{R_p^2}{[M]^3}. \qquad (3.49)$$

The large curvature in the plots with CHP or t-BHP initiator is due to both relatively low k_d and relatively high C_I, which makes the 3rd term in the left side of Eq. (3.49) important.

C_M of several monomers is given in Table 3.6. The values are generally small such that sufficiently high molecular weight polymers can be prepared for acceptable mechanical properties. This is not so for allyl monomers, which have quite large C_M. With styrene, methyl methacrylate, and vinyl acetate, monomer transfer occurs through hydrogen atom abstraction. The abnormally high C_M of allyl acetate and allyl chloride are due to the very high stability of allylic radicals formed by the removal of an alpha hydrogen atom from the

Table 3.6 Values of C_M for some selected monomers.

Monomer	Temperature (°C)	$C_M \times 10^4$	Ref.
Acrylamide	60	0.15	36, 37
Styrene	60	0.6	41, 42
Methyl methacrylate	60	1	38
Vinyl acetate	60	1.8	137
Vinyl chloride	100	50	138, 139
Allyl acetate	80	710	140
Allyl chloride	80	1600	140

former and chlorine atom from the latter.[140]

$$\dot{P}_n + CH_2=CH-CH_2OCOCH_3 \longrightarrow P_nH + CH_2\!=\!\dot{C}H\!=\!CH-OCOCH_3. \quad (3.50a)$$

$$\dot{P}_n + CH_2=CH-CH_2Cl \longrightarrow P_nCl + CH_2\!=\!\dot{C}H\!=\!CH_2. \quad (3.50b)$$

The allylic radical being too stable (Table 3.1) to add to monomer disappears by mutual bimolecular reaction. This results in abnormally low rate of polymerization and gradual stoppage of polymerization (autoinhibition). Accordingly, this kind of transfer is referred to as "degradative chain transfer".

Allyl hydrogen transfer from monomer has been suggested by Nozaki to be responsible for the failure of α-olefins to polymerize by radical means.[141] However, the absence of such transfers with monomers like methyl methacrylate and methacrylonitrile, which also have allylic hydrogens, has been attributed to the stabilizing effects of substituents. On the one hand, the substituent decreases the reactivity of the radical and hence chain transfer; on the other, it enhances the reactivity of the monomer and hence propagation, both factors contributing to make C_M small.[141]

3.8.1.3 *Initiator transfer*

C_I is conveniently determined using Eq. (3.51) in bulk polymerization ([S] = 0), which follows by rearranging Eq. (3.46) and considering polymer transfer to be negligible.

$$\frac{1}{\overline{DP}_n} - \frac{(1+x)k_t}{k_p^2}\frac{R_p}{[M]^2} = C_M + C_I\frac{[I]}{[M]}. \quad (3.51)$$

By plotting the left side of Eq. (3.51) *vs.* [I]/[M], C_I is obtained from the slope of the linear plot. The plot for CHP in the polymerization of methyl methacrylate is shown in Fig. 3.5.[38] The value of $(1+x)k_t/k_p^2$ required to make the plot is determined as discussed in the preceding section. In the polymerization of methyl methacrylate, its value at 60°C is 66.3.[38]

C_I of some initiators in the polymerization of styrene and methyl methacrylate are given in Table 3.7. AIBN has nearly zero transfer constant in the polymerization of both methyl methacrylate and styrene.[38,41] Benzoyl peroxide, however, behaves differently. It has $C_I = 0$ and 0.055 respectively in the polymerization of methyl methacrylate[38] and styrene.[42] The

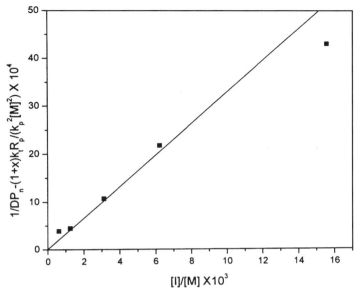

Figure 3.5 Plot of $1/DP_n - (1 + x)k_tR_p/(k_p^2[M]^2)$ vs. $[I]/[M]$ in the cumene hydroperoxide initiated bulk polymerization of methyl methacrylate at 60°C. "Reprinted with permission from Ref. 38. Copyright ©1962 John Wiley & Sons Inc."

Table 3.7   Initiator transfer constants.

| | Temperature | Monomer | |
Initiator	(°C)	Styrene	Methyl methacrylate
AIBN	60	~0[38,41]	~0[38]
Bz$_2$O$_2$	60	0.055[42]	~0[38]
t-BHP	60	0.035[41]	—
CHP	60	0.063[41]	0.33[38]

AIBN = 2,2′-azobisisobutyronitrile, Bz$_2$O$_2$ = benzoyl peroride, CHP = cumene hydroperoxide, t-BHP = t-butyl hydroperoxide.

transfer reaction (3.52) with benzoyl peroxide involves attack of the chain radical at the O–O bond of the peroxide, whereas with hydroperoxides it involves the abstraction of H atom from the ROO–H bond.[122]

$$P_n^\bullet + C_6H_5-\underset{O}{\overset{\|}{C}}-O-O-\underset{O}{\overset{\|}{C}}-C_6H_5 \longrightarrow P_n-O-\underset{O}{\overset{\|}{C}}-C_6H_5 + {}^\bullet O-\underset{O}{\overset{\|}{C}}-C_6H_5 \quad (3.52)$$

$$P_n^\bullet + ROOH \rightarrow P_nH + {}^\bullet OOR. \quad (3.53)$$

The transferred peroxide radical from hydroperoxide is of low reactivity such that it may get involved, to some extent, in the termination of chains.

Certain initiators have relatively high chain transfer constant. They are referred to as inifers (short of initiator and transfer agent). For example, tetraethylthiuram disulfide

(11) has a $C_I = 0.53$ in the polymerization of methyl methacrylate. Both initiation and transfer reactions introduce SC(= S)NEt$_2$ end group. While initiation occurs by homolytic dissociation of the disulfide bond, transfer occurs by the AFCT mechanism (Sec. 3.8.1.5) involving the addition of the chain radical to the S end of the thiocarbonyl bond followed by the fragmentation of the adduct radical.[142]

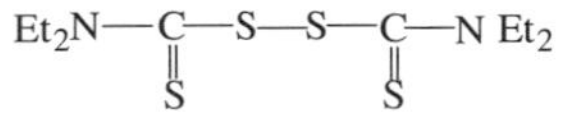

$$Et_2N-\underset{\underset{S}{\|}}{C}-S-S-\underset{\underset{S}{\|}}{C}-NEt_2$$

(11)

3.8.1.4 Regulator transfer

Regulators are chain transfer agents, which have high chain transfer constants (*ca.*, $C_T > 1$). Examples are mercaptans,[145] carbon tetrabromide,[144] and, to some extent, carbon tetrachloride[123] (Table 3.5), so that they are effective in small concentrations in controlling molecular weights. However, use of the Mayo equation (Eq. 3.48) for determining C_T requires [T]/[M] to remain constant during polymerization, which is ideally achieved when $C_T = 1$. For transfer agents with C_T not close to unity, this requirement is satisfied reasonably by keeping monomer conversion appropriately low. However, with $C_T \geq \sim 5$, conversion is to be restricted to impractically low levels.[145] The problem is overcome by using an alternative method, which requires measuring the consumption of the regulator as well as that of the monomer.[146] The rate of consumption of the regulator follows from Eq. (3.42) as

$$-\frac{d[T]}{dt} = k_{tr,T}[T][P^\bullet]. \qquad (3.54)$$

Dividing Eq. (3.54) by Eq. (3.19) gives

$$\frac{d[T]}{d[M]} = C_T[T]/[M] \qquad (3.55)$$

or

$$\frac{d\log[T]}{d\log[M]} = C_T. \qquad (3.56)$$

Thus, the slope of a plot of log [T] *vs.* log [M] gives C_T. The consumption of the regulator may be estimated from the analysis of the polymer for the regulator fragments, *e.g.*, Cl for CCl$_4$[124] and S for mercaptans.[143,145] Alternatively, the unreacted regulator is to be estimated.[145,146] Among sulfur compounds, apart from mercaptans, several disulfides and xanthogen disulfides also have high C_Ts.[147,148]

3.8.1.5 Addition-fragmentation chain transfer (AFCT)

In AFCT, the chain radical adds to the unsubstituted end of the C=C bond in an allylic compound having the general structure shown in reaction (3.57) where X may be Br, SR, SnR$_3$, SOR, SO$_2$Ar, P(O)(OR)$_2$, OOR or SiR$_3$. Addition is followed by the fragmentation of the adduct radical at the labile C–X bond.[149–151] The substituent Y (Y=Ph, CN, or CO_2R) facilitates addition.

$$P_n^\bullet + CH_2=\underset{Y}{\overset{CH_2X}{C}} \xrightarrow{\text{addition}} P_n-CH_2-\underset{Y}{\overset{CH_2-X}{C^\bullet}}$$

$$\downarrow \text{fragmentation} \qquad (3.57)$$

$$P_n-CH_2-\underset{Y}{C}=CH_2 + \dot{X}.$$

3.8.1.6 *Polymer transfer*

Chain transfer to polymer (polymer transfer) can occur both intermolecularly and intramolecularly. The former becomes important at high conversions when polymer concentration is high. It gives rise to long chain branches. The latter works right from the beginning of polymerization and gives rise to short chain branches. The site of transfer depends on the chemical structure of the polymer. It can be ascertained from the analysis of polymer microstructure by ^1H, ^{13}C NMR, and IR spectroscopy.

Intermolecular Polymer Transfer

Measurement of polymer transfer constant (C_P) using the Mayo method presents difficulties because polymerization in the presence of the same polymer yields a mixture of the unreacted polymer, the branched polymer from the reacted polymer, and the formed linear polymer. In order to determine C_p, one has to measure the combined $\overline{DP}_n$ of the formed linear polymer and of the branches of the branched polymer that formed, which is not ordinarily possible. However, a reasonable estimate of C_P may be obtained using model compounds as transfer agents.[152,153]

More than one kind of transfer sites may exist in a polymer molecule. For example, in the polymerization of vinyl acetate (VA) polymer transfer occurs through H abstraction from both the acetoxy methyl and the in-chain methine groups. Subsequent growth from the transferred radical sites gives rise to hydrolytically cleavable and noncleavable branches respectively.[154–156] The former, however, are abundant, which suggests that the transfer of hydrogen atom from the acetoxy methyl is faster than that from the backbone methine, even though a *primary* carbon radical is formed in the former site compared to the *tertiary* in the latter. This may be due to three times as many abstractable hydrogens in the former case and resonance stabilization of the primary radical by the adjacent C=O group.[156]

Branching may also lead to crosslinking through coupling of branch radicals under suitable conditions, *viz.*, slow initiation and negligible transfer to monomer, solvent, initiator or other agents, particularly in the polymerization of monomers with relatively high $k_p/(2k_t)^{1/2}$ values (Table 3.3).[157] High molecular weight polymers form under these conditions and higher is the molecular weight, greater is the degree of branching. Furthermore, slow initiation lessens chances of coupling branch radicals with linear radicals, whereas transfer reaction between branch radicals and small molecules eliminates chances of crosslinking.[158]

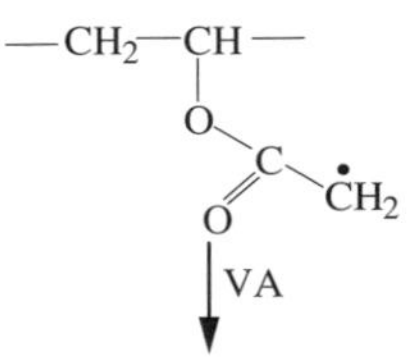

hydrolytically cleavable branches 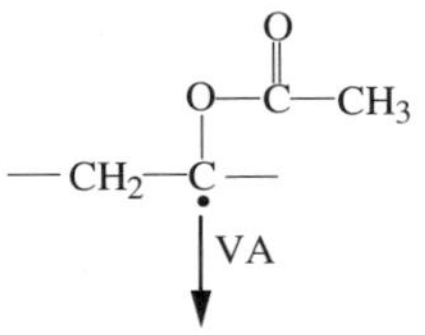   hydrolytically noncleavable branches

Intramolecular Polymer Transfer or Backbiting

Along with long chain branches, short chain ones also coexist in some polymers when the radicals are high in reactivity, *e.g.*, polyethylene, poly(vinyl acetate), poly(vinyl chloride), and polyacrylates. Branching is most abundant in polyethylene. The number of branches per 1000 carbon atoms in the polymer chain may vary from about 10 to 50 depending on polymerization conditions, the long chain branches constituting only about 10% of the total. Short chain branches comprise ethyl, butyl, pentyl, hexyl, and 2-ethylhexyl groups with butyl being usually the most abundant.[159–161] The proportion of the groups also changes with polymerization conditions. ^{13}Carbon NMR spectroscopy is the best analytical tool for the determination of the branch profile. However, it cannot distinguish between hexyls and longer branches.

Intramolecular polymer transfer reaction (backbiting) was proposed to explain short chain branching as shown for the formation of butyl branches[162]

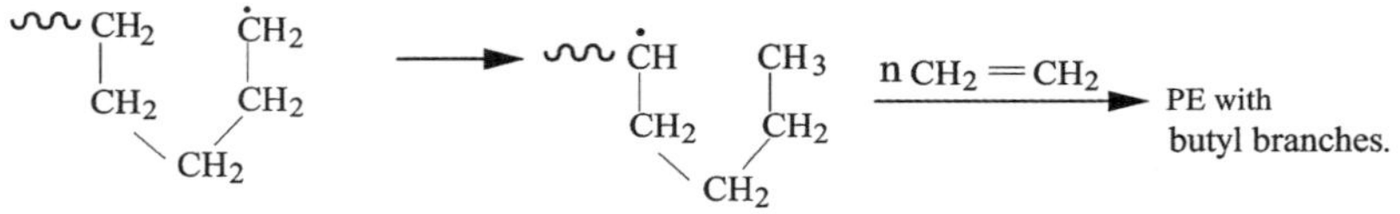

$$\text{(3.58)}$$

Backbiting also explains the presence of 2,4-dichlorobutyl and 2-chloroethyl branches in PVC[163] and of 2,4-diacetoxybutyl and 2-acetoxyethyl branches in poly(vinyl acetate).[164]

3.8.1.7 Redox transfer

Chain transfer reaction involving transfer of halogen atom from polyhalocarbons is facilitated when mediated by transition metal ions at their lower oxidation states, *e.g.*, Cu(I), and Fe(II).[165] The mechanism of Cu(I) mediated chain transfer to CCl$_4$ has been suggested to be as follows.

$$Cu^{I} + CCl_4 \longrightarrow Cu^{II}Cl + \overset{\bullet}{C}Cl_3.$$

$$CCl_3^{\bullet} + nM \longrightarrow P_n^{\bullet}.$$

$$P_n^{\bullet} + Cu^{II}Cl \longrightarrow P_n - Cl + Cu^{I}.$$

With CHCl$_3$ as the transfer agent, metal ion mediation brings about a change in the atom transferred. Ordinarily, the hydrogen atom is transferred, but in the presence of Cu(I) it is the Cl atom, which is transferred.

$$Cu^I + CHCl_3 \longrightarrow Cu^{II}Cl + \overset{\bullet}{C}HCl_2.$$

$$\overset{\bullet}{C}HCl_2 + nM \longrightarrow P_n^{\bullet}.$$

$$P_n^{\bullet} + Cu^{II}Cl \longrightarrow P_n - Cl + Cu^I.$$

3.8.1.8 Catalytic chain transfer (CCT)

CCT involves transfer agents which are regenerated so that each CCT agent (CCTA) participates in many hundreds of chain transfer reactions.[166] Low spin cobalt (II) complexes such as cobalt porphyrins,[166] cobaloximes and their derivatives[167] are very powerful CCTAs such that in AIBN initiated polymerization of alkyl methacrylates their use in catalytic quantities (ppm levels) produces oligomers. C_T is often greater than 40 000 in methacrylate polymerizations. The mechanism of action has been suggested to be as follows.[166,167]

$$P_n^{\bullet} + Co(II) \longrightarrow P_n - Co(III).$$

$$P_n - Co(III) \longrightarrow Co(III) - H + P_n^{=}.$$

$$(P_n^{=} = \text{polymer with unsaturated end group}).$$

$$Co(III) - H + M \longrightarrow P_1^{\bullet} + Co(II).$$

3.8.1.9 Instantaneous chain length distribution (CLD) and chain transfer

The CLD method was developed by Gilbert and coworkers as an alternative to the Mayo method of measuring the chain transfer constant.[168] It relates the high molecular weight slope (Λ_{high}) of the number molecular weight distribution, $P(M)$, to kinetic parameters as given in Eq. (3.59).

$$\Lambda_{high} = \lim_{M \to \infty} \frac{d \ln P(M)}{dM} = -\left(\frac{\langle k_t \rangle [P^{\bullet}]}{k_p[M]} + C_M + C_T \frac{[T]}{[M]}\right) \frac{1}{m_0}, \quad (3.59)$$

where Λ_{high} is the high molecular weight slope of a plot of $\ln P(M)$ vs. M, $\langle k_t \rangle$ is the average rate constant of termination, $[P^{\bullet}]$ is the total chain radical concentration, m_0 is the molar mass of monomer. All other symbols have been defined earlier. For the determination of C_T, one needs to measure Λ_{high} as a function of $[T]/[M]$ for low conversion polymerizations at a fixed initiator concentration. C_T is evaluated from the slope of the linear plot of $\Lambda_{high} m_0$ vs. $[T]/[M]$. Eq. (3.59) is analogous to Mayo equation (3.44) with $\Lambda_{high} \cdot m_0$ being the equivalent of $1/\overline{DP}_n$.

Although Eq. (3.59) is only valid for very high molecular weights, it has been shown that in polymerization systems completely dominated by chain transfer the most reliable value of chain transfer constant is obtained if the value of Λ_{high} is taken as the slope of the above-mentioned plot in the peak molecular weight region instead of the very high molecular weight region.[170]

The CLD method measures C_T as good as the Mayo method does. However, it is more reliable when C_T is high. This is due to the lower accuracy of $\overline{M}_n$ determined by GPC (particularly when $\overline{M}_n$ is low) arising out of the uncertainty in peak and base line selection. The problem is avoided in CLD where the higher molecular weight region of MWD is used. The CLD method is also more suitable for systems requiring analysis of contaminated polymer samples. The contaminant may affect MWD and $\overline{M}_n$. However, it has been suggested that if a region of MWD is identified, which is less affected by the contaminant, the CLD method can still be used.[169,170]

3.8.2 Factors influencing chain transfer

Consider the transfer of hydrogen atom from a series of transfer agents (substrates) to a given radical. In the absence of any polar effect, the rate constant of transfer increases with the decrease in the bond dissociation energy (BDE) of the C–H bond to be broken.[171–173]

$$\dot{R} + Y\underset{H}{\overset{X}{\diagup}}R' \longrightarrow R-H + \underset{Y}{\overset{X}{\diagup}}\dot{R}'$$

The BDE decreases with the increase in the resonance energy of the product radical and in the ground state strain in the substrate. The ground state of the substrate increasingly becomes strained due to the steric compression that occurs with the increase in the number and/or the size of the substituents on the tetrahedral carbon bearing the H atom to be abstracted. However, the substituents become separated to greater distances in a planar or nearly planar carbon radical (methyl, benzyl, and acceptor-substituted radicals known to be planar[9]) formed by breaking the C–H bond. The consequent release of strain is reflected in BDE. Larger is the strain smaller is the BDE.

The relative C_T of three alkylbenzenes, *viz.*, toluene, ethylbenzene, and *i*-propylbenzene, in the polymerization of each of the monomers, *viz.*, styrene, methyl methacrylate, vinyl acetate, and acrylonitrile, (Table 3.5) agrees with what would be predicted from the relative BDE. The C_T increases in the order: toluene < ethylbenzene < *i*-propylbenzene, although the number of abstractable benzylic hydrogens increases in the opposite order. This result is in line with the relative stability of the three product radicals

$$\underset{a}{\text{Ph–}\dot{C}H_2} < \underset{b}{\text{Ph–}\dot{C}HCH_3} < \underset{c}{\text{Ph–}\dot{C}(CH_3)_2}$$

All the three radicals are conjugated with the benzene ring and resonance stabilized. However, in *b* and *c* there is additional stabilization due to hyperconjugation of the radicals with the C–H bonds in the CH_3 groups and there are two such groups in *c*, one in *b*, and none in *a*. However, the hyperconjugation effect on the stability of small *secondary* or *tertiary* alkyl radicals has been proved to be small.[6]

On the other hand, the ground state strain increases in the following order for the transfer agents due to increasing methyl substitution at the benzylic carbon

$$\text{toluene} < \text{ethylbenzene} < i\text{-propylbenzene}.$$

Accordingly, the strain effect explains the C_T order.

When the C_Ts of any of the above hydrocarbon transfer agents with the four radicals entered in Table 3.5 are compared, the following order of decreasing reactivity is observed for the radicals. Although the comparison should be based on k_{tr} rather than on C_T, the same order obtains from both parameters for the radicals

$$-CH_2\overset{\bullet}{C}HOCOCH_3 > -CH_2\overset{\bullet}{C}HCN > -CH_2\overset{\bullet}{C}(CH_3)COOCH_3 > -CH_2\overset{\bullet}{C}HC_6H_5.$$

This order is the same as the general reactivity order of radicals (Sec. 3.1). However, the order is not maintained when the transfer agent is CCl_4, CBr_4, or a mercaptan (Table 3.5) with which the radical reactivity decreases in the order

$$-CH_2\overset{\bullet}{C}HOCOCH_3 \gg -CH_2\overset{\bullet}{C}HC_6H_5 \gg -CH_2\overset{\bullet}{C}(CH_3)COOCH_3 \approx -CH_2\overset{\bullet}{C}HCN.$$

This difference in the reactivity order of radicals towards the two sets of transfer agents suggests the role of polar effect in the latter set. As in radical addition (Sec. 3.1), the energy of the transition state of a radical transfer reaction is lowered due to the contribution of charge transfer resonance structures, when radicals and transfer agents of opposite polar character react.[127,143,144] The suggested structures are shown below with carbon tetrabromide and mercaptan as transfer agents

I. with CBr_4 (electron acceptor)[144]

$$\overset{\bullet}{R} + CBr_4 \longrightarrow [R^+Br.\overset{-}{C}Br_3 \longleftrightarrow R^+\overset{-}{Br}.CBr_3 \longleftrightarrow R^+\overset{\bullet}{Br}\,C\,Br_2 = \overset{-}{Br}],$$

II. with RSH (electron acceptor)[143]

$$\overset{\bullet}{R} + RSH \longrightarrow [R^+H.\overset{-}{S}-R \longleftrightarrow R^+H-\overset{-}{S}.R \longleftrightarrow R^+H-S.\overset{-}{R}\,].$$

Polystyrene radical is nucleophilic, whereas polyacrylonitrile and poly(methyl methacrylate) radicals are electrophilic, the latter being less so. On the other hand, conflicting views exist on the polarity of the poly(vinyl acetate) radical. For instance, Fuhrman and Mesrobian[144] considered it nucleophilic based on the negative Alfrey-Price e parameter of its monomer (Chap. 8), Whereas, Bamford et al.[174] considered it electrophilic based on the positive sign of the Hammett *para* σ parameter for the acetate group. Indeed, the acetate group exhibits electron releasing property in the monomer but electron withdrawing property in the radical.[8] This dual role is attributable to the interplay of both the resonance and the inductive effect. The resonance effect acts in opposition to the inductive effect.[7] Apparently, the resonance effect is stronger than the inductive effect in the monomer but not in the radical. However, the general reactivity of poly(vinyl acetate) radical is so high that it overwhelms the effect of polarity in determining the order of reactivity of radicals.

$$CH_2=CH-OCOCH_3 \longleftrightarrow \overset{-}{C}H_2-CH=\overset{+}{O}COCH_3 \cdot$$

$$RCH_2-\overset{\cdot}{C}H-OCOCH_3 \longleftrightarrow RCH_2-\overset{-}{C}H-\overset{+}{\overset{\cdot}{O}}COCH_3 \cdot$$

The substrates, CCl_4, CBr_4, and mercaptans are electron acceptors. They are also intrinsically highly reactive due to the low bond dissociation energies, in particular, the $Br_3C–Br$ and $RS–H$ bonds, and hence the C_Ts are high. The highest C_T with poly(vinyl acetate) radical is attributable to its highest general reactivity disregarding the polar effect. On the other hand, although polystyrene radical is the weakest of the four in general reactivity, its nucleophilic character enhances the rate of transfer so much that it is placed second in the reactivity order among the four radicals under discussion. The effect of the difference in the general reactivity of the other two radicals is largely counterbalanced by that of the difference in their electrophilicities so that they do not differ much in reactivity towards the above substrates.

Transfer reactions with alcohols also exhibit polar effect. This is evident from the C_T of a given alcohol being greater with polystyrene than with poly(methyl methacrylate) radical (Table 3.5) indicating that alcohols act as electron acceptors.[175]

A polar effect opposite to that observed with electron-acceptor substrates operates with electron-donor substrates. For example, with triethylamine as transfer agent, the radical reactivity order (vide Table 3.5) follows from the C_T values as[127]

$$-CH_2\overset{\cdot}{C}HCN \gg -CH_2\overset{\cdot}{C}HOCOCH_3 \gg -CH_2\overset{\cdot}{C}(CH_3)COOCH_3 > -CH_2\overset{\cdot}{C}HC_6H_5.$$

In this case, C_T of the electron-donor substrate triethylamine with polyacrylonitrile radical, the strongest electrophilic radical among the four under discussion, is the highest. The electrophilcity of poly(methyl methacrylate) radical is not strong enough to exert a large polar effect. The high C_T with poly(vinyl acetate) radical is attributable to its high general reactivity. The lowest C_T with polystyrene radical is due to the radical's nucleophilic character and lowest general reactivity.

3.8.2.1 Quantifying the resonance and polar effects

3.8.2.1.1 The Q, e Scheme

A quantitative treatment of the resonance and polar effects in radical addition reaction was provided by Alfrey and Price in their Q, e Scheme, which will be discussed in Chapt. 8.[2] In this scheme, a monomer and the corresponding polymer radical are assigned the general reactivity parameters Q and P respectively, and the same polarity parameter e for both. Fuhrman and Mesrobian extended the Q, e Scheme to chain transfer reaction assigning Q and e parameters to transfer agents also.[144] In analogy to the application of the Q, e scheme to copolymerization (Chap. 8) the C_T may be expressed as in Eq. (3.60)

$$C_T = \frac{k_{tr,T}}{k_P} = \frac{P_r Q_T \exp^{-e_T e_r}}{P_r Q_m \exp^{-e_m e_r}}, \qquad (3.60)$$

where P_r, Q_m and Q_T are the general reactivity parameters of the polymer radical, the monomer, and the transfer agent (TX) respectively, e_T is the polarity parameter of the transfer agent, e_r and e_m are the polarity parameters of the polymer radical and the monomer respectively. Setting $e_r = e_m$, as is assumed,

$$C_T = \frac{Q_T}{Q_m} \exp^{-e_m(e_T - e_m)}. \tag{3.61}$$

Since Q_m and e_m are known from copolymerization studies, Q_T and e_T of a transfer agent may be derived using Eq. (3.61), provided experimental values of C_T in the polymerizations of at least two monomers are available. Once Q_T and e_T are known, C_T in the polymerization of any other monomer can be theoretically calculated using the Q and e values of the latter in Eq. (3.61). Fuhrrman and Mesrobian thus calculated the C_T values of CBr_4 for a series of monomers and Katagari *et al.* determined Q, e values of a number of transfer agents.[176]

The weakness of the Q, e scheme lies in its lack of firm theoretical basis and in the arbitrarily chosen reference values. In addition, the assumption of the same e values for the monomer and the corresponding radical is not necessarily valid, as we have seen already in the vinyl acetate case.

3.8.2.1.2 The radical reactivity pattern scheme

Bamford, Jenkins, and Johnston provided an alternative method of separating the resonance and polar factors in radical reactions.[174] Like the Q, e scheme, the pattern scheme is applicable to both chain transfer and radical addition (copolymerization) reactions. However, unlike the Q, e scheme, the pattern scheme avoids the necessity of choosing arbitrary reference parameters. The reference parameter for general reactivity (resonance factor) was chosen to be the rate constant of transfer reaction of a radical with toluene, which is virtually free from polar effect

$$\overset{\bullet}{P_n} + \bigcirc\!\!-CH_3 \xrightarrow{k_{tr,t}} P_n-H + \bigcirc\!\!-\overset{\bullet}{C}H_2.$$

The relative order of general reactivity of radicals is, therefore, set from their $k_{tr,t}$ values. The rate constants ($k_{tr,HC}$) of the reactions of a series of radicals with another hydrocarbon are proportional to the $k_{tr,t}$ values of the respective radicals, *i.e.*,

$$k_{tr,HC} = \nu k_{tr,t}.$$

The factor ν is governed by the strength of the C–H bond in the hydrocarbon concerned relative to that in toluene and has therefore the same value applicable to all attacking radicals. However, when polar factors come into play the constancy of ν for any given substrate is lost. In such a situation the rate constant ($k_{tr,T}$) can be quantitatively related to $k_{tr,t}$ by Eq. (3.62)

$$\log k_{tr,T} = \log k_{tr,t} + \alpha\sigma + \beta, \tag{3.62}$$

where σ is the Hammett's σ parameter (the *para* σ value of the α-substituent in the radical and the algebraic sum of *para* σ values in the case of two α-substituents), α and β are constants for a given substrate. The first term in the right side of the equation refers to the general reactivity of the attacking radical; the second term is a measure of the contribution of the polar effect, α being analogous to ρ in the Hammett equation. The sign of α is negative for electron-acceptor substrates and positive for electron-donor substrates. The third term, *i.e.*, β, is determined by the general reactivity of the substrate. It follows from Eq. (3.62) that the parameters α and β for a given substrate can be evaluated from the rate constants of its reaction with two calibrated radicals, *i.e.*, radicals of which $k_{tr,t}$ and σ are known. In this way, the α and β parameters for various substrates were determined by Bamford *et al.*[174,177]

3.8.3 *Telomerization*

A telomer is an oligomer formed through chain transfer reaction with a regulator such that one fragment of the regulator makes one end of the oligomer, while the other fragment makes the other end. Regulators used for this purpose are referred to as telogens. Common telogens are polyhalocarbons, *viz.*, CCl_4, CBr_4, CCl_3Br, *etc.*, which are electron acceptors. C_T values of two of them, *viz.*, CCl_4 and CBr_4, in the polymerizations of four selected monomers are given in Table 3.5. From the discussion in Sec. 3.8.2, suitable monomers for telomerization should be those, which yield radical of high general reactivity and high nucleophilicity. Of the monomers given in Table 3.5 vinyl acetate fulfills the first criterion and styrene the second. However, ethylene, which gives nucleophilic radical of the highest general reactivity, is the monomer of choice. Besides, the low k_p in ethylene polymerization (*vide infra*) favors oligomer formation. Another requirement is that the initiator radical should react with the telogen instead of adding to the monomer. This condition is also fulfilled with ethylene as the monomer, it being the lowest in reactivity. Thus, the telomerization of ethylene with CCl_4 as telogen may be represented as follows

$$I \rightarrow 2\dot{R}.$$
$$\dot{R} + CCl_4 \rightarrow RCl + \dot{C}Cl_3.$$
$$\dot{C}Cl_3 + C_2H_4 \rightarrow CCl_3CH_2\dot{C}H_2.$$
$$CCl_3CH_2\dot{C}H_2 + nC_2H_4 \rightarrow CCl_3(CH_2CH_2)_n CH_2\dot{C}H_2.$$
$$CCl_3(CH_2CH_2)_n CH_2\dot{C}H_2 + CCl_4 \rightarrow CCl_3(CH_2CH_2)_n CH_2CH_2Cl + \dot{C}Cl_3.$$

No evidence of products with same fragments of the telogen, *e.g.*, $Cl(CH_2CH_2)_n Cl$ or $CCl_3(CH_2CH_2)_n CCl_3$, has been obtained,[178,179] which supports the reaction course shown and that termination is insignificant. Combination of redox initiation and redox transfer has been widely used in telomerization.[180–183] The reactions involved are as follows.

Redox Initiation

$$CCl_4 + CuCl \rightarrow \dot{C}Cl_3 + CuCl_2.$$
$$\dot{C}Cl_3 + M \rightarrow CCl_3\dot{P}_1^{\bullet}.$$

Propagation

$$CCl_3\dot{P}_1 + nM \rightarrow CCl_3\dot{P}_{n+1}.$$

Redox Transfer

$$CCl_3\dot{P}_n + CuCl_2 \rightarrow CCl_3P_nCl + CuCl.$$

Chloroform is also an effective telogen. As discussed in Sec. 3.8.1.7, it is the Cl atom and not the H atom, which is transferred from $CHCl_3$ to CuCl in the initiation step giving rise to the telomer $CHCl_2P_nCl$, and a mechanism analogous to the above operates.

3.9 Inhibition and Retardation of Polymerization

Certain substances when present in polymerization systems stop polymerization altogether until such times as are expended to convert them into inert products by way of reaction with chain radicals. These substances are referred to as inhibitors. Certain others reduce rate as well as degree of polymerization without stopping polymerization altogether. These are called retarders. A substance may inhibit polymerization of certain monomers but only retard those of certain others.

An inhibited polymerization shows an induction period following which the polymerization proceeds at almost the same rate as the uninhibited one provided the induction period (inhibition period) is not too long to effect a substantial consumption of the initiator. The polymer produced has almost the same degree of polymerization as the one produced in an uninhibited system. In some cases, however, the reaction products formed during the induction period may not be completely inert but act as retarders so that polymerization progresses at a reduced rate after the induction period is over producing polymer of reduced molecular weight.

An inhibitor reacts with a chain radical at an exceedingly fast rate producing either nonradical inert products or relatively stable radicals, which are incapable of initiating polymerization. Stable radicals like 2,2-diphenyl-1-picrylhydrazyl (DPPH), nitroxides, *e.g.*, 2,2,6,6-tetramethyl-1-piperidinyl-1-oxy (TEMPO), triphenylmethyl (TPM), galvinoxyl, verdazyl, *etc.*, (Fig. 3.6) are efficient inhibitors.[32–34,184–189]

Reaction with nitroxide radicals gives the combination product alkoxyamines in the main,[187] as shown in reaction (3.63) with TEMPO

$$\text{N}-\dot{\text{O}} + \dot{P}_n \rightleftharpoons \text{N}-\text{O}-P_n \cdot \quad (3.63)$$

Disproportionation products become significant with chain radicals having α methyl groups as in the PMMA radical. The combination reaction (3.63) is reversible specially at elevated temperatures. This property has been utilized to develop nitroxide mediated living radical polymerization (Sec. 3.15).

DPPH reacts with polystyrene and poly(vinyl acetate) radicals with a stoichiometry of 1:1. The induction period (i.p.) that occurs in its presence is proportional to its initial concentration, which suggests that the consumption is of zero order. The number of chain radicals formed during the induction period is, therefore, equal to the number of DPPH

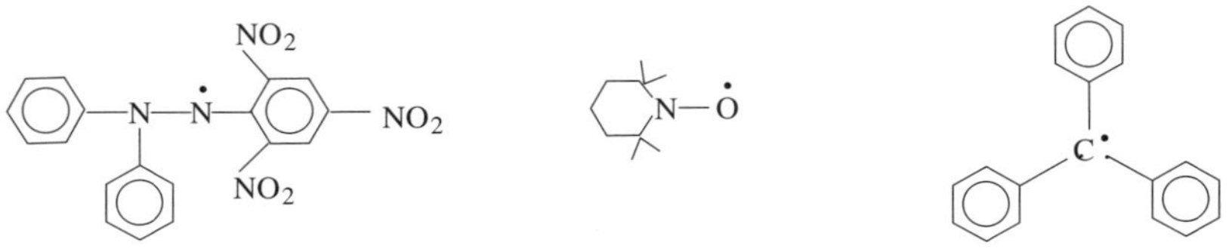

DPPH **(12)** TEMPO **(13)** TPM **(14)**

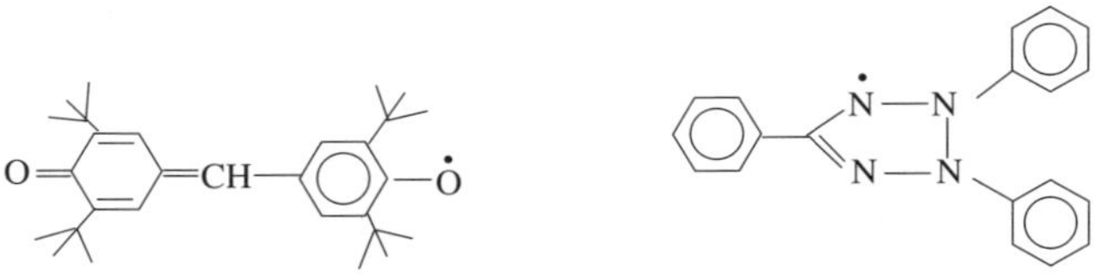

Galvinoxyl **(15)** Triphenylverdazyl **(16)**

Figure 3.6 Some stable radicals.

radicals initially used. Thus, the rate of initiation is given by

$$R_i = [DPPH]_0/i.p.,$$

where the subscript 0 refers to the initial concentration. In the inhibited polymerization of vinyl acetate, the post-inhibition polymerization proceeds at almost the same rate as the uninhibited polymerization, when DPPH concentration is low. However, at relatively high concentrations of DPPH the post-inhibition polymerizations are retarded. The retardation is presumably effected by the nitroaromatic moiety in DPPH (*vide infra*).

Nitroxide radicals also terminate chains with a stoichiometry of 1:1 (reaction 3.63) and do not initiate polymerization. On the other hand, although triphenylmethyl inhibits thermal polymerization of styrene, a part of the inhibitor is used up in initiation also.[189]

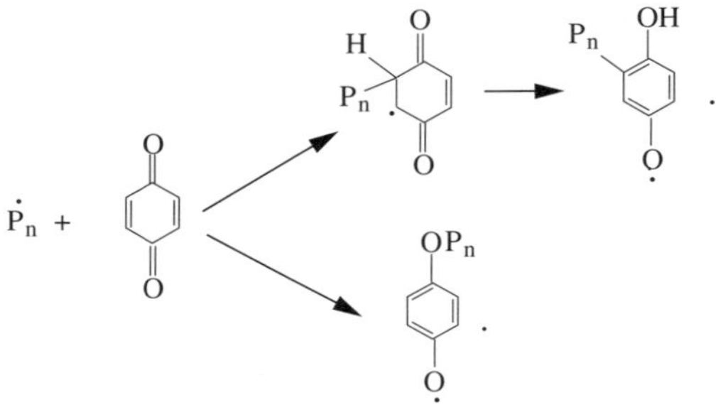

(3.64)

(3.65)

However, an inhibitor does not need necessarily be a stable radical; there are many nonradical inhibitors as well. Benzoquinone is one such inhibitor.[189–195] In inhibited polymerization, the length of the induction period is proportional to the initial benzoquinone concentration. However, it is not a universal inhibitor. For example, it inhibits polymerization of both styrene and vinyl acetate but only retards polymerization of methyl methacrylate and methyl acrylate.[190] The difference in reactivity is attributable to polar effect. Styrene radical is nucleophilic, while the two (meth)acrylate radicals are electrophilic.

Benzoquinone, being an electron acceptor, reacts with the latter two at much slower rates (*vide* Sec. 3.8.2). Possible reaction paths involve the addition of polymer radical to quinone nucleus (3.64) and to oxygen (3.65) respectively.[193,194]

Another possibility involves β-H transfer from the chain radical to quinone

$$P_n^{\bullet} + O=\!\!\bigcirc\!\!=O \longrightarrow \dot{O}-\!\!\bigcirc\!\!-OH + P_n^{=}.$$

The resultant semiquinone radical is resonance stabilized so that it fails to initiate polymerization of styrene except at rather high temperatures *ca.*, 100°C, when a fraction of the radicals take part in initiation.[11,189] Unlike benzoquinone, its reduction product hydroquinone is not an inhibitor.[192] However, it is air oxidized to benzoquinone (readily in alkaline medium).

In contrast to quinones, nitrobenzene and dinitrobenzenes only retard polymerization of styrene[195,196] but the dinitrobenzenes inhibit polymerization of vinyl acetate.[184] According to Price, the chain radicals add to aromatic nucleus in these compounds producing less reactive radicals[196]

$$\dot{P}_n + \bigcirc\!\!-NO_2 \longrightarrow \underset{H\ P_n}{\bigcirc\!\!-NO_2}.$$

However, Bartlett *et al.* found evidence, which suggests that addition of radical to nitro group also occurs.[184,186]

3.9.1 *Oxidative and reductive termination*

Transition metal halides capable of undergoing redox reactions with chain radicals also act as retarders or inhibitors of polymerization depending on the magnitude of the rate constants. The most studied of these halides are the ferric and cupric.[35,197–204] $FeCl_3$ inhibits polymerization of styrene but retards those of methyl methacrylate, acrylonitrile, methyl acrylate, and methacrylonitrile.[197] $CuCl_2$ is a stronger chain terminator than $FeCl_3$. It inhibits polymerization not only of styrene[201] but also of methyl methacrylate,[199] acrylonitrile,[200,35] and methyl acrylate.[35] The reaction involves chlorine atom transfer from metal halide to chain radical.[205,206]

$$P_n^{\bullet} + FeCl_3 \xrightarrow{k_x} P_nCl + FeCl_2.$$

$$P_n^{\bullet} + CuCl_2 \xrightarrow{k_x} P_nCl + CuCl.$$

The rate constants of the reactions in dimethylformamide are given in Table 3.8.

Kinetic studies of polymerizations retarded by metal halides under appropriate conditions allow determination of rate of initiation (R_i). Thus, when the metal halide concentration is sufficiently large such that the mutual bimolecular termination of chain radicals (reaction 3.6) is negligible compared to the unimolecular termination by metal halide, one obtains applying stationary state approximation[35,197]

$$R_i = k_x[P^{\bullet}][MtX_{n+1}].$$

Therefore, R_i may be estimated from the measurement of the rate of disappearance of MtX_{n+1} or of the rate of formation of MtX_n.

Table 3.8 Rate constants of oxidative or reductive termination of chain radicals in dimethylformamide at 60°C.

Monomer	k_p^a (l mol^{-1}s^{-1})	$k_x \times 10^{-4}$ l mol^{-1}s^{-1}		
		$CuCl_2$	$CuCl$	$FeCl_3^e$
VA	7940	—	—	500
St	340	350[b]	—	10.43
MMA	830	87[c]	—	0.34
MA	27800	68[d]	0.73[d]	5.4
AN	(2290)	(15)[d]	(1)[d]	(0.45)
MAN	58	14[d]		0.008

[a] New values by PLP-SEC method except for AN. These have been used to calculate k_x from the C_x values given in the literature; [b] Ref. 201; [c] Ref. 199; [d] Ref. 35; [e] Ref. 197.

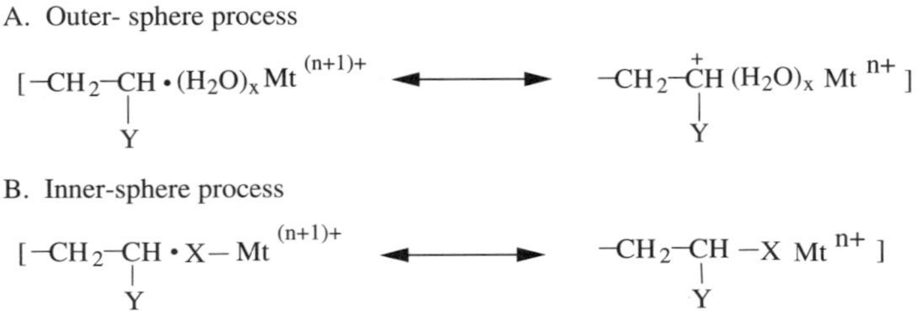

A. Outer-sphere process

$$[-CH_2-\overset{\bullet}{CH}\cdot(H_2O)_x Mt^{(n+1)+}] \longleftrightarrow [-CH_2-\overset{+}{CH}(H_2O)_x Mt^{n+}]$$
$$\quad\quad | \quad\quad\quad\quad\quad\quad\quad\quad\quad\quad\quad | $$
$$\quad\quad Y \quad\quad\quad\quad\quad\quad\quad\quad\quad\quad\quad Y$$

B. Inner-sphere process

$$[-CH_2-\overset{\bullet}{CH}\cdot X-Mt^{(n+1)+}] \longleftrightarrow [-CH_2-CH-X\ Mt^{n+}]$$
$$\quad\quad | \quad\quad\quad\quad\quad\quad\quad\quad\quad\quad\quad | $$
$$\quad\quad Y \quad\quad\quad\quad\quad\quad\quad\quad\quad\quad\quad Y$$

Scheme 3.4 Resonating structures in the transition states.[205]

Reductive termination of some chain radicals presumably also occurs. For example, polymerization of acrylonitrile and, to some extent, of methyl acrylate is retarded by CuCl.[200–203,35]

Metal ions such as Cu^{+2}, Fe^{3+}, and Ce^{4+} also retard polymerization by oxidative termination.[207] Oxidation by solvated free ion proceeds through outer-sphere electron transfer, which gives carbenium ion products. The transition states of the outer-sphere and inner-sphere electron transfer processes may be represented as shown in Scheme 3.4.[205]

Thus, in the oxidation using aquo complexes of Cu(II), polymer forms with terminal unsaturation ($P_n^=$) and hydroxyl group, which are carbenium ion products formed by β proton elimination and substitution reaction respectively following the outer-sphere electron transfer

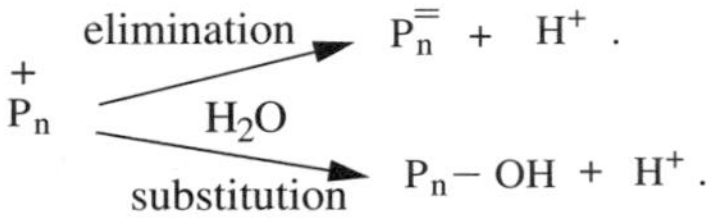

$$\overset{+}{P_n} \underset{\text{substitution}}{\overset{\text{elimination}}{\underset{H_2O}{\rightleftarrows}}} \begin{array}{l} P_n^= + H^+ \\ P_n - OH + H^+ \end{array}$$

The outer-sphere electron transfer either does not occur or occurs slowly when α-substituents of the radical are electron withdrawing. This is due to the destabilization of the cationic transition state by electron-withdrawing group. On the other hand, in the inner-sphere process, the charge development in the transition state is moderated through direct transference of an atom or ligand from the metal complex to the chain radical and hence electron-withdrawing effect of α-substituent(s) in the radical has lesser influence on the rate.

Ligands, such as H_2O, $RCOO^-$, acetylacetonate, pyridine, acetonitrile, phenanthroline, and bipyridine, among others, do not mediate electron transfer by the inner-sphere process. On the other hand, halides, *viz.*, chloride and bromide as well as pseudohalides, *viz.*, thiocyanate, cyanide, and azide, do.[206]

Oxidation of chain radical by solvated free ion is much slower than by the halide ion pair. However, although secondary radicals with electron withdrawing α substituents such as methyl acrylate and acrylamide undergo oxidation by hydrated Fe^{3+} ion, steric hindrance prevents oxidation of the *tertiary* PMMA radical to take place.[207–210]

3.9.2 Oxygen as an inhibitor

Oxygen inhibits radical polymerization. It reacts with carbon radicals (both primary and chain) forming peroxy radicals, which are relatively unreactive[11]

$$P_n^\bullet + O_2 \rightarrow P_nOO^\bullet.$$

The peroxy radical may eventually recombine with each other yielding peroxide and oxygen.

$$2P_nOO^\bullet \rightarrow P_nOOP_n + O_2.$$

Some of the peroxy radicals may also occasionally add to monomer to regenerate a carbon radical, which in turn reacts with O_2 to regenerate a peroxy radical.[211–213] The cycle continues to produce an approximately alternating copolymer of monomer and oxygen[212]

$$P_nOO^\bullet + M \rightarrow P_nOOP^\bullet.$$
$$P_nOOP^\bullet + O_2 \rightarrow P_nOOPOO^\bullet.$$

The peroxides, however, may decompose to generate reactive alkoxy radicals although at very slow rates.

At low concentrations, oxygen may act as an indirect initiator, the peroxides being the actual initiators. An important example is its use as an initiator in high pressure polymerization of ethylene.

Despite the inhibiting action of oxygen, polymerization of monomers exposed to air is possible under suitable circumstances. For example, in the ultrafast photocuring of surface coatings in air the dissolved oxygen is scavenged rapidly by initiator radicals, which are generated at very fast rates. It has been measured that the oxygen concentration is rapidly reduced from the original concentration of $\sim 10^{-3}$ mol/L[214] in multi-acrylate monomers to a stationary state level of $\sim 4 \times 10^{-6}$ mol/L, when the rate of oxygen consumption becomes equal to that of its diffusion through the film surface.[63]

3.10 Rate Constants of Propagation and Termination

Various methods have been used for the determination of k_p and k_t.[215,216] The earliest of these is the rotating sector method, that uses the sector for shining light intermittently from a

UV source on a polymerization reactor containing a photoinitiator.[32–34,217–220] As a result, a periodic change in radical concentration is produced in the polymerization medium. From the analysis of the results using nonstationary state kinetics k_p/k_t is evaluated.[11] For the determination of the individual rate constants, k_p and k_t, use is made of another parameter, viz., k_t/k_p^2, which is determined from a polymerization, observing stationary state kinetics as described earlier (Sec. 3.2).

In the shining phase of the rotating sector experiment, the rate of generation of radicals is given by $2fI_{abs}$ (where I_{abs} is the intensity of the absorbed light) and the rate of disappearance of radicals is given by $2k_t[P^\bullet]^2$. Thus,

$$\frac{d[P^\bullet]}{dt} = 2fI_{abs} - 2k_t[P^\bullet]^2. \tag{3.66}$$

If the light is continuously shined on the reactor, a stationary state of radical concentration $[P^\bullet]_s$ is built up the value of which follows from Eq. (3.66) putting $d[P^\bullet]/dt = 0$

$$[P^\bullet]_s = (fI_{abs}/k_t)^{1/2}. \tag{3.67}$$

Substituting from Eq. (3.67) for fI_{abs} in Eq. (3.66) one obtains for the nonstationary state

$$\frac{d[P^\bullet]}{dt} = 2k_t([P^\bullet]_s^2 - [P^\bullet]^2), \tag{3.68}$$

which gives on integration

$$\ln\{(1 + [P^\bullet]/[P^\bullet]_s)/(1 - [P^\bullet]/[P^\bullet]_s)\} = 4k_t[P^\bullet]_s(t - t_o), \tag{3.69}$$

where t_o is an integration constant; at $t = t_o$, $[P^\bullet] = 0$.

The $k_t[P^\bullet]_s$ factor in the right side of Eq. (3.69) is related to the average life time (τ) of the radical in the stationary state

$$\tau_s = \frac{[P^\bullet]_s}{2k_t[P^\bullet]_s^2} = \frac{1}{2k_t[P^\bullet]_s}. \tag{3.70}$$

Substituting from Eq. (3.70) for $k_t[P^\bullet]_s$ in Eq. (3.69) and rearranging one gets[11]

$$\tanh^{-1}([P^\bullet]/[P^\bullet]_s) = (t - t_o)/\tau_s. \tag{3.71}$$

Since R_p is proportional to the radical concentration, Eq. (3.71) may be written as

$$R_p/R_{p,s} = \tanh[(t - t_o)/\tau_s], \tag{3.72}$$

where $R_{p,s}$ is the rate of polymerization in the stationary state.

As has been noted above, at $t = t_o$, $R_p = 0$. When $R_p = 0$ at the start of illumination ($t = 0$) in a cycle, $t_o = t = 0$. In the event of $R_p > 0$ at the start of illumination in a cycle, $R_p = R_{p,0}$ at $t = 0$. Making use of these relations it may be derived easily from Eq. (3.72) that

$$\tanh^{-1}(R_p/R_{p,s}) - \tanh^{-1}(R_{p,o}/R_{p,s}) = t/\tau_s, \tag{3.73a}$$

or

$$\tanh^{-1}([P^\bullet]/[P^\bullet]_s) - \tanh^{-1}([P^\bullet]_0/[P^\bullet]_s) = t/\tau_s, \tag{3.73b}$$

where $R_{p,o}$ and $[P^\bullet]_0$ are the rate of polymerization and the radical concentration respectively at the start of illumination in a cycle.

Thus, measurement of the rate of polymerization as a function of time during the shining phase allows evaluating τ_s using Eq. (3.72) or (3.73a), whichever is applicable.

During the dark period following illumination the rate of decay of radical concentration is given by

$$\frac{d[P^\bullet]}{dt'} = -2k_t[P^\bullet]^2, \qquad (3.74a)$$

where t' refers to time in the dark period.

Integration of Eq. (3.74a) gives

$$\frac{1}{[P^\bullet]} - \frac{1}{[P^\bullet]_i} = 2k_t t', \qquad (3.74b)$$

where $[P^\bullet]_i$ is the radical concentration at the start of the dark period. Multiplying Eq. (3.74b) by $[P^\bullet]_s$ and substituting $2k_t[P^\bullet]_s$ with $1/\tau_s$ one gets

$$[P^\bullet]_s/[P^\bullet] - [P^\bullet]_s/[P^\bullet]_i = t'/\tau_s. \qquad (3.75)$$

Thus, following the decay of the polymerization rate with the start of the dark phase (photochemical after-effect) it is possible to determine τ_s. If the after-effect is studied following the establishment of the stationary state during the light period, the second term in the left side of Eq. (3.75) is reduced to 1. In that case, plotting $R_{p,s}/R_p$ against t' the value of $1/\tau_s$ is obtained from the slope of the resulting straight line.[217] Knowing τ_s, the value of $k_p/(2k_t)$ is obtained using Eq. (3.76), which follows from Eq. (3.70) by substituting $R_{p,s}/(k_p[M])$ for $[P^\bullet]_s$ (vide Eq. 3.19)

$$\tau_s = (k_p/2k_t)[M]/R_{p,s}. \qquad (3.76)$$

Determination of the value of the other kinetic parameter $k_p/(2k_t)^{1/2}$ from the kinetic studies of stationary state polymerization has already been discussed in Sec. 3.2. Having thus obtained the values of $k_p/(2k_t)$ and $k_p/(2k_t)^{1/2}$ the individual rate constants k_p and k_t are readily evaluated.

However, it is often not possible to determine τ_s following the approach of R_p to the stationary state with the start of the light period or the decay of R_p during the subsequent dark period. The rate changes may be too rapid to measure with reasonable accuracy. It has been possible to get rid of this problem by following a method which was originally used in gaseous systems.[218] The method uses a large number of illumination cycles, which give rise to uniform oscillation of radical concentration after several initial cycles. The average rates during the latter cycles are measured for different flashing frequencies and a fixed ratio (r) of dark to light period using an appropriately designed sector. A brief description of the method follows.[11]

We may refer the maximum radical concentration in a cycle at the end of the illumination period of duration t and the start of the dark period in the uniform oscillatory zone as $[P^\bullet]_{max}$, and the minimum radical concentration at the start of the illumination period and the end of the dark period of duration rt as $[P^\bullet]_{min}$. The individual concentration ratios $[P^\bullet]_{max}/[P^\bullet]_s$ and $[P^\bullet]_{min}/[P^\bullet]_s$ may be related to each other by Eqs. (3.77a) and (3.77b), which are obtained respectively by substituting $[P^\bullet]_{max}/[P^\bullet]_s$ for $[P^\bullet]/[P^\bullet]_s$ and $[P^\bullet]_{min}/[P^\bullet]_s$ for $[P^\bullet]_0/[P^\bullet]_s$ in

Eq. (3.73b), and $[P^\bullet]_s/[P^\bullet]_{min}$ for $[P^\bullet]_s/[P^\bullet]$ and $[P^\bullet]_s/[P^\bullet]_{max}$ for $[P^\bullet]_s/[P^\bullet]_i$ in Eq. (3.75).

$$\tanh^{-1}([P^\bullet]_{max}/[P^\bullet]_s) - \tanh^{-1}([P^\bullet]_{min}/[P^\bullet]_s) = t/\tau_s. \qquad (3.77a)$$

$$[P^\bullet]_s/[P^\bullet]_{min} - [P^\bullet]_s/[P^\bullet]_{max} = t'/\tau_s = rt/\tau_s. \qquad (3.77b)$$

The individual concentration ratios $[P^\bullet]_{max}/[P^\bullet]_s$ and $[P^\bullet]_{min}/[P^\bullet]_s$ can be theoretically determined solving Eqs. (3.77a) and (3.77b) for given values of r and t/τ_s, where t is the length of the illumination period in a cycle and $t' = rt$, r being constant for a given sector.

The average radical concentration ratio $[P^\bullet]_{av}/[P^\bullet]_s$ in the uniform oscillatory rate zone is obtained making use of $[P^\bullet]_{max}/[P^\bullet]_s$ and $[P^\bullet]_{min}/[P^\bullet]_s$ determined as above.[32,219]

$$[P^\bullet]_{av} = \left(\int_0^t [P^\bullet]dt + \int_0^{t'} [P^\bullet]dt' \right) / (t + t'),$$

where $[P^\bullet]$ in the first integral is given by Eq. (3.73b) and that in the second integral is given by Eq. (3.75). Evaluation of these integrals give the necessary equation for $[P^\bullet]_{av}/[P^\bullet]_s$.

$$[P^\bullet]_{av}/[P^\bullet]_s = (r+1)^{-1}\left[1 + (\tau_s/t)\ln\left(\frac{[P^\bullet]_{max}/[P^\bullet]_{min} + [P^\bullet]_{max}/[P^\bullet]_s}{1 + [P^\bullet]_{max}/[P^\bullet]_s}\right)\right].$$

Using $[P^\bullet]_{av}/[P^\bullet]_s$ thus obtained, a theoretical curve can be drawn for the variation of $R_{p,av}/R_{p,s}$ with $\log(t/\tau_s)$ for a fixed value of r. The experimental $R_{p,av}/R_{p,s}$ values for the same value of r detemined from experimentally measured polymerization rates at different flashing frequencies (different t values) and at constant illumination are plotted against log t. The theoretical curve is then superposed on the experimental points such as to obtain the best fit. The lateral shifting of one graph relative to the other done in the process gives a measure of $\log \tau_s$ inasmuch as the abscissa is $\log t - \log \tau_s$ for the theoretical curve, whereas it is log t for the experimental data points.[219,220]

The EPR method directly measures the radical concentration in a stationary state polymerization.[221,222] With the rate of polymerization also measured, k_p is easily obtained.

Emulsion polymerization obeying Smith-Ewart Case 2 kinetics (Ch. 9, Eq. 9.5) also provides a method. The method requires the measurement of the steady rate of polymerization in interval 2 and the number concentration of the latex particles. The latter is obtainable from the mass concentration of the polymer in the latex and the average mass of a particle, which can be determined from the average particle size measured by transmission electron microscopy with the polymer density known.[226]

In the late 1980s, methods based on pulsed laser polymerization began to emerge.[223–225] The method developed for k_p is considered superior to the others because it provides self-consistency checks for the reliability of the results. Besides, it is straightforward and almost free from any model-dependent assumption.[227] An IUPAC expert committee compiled 'benchmark' k_p values obtained through the use of the method for styrene,[227] various methacrylates,[228–230] butyl acrylate,[231] and methacrylic acid[232] till 2007.

3.10.1 *The PLP-SEC method*

In brief, the method uses UV laser pulses of nanoseconds duration to generate an intense burst of radicals from a photoinitiator (usually acetophenone or benzoin) in a polymerization medium. A train of evenly spaced pulses is used but the number of pulses is so chosen as to restrict the monomer conversion to a low level ca., 2 to 3%. The intense burst of initiator-derived primary radicals generated in each pulse effects a surge in the termination of chain radicals, which were initiated by primary radicals of the preceding pulses, besides initiating new chains. As a result, a periodic variation of radical concentration is brought about, which gives rise to a structured molecular weight distribution (MWD). However, structure-less MWD may result if rate of chain transfer or of mutual termination is high.[233] Hence appropriate reaction conditions are to be used. For example, low temperature helps to decrease k_{tr} to a sufficiently low level. Also, dilution of the reaction mixture with a solvent of zero chain transfer activity, e.g., CO_2,[234,235] or increase of laser pulse repetition rate[236] decreases the kinetic chain length and lowers the impact of chain transfer to monomer on the MWD structure. Similarly, the impact of mutual termination of chain radicals may be reduced by reducing the photoinitiator concentration besides other means.[236]

Figure 3.7 shows the log MWD of poly(decyl acrylate) prepared at $-4°C$ and 200 bar in CO_2 following the above principle.[233] The three maxima and one shoulder in increased order of molecular weights are due to chain radicals initiated by primary radicals in a pulse and terminated respectively by primary radicals generated in successive four pulses. Not all the chain radicals generated in the first pulse are terminated on the application of the second pulse; a part of the rest is terminated on the application of the third pulse, and so on. As a result, the chain lengths of the polymers formed due to the four successive pulses following a given pulse would be in the ratio 1:2:3:4 in that order.

However, studies on MWD suggested that the molecular weight corresponding to the point of inflection on low molecular weight side of the major peak in the MWD curve is to be used for evaluating k_p.[224,228,229] Equation (3.78) relates k_p with chain lengths (L_i) corresponding to the low molecular weight side inflection points (i) of the maxima

$$L_i = ik_p[M]t_0 \quad i = 1, 2, 3, \text{ etc.}, \tag{3.78}$$

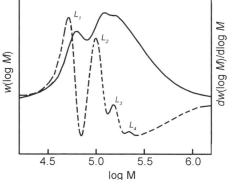

Figure 3.7 Log molecular weight distribution (full line) of poly(decyl acrylate) prepared in pulsed laser polymerization at $-4°C$ and 200 bar in 36 wt% CO_2 at a laser pulse repetition rate of 100 Hz and the corresponding first derivative curve of MWD (dotted line). "Reprinted with permission from Ref. 233. Rate coefficients of free-radical polymerization deduced from pulsed laser experiments. Copyright © 2002 with Elsevier Science Ltd."

where t_0 is the time interval between two successive pulses and [M] is the monomer concentration.[233] The inflection points are best determined from the peak positions in the first derivative of the MWD curve as shown in Fig. 3.7.[233] The intense peak (L_1) corresponding to the first inflection point is more accurately determined. This is another reason of using it to evaluate k_p. The other three inflection points (L_2, L_3 and L_4) are used for self-consistency checks, *i.e.*, $L_2 \approx 2L_1$, $L_3 \approx 3L_1$, and $L_4 \approx 4L_1$.

3.10.2 The single pulse-pulsed laser polymerization (SP-PLP) method for determining k_t

For k_t, no benchmark values are available even for very common monomers. An IUPAC task-group could not recommend a method, which should be superior to others in every respect.[238,239] Nevertheless, the SP-PLP method finds favor. It provides unparallel control over conversion so as to be routinely used to follow the change of k_t with conversion at increments of less than 1%.[240]

In SP-PLP method, a laser pulse typically of about 20 ns width, is applied to a monomer containing a photoinitiator. Conversion of monomer is followed with time *via* on-line infrared or near-infrared spectroscopy with a time resolution of 10 microseconds.[233,238] Signal-to-noise ratio in conversion-time trace is increased with increase in k_p/k_t, which is increased by conducting polymerization under high pressure since the activation volume of propagation is negative in sign and that of termination is positive.[233,238,239]

The time interval between pulses varies from milliseconds to seconds depending on the rate of polymerization of the monomer under investigation. The photoinitiator is so chosen that the pulses create the radicals instantaneously on the time scale of termination. The small conversion per pulse allows one to measure k_t at consecutive small conversion intervals by a series of successive SP-PLP experiments. To extract k_t from the conversion *vs.* time trace, non-stationary state kinetics of polymerization beginning with the application of each single pulse is used (Fig. 3.8). Eqs. (3.79) and (3.80) give the rate of decrease of monomer and

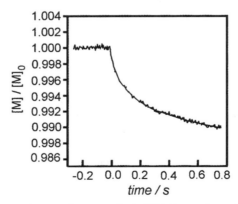

Figure 3.8 Monomer concentration *vs.* time trace in a SP-PLP experiment for the copolymerization of an equimolar mixture of methyl acrylate and dodecyl acrylate at 40°C, 1000 bar, and in the presence of 5 wt% initial copolymer concentration. "Reprinted in part with permission from Ref. 241. Copyright © 1999 American Chemical Society."

radical concentrations respectively within each SP-PLP experiment[240]

$$\frac{d[M]}{dt} = -k_p[P^\bullet][M], \qquad (3.79)$$

and

$$\frac{d[P^\bullet]}{dt} = -2k_t[P^\bullet]^2. \qquad (3.80)$$

With the assumption of chain-length-independent (hence time-independent) k_p and k_t integration of the above two coupled differential equations gives[233]

$$\frac{[M]}{[M]_0} = (2k_t[P^\bullet]_0 t + 1)^{-k_p/2k_t}, \qquad (3.81)$$

where [M] and $[M]_0$ are the monomer concentrations at time t and 0 respectively, and $[P^\bullet]_0$ is the instantaneously generated radical concentration at the application of a single pulse.

A fitting of the experimental conversion *vs.* time trace obtained for a SP-PLP experiment with Eq. (3.81) yields k_p/k_t and $k_t[P^\bullet]_0$ for that particular experiment. From the former k_t is obtained when k_p is known.

However, the assumption of chain-length independence is generally valid for k_p but not for k_t (*vide infra*). Hence, k_t obtained in the above-described way represents a chain-length averaged value. Equation (3.81) applies very well to many systems, which suggests a time-independent k_t. This is true not only when k_t^{ii} is relatively invariant with the chain length i but also when transfer reaction is sufficiently fast so that the chain length distribution of the growing radicals becomes broad.[238,239] However, in the latter case, one deals with k_t^{ij} (which is time independent) rather than k_t^{ii}.[233]

3.10.3 *Propagation rate constants*

Table 3.9 presents the values of k_p and Arrhenius parameters for various monomers. The latter are arranged in the order of decreasing E_p. Values of k_p at 50°C are given except for the acrylates, which are at 20°C. This is because with the acrylates the method could not be used above about 30°C due to the relatively high chain transfer to monomer, which gives structureless MWD.

A comparison of the E_p values with those of radical stabilization energy (available for some of the radicals and given in Table 3.1) shows a correlation except for the two poorly stabilized radicals, ethylene and vinyl acetate. Greater is the radical stabilization energy greater is the activation energy. For example, both radical stabilization energy and activation energy of propagation increase in the following order of monomers

$$MA < MMA < MAN < St < BD.$$

However, the examples of the nonconforming monomers suggest that apart from radical reactivity, monomer reactivity is also important. The low monomer reactivity of ethylene and vinyl acetate seems to be the dominating factor governing k_p for them.

The frequency factors for the 1,1-disubstituted monomers are about an order of magnitude lower than those for the monosubstituted monomers. This is attributable to the

Table 3.9 Rate constants and kinetic parameters of propagation for various monomers.[233]

Monomer	Temp. °C	k_p^a L mol^{-1}s^{-1}	$A_p \times 10^{-6}$ L mol^{-1}s^{-1}	E_p kJ/mol	$\Delta V^{\#}$ mL/mol
BD	50	135	80.5	35.7	—
Ethylene	50	53^d	18.8	34.3	-27^e
Styrene	50	240	42.7	32.5	-12.1^e
MAN	50	42	2.69	29.7	
ClPr	50	980	19.5	26.6	
EMA	50	670	4.06	23.4	
BMA	50	757	3.78	22.9	-16.5^f
MMA	50	648	2.67	22.4	-16.7^e
DMA	50	995	2.5	21	-16^f
VA	50	6300	14.7	20.7	-10.7^g
			10	19.8	
MAA (CH$_3$OH)^b	50	790	1.63	20.5	
MAA (H$_2$O)^c	50	5770	1.72	15.3	
MA	20	11600	16.6	17.7	-11.7^h
BA	20	14400	18.1	17.4	
DA	20	16700	17.9	17.0	-11.7^i

Key: MMA = methyl methacrylate, EMA = ethyl methacrylate, BMA = butyl methacrylate, DMA = dodecyl methacrylate, MAA = methacrylic acid, BD = butadiene, ClPr = chloroprene, MAN = methacrylonitrile, VA = vinyl acetate, MA = methyl acrylate, BA = butyl acrylate, DA = dodecyl acrylate. "Reprinted with permission from Ref. 233, Rate coefficients of free-radical polymerization deduced from pulsed laser experiments. Copyright © 2002 Elsevier Science Ltd."
^{a}At 1 atm; ^{b}30 wt% MAA; ^{c}15 wt% MAA; ^{d}Formal value at 1 atm; ^{e}30°C; ^{f}30°C; ^{g}25°C; ^{h}−15°C; ^{i}15°C.

larger steric effect arising out of the disubstitution in the former group of monomers and in the corresponding radicals.

3.10.4 Termination rate constants

Values of k_t for some monomers as determined by the SP-PLP method are given in Table 3.10. These values are obtained at low conversions where translational diffusion-control has not set in (*vide infra*). Also included in the table are the values of activation energy and activation volume. The former, however, does not necessarily represent the energy barrier of termination. For instance, it has been shown in the polymerization of styrene that both

Table 3.10 Kinetic parameters of termination.[233]

Monomer	k_t^a(40°C) L mol^{-1}s^{-1}	E_t kJ/mol	$\Delta V^{\#}$ mL/mol
Styrene	5.0×10^7	8.9 ± 3.2	14 ± 2.6
Methyl methacrylate	2.5×10^7	5.6 ± 2.6	15 ± 5
Butyl methacrylate	5.0×10^6		18
Dodecyl methacrylate	1.6×10^6		10.8
Methyl acrylate	1.1×10^8	9 ± 6	20 ± 6
Dodecyl acrylate	2.5×10^6	4 ± 5	21 ± 6

^{a}at 1000 bar

the activation energy and the activation volume are similar to those of the fluidity of styrene monomer ($\Delta V^{\#} = 14.6$ mL/mol and $E_A = 9.9$ kJ/mol).[233] This is what would be expected if the diffusion of polymer segments were rate-controlling at low conversions.

3.10.5 Variation of k_t with monomer conversion

Figure 3.9 shows the pattern of variation of k_t with conversion in the bulk polymerization of methyl methacrylate at different initiator concentrations and 0°C determined by rotating sector technique and after-effect studies (*vide supra*).[242] For the initial approximately 10–20% conversion depending on initiator concentration the change in k_t is relatively small. A drop in k_t by about 3 orders of magnitude occurs thereafter as the conversion increases to about 50%. Then a region of slow decrease follows to about 70% conversion, above which no experimental data are available. Similar pattern is observed also with k_t obtained from SP-PLP experiments.[233]

In Chap. 2, Sec. 2.1, we have explained why termination of polymer radicals becomes translational diffusion–controlled after the viscosity of the polymerization medium exceeds certain threshold. The initial plateau region of the k_t vs. conversion curve is due to segmental diffusion controlling the rate.[243,244] At the end of the plateau region, the viscosity of the medium becomes sufficiently high to slow down the translational diffusion and k_t starts to be controlled by it. This control continues until at sufficiently high conversion (~70%) the translational diffusion ceases. At still higher conversions, where there are no experimental data available, the radicals are brought close together only by the growth of the polymer chains through propagation.[245] The process is termed 'reaction diffusion'. In addition, theoretical treatment suggests that above about 80% conversion k_p can be diffusion-controlled too so that it drops as well.[233]

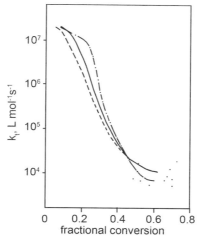

Figure 3.9 Variation of k_t with conversion in the polymerization of methyl methacrylate at ambient pressure and 0°C. "Reprinted with permission from Ref. 242. Copyright © 1988 American Chemical Society."

3.10.6 Chain length dependence of k_p and k_t

In view of the diffusion-controlled nature of termination k_t would be expected to be chain-length dependent. The effect of chain-length dependence (CLD) of k_t in SP-PLP experiments has been discussed already. However, the CLD may be expressed as the power law equation (3.82) in which the exponent α was determined to be about 0.2 at low conversions in the polymerization of styrene and methyl methacrylate.[246,247]

$$k_t^{ii} = k_t^{11} i^{-\alpha}. \tag{3.82}$$

Regarding k_p, the first few monomer additions are usually significantly faster than the rest. For example, in the polymerization of methyl methacrylate, the rate constants for the first-three monomer additions at 60°C are respectively 14 000, 3600, and 843 compared to the average k_p value of 820 L mol^{-1} s^{-1}.[248] Similarly, in the polymerization of methacrylonitrile, k_p for the first monomer addition and the average k_p are 340 and 55 L mol^{-1} s^{-1} respectively. However, this early chain-length dependence of k_p is not universally observed, methyl acrylate being one example.[249] However, in general, the rate constants of the addition of small model radicals to monomers are greater than k_p by more than an order of magnitude.[250]

3.10.7 Temperature dependence of rate of polymerization and molecular weight

From Eq. (3.20) the overall activation energy of polymerization is given by

$$E_{poly} = E_p + (E_i - E_t)/2, \tag{3.83}$$

where E_p, E_t and E_i are the activation energies of propagation, termination, and initiation respectively.

In the polymerization of a typical monomer, e.g., styrene, it follows from the values of E_p and E_t given respectively in Tables 3.9 and 3.10 that $E_p - 1/2E_t$ is approximately 28 kJ/mol. However, E_i changes with the type of initiator, being approximately 130, 50, and 0 kJ/mol for thermal, redox, and photoinitiators respectively. The corresponding E_{poly} values are, therefore, 93, 53, and 28 kJ/mol. Accordingly, the rate of polymerization would increase by approximately 2.5, 2, and 1.5 times per 10°C rise in temperature for thermal, redox, and photo initiated polymerizations respectively.

For the effect of temperature on molecular weight in the absence of chain transfer, it may be shown easily using Eq. (3.37) that the activation energy for $\overline{DP}_n^\upsilon$ (E_{DP}) equals $(E_p - 1/2E_t) - 1/2E_i$. With a thermal initiator, $1/2 E_i$ is larger than $(E_p - \frac{1}{2}E_t)$, as noted above. Hence, $\overline{DP}_n$ would decrease with increase in temperature. For redox initiation, $E_{DP} \approx 0$, hence, $\overline{DP}_n^\upsilon$ would be almost insensitive to change in temperature. For photoinitiation, E_{DP} is +ve, hence, $\overline{DP}_n^\upsilon$ would increase with increase in temperature.

The effect of temperature on transfer-controlled molecular weight depends on the variation of C_T with temperature. Thus, with transfer agents of low transfer constant ($C_T \ll 1$), E_{tr} is larger than E_p to the extent of about 10 to 60 kJ/mol, the value in the low side of the range applies to transfer agents with relatively high C_T.[11] Accordingly, molecular weight would decrease with increase in temperature. However, with transfer agents of high transfer constant ($C_T > 1$), *i.e.*, with regulators, E_{tr} may be smaller than E_p.[145] Accordingly, molecular weight would increase with temperature.

3.11 The Course of Polymerization and Gel Effect

Polymerization of monomers in bulk or in concentrated solution exhibits autoacceleration at some stage of polymerization, which varies with the monomer, solvent concentration, and, to some extent, the amount of initiator, and polymerization temperature.[45,242] In Fig. 3.10 are reproduced the monomer conversion curves in the photopolymerization of undiluted methyl methacrylate at 0°C using different initiator concentrations. At low conversions of monomer to polymer, below about 12%, polymerization progresses slowly and with nearly constant rate; it then undergoes autoacceleration, and becomes nearly explosive soon after. The phenomenon is called the 'gel effect'. The molecular weight of polymer increases sharply with the acceleration in rate, although not in proportion to the latter.[45,242]

The gel effect was attributed by Norrish and Smith[251] as well as Trommsdorff[252] to the marked decrease in the rate of termination occurring due to the viscosity of the medium becoming high. In effect, they considered k_t translation-diffusion-controlled (Sec. 3.10.5). Propagation is considered unaffected since unlike termination it is reaction-controlled. Besides, one of the reactants, *i.e.*, the monomer, being a small molecule, diffuses at fast enough rate. When a diluted monomer is polymerized, gel effect occurs at higher conversions; however, at sufficiently high dilution the effect may not be seen at all due to the viscosity not being high enough even as the polymerization is completed.[45] Similarly, using powerful chain transfer agents, the molecular weight may be so reduced as to prevent the gel effect.

The pattern of variation of k_t with monomer conversion, as determined experimentally and discussed above, shows that translational-diffusion-control in k_t starts at monomer conversion somewhat above 10% in the polymerization of undiluted methyl methacrylate at 0°C (Fig. 3.9), which is in fair agreement with the approximate conversion at which autoacceleration starts (Fig. 3.10).

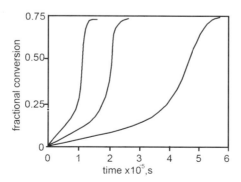

Figure 3.10 The course of the bulk photo polymerization of methyl methacrylate at 0°C at different photoinitiator concentrations. "Reprinted with permission from Ref. 242. Copyright © 1988 American Chemical Society."

Gel effect is observed also in the polymerizations of other monomers. With undiluted styrene[253] and vinyl acetate,[11] it is much less strong than with undiluted methyl methacrylate. This is in agreement with the onset of translational-diffusion-control on k_t in styrene polymerization occurring at about 30% conversion compared to about 10% in methyl methacrylate.[233] In the polymerizations of undiluted methyl acrylate or (meth) acrylic acids, autoacceleration may become so severe that explosive polymerization may occur.

An important aspect not related to gel effect is the stoppage of polymerization before being completed. For example, polymerization of undiluted methyl methacrylate stops at about 73% conversion (Fig. 3.10). This phenomenon is attributed to the polymerized mass containing the residual monomer becoming glassy at the polymerization temperature. In the glass, the monomer molecules become frozen leading to the stoppage of polymerization. Use of a solvent would decrease the glass temperature and prevent incomplete polymerization. Increase of polymerization temperature to above T_g also would have the same effect.

3.12 Popcorn Polymerization

Under certain conditions, usually under extremely slow initiation, some monomers give rise to polymers, which have a popcorn or cauliflower like appearance.[254–256] Polymerization starts after long induction periods with the formation of small nodules of insoluble polymer, which once formed proliferate rapidly until all the monomer is consumed. Large increase in volume occurs since the density of the polymer is very low due to the presence of many voids. This poses a potential hazard to monomer storage since the popcorn growth may cause bursting of storage vessels.[255]

The polymer is only lightly crosslinked, yet it swells very little in solvents. Monomers such as diolefins or mixtures of mono- and diolefins, which on polymerization yield crosslinked polymers exhibit popcorn polymerization more often.[255,256] However, it is also encountered with some monoolefins, *e.g.*, methyl acrylate and acrylic acid, under suitable conditions. Thus, highly purified methyl acrylate yields popcorn polymer when it is polymerized in bulk at low temperature (*ca.*, 18°C) using a very low initiator concentration *ca.*, 10^{-4} mol/L benzoyl peroxide. At higher initiator concentrations, normal polymerization occurs. Acrylic acid also forms popcorn polymer when polymerized in aqueous solution, (*ca.*, 0.4 mol fraction of monomer) using low initiator concentrations, but not in bulk.[257]

The popcorn structure is believed to arise through the formation of the crosslinked network polymer on which branched chains grow through polymer transfer. In the case of monoenes, the network structure may develop through coupling of branch radicals,[157] as has been discussed in Sec. 3.8.1.6. It has also been suggested that physical crosslinking occurring through extensive chain entanglement between the extremely long polymer molecules may be responsible.

3.13 Dead End Polymerization

Dead end polymerization refers to polymerization, which is stopped short of completion due to the exhaustion of initiator. It often occurs with polymerizations initiated by redox initiators since the latter undergo fast decay.

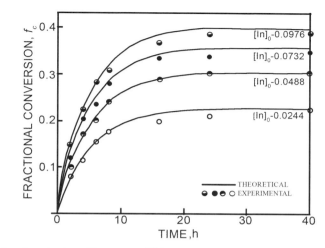

Figure 3.11 Bulk polymerization of isoprene at 60°C using AIBN initiator (concentrations in mol/L shown on curves). The points are experimental; the drawn curves are theoretical. "Reprinted with permission from Ref. 259. Copyright © 1962 John Wiley & Sons Inc."

However, even when a thermal initiator decaying at a slower rate is used, polymerization may reach dead end with a monomer, which polymerizes slowly due to a low $k_p/(2k_t)^{1/2}$ value. An example of the latter type is the polymerization of isoprene using AIBN initiator. Figure 3.11 shows the conversion-time curves in this polymerization. Incomplete conversion is evident from the curves. The case was mathematically treated by Tobolsky, providing a method for the determination of R_i.[258,259]

3.14 Molecular Weight Distribution

The molecular weight distribution in radically prepared polymers has been deduced theoretically using statistical methods for two different types of chain breaking reactions.[260,261] One of these involves chain transfer and/or termination by disproportionation, which yields a molecule of the same chain length as that of the chain radical. The other involves termination by combination, which yields a molecule of chain length equal to the sum of the chain lengths of the two coupled chain radicals.

In the first case, the probability (p) that a chain radical undergoes a propagation step rather than a growth interruption by transfer and termination by disproportionation is given by

$$p = \frac{R_p}{R_p + R_{tr} + R_{td}} = \frac{k_p[M]}{k_p[M] + k_{tr}[T] + 2k_{td}[P^\bullet]}. \tag{3.84}$$

However, if one or the other chain breaking reaction is absent, the corresponding term in the denominator is to be omitted.

The rate constants in Eq. (3.84) are essentially independent of the chain length of the radical as discussed in Sec. 3.2, so should be p. Thus, the probability that a primary or transfer radical undergoes $r - 1$ successive propagation steps without undergoing transfer and/or termination by disproportionation is given by the product of individual probabilities of $r-1$ independent propagation steps, i.e., p^{r-1}. Now, the probability that a chain molecule containing exactly r units (counting the primary or transfer radical starting a chain as a unit) has been formed is given by

$$p_r = p^{r-1}(1-p). \tag{3.85}$$

However, ordinarily, p does not remain constant since both [M] and [T] decrease with increase in conversion even as [P$^\bullet$] and the various rate constants remain constant in a stationary state polymerization. Thus, Eqs. (3.84) and (3.85) may be used only in low conversion polymerization or over a small range of higher conversions.[261]

Now, the number average degree of polymerization is given by the number of monomer molecules polymerized per chain breaking event (with the primary or transfer radical providing a unit).

$$\overline{DP}_n = \bar{r}_n = 1/1-p, \tag{3.86}$$

If the number of polymerized units in the increment of polymer formed within a small range of conversions is N_0, the number of polymer molecules in that increment is given by

$$N = N_0(1-p). \tag{3.87}$$

Equations (3.85) through (3.87) are equivalent to the corresponding equations in linear step polymerization where p is equal to the extent of reaction (p). Thus, the most probable molecular weight distribution, which holds in linear step polymerization (Chap. 2, Sec. 2.5), should also hold here giving

$$n_r = (1-p)p^{r-1}, \tag{3.88}$$

and

$$w_r = r(1-p)^2 p^{r-1}. \tag{3.89}$$

In addition, as discussed in Chapter 2, the distribution gives rise to

$$\overline{DP}_n = \bar{r}_n = 1/(1-p), \tag{3.90}$$

$$\overline{DP}_w = \bar{r}_w = (1+p)/(1-p), \tag{3.91}$$

and

$$PDI = 1 + p. \tag{3.92}$$

A high polymer is formed when $p \to 1$ and PDI has an upper limiting value of 2.

However, chain transfer and bimolecular chain termination are not the only chain breaking reactions, which give rise to the most probable molecular weight distribution. Unimolecular chain termination also would yield the same distribution.

A different distribution arises when chain transfer is absent and termination occurs entirely by combination. In such case, the chain length of the polymer is given by the sum of the chain lengths of two independently grown chain radicals that couple together to form the polymer. However, since the probability of coupling is independent of radical size, the molecular weight distribution is narrower than the most probable one. In this respect, the polymer resembles the dichain step polymer, which is prepared by complete polymerization of an A-B monomer with small quantity of a RA_2 monomer that links two independently grown A-B polymer chains.

In the present case, p is defined as the probability of a chain radical undergoing a propagation step rather than termination by coupling

$$p = \frac{R_p}{R_p + R_{tc}} = \frac{k_p[M]}{k_p[M] + 2k_{tc}[P^\bullet]}. \tag{3.93}$$

The number average degree of polymerization is given by

$$\overline{DP}_n = \bar{r}_n = 2/(1-p) = 2 + k_p[M]/k_{tc}[P^\bullet], \tag{3.94}$$

where $\bar{r}_n$ counts also two initiator derived end groups as two units. Thus, a polymer chain of degree of polymerization r is formed by coupling two chain radicals, which together have undergone $r - 2$ propagation steps starting from two primary radicals. However, the two coupling radicals may individually go through $x - 1$ and $y - 1$ propagation steps maintaining $x + y = r$.

The probability that a radical has grown through $x - 1$ propagation steps and the growth is limited to exactly x units is given by

$$p_x = p^{x-1}(1-p), \tag{3.95}$$

where p is defined by Eq. (3.93).

Similarly, for the other radical which has grown through $y - 1$ propagation steps and the growth is limited to exactly y units, the probability is

$$p_y = p^{y-1}(1-p). \tag{3.96}$$

Now the probability that the two radicals have coupled to form the r-mer is given by

$$p_{x+y} = p_x \cdot p_y = p^{x+y-2}(1-p)^2 = p^{r-2}(1-p)^2. \tag{3.97}$$

Since x and y may both vary from 1 to $r - 1$ holding $x + y = r$, there can be $r - 1$ pairs of chains that give rise to r-mers. The total probability for the formation of an r-mer is then

$$P_r = P_{x+y} = \sum p_{x+y} = (r-1)p^{r-2}(1-p)^2. \tag{3.98}$$

However, P_r equals the mol fraction x_r (*vide* Chap. 2, Sec. 2.5.1). Thus, the mol fraction distribution follows as

$$x_r = (r-1)p^{r-2}(1-p)^2. \tag{3.99}$$

Representing the number of monomer units in the increment of polymer formed within a small range of conversion as N_0, the number of polymer molecules in that increment is given by

$$N = N_0/\bar{r}_n. \tag{3.100}$$

Substituting Eq. (3.94) for $\bar{r}_n$ in Eq. (3.100) one gets

$$N = \frac{N_0(1-p)}{2}. \tag{3.101}$$

The number of r-mers in the increment of polymer is obtained by multiplying Eqs. (3.99) with (3.101)

$$N_r = Nx_r = \frac{N_0}{2}(1-p)^3(r-1)p^{r-2}. \tag{3.102}$$

The weight fraction of r-mer is given by

$$w_r = \frac{rN_r}{N_0}. \tag{3.103}$$

Substituting Eqs. (3.102) for N_r in Eq. (3.103) one gets

$$w_r = \frac{r}{2}(r-1)p^{r-2}(1-p)^3. \tag{3.104}$$

Proceeding as with the most probable distribution (*vide* Chap. 2) the number- and weight-average degrees of polymerization and the polydispersity follow as

$$\overline{DP}_n = \bar{r}_n = \frac{2}{1-p}. \tag{3.105}$$

$$\overline{DP}_w = \bar{r}_w = \frac{2+p}{1-p}. \tag{3.106}$$

$$PDI = \frac{2+p}{2}. \tag{3.107}$$

From Eq. (3.107), it follows that PDI has an upper limit of 1.5 inasmuch as the upper limit of p is 1. Thus, termination by combination leads to a narrower distribution than that obtained when chain growth is interrupted either by transfer and/or by termination by disproportionation for which the upper limit of PDI is 2.

In a real polymerization, chain breaking takes place by all three processes, *viz.*, chain transfer and termination by disproportionation as well as combination. Accordingly, the corresponding distributions may be considered to coexist. However, to all of these the same probability parameter p defined by Eq. (3.108) applies[261]

$$p = k_p[M]/(k_p[M] + k_{tr}[T] + 2k_t[M^\bullet]). \tag{3.108}$$

The resultant distribution is given by the weighted average of the component distributions.

3.15 Living Radical Polymerization (LRP)

As already introduced in Chap. 1, living polymerization is defined as a chain polymerization, which is free from chain termination and irreversible chain transfer.[262] Given the propensity of mutual bimolecular termination of radicals, it was inconceivable that living radical polymerization (LRP) could ever be achieved. However, in the mid 1990s, several radical polymerization methods were discovered where termination and irreversible chain transfer are overwhelmed by reversible deactivation through either reversible termination or reversible (degenerative) chain transfer (Scheme 3.5) yielding dormant polymers that constitute about 90 to 99 mol % of the total polymers and have all the attributes of living polymers.[263–270] For example, they can be chain-extended with their own or different monomers and have defined chain ends, predicted molecular weights, and narrow molecular weight distributions.

However, due to the inevitable occurrence of chain termination and chain transfer (howsoever small these may be), the polymerization does not conform to the definition of

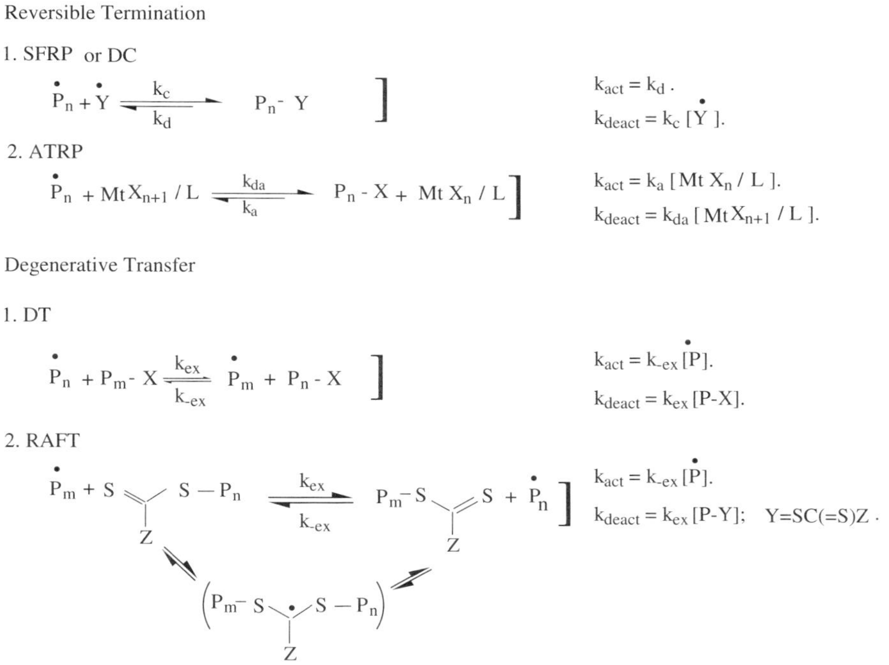

Reversible Termination

1. SFRP or DC

$$\overset{\bullet}{P_n} + \overset{\bullet}{Y} \underset{k_d}{\overset{k_c}{\rightleftharpoons}} P_n-Y \quad]$$

$k_{act} = k_d$.
$k_{deact} = k_c [\overset{\bullet}{Y}]$.

2. ATRP

$$\overset{\bullet}{P_n} + MtX_{n+1}/L \underset{k_a}{\overset{k_{da}}{\rightleftharpoons}} P_n-X + MtX_n/L \quad]$$

$k_{act} = k_a [MtX_n/L]$.
$k_{deact} = k_{da} [MtX_{n+1}/L]$.

Degenerative Transfer

1. DT

$$\overset{\bullet}{P_n} + P_m-X \underset{k_{-ex}}{\overset{k_{ex}}{\rightleftharpoons}} \overset{\bullet}{P_m} + P_n-X \quad]$$

$k_{act} = k_{-ex} [\overset{\bullet}{P}]$.
$k_{deact} = k_{ex} [P-X]$.

2. RAFT

$$\overset{\bullet}{P_m} + S=C(Z)-S-P_n \underset{k_{-ex}}{\overset{k_{ex}}{\rightleftharpoons}} P_m-S-C(Z)=S + \overset{\bullet}{P_n} \quad]$$

$$\rightleftharpoons (P_m-S-C(Z)(\bullet)-S-P_n)$$

$k_{act} = k_{-ex} [\overset{\bullet}{P}]$.
$k_{deact} = k_{ex} [P-Y]$; Y=SC(=S)Z .

Scheme 3.5 Various methods of reversible deactivation.

living polymerization, as given above. Hence, the term "controlled radical polymerization" was suggested.[271] However, both the terms "living radical polymerization" (LRP) and "controlled radical polymerization" are used in the polymer literature interchangeably or even together. The discoveries rejuvenated the field of radical polymerization, which was reaching a dead end. A large number of vinyl monomers, which were not amenable to already-existing living ionic polymerizations, became polymerizable in a living manner with the experimental ease that goes with radical polymerization.

As shown in Scheme 3.5, the reversible termination methods of deactivation are of two types. One of these operating in the stable free radical polymerization (SFRP) involves dissociation-combination (DC) equilibrium between a chain radical and a stable radical (Y•). The latter may be a nitroxide radical[263] or a metal-centered radical, e.g., a cobalt (II) metalloradical.[264] The specific polymerizations are known by the names of the stable radicals, e.g., 'nitroxide mediated polymerization' (NMP) for the former and 'cobalt mediated polymerization' (CMP) for the latter. The other involves reversible halide transfer oxidation of a chain radical by a transition metal halide complex (Mt X_{n+1}/L). The chain radical is oxidized to form the dormant polymer halide, while the metal is reduced to an oxidation state, one unit lower. Since atom transfer is involved in the deactivation process, the polymerization is referred to as 'atom transfer radical polymerization' (ATRP).[267,268]

Degenerative transfer has been defined as a "*chain-transfer* reaction that generates a new *chain carrier* and a new *chain-transfer agent* with the same reactivity as the original *chain carrier* and *chain-transfer agent*".[262] It is also of two types as will be found in

Scheme 3.5. One type, *viz.*, DT, involves reversible transfer of an atom or a group of atoms (X) such as I, TeR, BiR$_2$, SbR$_2$, and ASR$_2$, from the chain end of a dormant polymer to a chain radical.[269,280–283] The other type, *viz.*, RAFT, involves reversible addition of a chain radical to the S end of the C=S group of a polymeric thiocarbonylthio compound. This is followed by reversible fragmentation of the resulting adduct radical (shown in parenthesis) generating a new chain radical and a new polymeric thiocarbonylthio compound, both having the same reactivity as their original counterparts (*vide infra*). This variant of DT is referred to specifically as 'reversible addition–fragmentation chain transfer' (RAFT).[278,279] Of all the methods introduced above, NMP, ATRP, and RAFT have been widely studied and we shall confine our discussion principally to these three methods.

Also shown in the scheme in the right is the rate constant of deactivation, k_{deact}, which is pseudo first order, and the rate constant of activation, k_{act}, which is first order or pseudo first order according as activation is unimoleular or bimolecular respectively.

Deactivation involves a small concentration (*ca.*, 10^{-8} to 10^{-7} mol/L) of active chains existing in dynamic equilibrium with a large concentration (*ca.*, 10^{-3} to 10^{-2} mol/L) of dormant chains,[272–279] the former being of the same order as in a conventional radical polymerization (CRP). In fact, the concentration of the active chains would be the same and the dead polymer chains would be formed in comparable numbers in the two polymerizations, LRP and CRP, proceeding at the same rate. However, in the living system, the number fraction of the dead chains would be reduced to about 0.01 to 0.1 as they would be swamped by dormant chains with molecular weights only a few hundredths as large as those of the dead chains that would be formed in the conventional polymerization. In the living system, the chains grow intermittently through a large number of activation-deactivation cycles during the whole time of polymerization but the average total time of growth of any chain is much less than the average time of growth of a chain in the conventional system. The latter is of the order of a second, being equal to $(2k_t[P^\bullet])^{-1}$.

The first attempted living radical polymerization is due to Otsu and coworkers who used iniferters for the purpose.[284,285] These are compounds, which act in all three capacities, *viz.*, initiator, transfer agent, and terminator, in a chain polymerization. The iniferters ensure the initiator fragments to be present at both ends of the polymer molecule. With suitably chosen iniferters, the polymer may act as a macroinitiator. For example, using benzyl N, N-diethyldithiocarbamate (BDC) as the iniferter in the photo polymerization of styrene, the obtained polymer is capable of polymerizing styrene resulting in chain extension.

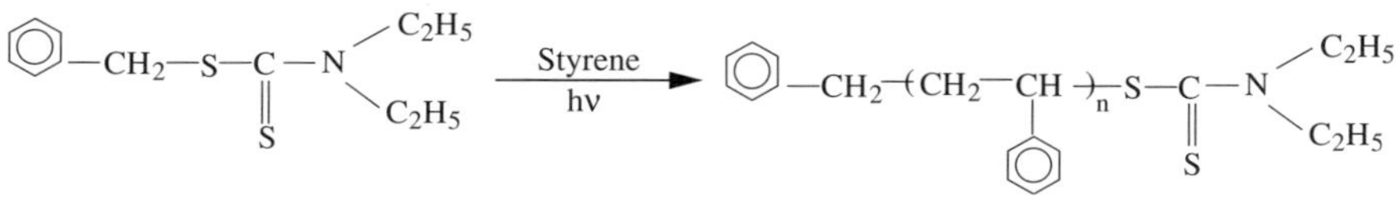

$$\text{Ph}-CH_2-S-\underset{\underset{S}{\|}}{C}-N\begin{array}{c}C_2H_5\\C_2H_5\end{array} \xrightarrow[h\nu]{\text{Styrene}} \text{Ph}-CH_2-(CH_2-\underset{\text{Ph}}{CH})_n-S-\underset{\underset{S}{\|}}{C}-N\begin{array}{c}C_2H_5\\C_2H_5\end{array}$$

The mechanism proposed did indeed consider reversible deactivation between radical chain growths[285]

$$R-Y \rightleftharpoons \overset{\bullet}{R} + \overset{\bullet}{Y} \xrightarrow{nM} P_n-Y \rightleftharpoons \overset{\bullet}{P}_n + \overset{\bullet}{Y} \xrightarrow{mM} P_{n+m}-Y \rightleftharpoons \overset{\bullet}{P}_{n+m} + \overset{\bullet}{Y}$$

and so on, where $\overset{\bullet}{R} = C_6H_5 \overset{\bullet}{C}H_2$ and $\overset{\bullet}{Y} = \overset{\bullet}{S} - C (=S) N (C_2H_5)_2$.

However, these initial attempts fell far short of the target. For example, the polydispersities of the polymers were high (PDI = 1.7 − 5.2) and the block copolymers (both AB and ABA) had large contamination of both homopolymers A and B.

The lack of success in Otsu's work has been attributed to the loss of the dithiocarbamate radical in self-coupling and initiation of polymerization, which lowers the selectivity of the formation of the dithiocarbamate ended dormant polymer (*vide* Sec. 3.15.1). Besides, the iniferter is not a suitable RAFT agent either due to the slowness of addition of the polystyrene radical to the dithiocarbamate group (*vide* Sec. 3.15.7.1).[286]

3.15.1 *Persistent radical effect in reversible deactivation*

Reversible termination type of deactivation is governed by the persistent radical effect discussed below. Consider a living radical polymerization controlled by reversible termination by a stable radical, which involves dissociation-combination equilibrium. The stable radical is referred to as the persistent radical. The kinetic scheme for polymerization may be written as shown in Scheme 3.6.

In order to avoid unnecessary complexity in the discussion a low molar mass analog of the dormant polymer has been chosen as the initiator so that $P_0^\bullet$ is deemed to be kinetically

Initiation

$$P_0-Y \underset{k_c^o}{\overset{k_d^o}{\rightleftarrows}} \dot{P}_0 + \dot{Y} \qquad K^o = k_d^o/k_c^o \qquad (3.109)$$

$$\dot{P}_0 + M \xrightarrow{k_i} \dot{P}_1 \qquad (3.110)$$

Propagation

$$\left.\begin{aligned}\dot{P}_1 + M &\xrightarrow{k_p} \dot{P}_2 \\ \dot{P}_2 + M &\xrightarrow{k_p} \dot{P}_3 \\ - - - - - - - - - - \\ \dot{P}_n + M &\xrightarrow{k_p} \dot{P}_{n+1}\end{aligned}\right\} \qquad (3.111)$$

Reversible Deactivation

$$\dot{P}_n + \dot{Y} \underset{k_d}{\overset{k_c}{\rightleftarrows}} P_n-Y \qquad K = \frac{k_d}{k_c} \qquad (3.112)$$

Termination

$$\dot{P}_n + \dot{P}_m \xrightarrow{k_t} \text{Dead Polymer} \qquad (3.113)$$

Side Reactions

$$\dot{Y} + \dot{Y} \xrightarrow{\quad\not\quad} \text{Waste Products} \qquad (3.114)$$

$$\dot{Y} + M \xrightarrow{\quad\not\quad} \dot{P}_1 \qquad (3.115)$$

Scheme 3.6 Kinetic scheme for living radical polymerization involving dissociation-combination type of reversible termination.

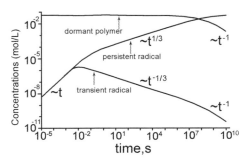

Figure 3.12 Evolution of the transient and persistent radicals and dormant chains determined by numerically analyzing the kinetics according to the Scheme 3.6. Parameters used are; $k_d^0 = k_d = 10^{-2}\,s^{-1}$, $k_c^0 = k_c = 10^7\,L\,mol^{-1}\,s^{-1}$, $k_t = k_t^0 = 10^8\,L\,mol^{-1}\,s^{-1}$, $k_p = k_i$, $I_0 = [P-Y]_0 = 5 \times 10^{-2}\,mol\,L^{-1}$. Equilibrium constant $K = k_d/k_c = 10^{-9}\,mol\,L^{-1}$. The time dependence of the evolution of [P•] and [Y•] in the various zones is shown on the respective curves. "Reprinted with permission from Ref. 272. Copyright © 1999 John Wiley & Sons, Inc."

equivalent to $P_n^\bullet$. This allows denoting the transient radical as P• without distinction as to its chain length. Y• is a persistent radical; it can neither self-terminate nor initiate. It disappears only by coupling with the transient radical. Although not a radical, the transition metal halide in a higher oxidation state is referred to as the persistent radical in ATRP since it is capable of reversibly deactivating the transient radical forming the dormant polymer P-X (Scheme 3.5).

The evolution of P• and Y• determined by numerically analyzing the kinetics according to the Scheme 3.6 is shown in Fig. 3.12 with monomer omitted.[272] However, the result should be the same with monomer present if $k_d = k_d^0$, $k_c = k_c^0$, $k_p = k_i$, and $k_t = k_t^0$. To start with, both P• and Y• form in equal concentrations, which build up linearly with time from zero, both being formed at the same rate, $k_d[P-Y]$. As their concentrations increase, self-termination of P• occurs in preference to cross-coupling with Y• since k_t is 10 to 1000 times larger than k_c.[289] As a result, within a fraction of a second the growth of concentration of P• is halted and [P•] would have been stationary but for the termination by Y•. Thus, the concentration of P• begins to decrease passing through a transient stationary period. Meanwhile, Y• accumulates due to the absence of self-termination but its rate of accumulation is retarded due to the cross-coupling with P• taking place. Eventually, [Y•] becomes high and [P•] low enough that self-termination of the latter becomes negligible in comparison to cross-coupling.[272]

Soon, the rate of cross-coupling (deactivation) becomes virtually equal to that of the dissociation of the dormant polymer establishing the quasi-equilibrium (3.112). Thus, a certain fraction of P-Y, depending on the relative values of k_c and k_t, is converted to dead polymer resulting in [Y•] ≫ [P•] before the equilibrium is reached. This phenomenon of self-adjustment of the persistent radical concentration and consequent near absolute dominance of cross-coupling is known as the Persistent Radical Effect (PRE).[287,288] It is obvious that if Y• were continually lost in self-termination (reaction 3.114) or reinitiation (reaction 3.115), or disproportionation with P•, the effect would not have been observed.

However, the concentrations of both P• and Y• continue to drift with time, P• being continually lost by the ever-present self-termination and Y• increasing by that amount. Eventually, at infinite time, all P − Y is converted to dead polymer and [Y•] becomes equal to the initial [P₀ − Y] even as [P•] is reduced to zero.

3.15.2 *Kinetic equations for LRP with reversible deactivation*

The rate of monomer disappearance at any time t in a radical polymerization is given by

$$\frac{-d[M]}{dt} = k_p[P^\bullet][M], \tag{3.19}$$

where k_p is the propagation rate constant, [P•] is the chain radical concentration, and [M] is the monomer concentration. In a living polymerization, the concentration of active centers remains constant throughout the course of polymerization. Therefore, integration of Eq. (3.19) between limits, [M] = 0 at t = 0, and [M] = [M] at t = t, gives

$$\ln \frac{[M]_0}{[M]} = k_p[P^\bullet]t. \tag{3.116}$$

Thus, the monomer disappearance follows first order kinetics. In LRP, with the approximation that termination is negligible, we may obtain [P•] from the equilibrium (3.112)

$$[P^\bullet] = \frac{k_d[P-Y]}{k_c[Y^\bullet]}, \tag{3.117}$$

where the symbols are as defined earlier. Substituting Eq. (3.117) for [P•] in Eq. (3.116) gives

$$\ln \frac{[M]_0}{[M]} = k_p \frac{k_d[P-Y]t}{k_c[Y^\bullet]} = k_p \frac{K[P-Y]t}{[Y^\bullet]}. \tag{3.118}$$

Thus, the first order kinetics follows if [P−Y]/[Y•] remains constant during polymerization. This may be approximately achieved by adding the persistent radical into the system at the beginning of polymerization in sufficient amount (*vide infra*). Otherwise, both [Y•] and [P•] and, consequently, [P−Y] change with time. Fischer as well as Fukuda *et al.* independently derived equations for the time variation.[272–274] These equations read as

$$[Y^\bullet] = (6k_t K^2 I_0^2)^{1/3} t^{1/3}, \tag{3.119}$$

and

$$[P^\bullet] = \left(\frac{KI_0}{6k_t}\right)^{1/3} t^{-1/3}, \tag{3.120}$$

where I_0 is the initial initiator (P₀ − Y) concentration. The equations were derived assuming that [P − Y]₀ ≅ [P − Y], *i.e.*, dead polymer molecules constitute only a small fraction (<10%) of the total polymer molecules. Furthermore, Fisher's treatment requires that $K \ll I_0 k_c/16k_t$.[289] However, the equations predict that the persistent radical concentration increases as 1/3 power of time, while the transient radical concentration decays as inverse

1/3rd power of time. Substituting Eq. (3.120) for [P$^\bullet$] in Eq. (3.19) and integrating gives the equation for the rate of polymerization

$$\ln \frac{[M]_0}{[M]} = \frac{3}{2} k_p \left(\frac{KI_0}{6k_t} \right)^{1/3} t^{2/3}. \tag{3.121}$$

The experimentally determined kinetics of monomer disappearance in several nitroxide mediated polymerization agrees well with Eq. (3.121).[272–274]

Fukuda et al., however, showed that when the persistent radical is added initially into the system in an amount given by Eq. (3.122), the first order monomer disappearance rate law (Eq. 3.118) is obeyed with concentrations in the equation changed to their initial values

$$[Y^\bullet]_0 \gg (6k_t K^2 I_0^2 t)^{1/3}. \tag{3.122}$$

They also showed that when K is extremely small, use of a conventional initiator gives rise to the stationary state kinetics (Eq. 3.123), the rate of polymerization being controlled by it and independent of I_0[273,274]

$$\ln([M]_0/[M]) = k_p (R_i/2k_t)^{1/2} t, \tag{3.123}$$

where R_i is the rate of initiation effected by a conventional initiator.

However, a general treatment of the kinetics, which is applicable to all systems including those where the dead polymer fraction is greater than even 10%, was given by Fukuda and Matyjaszewski et al.[289] It gives Eqs. (3.124) and (3.125) respectively for time dependence of the persistent radical concentration and of the monomer concentration.

$$\frac{I_0^2}{I_0 - [Y^\bullet]} + 2I_0 \ln \left(\frac{I_0 - [Y^\bullet]}{I_0} \right) - (I_0 - [Y^\bullet]) = 2k_t K^2 t. \tag{3.124}$$

$$\ln \frac{[M]_0}{[M]} = \frac{k_p}{2k_t K} \left(I_0 \ln \left(\frac{I_0}{I_0 - [Y^\bullet]} \right) - [Y^\bullet] \right). \tag{3.125}$$

3.15.2.1 Gel effect in living radical polymerization

The gel effect in conventional radical polymerization (CRP) discussed in Sec. 3.11 arises because of a sharp decrease in k_t, which occurs as the viscosity of the polymerization medium exceeds certain threshold. In LRP, this effect is not observed even in systems operating with conventional initiators and exhibiting stationary state kinetics as in CRP (Eq. 3.123). This is attributable to viscosity not being high enough due to the chains produced in LRP being 10–100 times shorter in length. The situation is similar to the elimination of gel effect in CRP by lowering chain lengths of polymers using chain transfer agents. However, in LRP operating without the aid of conventional initiation, termination has either no role (Eq. 3.117) or only a weak one (Eq. 3.120) to play in determining the chain radical concentration and, consequently, the rate of polymerization. In the former case, [P$^\bullet$] is determined by the balance between rates of activation and deactivation of which the latter is more likely to be diffusion-controlled in view of its high rate constant ca., 10^7–10^8 L mol^{-1}s^{-1}. However, one of the reacting species, viz., the deactivator, being a low molar mass substance, its mobility is not affected by the increase in viscosity, and, accordingly, k_c is not affected. In the other case, [P•] depends more weakly on k_t than it does in CRP [compare Eq. (3.120) with Eq. (3.18)]. This added with the fact that polymer of much lower DP is formed in LRP than in CRP prevents the occurrence of gel effect.

3.15.3 *Molecular weight*

In living polymerization, the degree of polymerization is given by

$$\overline{DP}_n = \frac{[M]_0}{I_0} \cdot f_c, \qquad (3.126)$$

where $[M]_0$ and I_0 are the initial monomer and initiator concentrations respectively and f_c is the fractional conversion. Thus, molecular weight should increase linearly with conversion. In living radical polymerization, initiation is not instantaneous. The equation is not applicable until initiation is completed. Hence, large positive deviation from the theoretical $\overline{DP}_n$ occurs in the low conversion region, as shown below. Slower is the initiation longer is the positive deviation region.

A typical theoretical $\overline{DP}_n$ evolution curve is shown in Fig. 3.13, which was determined by numerical simulation based on the kinetic Scheme 3.6 using typical values of the various rate constants.[272] In this case, the positive deviation disappears at conversions above *ca.*, 20% with the completion of initiation. The higher molecular weight polymer formed below this conversion is due to polymerization in the pre-equilibrium regime. Figure 3.14 shows the experimentally determined evolution of molecular weights in ATRP along with the MWD curves at 90% conversion.[290] Two initiators were used; one is a monomeric model initiator and the other is a dimeric one. The former is slower initiating as is evident from the broader MWD polymer it produces. The deviation in molecular weights from the theoretical values is eliminated at much lower conversion with the faster initiating dimeric initiator (*vide infra*).

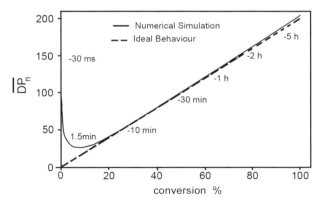

Figure 3.13 Numerical simulation of the evolution of number average degree of polymerization with monomer conversion. Parameters used are $k_d^0 = k_d = 10^{-2}\,\text{s}^{-1}$, $k_c^0 = k_c = 10^7\,\text{L mol}^{-1}\,\text{s}^{-1}$, $k_t = 10^8\,\text{L mol}^{-1}\,\text{s}^{-1}$, $I_0 = [P - Y]_0 = 5 \times 10^{-2}\,\text{mol L}^{-1}$. $[M]_0 = 10\,\text{mol L}^{-1}$ and $k_p = k_i = 5000\,\text{L mol}^{-1}\,\text{s}^{-1}$. Equilibrium constant $K = k_d/k_c = 10^{-9}\,\text{mol L}^{-1}$. "Reprinted with permission from Ref. 272. Copyright © 1999 John Wiley & Sons, Inc."

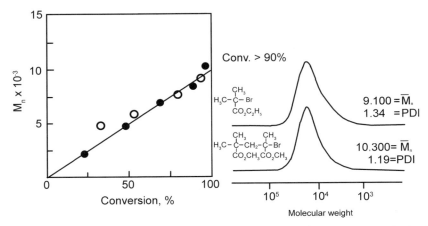

Figure 3.14 Experimentally determined molecular weight evolution and molecular weight distribution (>90% conversion) curves in the ATRP of methyl methacrylate initiated by a monomeric model bromide (open circle) and a dimeric bromide (filled circle) initiator (chemical structures of initiators shown on MWD curves). "Reprinted with permission from Ref. 290. Design of initiators for living radical polymerization of methyl methacrylate mediated by ruthenium (II) complex. Copyright © 1997 Elsevier Science Ltd."

When a conventional initiator is also used, as in the case of NMP with exceedingly low K (Sec. 3.15.4.2), $\overline{DP}_n$ becomes somewhat lower than that predicted by Eq. (3.126) since the total number of chains formed exceeds the number of the alkoxyamine initiator molecules by the number of chains formed by conventional initiation. An analogous situation arises in DT controlled LRP.[278]

3.15.3.1 Molecular weight distribution and polydispersity index

Near monodisperse polymer is obtained in a living polymerization if, in addition to the absence of termination and transfer, initiation is not slower than propagation and the chains get equal opportunity to grow. These conditions ensure that all chains start early in the polymerization and they grow to almost the same lengths. The number or mol fraction distribution of the degree of polymerization in such polymers is given by that of Poisson[261]

$$n_x = \frac{N_x}{N_0} = e^{-\nu}\nu^{x-1}/(x-1)!, \qquad (3.127)$$

where N_x is the number of x-mers with the initiator residue counted as one unit, N_0 is the total number of chains and ν is the number of monomer molecules added per initiator molecule i.e., $\nu = x - 1$. However, the weight fraction distribution is represented by a slightly modified Poisson distribution

$$w_x = (\nu/\nu + 1)[e^{-\nu}x\nu^{x-2}/(x-1)!]. \qquad (3.128)$$

Using these distributions Flory derived Eq. (3.129) for the polydispersity index (PDI).[291]

$$\text{PDI} = \bar{x}_w/\bar{x}_n = 1 + \frac{\nu}{(\nu+1)^2}. \qquad (3.129)$$

$$\approx 1 + \frac{1}{x}. \qquad (3.130)$$

From Eq. (3.130) it follows that polydispersity decreases as the chains become longer. A virtually monodisperse polymer results when $\overline{DP}_n$ is high ca., 100.

The effect of slow initiation on PDI in a termination- and transfer-free polymerization was examined by Gold.[292] A modified Poisson type distribution was derived. It reveals that even if propagation is 10^6 times faster than initiation, PDI lies in the range 1.3–1.4.

However, if the active centers in a chain polymerization exist in dynamic equilibrium between two states differing in reactivity, e.g., an active state and a dormant state as in a living radical polymerization, the sharpness of distribution depends on the frequency of exchange (frequency of deactivation) relative to that of propagation. Referring to the LRP equilibriums presented earlier in Scheme 3.5, an activated chain, on average, remains active for $(k_{deact})^{-1}$ second(s), k_{deact} being the pseudo first order rate constant of deactivation in s^{-1} units. In this period, the chain adds $k_p[M]/k_{deact}$ number of monomer units. More is this number, fewer is the number of activation-deactivation cycles required on a chain to reach a given chain length and lesser is the opportunity for chains to go through equal number of cycles and make even the statistical variation in the lengths of the individual activation periods. Accordingly, broader is the MWD broadening.[293] In such systems, the polydispersity index may be expressed as

$$\text{PDI} = (\overline{M}_w/\overline{M}_n)_{\text{Poisson}} + U_{ex}, \quad (3.131)$$

where U_{ex} is the polydispersity in excess of that given by Poisson distribution.

U_{ex} has been theoretically related to the inverse of the average number of activation-deactivation cycles on a chain as[294–298]

$$U_{ex} = \frac{2\langle n \rangle}{\overline{DP}_n}, \quad (3.132)$$

where $\langle n \rangle$ is the average number of monomer molecules added on a chain per cycle of activation-deactivation averaged over the period of polymerization. Substituting Eq. (3.132) for U_{ex} in Eq. (3.131) one gets

$$\text{PDI} = (\overline{M}_w/\overline{M}_n)_{\text{Poisson}} + 2\langle n \rangle/\overline{DP}_n, \quad (3.133)$$

$$= 1 + \frac{1}{\overline{DP}_n} + \frac{2\langle n \rangle}{\overline{DP}_n}, \quad (3.134)$$

$$\approx 1 + \frac{2\langle n \rangle}{\overline{DP}_n}. \quad (3.135)$$

As already discussed, the average number ($\overline{n}$) of monomer units added to a chain per cycle of activation-deactivation at any given instant is given by

$$\overline{n} = \frac{k_p[M]}{k_{deact}}, \quad (3.136)$$

where [M] is the monomer concentration at that instant. Averaging $\overline{n}$ over monomer concentrations which reduce from $[M]_0$ to $[M]$ in a batch polymerization[295]

$$\langle n \rangle = \frac{\int_{[M]_0}^{[M]} \overline{n} \, d[M]}{\int_{[M]_0}^{[M]} d[M]}. \quad (3.137)$$

Substituting Eq. (3.136) for $\bar{n}$ in (3.137) and integrating one obtains

$$\langle n \rangle = \frac{k_p}{2k_{deact}}([M]_0 + [M]) \qquad (3.138)$$

$$= \frac{k_p[M]_0(2-f_c)}{2k_{deact}}, \qquad (3.139)$$

where f_c is the fractional conversion.

Substituting Eqs. (3.126) and (3.139) for $\overline{DP}_n$ and $\langle n \rangle$ respectively in Eq. (3.135) gives the general PDI equation

$$PDI = 1 + \frac{k_p I_0}{k_{deact}}\left(\frac{2}{f_c} - 1\right). \qquad (3.140)$$

Thus, with initiation completed very early in the polymerization, polydispersity index depends on the ratio of the rate constant of propagation to that of deactivation. Lower is this ratio lower is the PDI. In addition, PDI decreases as conversion increases.

The PDI equations for individual living radical polymerizations follow by substituting the respective expressions from Scheme 3.5 for k_{deact} in Eq. (3.140).

In NMP,

$$PDI = 1 + \left(\frac{2}{f_c} - 1\right)\frac{k_p I_0}{k_c[Y^\bullet]}, \qquad (3.141)$$

where I_0 is the initial concentration of alkoxyamine and $[Y^\bullet]$ is the equilibrium concentration of the free nitroxide radical.

In ATRP,

$$PDI = 1 + \left(\frac{2}{f_c} - 1\right)\frac{k_p I_0}{k_{da}[MtX_{n+1}]}, \qquad (3.142)$$

where I_0 is the initial concentration of the organic halide initiator and $[MtX_{n+1}]$ is the equilibrium concentration of the oxidized metal halide complex.

In DT,

$$PDI = 1 + \left(\frac{2}{f_c} - 1\right)\frac{k_p I_0}{k_{ex}[P-Y]}. \qquad (3.143)$$

Since $[P-Y] \approx I_0$, Eq. (3.143) reduces to

$$PDI = 1 + \left(\frac{2}{f_c} - 1\right)\frac{k_p}{k_{ex}}. \qquad (3.144)$$

3.15.4 Nitroxide mediated polymerization

Nitroxide radicals are inhibitors of radical polymerization (Sec. 3.9). Solomon et al. recognized that the coupling reaction of a nitroxide radical with a carbon-centered radical is reversible at elevated temperatures[299,300]

$$R_2N\dot{O} + \dot{R}' \rightleftharpoons R_2N-O-R'. \qquad (3.145)$$

They used model alkoxyamines (model-nitroxides) in which R' is the monomeric model of the polymer to be prepared for initiating polymerization of certain monomers (mostly acrylates) at 80–100°C and ended up with polydisperse low molecular weight polymers, which were end-capped with nitroxides. Block copolymers were prepared using the latter as macroinitiators. Thus, the polymerization exhibited living character although not quite inasmuch as the polymers were polydisperse. The deficiency is attributable to relatively low temperature and inappropriate combinations of monomers and nitroxides used (*vide infra*).[300]

Several years later, Georges *et al.* succeeded in producing narrow disperse polystyrene using basically the same method. In particular, they polymerized styrene at 120°C using benzoyl peroxide initiator in the presence of the stable radical, 2,2′,6,6′-tetramethyl-1-piperidinyl-1-oxy (TEMPO), which produces the model-nitroxide (S-TEMPO) initiator *in situ*.[263] The molecular weights of the polymers increase linearly with conversion and the polydispersities are lower than 1.3.

Figure 3.15 shows some of the most commonly used nitroxide radicals.[273,274,301,302] Rate constants as well as equilibrium constants of reversible dissociation of some model- or polymer-nitroxides represented by Eq.(3.145) or (3.146) respectively are given in Table 3.11.

$$P-O-NR_2 \underset{k_c}{\overset{k_d}{\rightleftharpoons}} \dot{P} + R_2N\dot{O}, \quad K = k_d/k_c. \tag{3.146}$$

The method of determination of these constants is described later (Sec. 3.15.4.4). K as well as k_d of model- or polymer-nitroxides increases in the order of nitroxide radicals as[303–309]

$$\text{TEMPO} < \text{TIPNO} < \text{DEPN} < \text{DBN},$$

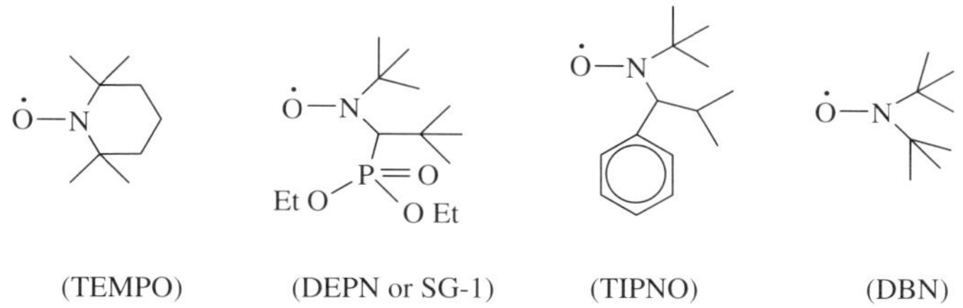

(TEMPO) (DEPN or SG-1) (TIPNO) (DBN)

Figure 3.15 Some examples of stable nitroxide radicals[273,274,301,302]: TEMPO = 2,2,6,6-tetramethyl-1-piperidinyl-1-oxy, DEPN = N-*tert*-butyl-N-[1-diethylphosphono(2,2-dimethylpropyl)]-N-oxy, TIPNO = 2,2,5-trimethyl-4-phenyl-3-azahexane-3-oxy, DBN = di-*tert*-butyl nitroxide.

Table 3.11 Values of rate constants and equilibrium constants for some model- and polymer-nitroxides.

Polymer-nitroxide or Model-nitroxide	T°C	k_d or k_d^0, s^{-1}	k_c, L mol^{-1}s^{-1}	K, mol L^{-1}	Ref.
PS-TEMPO[a]	120	1×10^{-3}	7.7×10^7	2.1×10^{-11}	307, 308
PS-DEPN[a]	120	11×10^{-3}	0.18×10^7	6×10^{-9}	303, 305
PS-DEPN	90			4×10^{-10}	303
PMMA-DEPN	90			1×10^{-7}	304
PBA-DEPN[a]	120	7.1×10^{-3}	4.2×10^7	1.7×10^{-10}	303
PAN-TEMPO[b]	120	0.34×10^{-3}			307
PS-DBN[a]	120	42×10^{-3}			305
S-TIPNO	120	3.3×10^{-3}			309

[a]In monomer; [b]In *t*-butylbenzene.

and in the order of monomers corresponding to the alkyl group as

butylacrylate < acrylonitrle ≈ styrene < methyl methacrylate.

Following the discussion in Sec 3.8.2 the C–N bond dissociation energy (BDE) in alkyl nitroxides having a common nitroxide part should be smaller the larger is the resonance energy of the alkyl radical to be formed and larger is the steric bulk of the substituents on the carbon atom. Besides, the polarity of the bond also influences the BDE. In the present case, electron-withdrawing substituents on the carbon of the C–O bond in polymer-nitroxide decrease the electronegativity difference between the bonded atoms, which makes the bond weaker.[310,5] As a result, the ground state energy is increased, and BDE is decreased. The lower the BDE is the faster the bond dissociation is. The combined effect of reactivity, polarity, and the bulk of the radicals on k_d of alkylnitroxides has been quantitatively treated using a Taft–Ingold approach.

The equation relating k_d with parameters representing the above effects is given by[5]

$$\log k_d = \log k_{d,0} + \rho_{RS}\sigma_{RS} + \rho_u\sigma_u + \delta v, \tag{3.147}$$

where σ_{RS}, σ_u, and v are parameters representing radical stabilization energy, polarity (inductive/field), and bulkiness respectively; the factors ρ_{RS}, ρ_u, and δ represent respectively the sensitivity of the rate constant to these parameters and $k_{d,0}$ is the value of k_d in the absence of the effects. The parameters for various small radicals estimated by Marque et al.[5] are given in Table 3.12.

The k_d values of alkylnitroxides having a common nitroxide (TEMPO or DEPN) part but different alkyl groups some of which are monomeric models of polymers fit well with Eq. (3.147). Figure 3.16 shows the fit obtained for the k_d of alkyl-TEMPO using linear regression analysis.[5] An evaluation of the individual terms comprising the abscissa in Fig. 3.16 reveals that the steric factor plays the dominant role in determining k_d. Principally for this reason, the k_d is the largest for PMMA-nitroxides.

Table 3.12 Radical stability (σ_{RS}), polar (σ_u), and steric (v) parameters of some radicals.[5]

Radical	σ_{RS}	σ_u	v
$CH_3C^\bullet HCH = CH_2$	0.46	0.03	0.86
$C_6H_5C^\bullet(CH_3)_2$	0.36	0.05	1.28
$C_6H_5C^\bullet H\ CH_3$	0.31	0.03	0.70
$CH_3\ C^\bullet(CH_3)CN$	0.22	0.14	1.20
$CH_3\ C^\bullet(CH_3)COOCH_3$	0.20	0.07	1.43
$CH_3\ C^\bullet H\ CN$	0.19	0.17	0.79
$CH_3\ C^\bullet H\ COOCH_3$	0.18	0.09	1.00
$CH_3\ C^\bullet H\ COOH$	0.18	0.09	0.83
$C^\bullet(CH_3)_3$	0.12	−0.01	1.24
$C^\bullet H_3$	0.00	−0.01	0.52

"Reprinted in part with permission from Ref. 5. Copyright © 2005 American Chemical Society."

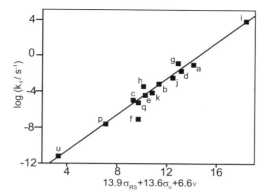

Figure 3.16 Plot of log k_d at 120°C of adducts of various alkyl/model radicals and TEMPO against resonance, polar, and steric parameters according to Eq. (3.147). The alkyl/model radicals are: [a]α-methylstyrene; [b]styrene; [c]benzyl; [d]methyl methacrylate; [e]methyl acrylate; [f]CH_2COOR; [g]methacrylonitrile; [h]acrylonitrile; [i]CH_2CN; [j]butadiene; [k]allyl; [p]cyclohexyl; [q]trimethylmethyl; u, methyl. "Reprinted with permission from Ref. 5. Copyright © 2005 American Chemical Society."

The above discussion considers reversible coupling of chain radical with nitroxide radical as the sole reaction between the two. However, radicals may also undergo disproportionation to various degrees depending on their structures forming dead chains. Radicals with greater number of α hydrogens, such as PMMA•, are prone to disproportionation yielding a macromonomer and a hydroxylamine as shown in reaction (3.148).[311]

$$\text{wwCH}_2-\underset{\underset{\text{COOCH}_3}{|}}{\overset{\overset{\text{CH}_3}{|}}{\text{C}^\bullet}} + \overset{\bullet}{\text{O}}-\text{N}\underset{}{\bigcirc} \longrightarrow \text{wwCH}=\underset{\underset{\text{COOCH}_3}{|}}{\overset{\overset{\text{CH}_3}{|}}{\text{C}}} + \text{HO}-\text{N}\underset{}{\bigcirc} \cdot \quad (3.148)$$

The hydroxylamine acts as an inhibitor by transferring the hydroxyl hydrogen to a chain radical, itself being converted to the corresponding nitroxide radical in the process. Thus, essentially two growing chains become dead.

3.15.4.1 *Selection of initiator*

As discussed earlier, one of the requirements for achieving low polydispersity is that initiation should not be slower than propagation. This ensures that initiation is completed early in the polymerization. Thus, all chains are set for growing from the early stage of polymerization. Referring to the kinetic Scheme 3.6, the requirement is fulfilled when

$$K^0 k_i \geq K k_p. \quad (3.149)$$

With model alkoxyamines the inequality (3.149) is satisfied, since $K^0 \approx K$ and as already discussed, $k_i > k_p$ (Sec. 3.10.6). Hence, they make satisfactory initiators. The alkyl moiety in the model initiator usually represents the hydrogenated monomer *e.g.*, in the TEMPO mediated styrene polymerization the model initiator used is 1-phenylethyl-TEMPO (S-TEMPO, S representing styrene). It may be either preformed or prepared *in situ* at the start of polymerization using a conventional initiator along with TEMPO. For example, in the work of Georges *et al.* benzoyl peroxide (Bz_2O_2) was used as the initiator and TEMPO as the nitroxide radical to conduct NMP of styrene at 123°C.[265] Bz_2O_2 has a half life of about 1.5 min at this temperature.[312] Thus, within a few minutes the formation of the alkoxyamine initiator, which is predominantly S-TEMPO, is completed.

3.15.4.2 *Polymerization of styrene*

TEMPO mediated polymerization of styrene at 120°C occurs very slowly proceeding at the same rate as the thermal uncatalyzed polymerization and independent of S-TEMPO concentration. Refering to Eq. (3.123) this result suggests that the equilibrium constant for reversible dissociation of PS-TEMPO is extremely small, which indeed is found to be true (Table 3.11).[313–315]

$$PS-TEMPO \rightleftharpoons PS^\bullet + TEMPO, \; K = 2 \times 10^{-11} \; mol/L \; at \; 120°C.$$

The rate, however, can be boosted by small amounts of certain additives, *e.g.*, camphorsulfonic acid,[316,317] 2-fluoro-1-methylpyridinium *p*-toluenesulfonate (FMPTS)[318] or acetic anhydride.[319] These additives react with TEMPO and reduce its concentration resulting in the increase in the rate of polymerization. Camphorsulfonic acid additionally prevents the thermal self-polymerization of styrene by destroying the Diels–Alder dimer of styrene (Section 3.7).[118,119] This has the beneficial effect of narrowing the molecular weight distribution. However, the concentration of free TEMPO may be reduced even without additives by using appropriate ratios of conventional initiator to TEMPO. Thus, using [TEMPO]/[Bz_2O_2] = 1.1, it takes 5 h to reach about 70% conversion at 125°C,[320] whereas a ratio of 1.2 takes 45 h to reach 76% conversion at 123°C[265] using about 1.8 times as much Bz_2O_2. Free TEMPO concentration may also be reduced by using a conventional initiator, *e.g.*, dicumyl peroxide (DCP) or t-butyl hydroperoxide (*t*-BHP), each having a long half life. The free radicals produced by these initiators remove some TEMPO by coupling resulting in an increase in the propagating radical concentration and, in turn, the rate of polymerization.[296,321]

Kinetics

Figure 3.17 shows the first order plots for monomer disappearance in TEMPO mediated polymerization of styrene at 120°C using the conventional initiator DCP.[321] The rate increases with the increase in DCP concentration as would be expected from Eq. (3.123).

The polymers are of much lower molecular weights compared to those prepared in the absence of S-TEMPO. In fact, molecular weights are controlled essentially by the concentration of S-TEMPO. However, as DCP concentration increases, molecular weight decreases (Fig. 3.18). This is because the total number of chains increases with the increase in DCP concentration (Sec. 3.15.3).

Polydispersities are low, decreasing with increase in conversion (Fig. 3.19), as would be expected from Eq. (3.141). However, at any given conversion, PDI increases with increase in DCP concentration. This is because the greater rate of radical injection at higher

Radical Polymerization 163

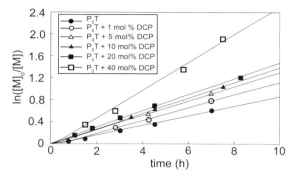

Figure 3.17 First order plots for monomer disappearance in bulk polymerization of styrene at 120°C in the presence of (1-phenylethyl)-TEMPO ([P_0T] = 0.01 mol L^{-1}) and varying amounts of dicumyl peroxide (DCP) as indicated in the inset. The lines are drawn to serve as visual guide. "Reprinted with permission from Ref. 321. Copyright © 1997 John Wiley & Sons, Inc."

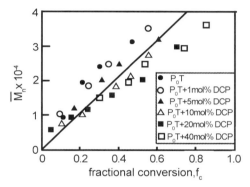

Figure 3.18 Molecular weight evolution with conversion for bulk polymerization of styrene at 120°C in the presence of (1-phenylethyl)-TEMPO (P_0T, 0.01 mol L^{-1}) and varying amounts of dicumyl peroxide (DCP) as indicated in the inset. The line is drawn to serve only as visual guide. "Reprinted with permission from Ref. 321. Copyright © 1997 John Wiley & Sons, Inc."

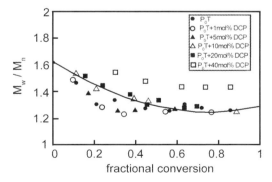

Figure 3.19 Evolution of polydispersity index with conversion for bulk polymerization of styrene at 120°C in the presence of (1-phenylethyl)-TEMPO (P_0T, 0.01 mol L^{-1}) and varying amounts of dicumyl peroxide as indicated in the inset. The line serves as visual guide. "Reprinted with permission from Ref. 321. Copyright © 1997 John Wiley & Sons, Inc."

DCP concentration decreases the free TEMPO (deactivator) concentration but increases the radical concentration, both factors contributing to making dead chain fraction larger.

In contrast to the TEMPO mediated polymerization, which follows stationary state radical polymerization kinetics, that mediated by DEPN at 120°C follows power law kinetics (Eq. 3.121).[303,322] This is due to the three hundred times larger equilibrium constant in the latter system (Table 3.11), which makes thermal initiation unimportant. However, subsequent simulation studies showed that this system does not perfectly follow power law kinetics due to the dead polymer fraction being relatively large (> 0.1). The kinetics, however, conforms, to the general kinetic Eq. (3.125).[289]

However, for such systems better control is achieved with the addition of an appropriate amount of nitroxide radical along with the alkoxyamine, to start with.[303] The high equilibrium constant, of course, offers the advantage of performing DEPN mediated polymerization at significantly lower temperatures ca., 90°C.[323]

3.15.4.3 *Polymerization of (meth)acrylates and other monomers*

Controlled polymerization of butyl acrylate is not successful with TEMPO as the mediator due to the extremely small value of the dissociation–association equilibrium constant.[324] However, it is successfully done at 120°C with DEPN as the mediator primarily due to the relatively large value of the equilibrium constant (Table 3.11).[303] Besides, irreversible termination by disproportionation with the DEPN radical is not significant unlike that with the TEMPO radical. TIPNO is another mediator using which successfully controlled polymerization has been reported even at temperatures lower than 100°C.[325] In fact, TIPNO is considered a universal mediator for NMP.[325,326]

NMPs of methacrylates have not been successful. This is mainly due to the large K of the nitroxides, e.g., PMMA-DEPN (Table 3.11), which results in significant termination.[304,306] Although with TEMPO as the mediator K may not be too large, disproportionation between TEMPO and PMMA radical dominates over their reversible combination.[326] In contrast, with DEPN or TIPNO as the mediator, even though disproportionation does not occur, K is too large for controlled polymerization,. However, good control is achieved by using small amount of a comonomer, ca., <10 mol% styrene, for which K is very low, e.g., K of PS-DEPN is about 250 times smaller than that of PMMA-DEPN at 90°C (Table 3.11). Thus, at equilibrium, the majority of the dormant chains will have styrene as terminal units and the over all K for the system will be close to that in styrene polymerization (*vide* Chap. 8, Sec. 8.3). However, due to cross-propagation equilibrium the ratio of the concentrations of the two types of radicals will be the same as that in a conventional copolymerization. Nevertheless, because of the lowering of the overall K, the equilibrium radical concentration is sharply reduced and controlled polymerization occurs. However, the polymer formed is not a pure PMMA but a copolymer of MMA and styrene.

TIPNO and DEPN based alkoxyamines permit also polymerizations of a wide variety of other monomers besides styrene and acrylates, *e.g.*, acrylamides, 1,3-dienes, and acrylonitrile with accurate control of molecular weights. In some systems, polymers with PDI as low as 1.06 are obtained.[325,326]

3.15.4.4 Rate constants and equilibrium constants

Equilibrium constant, $K = [P^\bullet][Y^\bullet]/[P-Y]$, may be determined by measuring $[Y^\bullet]$ and $[P^\bullet]$ in a polymerization system. $[Y^\bullet]$ may be measured using ESR spectroscopy and $[P^\bullet]$ determined from the rate of polymerization using Eq. (3.19) with k_p obtained from the literature.

For the determination of k_d, Fukuda *et al.* developed a GPC curve resolution method in which a polymer-nitroxide is heated with the corresponding monomer in the presence of different concentrations of a conventional initiator for a predetermined time. During polymerization, some of the polymer molecules are activated, on average, only once; others remain unactivated.[308] For example, from the value of k_d of PS-TEMPO given in Table 3.11, it follows that a PS-TEMPO molecule becomes activated once in about 17 min (value of k_d^{-1}), on average, at 120°C. When PS-TEMPO is heated in styrene with varying concentrations of t-BHP for 10–60 min at 110°C, the resultant polymers show bimodality in MWD. Figure (3.20) shows this for the polymers obtained after 10 min reaction at each t-BHP concentration written on the curves. The low molecular weight component comprises the unactivated polymer, while the high molecular weight component comprises the once-activated polymer and minor amounts of other polymers originated from initiation by t-BHP as well as twice activated polymers. Increase in the t-BHP concentration decreases the free TEMPO concentration resulting in the decrease in the deactivation frequency. As a result, the weight fraction of the high molecular weight component increases.

Similar experiments are conducted by varying the time of polymerization. From the resolution of the bimodal curves into two components, a first order plot for the decrease of

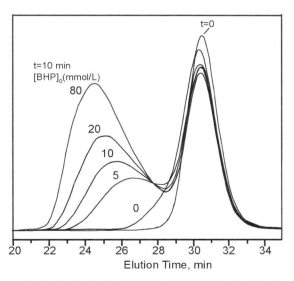

Figure 3.20 GPC traces of the chain extended PS-TEMPO obtained after heating PS-TEMPO in styrene at 110°C for 10 min in the presence of various concentrations of t-butyl hydroperoxide in mmol/L indicated on the curves. "Reprinted with permission from Ref. 308. Copyright © 1997 American Chemical Society."

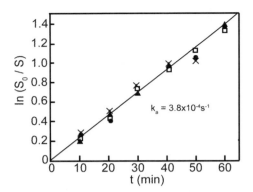

Figure 3.21 First order plot for the consumption of PS-TEMPO used in chain extension experiments at 110°C in the absence or presence of various concentrations of t-butyl hydroperoxide. [PS-TEMPO]$_o$ = 23 mmol L^{-1} and t-BHP concentration ranged from 0 to 80 mmol L^{-1}. "Reprinted with permission from Ref. 308. Copyright © 1997 American Chemical Society."

the starting PS-TEMPO concentration is made. The plot gives a straight line according to the first order rate law (Fig. 3.21)

$$\ln \frac{S_0}{S} = k_{act} t,$$

where S_0 and S are the concentrations of the starting PS-TEMPO at time 0 and t respectively. From the slope of the straight line, k_{act} is evaluated.[308]

Alternatively, k_d^0 is determined by measuring the rate of decomposition of model alkoxyamines under conditions where the transient radical is scavenged as fast as it is formed

$$P_0\text{-}Y \xrightarrow{k_d^0} \dot{P}_0 + \dot{Y} \xrightarrow{scavenger} \text{inert products}.$$

The scavenger may be a different nitroxide, oxygen, or deuterated styrene.[273] A large excess of the scavenger is to be used to prevent recombination of $P_0^{\bullet}$ and $Y^{\bullet}$. From the first order decay rate of $P_0 - Y$ measured by HPLC or NMR, or from the rate of accumulation of $Y^{\bullet}$ measured by ESR, k_d^0 is evaluated.

With K and $k_d^0 (\approx k_d)$ measured as described above, k_c is readily evaluated. Values of these constants for some polymer-nitroxides are given in Table 3.11, which have been already discussed.

3.15.4.5 Synthesis of block copolymers

A nitroxide ended dormant polymer (P$_1$NO) may be used as a macroinitiator for the polymerization of a second monomer (M$_2$) resulting in a block copolymer. However, for this purpose, a more active nitroxide such as DEPN or TIPNO should be used as the mediator, since it allows polymerization at lower temperatures where thermal initiation is negligible and no conventional initiator is needed.[301,302,323,325] Conventional initiation including thermal initiation leads to contamination with homopolymers. Besides, the sequence of monomers should be chosen judiciously so that all the macroinitiator is used up early in the polymerization of M_2. This is assured if

$$K_1 k_i > K_2 k_{p(2)},$$

where K_1 is the equilibrium constant for the dissociation of the macroinitiator, k_i is the rate constant of initiation of M_2 by the macroinitiator radical ($P_1^\bullet$), K_2 is the equilibrium constant for the dissociation of P_2–NO, and $k_{p(2)}$ is the rate constant of propagation in the polymerization of M_2.

Thus, TIPNO mediated block copolymerization of styrene with butyl acrylate successfully occurs when poly(butyl acrylate)-TIPNO is used to initiate polymerization of styrene but not the other way round.[301,302,309,325] In this case, $K_1 < K_2$, but $K_1 k_i > K_2 k_{p(2)}$.

3.15.5 Metalloradical mediated polymerization

Cobalt mediated living radical polymerization has been successful particularly with acrylate monomers.[266,327] The polymerization is initiated with an organometallic derivative of cobalt tetramesitylporphyrin ((TMP)Co–R), which reversibly dissociates at the polymerization temperature (ca., 60°C) to give the transient radical R$^\bullet$ and the persistent metalloradical ((TMP)Co^{II}). The latter mediates living radical polymerization of acrylates.[327] However, the method has failed to attract much interest because of the difficulty and impracticality of preparing the cobalt complex.

An improved and practical method prepares the model initiator *in situ*.[328] It uses a fast decomposing azo initiator, *viz.*, V-70 ($t_{1/2}$ = 11 min at 60°C). The radicals generated at a high concentration react with (TMP)Co^{II} by disproportionation forming cobalt hydride (TMP)Co-H, which reversibly adds to monomer to give 1:1 adduct

$$(TMP)Co-H + CH_2=CHX \longrightarrow (TMP)Co-CH(X)CH_3.$$

The adduct reversibly dissociates to give the transient radical $\overset{\bullet}{C}H(X)CH_3$ and the persistent radical (TMP)Co^{II}. Apart from cobalt metalloradicals, other metalloradicals of transition metals like Mo,[329] Os,[330] and Fe[331] may also be used.

Importantly, cobalt (acetylacetonate)$_2$ derivatives have been used to control polymerizations of unconjugated monomers, like vinyl acetate and N-vinyl pyrrolidone, which are difficult to control by many of the living radical polymerization methods.[332–335]

3.15.6 Atom transfer radical polymerization

ATRP has its roots in atom transfer radical addition (ATRA) long used in organic synthesis of 1:1 adducts of olefins and alkyl halides using transition metal halides as catalysts. The reaction cycle involved in ATRA is presented in Scheme 3.7.[336]

The cycle starts with the generation of a radical R$^\bullet$ through the transfer of a halogen atom from an alkyl halide (RX) to a transition metal halide, *e.g.*, Cu(I), Fe(II), or Ru(II) halide, resulting in the increase in the oxidation state of the metal by one unit. The radical adds to an olefin forming the adduct radical, which is converted to the adduct halide by

$$RX + Mt\ X_n \longrightarrow \dot{R} + Mt\ X_{n+1} \quad (a)$$

$$\dot{R} + CH_2 = CHY \longrightarrow R\ CH_2 - \dot{C}HY \quad (b)$$

$$R\ CH_2\dot{C}HY + Mt\ X_{n+1} \longrightarrow R\ CH_2\ CHY\ X + Mt\ X_n \quad (c)$$

$$2R\ CH_2\dot{C}HY \longrightarrow \text{Non adducts (minor products)} \quad (d)$$

Scheme 3.7 Reaction scheme for ATRA.

reverse transfer of a halogen atom from the oxidized metal halide. In the process, the metal halide in the reduced state is regenerated, which starts a new cycle. Thus, only a catalytic amount of metal halide is required to complete addition. Termination of the adduct radical by the oxidized metal halide occurs in preference to self-termination (persistent radical effect).

Only easily reducible organic halides, *e.g.*, polyhalides, sulfonyl chlorides, and N-haloamines, may be used as the addenda when Cu(I) halides or their complexes with monoamines are used as catalysts.[336–339] However, the use of chelating amine ligands to complex the copper halide makes the addition of difficultly reducible halides also possible.[340] The chelating ligand makes the catalyst more reducing.

The progress from ATRA to ATRP materialized in 1995 through the pioneering works of Sawamoto[265] and Matyjaszewski[266] who independently used ATRA catalysts under appropriate conditions to polymerize methyl methacrylate and styrene respectively obtaining dormant polymers of predicted molecular weight and narrow MWD. While Sawamoto used the $RuCl_2/(PPh_3)_3$ complex as catalyst along with an aluminum-based promoter, Matyjaszewski used CuCl/bipyridine complex, which does not require a promoter to be active. Subsequently, complexes of many other metals have been used such as those of Ti, MO, Re, Mn, Fe, Rh, Ni, Pd, Co, and Os.[275–277,341–345] However, copper(I) complexes are the most versatile and widely used in ATRP. In addition, although, halides (Cl and Br) are the commonly used metal salts, others such as pseudohalides, hexafluorophosphates, and perchlorates may also be used. Proper selection of monomer, initiator, catalyst, and ligand and their ratios, solvent, temperature, *etc.*, are important for successful ATRP.

The mechanism of ATRP is shown in Scheme 3.8 with copper complex as the catalyst.[275–277,341–345] The superscript 0 on the rate constants refers reactions involving the initiator. The mechanism is similar to that of NMP discussed earlier (Scheme 3.6) except that the activation is bimolecular in ATRP but unimolecular in NMP. In both, the deactivation equilibrium is established by the persistent radical effect. Hence, all rate equations (3.118) through (3.122) derived for NMP are applicable to ATRP as well, substituting $([P-X]_0[Mt\ X_n]_0)$ for I_0, *i.e.*, $[P-Y]_0$, and $[Mt\ X_{n+1}]$ for $[Y^\bullet]$ in the equations.

3.15.6.1 *Initiators*

As discussed earlier (Sec. 3.15.4.1), initiation must not be slower than propagation, *i.e.*,

$$K^0 k_i \geq K k_p. \quad (3.150)$$

Scheme 3.8 Reaction scheme for ATRP.

Initiation

$$RX + Cu(I)/L \underset{k_{da}^0}{\overset{k_a^0}{\rightleftharpoons}} \overset{\bullet}{R} + Cu(II) X/L, \qquad K^0 = k_a^0 / k_{da}^0$$

$$\overset{\bullet}{R} + M \xrightarrow{k_i} \overset{\bullet}{P_1}$$

Propagation

$$\overset{\bullet}{P_1} + M \xrightarrow{k_p} \overset{\bullet}{P_2}$$

$$\overset{\bullet}{P_2} + M \xrightarrow{k_p} \overset{\bullet}{P_3} \quad \text{and so on}$$

Reversible deactivation

$$\overset{\bullet}{P_n} + Cu(II) X/L \underset{k_a}{\overset{k_{da}}{\rightleftharpoons}} P_n-X + Cu(I)/L, \qquad K = k_a / k_{da}$$

Termination

$$\overset{\bullet}{P_n} + \overset{\bullet}{P_m} \xrightarrow{k_t} \text{Dead Polymer}$$

A low molar mass model of the dormant polymer, *e.g.*, the 1:1 adduct of the monomer and hydrogen halide (hereafter referred to as unimer halide or model initiator) is usually satisfactory as an initiator. However, this is not always true. For example, the unimer bromide initiator, ethyl 2-bromoisobutyrate, produces poly(methyl methacrylate) with relatively broad MWD, as discussed earlier (Fig. 3.14).[290,347,348] This unusual result is attributed to the B-strain effect.[349] The B-strain released during radical formation due to the change in hybridization from sp^3 to sp^2 is greater from the polymer halide than from the initiator. This makes k_a larger than k_a^0 or $K > K^0$ and, accordingly, propagation faster than initiation. The effect is minimized when the initiator is a dimer halide instead of the unimer one. Thus, the dimer bromide, dimethyl 2-bromo-2,4,4-trimethylglutarate (**17**), proves to be an efficient initiator, as already seen (Fig. 3.14).[290] In fact, it is activated by Cu(bpy)$_2$Br in acetonitrile at 35°C nearly 7.4 times as fast as the unimer bromide is activated, the respective k_a^0 values being 1.92 and 0.26 L mol^{-1}s^{-1}.[350]

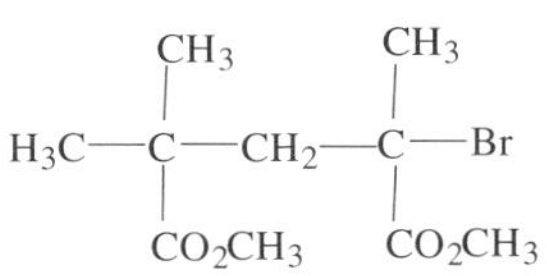

$$H_3C-\underset{\underset{CO_2CH_3}{|}}{\overset{\overset{CH_3}{|}}{C}}-CH_2-\underset{\underset{CO_2CH_3}{|}}{\overset{\overset{CH_3}{|}}{C}}-Br$$

(**17**)

Similarly, the dimer chloride is a more efficient initiator than the unimer chloride.[347]

We shall see similar B-strain effect later in this Chapter in the initiation of RAFT polymerization of methyl methacrylate and in Chapter 6 in the initiation of cationic polymerization of isobutylene by its model initiator, *t*-butyl chloride.

The B-strain effect is absent with initiators generating secondary or primary radicals. For example, 1-phenylethyl bromide (PEBr), is an efficient initiator for the ATRP of styrene. Indeed, measurements of rate constants of activation of polystyrene bromide (PSBr) and PEBr revealed that they have nearly equal reactivity, k_a and k_a^0 at 110°C being 0.45 and 0.42 L mol^{-1}s^{-1} respectively with CuBr/dHbpy catalyst in nonpolar solvents.[346]

Initiators other than the models may be selected from a consideration of the relative equilibrium constant (K_{BD}) values of the homolytic dissociation of C–X bonds, which have been theoretically estimated from the Density Functional Theory (DFT)-calculated free energies[351]

$$R-X \xrightleftharpoons{K_{BD}} \dot{R} + \dot{X}.$$

The relative K_{BD}s agree reasonably well with the experimentally determined relative K^0s, which are available for some initiators.[352] In general, K_{BD} increases with decrease in bond dissociation energy (BDE). As discussed under Sec. 3.15.4, the BDE decreases with increase in one or more of the following factors, viz., the resonance energy of the carbon radical to be formed, the bulkiness, and electron-withdrawing ability of the α substituents of the radical. The relative K_{BD} values of bromide initiators at 90°C as obtained from DFT calculations are shown in Fig. 3.22 (setting $K_{BD} = 1$ for methyl 2-bromopropionate).

Excepting benzyl, trichloromethyl, and p-toluenesulfonyl bromide, the other bromides in the Figure are model initiators (unimer bromides) of the corresponding monomers. For a given monomer, the initiators that have K_{BD}s larger than the model are likely to be suitable. Indeed, p-toluenesulfonyl bromide and trichloromethyl bromide, which have the largest K_{BD}s, are powerful initiators.[353–360] An important feature of the arenesulfonyl halide initiators is that the ArṠO$_2$ radicals recombine and disproportionate only slowly. Therefore, almost all the generated radicals are used up in initiation. However, with some activators radical generation may be too fast to make termination negligible resulting in loss of control and low conversion.[361]

Significantly, the electron-withdrawing property of the substituents in the leaving radical exerts a dominating influence on K_{BD} of the initiator in ATRP. Thus, whereas methyl methacrylate radical has comparable stability and larger steric bulk but weaker electron-withdrawing ability vis-à-vis acrylonitrile radical (Table 3.12); the K_{BD} of the unimer

$CH_3CH_2Br < CH_3CH(Br)OCOCH_3 < CH_3CH(Br)Cl < CH_3CH(Br)CO\ N(CH_3)_2$

K_{BD} (relative) 1.6×10^{-7} 2.7×10^{-5} 4×10^{-3} 5.8×10^{-2}

$< C_6H_5CH_2Br < CH_3CH(Br)COOCH_3 < C_6H_5CH(Br)CH_3 <$
  0.4   1   5.6

$(CH_3)_2C(Br)COOCH_3 < CH_3CH(Br)CN < CCl_3Br < p\text{-}CH_3C_6H_5SO_2Br$
  28   730 1.6×10^5 2×10^7

Figure 3.22   Relative K_{BD} values at 90°C for some alkyl bromides as derived by DFT calculations. "Adapted with permission from Ref. 351. Copyright © 2003 American Chemical Society."

bromide is smaller for methyl methacrylate than for acrylonitrile. This is opposite to that observed for the equilibrium constant (K) of the dissociation of the nitroxide initiator in NMP, which has a larger value for MMA-nitroxide, as discussed earlier.

The rate of activation also depends on the nature of the halogen atom. The C–F bond is too strong to make fluorides initiators. However, grafting of poly(alkyl methacrylate)s from secondary fluorine sites in poly(vinylidene fluoride) using ATRP initiators has been reported.[362,363] Because of the strong C–F bonds initiation is very slow compared to propagation limiting the living nature of the polymerization. Only about 0.1% of the secondary fluorines react in the graft copolymerization with t-butyl methacrylate at 90°C using CuCl/4,4′-dimethyl-2,2′-bipyridine catalyst. In the other extreme, organic iodides having the weakest C–I bonds are not preferred as initiators either, since the oxidized metal iodide complexes suffer from a solubility problem. Besides, the iodides induce degenerative transfer reaction.[344] However, they have been successfully used with Ru and Re based catalysts.[364,365] The remaining two classes of halides, viz., chlorides and bromides, work well as initiators. However, initiation is faster with bromides because of their lower C–Br bond dissociation energies.

3.15.6.1.1 Halide exchange effect

Another effective method of achieving initiation faster than propagation is to use a bromide initiator and a chloride catalyst instead of a bromide initiator and a bromide catalyst.[347,366–368] The method leads to halide exchange giving predominantly Cl-ended polymer and bromide catalyst by a mechanism shown in Scheme 3.9. The same effect also may be brought about using a bromide catalyst in conjunction with a soluble chloride salt, e.g., a quaternary ammonium chloride.[361]

The persistent radical effect sets up two activation-deactivation equilibria involving P$^\bullet$ with equilibrium constants K_1 and K_2, K_2 being greater than K_1 since $K_{ex.} \gg 1$.[369] However, the halide exchange is not as great with the initiator as with the polymer. This is because the concentration of R$^\bullet$ is much smaller than that would have been established by the regular persistent radical effect due to capture by the monomer. Thus, the initiator

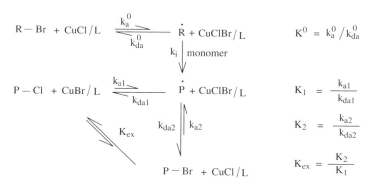

Scheme 3.9 Halide exchange using bromide initiator and chloride catalyst.

remains a bromide, while the polymer becomes a chloride predominantly. The weaker C–Br bond in the initiator compared to the stronger C–Cl bond in the polymer makes initiation faster than propagation.[366–368] However, even though most of the polymers are Cl ended, most of the deactivation events occur by the faster process of Br atom transfer, inasmuch as $k_{da2} > k_{da1}$, resulting in narrow MWD polymers. Nevertheless, P–Cl builds up due to its much slower frequency of activation. An additional benefit derived from the halide exchange is that P–Cl being more stable than P–Br, the end group loss at higher conversions due to side reactions becomes much less.[366,367,371,372]

3.15.6.2 Ligands

Phosphorous-based ligands, usually PPh$_3$, are used with most transition metals other than copper. With the latter, nitrogen based multidentate ligands work well. A wide variation in ligand structure is possible so that the rate of activation may be changed over a million fold.[373] This is effected through tuning the redox potential of the complex by the appropriate ligand.

Lowering the reduction potential increases the rate of activation. The ligands vary in the number of nitrogen donor atoms and their types (*viz.*, pyridine, aliphatic amine, and imine) and in the skeletal motifs, *e.g.*, aliphatic (linear, branched, cyclic, or multipodal), aromatic (fused or substituted rings), *etc*. Some of the representative ligands are shown in Figure 3.23.

Based on the relative K^0 values determined at 22°C for the activation of ethyl 2-bromoisobutyrate initiator (EBrIB) by CuBr complexed with some selected ligands, the activity of the complexes decreases in the following order of ligands[352]

$$K^0_{(relative)} \quad \begin{array}{ccccc} Me_6TREN & > TPMA & > TPEN & > PMDETA & > bpy \\ 39\,000 & 2400 & 500 & 20 & 1 \end{array}$$

However, the highest K^0 does not necessarily make an automatic choice for the best ligand. This is because too large a K^0 or K results in high radical concentration causing irreversible termination to be important.

Thus, with the *tertiary* bromide, EBrIB, or the corresponding PMMA–Br as the halide and the CuBr/Me$_6$TREN as the activator, the value of K^0 or K is so high that uncontrolled polymerization of MMA occurs.[381] On the other hand, using the *secondary* bromide initiator, ethyl 2-bromopropionate, controlled polymerization of methyl acrylate occurs at fast rates at ambient temperature.[378]

Large alkyl substitution in the aromatic rings of bpy increases activation rate. This also makes the catalyst complex soluble in nonpolar solvents and greatly improves control. Thus, using dNbpy as the ligand homogeneous polymerization occurs yielding polystyrene with PDI = 1.05.[382]

The catalytic activities of the complexes of a given transition metal with various ligands correlate well with the reduction potentials of the complexes.[383] With the decrease in the reduction potential k^0_a increases and usually k^0_{da} decreases.[344] However, Me$_6$TREN is exceptional. It makes a very active Cu based activator as well as an efficient deactivator.

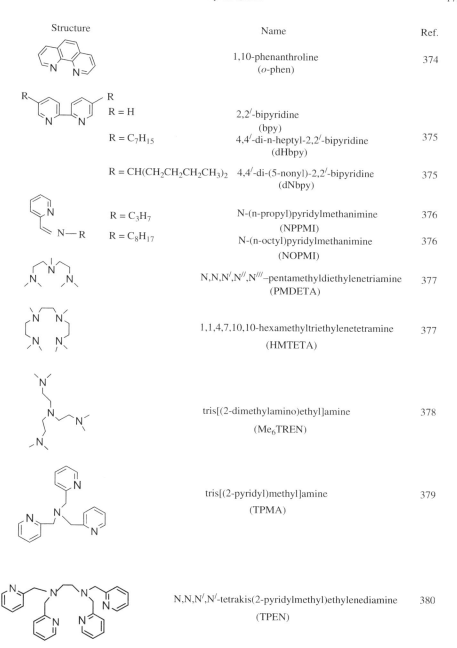

Figure 3.23 Examples of some ligands used in copper mediated ATRP.

Reduction potential refers to outer-sphere electron transfer

$$R-X + e \rightleftharpoons [R \overset{\bullet}{-} X]^- \rightleftharpoons \overset{\bullet}{R} + X^-,$$

whereas, activation proceeds through inner-sphere electron transfer from the reduced metal complex to the alkyl halide in a concerted one step process mediated by the halogen atom[205,344]

$$R-X + Mt\ X_n/L \rightleftharpoons [R \cdots X \cdots Mt\ X_n/L] \rightleftharpoons \dot{R} + Mt\ X_{n+1}/L\ .$$

The rate of activation, therefore, depends not only on the reduction potential but also on the halogenophilicity of the complex. For example, ruthenium complexes have much higher reduction potentials ($\sim +300$ mV) than copper complexes (~ -100 mV), but their halogenophilicities are also higher such that they become comparable in activity with the copper complexes.[344,384]

However, as we shall see in the section immediately below, activation may also be effected by an outer-sphere single electron transfer (SET) from Cu(0) to halide initiators or dormant polymer halides.

3.15.6.3 Solvents

Solvent properties like polarity, ion coordinating power, and ability of dissolving the activator as well as the deactivator greatly influence ATRP.[386,387] For example, remarkable rate increase occurs in the solution ATRP of butyl acrylate activated by CuBr/bpy when a lowly polar solvent such as anisole, diphenyl ether, 1,4-dimethoxybenzene, p-dioxane, propylene carbonate, dimethyl carbonate, or benzene is replaced with strongly polar ethylene carbonate.[385]

Similarly, rate of solution ATRP of methyl methacrylate activated by CuCl/bpy or CuCl/o-phenanthroline is approximately doubled on changing the solvent from acetone to more polar ethanol.[388,389] The polydispersity does not increase in spite of the increase in the rate of polymerization indicating that the deactivation rate is faster too. Use of aqueous ethanol as solvent increases the rate of polymerization even further, although with decrease in control.[388,390–396] The rate increase is attributable to the increase in k_a. The decrease in control is due to the oxidative termination caused by outer-sphere electron transfer to the aquo complexes formed by the displacement of the halide ligand from the Cu (II) complexes by water (*vide* Sec. 3.9.1). The corresponding displacement by ethanol is less efficient

$$[CuLn_xX]^+ + H_2O \rightleftharpoons [CuLn_x(H_2O)]^{+2} + X^- (X = Cl, Br).$$

The extraneous addition of some amount of the cupric complex may establish enough deactivator concentration for effective control. Nevertheless, the aquo complex present in equilibrium may cause termination by outer-sphere electron transfer. Of course, its formation may be prevented or minimized utilizing the common ion effect by adding common ion halide salts extraneously. Hence, the addition of both the deactivator and a common ion salt may be necessary to exert effective control.[397–399] However, the rate enhancing effect of water with Cu based activators may be advantageously utilized through the optimum use of water in aqueous organic solvents keeping hydrolysis to a tolerable level for reasonably good control. With the use of lowly oxophilic activator, *e.g.*, Ru(PPh$_3$)$_3$Cl$_2$, the hydrolysis problem is avoided altogether.[358-360]

Besides decreasing the rate of deactivation, use of water or a coordinating compound presents another problem in copper mediated ATRP. For example, some Cu(I) complexes undergo disproportionation in water or aqueous organic media due to the much greater stability of Cu(II) complexes relative to Cu(I) complexes in these media. For example, Cu(I) complexes of PMDETA and Me$_6$TREN disproportionate in water or in aquated organic solvents, while those of bpy and HMTETA do not.[345] Hence, the former two complexes are unsuitable as activators. Thus, ATRP of methyl methacrylate using CuX/PMDETA or CuX/TREN in aqueous ethanol[388] or aqueous acetone[389] containing less than even 20 volume percent water at 35°C gives low yield and is uncontrolled, whereas with the use of bpy or HMTETA as the ligand, controlled polymerization occurs (*vide* next section). However, some monomers can prevent disproportionation by taking part in the complexation of Cu(I) and stabilizing it relative to Cu(II) (acrylamide being an example).[398,399]

Measurement of formation constants of [Cu(PMDETA)(π-M)]$^+$BPh$_4^-$ (where π-M represents the π-coordinated monomer) from [Cu(PMDETA)]$^+$ and monomer (M) in acetone revealed that methyl methacrylate binds with Cu(I) only poorly. For example, the formation constant with MMA as the π ligand is about 100 times lower than that with methyl acrylate at room temperature.[400]

Importantly, the disproportionation of CuX in polar solvents has been taken advantage of in the so-called single electron transfer-ATRP (SET-ATRP).[401] In this method of ATRP, the use of cuprous halide complex as activator is obviated. Instead, activation is effected by outer-sphere single electron transfer from Cu(0) powder or wire to organic halide initiators or to the dormant polymer halides

$$Cu(0) + RX = CuX + R^\bullet.$$

Although a chelating amine ligand is also used in the formulation, it is of the kind that does not stabilize Cu(I) relative to Cu(II). As a result, the CuX generated in the activation process instantly undergoes disproportionation generating the nascent and extremely reactive Cu(0) and CuX$_2$ species. The latter becomes complexed with the ligand and acts as the deactivator, while the former acts as the activator. The deactivator concentration does not depend on the persistent radical effect. The method yields polymer with much lower copper contamination than a usual ATRP does.

Another side reaction, which is more important in protic and aqueous based medium, is the substitution of halogen atom or elimination of hydrogen halide from initiator as well as from polymer chain end by nucleophilic or basic groups in monomer and polymer. The problem is much less or absent when the halide is chloride instead of bromide.[402]

3.15.6.4 *Kinetics of ATRP*

Well-controlled ATRP, unaided by thermal self-initiation or conventional initiation, exhibits first order kinetics with respect to monomer concentration. This is evident from first order plots for monomer disappearance, shown in Fig. 3.24,[388] in the ATRP of methyl methacrylate

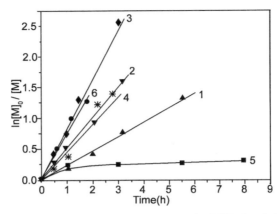

Figure 3.24 First order kinetic plot for monomer disappearance in the ATRP of methyl methacrylate at 35°C in ethanol (neat or aqueous) medium. [MMA] = 4.67 mol L^{-1}, [ethyl 2-bromoisobutyrate] = [CuBr] = 0.5[ligand] = 23.3 mmol L^{-1}. Ligands and solvent systems are for lines: 1, o-phen and ethanol; 2, o-phen and ethanol:water (5:1); 3, o-phen and ethanol:water (2:1); 4, bpy and ethanol:water (5:1); 5, PMDETA and ethanol:water (5:1); 6, HMTETA and ethanol:water (5:1). "Reprinted with permission from Ref. 388. Copyright © 2004 American Chemical Society."

at 35°C in ethanol and ethanol-water mixtures using ethyl 2-bromoisobutyrate as initiator and various CuBr complexes as catalysts. In all cases but one, the plots are linear. The linear plots indicate constancy of chain radical concentration throughout the course of polymerization. The nonlinear plot with the CuBr/PMDETA catalyst is due to the catalyst loss through disproportionation, as discussed above. The relative rates show that the catalyst activity decreases in the following order for ligands

$$\text{HMTETA} > o\text{-phen} \simeq \text{bpy}.$$

However, according to the theory, the concentration of the chain radical should decrease with time inasmuch as $[P^\bullet] \propto t^{-1/3}$ (Eq. 3.120) and, consequently, the first order plots for monomer disappearance should be nonlinear (Eq. 3.121).

This discrepancy is observed commonly in ATRP. It has been attributed to various factors:[273]

(i) inadvertent presence of deactivator at the very start of polymerization due to air oxidation of activator occurring, while handling and in the reaction vessel, by residual oxygen,
(ii) precipitation of deactivator with progress of reaction in some systems,
(iii) decrease of k_t with increase of viscosity and of chain lengths of radicals with time.

Factor (i) leads to kinetic equation (3.118) with the concentrations at the initial values provided the inadvertent deactivator concentration well exceed certain value as given by Eq. (3.122). Factor (ii) increases R_p with conversion and may compensate for the decrease in rate arising out of the decrease in chain radical concentration with time. Factor (iii) decreases the rate of decrease of chain radical concentration with time (Eq. 3.120).

Indeed, when proper measures are taken to eliminate the above factors, the power law kinetic equation (Eq. 3.121) is obeyed.[403–406] Thus; factor (i) may be eliminated using proper experimental procedure, factor (ii) by using an appropriate catalyst system, and factor (iii) by using a polymeric initiator and studying the kinetics at low conversions.[273]

However, adapting Eq. (3.118) developed for NMP to ATRP gives

$$\ln \frac{[M]_0}{[M]} = \frac{k_p K [Cu(I)][P-X]t}{[Cu(II)X]} = k_{app} t. \qquad (3.151)$$

Extraneous addition of the persistent radical, which is the cupric complex in the present case, well above certain amount as prescribed by (Eq. 3.122), results in a kinetic equation which is of the same form as (3.151) with concentrations changed to initial ones (vide Sec. 3.15.2)

$$\ln \frac{[M]_0}{[M]} = \frac{k_p K [Cu(I)]_0 [P-X]_0 t}{[Cu(II)X]_0} = k_{app} t. \qquad (3.152)$$

Thus, in ATRP conducted with sufficient cupric complex initially added, Eq. (3.152) may be used to determine k_{app} from the slope of the linear first order plot and K may be evaluated from k_{app} with the knowledge of k_p and initial concentrations of copper species and the initiator used.

Values of K in the ATRP of some monomers determined in this way are given in Table 3.13, which also includes values of the enthalpy change. It follows from the results that K decreases in the following order of monomers: methyl methacrylate > styrene > methyl acrylate. This order for the three monomers is the same as that observed for K^0s of the corresponding unimer bromides, which have been determined by DFT calculation (Fig. 3.22).[351] The relative $k_p K$ determines the relative rates of polymerization, which at 90°C is[351,366]

Monomer: methyl methacrylate > methyl acrylate > styrene.
Relative R_p(ATRP):   42   1   0.33.

This is different from the relative rate observed in conventional radical polymerization where the rate is determined by the kinetic parameter $k_p/(2k_t)^{1/2}$ (vide Sec. 3.2), the relative R_p

Table 3.13 ATRP equilibrium constants and enthalpy changes.

Monomer	Temp.°C	Catalyst	K	ΔH kJ/mol	k_p Lmol^{-1}s^{-1}	$K k_p$ Lmol^{-1}s^{-1}	Ref.
Methyl acrylate[a]	90	CuBr/dNbpy	1.2×10^{-9}	96	47100	5.4×10^{-5}	407
Styrene[b]	90	CuBr/dNbpy	2×10^{-8}	20	910	1.82×10^{-5}	375
Styrene[b]	90	CuCl/dNbpy	1×10^{-8}	26	910	0.9×10^{-5}	375
Methyl methacrylate[b]	90	CuCl/dNbpy	7×10^{-7}	40	1620	113×10^{-5}	408
Butyl acrylate[c]	110	CuBr/dNbpy	6.3×10^{-10}	—	78450	4.94×10^{-5}	404

[a]In bulk; [b]in diphenyl ether (50%); [c]in xylene (50%).

in conventional polymerization at 60°C being

Monomer: methyl acrylate > methyl methacrylate > styrene
Relative R_p(conventional):　　75　　　　　　4　　　　　　1.

The rate constant of activation and that of deactivation in polymeric or model systems have been determined by methods analogous to those discussed in the case of NMP.[346,409]

ATRP of lowly reactive monomers such as ethylene and vinyl acetate is not feasible using the same catalysts as are effective with the reactive monomers such as styrene, (meth)acrylates, or acryloniotrile. This would be obvious from the relative K_{BD} values of the model bromides given in Fig. 3.22. However, it may be possible to find highly active catalysts for the lowly reactive monomers with which their polymerizations become controlled and reasonably fast, whereas polymerizations of the reactive monomers are terminated immediately after initiation due to the extremely high radical concentration since K^0s are about 10^5–10^7 times larger for them (Fig. 3.22). Indeed, it has been possible to conduct ATRP of vinyl acetate at practicable rates at the moderate temperature of 60°C using a highly active catalyst, $[Fe(Cp)(CO)_2]_2$, and a model initiator with the weakest C–X bond, viz., the unimer iodide. Certain additives, such as aluminum or titanium isopropoxide, aluminum triisobutyl, and dibutyl amine, are however needed.[410] The additives increase the rates substantially; however their roles are not clear.

Although homopolymerization of ethylene is not feasible, as noted above, its copolymerization with methyl acrylate or methyl methacrylate proceeds in controlled fashion using a conventional Cu(I) catalyst, viz., Cu(PMDETA)Br.[411–413] This has been attributed to the rate of irreversible deactivation of the copolymer radical with terminal ethylene unit by Cu(PMDETA)Br_2 being 10^3 to 10^4 times slower than cross-propagation.[412]

3.15.6.5 Molecular weight and polydispersity

Molecular weight increases linearly with conversion according to Eq. 3.126 excepting some positive deviation in the lower conversion region (Fig. 3.14), as discussed already in Sec. 3.15.3. The positive deviation continues to larger conversions with slower initiators.

The polydispersity index decreases with increase in conversion (Fig. 3.24) as is theoretically predicted (Eq. 3.142).

3.15.6.6 Effect of temperature

Since the ATRP equilibrium is endothermic (Table 3.13), the equilibrium constant K increases with increase in temperature. In addition, k_p increases with temperature. Thus, it follows from Eq. (3.152) that the rate of ATRP should increase with temperature. Indeed, this is found, for example, in the CuX/2dNbpy catalyzed polymerizations of styrene,[375] methyl methacrylate,[408] and methyl acrylate.[407] The effect of temperature on PDI may be understood from Eq. (3.142). The parameter k_p/k_{da} increases with temperature, the activation energy of propagation being greater than that of deactivation. Thus, PDI would increase with increase in temperature. However, should a change in the structure of the deactivator occur, k_{da} would change unpredictably, and so would PDI. In addition, loss of end groups increases with increase in temperature due to side reactions leading to an increase in dead chain fraction

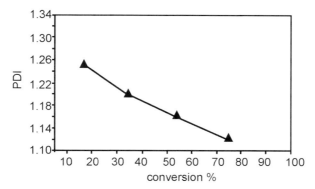

Figure 3.25 Evolution of PDI with conversion in the ATRP of methyl methacrylate at 35°C in ethanol using CuBr/o-phenanthroline as catalyst and ethyl 2-bromoisobutyrate as initiator. "Reprinted in part with permission from Ref. 388. Copyright © 2004 American Chemical Society."

and PDI. Besides, due to the increase in K the radical concentration increases leading also to an increase in dead chain fraction and PDI. To cite an example, CuBr/PMDETA catalyzed polymerization of methyl methacrylate in anisole gives polymer of lower PDI at ambient temperature than that obtained at a higher temperature *ca.,* 60°C.[414]

3.15.6.7 ATRP with activator generated in situ

In normal ATRP, a part of the activator is oxidized by air during its handling or by residual oxygen in the deaerated polymerization medium. While it is possible to restrict the oxidation to a low level in laboratory scale experiments, it is very difficult to do so in industrial scale. The problem is more acute with highly active activators, which may be required to be used at low concentrations lest too much radicals are formed favoring termination or when low concentrations of initiator and activator are used to get high molecular weight polymer. A solution to this problem lies in using the oxidatively stable deactivator in the polymerization recipe and generating the activator *in situ* with a reducing agent. Various methods of reduction have been devised, which are discussed in the following subsections.

3.15.6.7.1 Reverse ATRP

In this method, the deactivator is reduced by chain radicals initiated by a conventional initiator such as AIBN. The reduction process generates the activator and the halide initiator as shown in Scheme 3.10.

Both copper and iron mediated reverse ATRP has been performed successfully.[415,416] A high temperature of polymerization (*ca.,* ≥100°C) at which AIBN has a short half life allows the formation of the activator and the halide initiators early in the polymerization so that ATRP equilibrium is set up quickly. This reduces broadening of the MWD from continuous new chain formation from the coexisting AIBN, present otherwise. For example, using $CuBr_2$/dNbpy as the catalyst, polymerizations of styrene and methyl acrylate were successfully carried out at 110°C. The Cu(II) : AIBN ratio of 1.6 : 1 is optimum. Given

$$\text{AIBN} \longrightarrow 2\dot{\text{R}}$$
$$\dot{\text{R}} + \text{M} \longrightarrow \dot{\text{P}}_1 \xrightarrow{\text{M}_2} \dot{\text{P}}_2, \text{etc}.$$
$$\dot{\text{P}}_\text{m} \xrightarrow{\text{Mt X}_{n+1}/\text{L}} \text{P}_\text{m}\text{X} + \text{Mt X}_n/\text{L} \quad \text{(m being small)}$$

Scheme 3.10 Generation of halide initiator and activator in reverse ATRP.

the initiator efficiency of 0.76 for AIBN at 70°C (Sec. 3.6), this ratio is about enough to reduce all Cu(II) to Cu(I). Some of the Cu(I) will be reoxidized to Cu(II) by residual oxygen present in the system but they will be reduced back to Cu(I) by chain radicals. Subsequently, Cu(II) deactivator required for the control is generated by the persistent radical effect. With benzoyl peroxide initiator, reverse ATRP of styrene is not successful since copper bromide is converted to copper benzoate, which is not ATRP-active.[417] Reverse ATRP has the disadvantage that it is not suitable for block copolymer synthesis.

3.15.6.7.2 SR and NI ATRP

With the use of very active catalysts (large K^0), lower temperature ($\leq 60°C$) of polymerization and lower catalyst concentration ($[\text{Cu(I)}]_0/[\text{RX}]_0 \leq 0.1$) are required for obtaining satisfactory control in normal ATRP. Reverse ATRP is not effective since at the relatively high temperature of polymerization ($\geq 100°C$), which is required to generate the initiator *in situ*, the radicals are generated from the initiator by these highly powerful catalysts so fast that uncontrolled polymerization takes place. A solution to these problems was found in the simultaneous reverse and normal initiation (SR and NI) method.[418,419] In this method, normal and reverse ATRP are combined. The initiator concentration is independently controlled using an alkyl halide (RX) initiator as in normal ATRP, while the Cu(I) is generated as in reverse ATRP but at a lower temperature (*ca.*, 60 to 80°C) than that usually used. Typical concentrations of CuBr$_2$/L and AIBN for generating CuBr/L are about 10 and 6 mol % respectively of RX. Successful SR and NI ATRPs of styrene and acrylates were conducted using the deactivator CuBr$_2$/Me$_6$TREN that generates the very active CuBr/Me$_6$TREN catalyst *in situ*.[418]

3.15.6.7.3 AGET ATRP

In this method, the activator is generated by electron transfer to the deactivator by a reducing agent without producing an alkyl halide initiator unlike that in the reverse ATRP.[420–422] Thus, the activator CuCl/L is produced *in situ* by the reduction of CuCl$_2$/L with stannous 2-ethylhexanoate, Sn(COOR)$_2$ (reaction 3.153).

$$\text{Sn(COOR)}_2 + 2\text{CuCl}_2/\text{L} \rightleftharpoons (\text{Sn(COOR)}_2\text{Cl}_2 + 2\text{CuCl/L} \tag{3.153}$$

The tin compound is used at a concentration short of the stoichiometric requirement to achieve better control. For example, use of a molar ratio of 1: 0.45 for Cu : Sn leaves about 10% of Cu(II) unreduced, which effects better control. Although the tin compound is highly reducing, it itself is a poor activator of ATRP. This is presumably due to the activation being effected by it through outer-sphere electron transfer.[420] An alkyl halide initiator is used as in a normal ATRP. Other reducing agents, e.g., ascorbic acid[421] or *tertiary* alkyl amines,[422] besides Sn(COOR)$_2$, have been used. Glucose or copper powder may also be used.[423]

3.15.6.7.4 ARGET ATRP

Activator regenerated by electron transfer (ARGET) ATRP provides the best solution so far for reducing metal salt contamination in polymer to an acceptable level without going through the expensive process of polymer purification. Hence, the method has great potential for industrial utilization. It works by continuously regenerating the activator (CuX/L) using excess reducing agent so that ATRP can be conducted in the reverse mode starting with a very small amount of CuX$_2$/L (*ca.*, 10 to 50 ppm of Cu).[424-429]

Matyjaszewski *et al.* reasoned that copper concentration can be reduced without affecting R$_p$ inasmuch as the latter is dependent on the ratio [Cu(I)]/[Cu(II)X] rather than the total copper concentration (*vide* Eq. (3.151).[424] However, the absolute concentration of Cu(II)X does matter in determining PDI (Eq. 3.142) but the requirement can be very low. For example, only ~2 ppm of Cu(II) is required to produce polystyrene of DP = 200 and PDI = 1.2 at 90% conversion with k$_p$ ~ 10^3 L mol^{-1} s^{-1} and k$_{da}$ ~ 1×10^7 L mol^{-1} s^{-1}.[424] However, since at equilibrium, [Cu(I)/[Cu(II)X] is typically about 10, total copper requirement would be about 20 ppm. However, a normal ATRP with so low Cu(I) concentration is not feasible due to the oxidation of Cu(I) by the residual oxygen as well as the persistent radical effect so that all Cu(I) will be oxidized to Cu(II) and initiation will be stopped.

ARGET ATRP was devised to provide a solution to the problem. It starts with CuX$_2$/L (10–50 ppm Cu depending on the magnitudes of k$_p$ and k$_{da}$) and uses a reducing agent such as stannous 2-ethylhexanoate, ascorbic acid, glucose, hydrazine, or phenyl hydrazine at a concentration of about 10 times that of [CuX$_2$/L]$_0$.[424-429] Accordingly, Cu(II) produced by the persistent radical effect as well as by the oxidation of Cu(I) by the residual oxygen is reduced continuously by the reducing agent maintaining an equilibrium ratio of Cu(I) : Cu(II). A very active catalyst (high K^0), *e.g.* one with the Me$_6$TREN or TPEN ligand, is required to be used so that the equilibrium [Cu(II)] is large enough at such low total copper concentration to effect control. Furthermore, these ligands produce copper complexes with higher stability such that control is not lost due to the dissociation of the complexes at the low concentrations.[429] TPEN is better than Me$_6$TREN in this respect. Figure 3.26 shows the linear first order plot obtained in an ARGET ATRP of styrene using only 15 ppm copper.

Apart from eliminating the purification cost and lessening the environmental hazard with very low amount of copper in ARGET ATRP, the low Cu(II) concentration provides an additional benefit. It sharply reduces end group loss by Cu catalyzed outer-sphere electron transfer reaction so that polymerization can be carried to high conversion without loss of control.[88,371,372] Thus, the method also allows the synthesis of high molecular weight polymers.[426,428]

However, even though the total copper concentration is reduced, the ligand concentration should not be reduced proportionally lest the ligand gets dissociated from the metal ions, which is facilitated at higher temperatures or in coordinating solvents such as water, alcohols, DMF, *etc.*, or by complexation of Cu(I) (soft acid) with monomer (soft base).

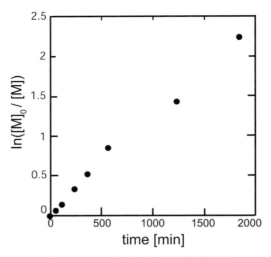

Figure 3.26 First order monomer disappearance plot in the ARGET ATRP of styrene at 120°C with 15 ppm of copper. Recipe: S/EBrIB/Cu(II)/Me$_6$TREN/Sn(oct)$_2$ = 200/1/0.003/0.1/0.1, [Styrene] = 5.82 mol/L in anisole. "Reprinted with permission from Ref. 424. Copyright © 2006 American Chemical Society."

Ligand dissociation decreases catalyst activity.[424] Furthermore, when Sn(COOR)$_2$ is used as the reducing agent sufficient amount of ligand is required to be used to complex the SnCl$_2$(COOR)$_2$ formed as well (reaction 3.153) so that copper ions are not denuded of the ligands by it due to its stronger Lewis acidity.[424]

3.15.6.7.5 ICAR ATRP

Initiators for continuous activator generation (ICAR) ATRP also allows controlled polymerization like ARGET ATRP using very small amounts of copper (<ca., 50 ppm).[429] The principle is the same as that in ARGET ATRP. The difference between the two lies in the method of regenerating Cu(I) from Cu(II). In ARGET ATRP, it is done by various organic reducing agents, whereas in ICAR ATRP by carbon radicals produced by a thermally dissociable initiator, e.g., AIBN, used at a concentration of about one tenth of the organic halide initiator at 60–70°C. A ligand yielding a highly active catalyst is to be used like in ARGET ATRP so that enough Cu(II) is available to deactivate the radicals at the low total Cu used. At higher temperatures, thermal self-initiation may be enough to provide the required concentrations of carbon radicals. The rate of polymerization is predominantly controlled by the conventional initiator. Since new chains continue to form throughout the course of polymerization, pure block copolymer synthesis cannot be done using this method. However, since the conventional initiator is used at about 10 mol percent of the halide initiator the homopolymer contamination is small.

3.15.6.8 *Synthesis of high molecular weight polymers*

Polymerization needs to be carried to high conversion for the synthesis of high molecular weight polymers. However, at high conversions (ca., >90%) in normal bulk ATRP a very high increase in viscosity takes place. As a result, deactivation, which has small activation energy, becomes diffusion-limited resulting in loss of control. This is more so when high molecular weight is targeted. The problem may be overcome by resorting to solution instead of bulk polymerization. In fact, poly(2-(dimethyl amino)ethyl methacrylate) has been reportedly prepared in aqueous methanol to a molecular weight of about 1 million with PDI = 1.23 using chloride based initiator and catalyst.[430]

However, with various other monomers, side reactions take place resulting in the end group loss, which becomes disproportionately large as conversion increases.[371] Thus, the upper limit of molecular weight with reasonably low polydispersity in ATRP of both styrene[88,431] and acrylonitrile[426,432] is about 30 000. Side reactions in these polymerizations involve outer-sphere electron transfer, which results in oxidative termination of nucleophilic polystyrene radical by $CuBr_2$/L and reductive termination of electrophilic polyacrylonitrile radical by CuBr/L (Sec. 3.9.1).

Oxidative termination

$$\dot{P}_n + CuBr_2/L \xrightarrow{k_x} CuBr/L + Br^- + P_n^+ \xrightarrow{-H^+} P_n^=$$

Reductive termination

$$\dot{P}_n + CuBr/L \xrightarrow{k_x'} P_n^- + Cu(II)Br/L \xrightarrow{Solvent\ (SH)} P_nH + S^-$$

Scheme 3.11 Oxidative and reductive termination by outer sphere electron transfer.

Apart from the terminally unsaturated polymer obtained in the oxidation reaction (Scheme 3.11), polystyrene radical gives yet another carbenium ion product, *viz.*, the polymer with an indanyl end group, which is formed by intramolecular electrophilic aromatic substitution (*vide* Chap. 6)[88,431]

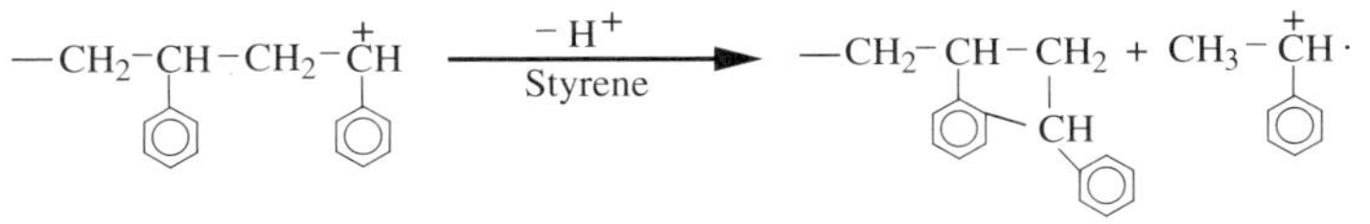

Yet another means of end group loss is the Cu(II) catalyzed β-H elimination from polystyrene like polymers[371,431]

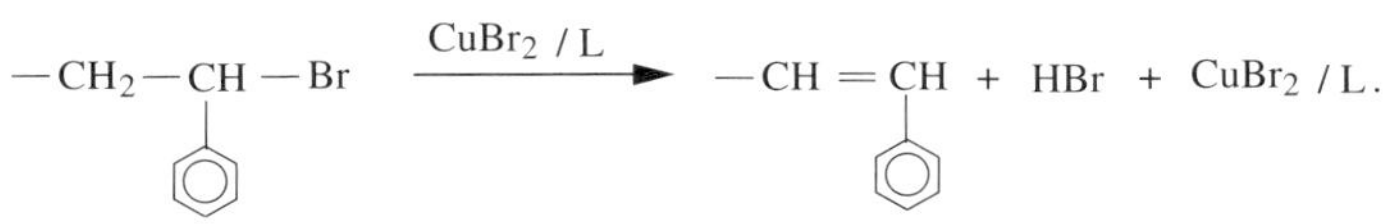

The importance of side reactions increases as polymerization is carried to high conversions (>90%) inasmuch as their rates are independent of monomer concentrations, while the rate of propagation decreases with decrease in monomer concentration. As a result, relatively more polymers lose end groups as conversion is increased.[371]

However, the problem can be eliminated by resorting to ARGET ATRP. As because the copper concentration is very small in this mode of ATRP, the side reactions are eliminated. Thus, polystyrene of $\overline{M}_n$ =185 000 with PDI = 1.35 was prepared using ARGET ATRP.[431] Similar results were obtained with polyacrylonitrile[426] and poly(styrene-co-acrylonitrile).[428]

End group loss also takes place by β-H abstraction in NMP mediated polymerization.[321,433] The effect is larger, larger is the target molecular weight. Thus, in the TEMPO mediated polymerization of styrene the chain end functionality is about 0.91 at $\overline{M}_n = 8500$ but only about 0.65 at $\overline{M}_n = 30\,000$.

3.15.6.9 Synthesis of block copolymers

In general, in sequential block copolymerization, the sequence of monomers should be in decreasing order of ATRP equilibrium constants: acrylonitrile > methacrylates > acrylates > (meth)acrylamides (cf., Fig. 3.22); otherwise, a complete change over from one block to another will not occur due to slower and incomplete initiation.

However, to be precise, for the complete initiation of polymerization of monomer (M_2) of the succeeding block in the block polymer by the preceding block P_1-X (X=Cl, Br), referred to as the macroinitiator, the following condition must satisfy

$$K_1 k_{12} \geq K_2 k_{p(2)},$$

where K_1 is the ATRP equilibrium constant for P_1-X, k_{12} is the rate constant of initiation of polymerization of M_2 by $P_1^\bullet$, K_2 is the ATRP equilibrium constant for P_2-X, and $k_{p(2)}$ is the propagation rate constant for monomer 2 (cf., Eq. 3.149). Thus, departure may occur from the monomer sequence given above designed solely based on the relative values of K_1 and K_2. For example, between styrene (S) and acrylates (A), $K_S > K_A$ but $K_S k_{sA} < K_A k_{p(A)}$ (vide Sec. 3.15.4.5). Similar situation is also encountered in NMP as we have seen already.

However, in ATRP a change in monomer sequence may be made possible by utilizing the halide exchange technique (Scheme 3.9).[366,434,435] For example, when poly(methyl acrylate)–Cl (PMA–Cl) is used as a macroinitiator in block copolymerization with methyl methacrylate using CuCl/dNbpy catalyst, block copolymer with bimodal MWD is obtained as shown in the GPC traces of Fig. 3.27.[366] The low molecular weight peak is due to PMA–Cl remaining unactivated indicating incomplete initiation. Thus, the block copolymer is contaminated with PMA homopolymer. This result is due to K(ATRP) of methyl acrylate being much lower than that of methyl methacrylate (Table 3.13).

On the other hand, utilizing the halide exchange technique, i.e., by using PMA–Br as the macroinitiator and CuCl/dNbpy as the catalyst, complete initiation takes place, as is evident from the monomodal GPC trace of the block copolymer shown in Fig. 3.28.[366]

Amongst the ATRP methods, which generate the activator in situ, only the AGET and ARGET give pure block copolymer. However, ARGET is superior due to the low amount of copper used. Successful block copolymerization with only 50 ppm or less of copper has been reported using the ARGET ATRP. However, the halide exchange technique, which is a very effective tool in making crossover from a secondary bromide macroinitiator to a tertiary chloride, is not suitable with ARGET ATRP. This is because only a small fraction of the Br end groups of macroinitiator changes to Cl since [CuCl]/[macroinitiator] is very low,

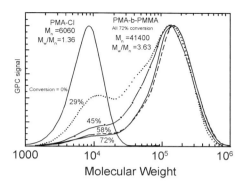

Figure 3.27 Molecular weight distributions of poly(methyl acrylate)-Cl (PMA–Cl) macro initiator and poly(methyl acrylate-*b*-methyl methacrylate) (PMA-b-PMMA) diblock copolymer at various conversions of monomer as indicated in the figure. The catalyst is CuCl/dNbpy in the synthesis of both the macroinitiator and the block copolymer. "Reprinted with permission from Ref. 366. Copyright © 1998 American Chemical Society."

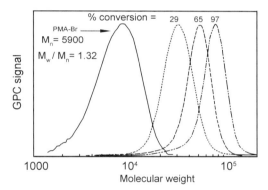

Figure 3.28 Molecular weight distributions of poly(methyl acrylate)-Br (PMA–Br) macroinitiator and poly(methyl acrylate-*b*-methyl methacrylate) (PMA-*b*-PMMA) at various conversions of methyl methacrylate. The diblock copolymer was prepared using CuCl/dNbpy as catalyst. "Reprinted with permission from Ref. 366. Copyright ©1998 American Chemical Society."

typically 0.01. However, the crossover can be efficiently effected without taking recourse to the halide exchange method by using a small amount (*ca.*, <10%) of a less active comonomer (low K) along with the more active one (high K), with which chain extension is desired. Thus, a polyacrylate or polystyrene macroinitiator can be chain-extended with methyl methacrylate containing about 10 vol. % styrene.[436] Such a technique is used in the NMP of methyl methacrylate, which has a high K, as discussed earlier (Sec. 3.15.4.3).[437,438] Although the ICAR method may also be used in place of the ARGET method for block copolymer synthesis, some homopolymer contamination can not be avoided.[426,428,436]

Multiblock copolymers may be prepared by multi-stage sequential polymerizations of monomers starting with a monofunctional macroinitiator, isolating block copolymers formed in successive stages and using a block copolymer formed in a previous stage as the macroinitiator for the next stage. In this method, as many stages of polymerizations are required as the number of blocks there is. However, the number of stages may be curtailed depending on the block composition. For example, an A–B–A triblock copolymer may be synthesized in two stages using a difunctional initiator to synthesize the center block (B) in the first stage. The latter is then used to construct the end A blocks. Thus, the thermoplastic elastomer poly(methyl methacrylate)-*b*-poly(*n*-butyl acrylate)-*b*-poly(methyl methacrylate) was prepared by polymerizing *n*-butyl acrylate in the first stage using the difunctional 1,2-bis(bromopropionyloxy)ethane as initiator and CuBr complexed with a dialkylsubstituted bipyridine as catalyst to yield α, ω-dibromo-poly(*n*-butyl acrylate). In the second stage, the halide exchange technique was used to chain extend poly(*n*-butyl acrylate) with poly(methyl methacrylate) from both ends of the former.[366]

Similarly, star block copolymers may be prepared in two stages using a multifunctional initiator (with the halide functions radiating from a nucleus) in the first stage polymerization of a monomer to form the inner core of the star block. The macroinitiator so formed is then used to polymerize a different monomer to get the desired star block copolymer. For example, trifunctional initiators like 1,3,5-(2′-bromo-2′-methylpropionato) benzene[439] (**18**) and 1,1,1-tris(4-(2-bromoisobutyryloxy)phenyl) ethane (**19**)[440,441] have been used to synthesize 3-arm star block copolymers.

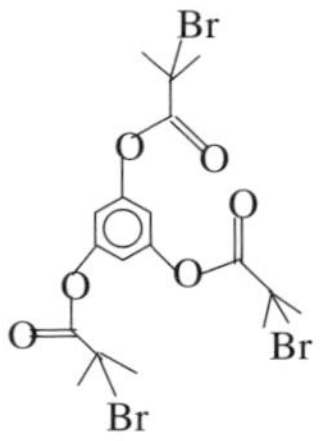

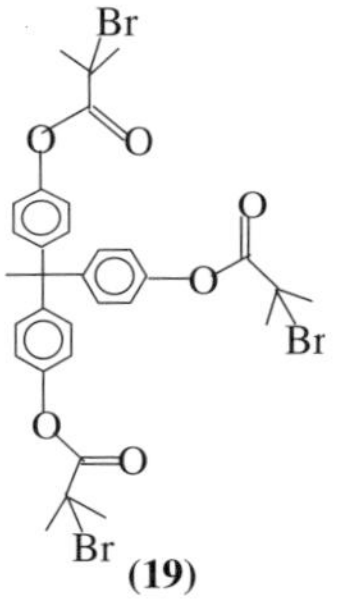

3.15.7 *Reversible addition–fragmentation chain transfer (RAFT) polymerization*

The discovery of RAFT polymerization[442,443] has its origin in the addition fragmentation chain transfer (AFCT) chemistry revealed in 1986 using poly(methyl methacrylate) macromonomer as the AFCT agent (Sec. 3.8.1.5).[444] As already introduced, reversible deactivation in RAFT polymerization is effected through degenerative transfer (Scheme 3.5). The mechanism of polymerization is shown in Scheme 3.12.[445–448] Since the RAFT agent does not produce any radical and termination is unavoidable, a conventional initiator is used in order to initiate and sustain polymerization.

The mechanism of polymerization involves reversible addition of a chain radical ($P_n^{\bullet}$) to the RAFT agent (A) forming the intermediate radical (A$^{\bullet}$), which fragments reversibly producing a polymeric RAFT agent (B) and a new radical (R$^{\bullet}$). This stage of the polymerization is referred to as pre-equilibrium.[448] The radical R$^{\bullet}$ adds on monomer generating a new chain radical $P_m^{\bullet}$. The latter participates in a chain equilibration process at a

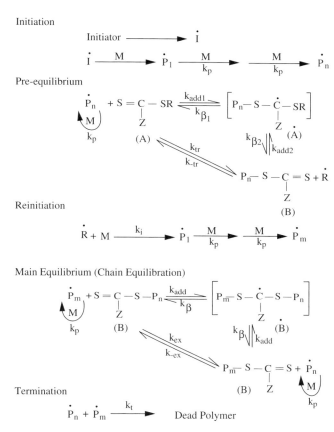

Scheme 3.12 Mechanism of RAFT Polymerization.[445–448]

fast rate ($k_{ex} \gg k_p$) mediated by the polymeric RAFT agent produced in the pre-equilibrium stage. This stage of the polymerization is referred to as 'the main equilibrium'[448] or 'chain equilibration'.[445]

The number of dormant polymer molecules equals the number of molecules of the RAFT agent used. There exists termination as well as a stationary concentration of chain radicals as in a conventional polymerization. However, the number of dead polymer molecules formed is insignificant compared to the number of dormant polymer molecules.

The mechanism shown in Scheme 3.12 finds support in the following evidences. The intermediate radicals A• and B• were identified using ESR spectroscopy[445,449] as were the thiocarbonylthio end groups in polymers using both NMR and MALDI–TOF or electrospray ionization mass spectrometry.[445]

A wide variety of thiocarbonylthio compounds (RSC(=S)Z) has been devised for efficient RAFT polymerization of a wide range of monomers: (meth)acrylates, styrenics, (meth)acrylamides, vinyl esters, *etc.*, in organic or aqueous solvents.[445–448] These are dithioesters (Z = alkyl, aryl), trithiocarbonates (Z = SR), xanthates (Z = OR), dithiocarbamates ($Z=NR_2$), and others. Xanthates (belonging to the family of thiocarbonylthio compounds) were independently developed to mediate living radical polymerization, which was referred to as MADIX (macromolecular design by interchange of xanthates).[450] For the sake of generality, however, the term RAFT will be used in subsequent discussion for living polymerization mediated by all kinds of thiocarbonylthio compounds.

3.15.7.1 Tuning the RAFT agent

In order to achieve success in making RAFT polymerization living the following requirements must be met.

1) The RAFT agent must be consumed very early in the polymerization by undergoing rapid transfer reaction with the chain radical and the transferred radical rapidly reinitiating polymerization. For this to achieve, reinitiation must be at least as fast as propagation, *i.e.*,

$$K_{tr}k_i \geq K_{ex}k_p, \quad (3.154)$$

where

$$K_{tr} = k_{tr}/k_{-tr}, \quad \text{and} \quad K_{ex} = k_{ex}/k_{-ex}.$$

Since $P_n^\bullet$ and $P_m^\bullet$ are indistinguishable, $K_{ex} = 1$ (k_{ex} being equal to k_{-ex}). Eq. (3.154) therefore reduces to

$$k_{tr}k_i \geq k_{-tr}k_p, \quad (3.155)$$

or

$$C_{tr}k_i \geq C_{-tr}k_p, \quad (3.156)$$

where $C_{tr} = k_{tr}/k_p$, and $C_{-tr} = k_{-tr}/k_p$.

2) The rate constant of exchange (main equilibrium in Scheme 3.12) must be much faster than propagation so that all chains grow to nearly equal lengths (*vide* Eq. 3.144). This is achieved with fast addition of chain radical to the polymeric RAFT agent B as well as fast fragmentation of the intermediate radical $B^\bullet$ so that

$$C_{ex} = C_{-ex} \gg 1, \quad (3.157)$$

where $C_{ex} = k_{ex}/k_p$, and $C_{-ex} = k_{-ex}/k_p$.

3) The intermediate radicals $A^\bullet$ and $B^\bullet$ must not undergo side reactions lest retardation occurs and dead chain fraction increases.

3.15.7.1.1 Selection of Z group

The Z group can activate or deactivate the thiocarbonyl double bond as well as influence the stability of the intermediate radicals $A^\bullet$ and $B^\bullet$. In xanthates and dithiocarbamates, the double bond character of the $C=S$ bond is decreased as a result of its conjugation with lone pairs of electrons on O and N centers respectively giving rise to delocalization of the π electrons.[451]

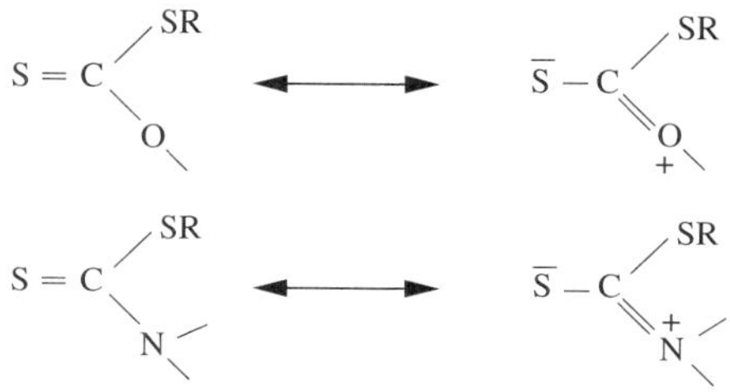

$$S=C\overset{SR}{\underset{O}{\diagdown}} \longleftrightarrow \overline{S}-C\overset{SR}{\underset{\overset{+}{O}}{\diagdown}}$$

$$S=C\overset{SR}{\underset{N}{\diagdown}} \longleftrightarrow \overline{S}-C\overset{SR}{\underset{\overset{+}{N}}{\diagdown}}$$

Deactivation of the C=S bond decreases to a relatively large extent the rate of addition of a relatively stable chain radical to it. Hence, alkyl xanthates or N,N-dialkyl dithiocarbamates are unsuitable as RAFT agents in the polymerizations of strongly reactive monomers such as, styrene, (meth)acrylates, and (meth)acrylamides, but not of weakly reactive ones such as vinyl esters.[451,452]

However, delocalization is hindered by electron-withdrawing (inductive as well as mesomeric) substituents such as carbonyl, phenyl, and substituted phenyls bonded to O or N atoms in xanthates or dithiocarbamates respectively or by incorporating these atoms in aromatic rings. As a result, the rate of radical addition to C=S bond is increased. Appropriate substitution makes these thiocarbonylthio compounds suitable mediators of living polymerization of the strongly reactive monomers as well.[451-453] For example, in methyl methacrylate polymerization satisfactory control is obtained using a dithiocarbamate in which the nitrogen atom is part of an aromatic ring, e.g., -N-pyrrolo

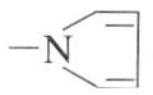

$$-N\!\!\diagup\!\!\!\diagdown\;.$$

Moad, Rizzardo, Thang, and coworkers determined quantitatively the effect of the Z group by measuring the chain transfer coefficients (C_{tr}) of a series of RAFT agents having the same R group, viz., benzyl, but different Z groups in the polymerizations of styrene.[453] The C_{tr} values are given in Table 3.14. C_{tr} decreases in the order of RAFT agents as

dithiobenzoates > trithiocarbonates > xanthates > dithiocarbamates.

The effect of electron-withdrawing substituents bonded to O and N atoms respectively in xanthates and dithiocarbamates in increasing C_{tr} is also evident. For examples, for the xanthates, C_{tr} decreases in the order of Z as

$-OC_6F_5 > -OC_6H_5 > -OC_2H_5.$

Similarly, for the dithiocarbamates, C_{tr} decreases in the order of Z as

$-$N-pyrrolo $> -$N-lactam $> -$N, N-diethyl.

Only the top four RAFT agents in the Table with $C_{tr} \geq \sim 10$ provide polymers with low polydispersity (PDI $\leq$1.2). Similarly, using another series of four RAFT agents for which R is cyano*iso*propyl (CMe_2CN) and Z is the same or similar to the first four entries in Table 3.14, i.e., Z $= -C_6H_5$, $-SCH_3$, $-CH_3$ and -N-pyrrolo, polymers with low polydispersity are also obtained.

Table 3.14 Chain Transfer coefficients of benzyl RAFT agents ($C_6H_5CH_2SC(=S)Z$) in the polymerization of styrene at $110°C$.[453]

Z	C_{tr}^a
$-C_6H_5$	26, 29
$-SCH_2C_6H_5$	18, 20
$-CH_3$	9.2, 10
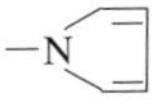 -N(pyrrole)	7.8, 11
$-OC_6F_5$	2.0, 2.3
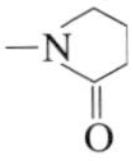 -N(2-pyridone)	1.6
$-OC_6H_5$	0.72
$-OC_2H_5$	0.105
$-N(C_2H_5)_2$	0.007, 0.009

[a]Two different values for the same RAFT agent are due to two different methods used for C_{tr} determination. "Reprinted with permission from Ref. 453. Copyright © 2003 American Chemical Society."

3.15.7.1.2 Selection of R group

For the realization of high chain transfer coefficient not only the rate of addition of a chain radical to the C=S bond but also the rate of forward fragmentation of the intermediate radical (A•) relative to that of the backward fragmentation should be high, i.e., R• should be a better leaving radical than the chain radical. Increase in one or more of the following properties of R•, viz.,[454] (i) resonance stabilization energy (ii) electron-withdrawing ability, and (iii) bulkiness of its α-substituents, increases the rate of forward fragmentation as in the case of dissociation of alkoxyamines or alkyl chlorides discussed earlier (Secs. 3.15.4 and 3.15.6.1). However, R• should be neither too stable nor sterically hindered relative to monomer radical as to be inefficient in reinitiation. For example, R should not be cumyl or 1-phenylethyl or benzyl for vinyl ester polymerization. The chain transfer coefficients of dithiobenzoates (S=C(Ph)S-R) with various R groups in the polymerizations of methyl methacrylate are given in Table 3.15. The coefficient decreases in the following order of R

$$R = -C(CH_3)_2CN \sim -C(CH_3)_2Ph > -C(CH_3)_2C(=O)OCH_3$$
$$> -C(CH_3)_2C(=O)NHR > -C(CH_3)_2CH_2C(CH_3)_3 \geq -CH(CH_3)Ph$$
$$> -C(CH_3)_3 \sim -CH_2Ph.$$

Significantly, only the first two members of the series, viz., $R = -C(CH_3)_2CN$ and $R = -C(CH_3)_2Ph$, which have C_{tr}s ~10 provide efficient control on polymerization of methyl methacrylate. In contrast, all the RAFT agents in the above series effectively control polymerization of styrene and methyl acrylate. Obviously, the corresponding C_{tr} s are relatively large. This is evident from C_{tr} of benzyl dithiobenzoate (entry 5) in the

Table 3.15 Chain transfer coefficients at 60°C for some dithiobenzoates (S=C(Ph)SR) differing in R group in the polymerizations of methyl methacrylate, styrene, and methyl acrylate.[454]

Entry	R	C_{tr} MMA	C_{tr} S	C_{tr} MA
1	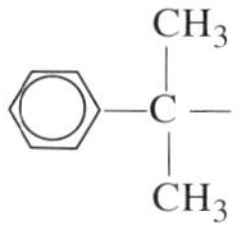 Ph–C(CH$_3$)$_2$–	5.9[a], 56[b]		
2	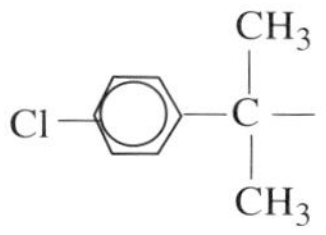 Cl–C$_6$H$_4$–C(CH$_3$)$_2$–	6.6[a]		
3	Ph–CH(CH$_3$)–	0.15[c]	>150[a]	
4	PSt	—	6000 ± 2000[d]	
5	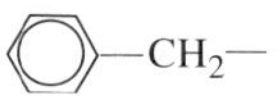 Ph–CH$_2$–	<0.03[c]	~50[a]	105[a]
6	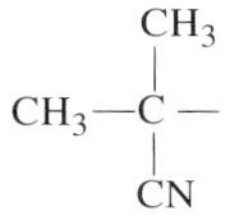 CH$_3$–C(CH$_3$)(CN)–	6.8[a], 25[b]		
7	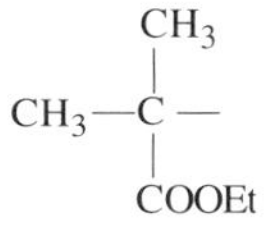 CH$_3$–C(CH$_3$)(COOEt)–	1.7[a]		
8	PMMA	—	140[d]	
9	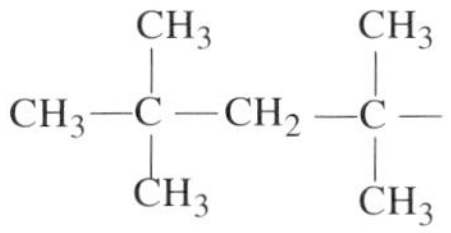 CH$_3$–C(CH$_3$)$_2$–CH$_2$–C(CH$_3$)$_2$–	0.4[c]		

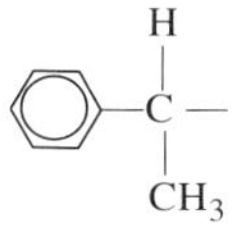

[a]Using Eq. 3.168; [b]by numerically solving Eq. 3.166 which also gives C_{-tr} = 2500 and 450 for R = cumyl and cyanoisopropyl respectively; [c]using Mayo method; [d]by GPC curve resolution method (ref. 455), [e]40°C. "Reprinted in part with permission from Ref. 454. Copyright © 2003 American Chemical Society."

polymerizations of these three monomers. The much smaller C_{tr} of this RAFT agent in the polymerization of methyl methacrylate is attributable to the relatively low rate constant of addition of the poly(methyl methacrylate) radical to RAFT agents (*vide infra*, Table 3.16).

The dramatic increase in C_{tr} on replacing R from monomeric to polymeric alkyl corresponding to methyl methacrylate (entry 8 *vs*. 7) is attributable to the B-strain effect,

Table 3.16 Values of k_{act}, C_{ex}, and k_{add} for some polymeric RAFT and DT agents.

Entry No.	RAFT/DT agent	Monomer	$10^3 k_{act}$, s^{-1}	C_{ex}^a	$k_{add} \times 10^{-3}$ L mol^{-1}s^{-1}	Temp. (°C)	Ref.
1	PS-SC(S)Ph	Styrene	360±120	6000±2000	1930±640	40	455
2	PMMA-SC(S)Ph	MMA	8.4	140	233	60	455
3	PS-SC(S)CH$_3$	Styrene	13	220	70	40	455
4	PS-SC(S)CH$_3$	MMA	0.05	0.83	0.83	40	468
5	PMMA-SC(S)CH$_3$	Styrene	25	420	135	40	468
6	PMMA-SC(S)CH$_3$	MMA	2.4	40	40	40	468
7	PMMA-Macromonomer	MMA	0.013	0.22	0.58	80	465, 466
8	PS-I	Styrene	0.22	3.6	—	80	469

^{a}For $R_p = 4.8 \times 10^{-4}$ mol L^{-1} s^{-1} (see Eq. 3.178).

as discussed earlier in the activation of model initiator *vs.* polymeric halide initiator in the ATRP of this monomer (Sec. 3.15.6.1). The larger B-strain in the polymeric alkyl makes it a better leaving group than the monomeric alkyl. Because of the much lower C_{tr}, the model RAFT agent is not effective in the RAFT polymerization of methyl methacrylate just as in the ATRP of this monomer, the model halide is a relatively inefficient initiator. However, when the alkyl is *secondary*, the C_{tr} for the model RAFT agent is high enough to be effective in the RAFT polymerization of styrene,[454,455] acrylates,[456] or N,N-dimethylacrylamide.[457] This is substantiated, for example, by C_{tr} = 150 for 1-phenylethyl dithiobenzoate in the polymerization of styrene. Remarkably, the C_{tr} of this model RAFT agent is much smaller than that of the corresponding polystyrene-RAFT agent (entries 3 and 4 in Table 3.15) indicating a strong B-strain effect. However, it is noted that the C_{tr} of the model RAFT agent is highly underestimated due to the method used in its determination (*vide infra*).

Although it is possible to fine tune the RAFT agent with appropriate choice of R and Z groups Moad *et al.* observed that "just two RAFT agents" are suitable for universal applications to two groups of monomers.[445] One of these is a *tertiary* cyanoalkyl trithiocarbonate, which is suitable for (meth)acrylates, (meth)acrylamides, and styrenics. The other is a cyanoalkyl xanthate suitable for monomers of relatively low reactivity, *e.g.*, vinyl esters.

3.15.7.2 Kinetics of RAFT polymerization

Since the RAFT polymerization is like a conventional radical polymerization (CRP) carried out in the presence of a RAFT agent, which is essentially a degenerative transfer agent, the kinetics of polymerization should be the same as that of a conventional radical polymerization involving biomolecular termination of chain radicals (Sec. 3.2)

$$R_p = \frac{-d[M]}{dt} = \frac{k_p R_i^{1/2}[M]}{(2k_t)^{1/2}}, \quad (3.20)$$

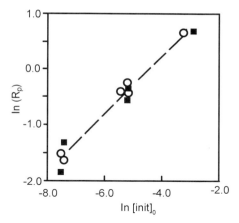

Figure 3.29 Plot of log (initial R_p) vs. log (initial initiator concentration) in bulk RAFT polymerization of methyl methacrylate at 60°C with AIBN initiator (0.0005–0.04 mol L^{-1}) and RAFT agent (0.006 to 0.03 mol L^{-1}), (■) cumyl dithiobenzoate, (o) 2-cyanoprop-2-yl dithiobenzoate. "Reprinted with permission from Ref. 454. Copyright © 2003 American Chemical Society."

and

$$\ln \frac{[M]_0}{[M]} = k_p \left(\frac{R_i}{2k_t} \right)^{1/2} t. \tag{3.123}$$

Indeed, the above kinetics is observed in RAFT polymerizations, which exhibit no retardation. For example, a plot of log (initial R_p) vs. log (initial initiator concentration) is shown in Fig. 3.29 in the bulk RAFT polymerization of methyl methacrylate at 60°C using either cumyl- or cyanoisopropyl dithiobenzoate as the RAFT agent.[454] The least square straight line through the data points has a slope of 0.507 proving adherence to Eq. (3.20). Remarkably, the data points are independent of the RAFT agent concentration over the range indicated except for cumyl dithiobenzoate, which showed slight retardation at the uppermost concentrations of the range.[454]

3.15.7.2.1 Retardation

Retardation as well as inhibition followed by retardation is observed rather commonly in RAFT polymerization at relatively high concentrations of RAFT agents.[458–460] Inhibition is eliminated when the leaving R group is an oligomer of the same monomer that is being polymerized.[447,459,460] This is shown in Fig. 3.30 in the RAFT polymerization of methyl acrylate.[459] Thus, the induction period (inhibition period) is associated with a slower rate of chain transfer in the pre-equilibrium and/or reinitiation stage. However, although the induction period is eliminated using an oligomeric model RAFT agent, retardation is not, as found in the RAFT polymerization of styrene mediated by polystyrene dithiobenzoate (Fig. 3.31).[461] Moad, Rizzardo, Thang, and coworkers attributed the induction period to slow reinitiation by the leaving radical R• and the retardation to slow fragmentation of the polymeric intermediate radical (B•) (Scheme 3.12)

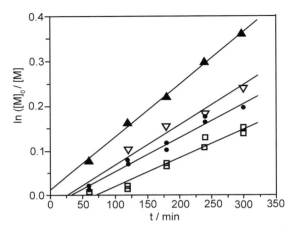

Figure 3.30 First order kinetic plot for monomer disappearance in the bulk polymerization of methyl acrylate at 60°C mediated by RAFT agents, 1-methoxycarbonyl ethyl dithiobenzoate (▽), 2-(2-cyanopropyl) dithiobenzoate (●), 1-phenylethyl dithiobenzoate (□) and poly(methyl acrylate) macro RAFT agent (▲) at an initial concentration of 7.7×10^{-3} mol L^{-1} for each. "Reprinted with permission from Ref. 459. Copyright © 2002 American Chemical Society."

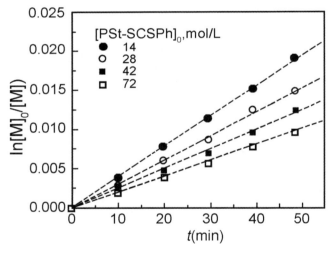

Figure 3.31 First order kinetic plot for monomer disappearance in the AIBN (10 mmol L^{-1}) initiated bulk polymerization of styrene at 60°C mediated by polystyrene dithiobenzoate at concentrations indicated in the figure. Total added polystyrene concentration was kept constant at 71 mmol L^{-1} using polystyrene bromide wherever needed. "Reprinted with permission from Ref. 461. Copright © 2004 American Chemical society."

However, Monteiro and de Brouwer put forward an alternative explanation. They postulated that both the intermediate radicals (A• and B•) partially undergo irreversible termination with chain radicals.[461] However, kinetic simulations of the rate of polymerization and the molecular weight resulted in six orders of magnitude difference in the fragmentation rates between the two models.[462,463]

3.15.7.2.2 Chain transfer constant

The chain transfer reaction refers to the pre-equilibrium stage (Scheme 3.12) and involves the addition of a propagating radical to a RAFT agent (A) followed by fragmentation of the adduct radical (A$^\bullet$).

Assuming a stationary concentration of A$^\bullet$ in the pre-equilibrium stage, it follows that

$$k_{add1}[P^\bullet][A] + k_{add2}[R^\bullet][B] = (k_{\beta 1} + k_{\beta 2})[A^\bullet], \tag{3.158}$$

or

$$[A^\bullet] = \frac{k_{add1}[P^\bullet][A] + k_{add2}[R^\bullet][B]}{k_{\beta 1} + k_{\beta 2}}. \tag{3.159}$$

For $k_{add1} \gg k_{add2}$, Eq. (3.159) reduces to

$$[A^\bullet] = k_{add1}[P^\bullet][A]/(k_{\beta 1} + k_{\beta 2}). \tag{3.160}$$

No reactions of A$^\bullet$ other than fragmentation are assumed in deriving the above equations.

Now, the rate of transfer is given by

$$R_{tr} = k_{tr}[P^\bullet][A] = k_{\beta 2}[A^\bullet]. \tag{3.161}$$

Substituting Eq. (3.160) for [A$^\bullet$] in Eq. (3.161) one obtains solving for k_{tr}

$$k_{tr} = \frac{k_{add1}k_{\beta 2}}{k_{\beta 1} + k_{\beta 2}} \quad (\text{for } k_{add1} \gg k_{add2}). \tag{3.162}$$

In RAFT polymerization, C_{tr} is referred to as chain transfer coefficient rather than chain transfer constant.[454] This is because C_{tr} is experimentally determined using equations that neglect reverse transfer, as will be discussed in the following. This approximation results in an underestimate of C_{tr}.

Less active RAFT agents are consumed at relatively slow rates so that the decrease of RAFT agent concentration from its initial value is negligible at low conversions of monomer to polymer. In such cases, Mayo plot may be used to obtain C_{tr}. On the other hand, with the use of more active RAFT agents, molecular weight increases with conversion even at very low levels of the latter (as is the characteristic of a living polymerization). This creates difficulty in the use of the Mayo equation. Accordingly, alternative methods have been developed as discussed below.[453,454]

Taking into consideration the reversible transfer in the pre-equilibrium stage (Scheme 3.12) and the consumption of monomer during this stage one gets for the rate of consumption of the RAFT agent and the monomer respectively as

$$-\frac{d[A]}{dt} \approx k_{tr}[P^\bullet][A] - k_{-tr}[R^\bullet][B], \tag{3.163}$$

and

$$-\frac{d[M]}{dt} \approx k_p[P^\bullet][M] + k_i[R^\bullet][M]. \tag{3.164}$$

The approximately equal signs used in the above equations are due to the neglect of the other minor pathways of consumptions of A and M such as reactions with initiator-derived radicals.

Dividing Eq. (3.163) by Eq. (3.164) gives[453]

$$\frac{d[A]}{d[M]} \approx \frac{k_{tr}[P^\bullet][A] - k_{-tr}[R^\bullet][B]}{k_p[P^\bullet][M] + k_i[R^\bullet][M]}. \tag{3.165}$$

Eliminating $[R^\bullet]$ and $[P^\bullet]$ in Eq. (3.165) using the stationary state approximation[453,454,464]

$$k_{tr}[P^\bullet][A] - k_{-tr}[R^\bullet][B] = k_i[R^\bullet][M],$$

one obtains using the further approximation $k_i \approx k_p$

$$\frac{d[A]}{d[M]} \approx C_{tr} \frac{[A]}{[M] + C_{tr}[A] + C_{-tr}[B]}. \tag{3.166}$$

Equation (3.166) may be solved numerically to yield values of C_{tr} and C_{-tr}. However, if the second and third terms in the denominator of the equation are negligible in comparison with the first term (which is valid when C_{tr} is ca., $<\sim 10$, and also C_{-tr} is negligible), Eq. (3.166) reduces to[465,466]

$$\frac{d[A]}{d[M]} \approx C_{tr} \frac{[A]}{[M]}, \tag{3.167}$$

or

$$\frac{d \ln[A]}{d \ln[M]} \approx C_{tr}. \tag{3.168}$$

Thus, the slope of a plot of ln [A] vs. ln [M] yields the value of C_{tr}. As discussed earlier, the same method is used to determine chain transfer constant in conventional radical polymerization when the value of the constant exceeds unity (Sec. 3.8.1.4). However, if C_{-tr} is not negligible, the method gives an underestimate of C_{tr}.

The decrease of concentrations of the RAFT agent (A) and the monomer may be followed using NMR spectroscopy.[453] An alternative method suggested by Moad et al. is based on a comparison of the experimental molecular weight with the theoretical one before all the RAFT agent is consumed (Eq. 3.169)[454]

$$f_{c,A} = \frac{[A]_0 - [A]_t}{[A]_0} = \left(\frac{[M]_0 - [M]_t}{[A]_0}\right) \Big/ \left(\frac{[M]_0 - [M]_t}{[A]_0 - [A]_t}\right) = \frac{\overline{DP}_{n(theory)}}{\overline{DP}_{n(exp\,tl)}}, \tag{3.169}$$

where $f_{c,A}$ is the fractional conversion of A, $\overline{DP}_{n(theory)}$ is the theoretical degree of polymerization assuming complete consumption of the RAFT agent and $\overline{DP}_{n(exp\,tl)}$ is the experimentally determined value at time t, the subscripts 0 and t on the concentration symbols refer to values at $t = 0$ and $t = t$ respectively. Obviously, the method can be used only when the RAFT agent is not completely consumed before polymers are isolated for molecular weight measurements.

C_{tr} may also be estimated from the rate of decrease of polydispersity or the narrowing of molecular weight distribution with consumption of monomer/RAFT agent.[453,455,467]

3.15.7.2.3 Rate constant for exchange and activation

The exchange reaction refers to the main equilibrium (Scheme 3.12) in which $P_n^\bullet$ and $P_m^\bullet$ are indistinguishable. Therefore, for a stationary concentration of the intermediate radical ($B^\bullet$) one obtains

$$\frac{d[B^\bullet]}{dt} = k_{add}[P^\bullet][B] - 2k_\beta[B^\bullet] = 0. \qquad (3.170)$$

$$\therefore [B^\bullet] = \frac{k_{add}[P^\bullet][B]}{2k_\beta}. \qquad (3.171)$$

The rate of exchange is given by

$$R_{ex} = k_{ex}[P^\bullet][B] = k_\beta[B^\bullet]. \qquad (3.172)$$

Substituting from Eq. (3.171) for [$P^\bullet$] [B] in Eq. (3.172) and solving for k_{ex} one gets

$$k_{ex} = \frac{k_{add}}{2}. \qquad (3.173)$$

Also, since $P_n^\bullet$ and $P_m^\bullet$ are indistinguishable, $k_{ex} = k_{-ex}$.

$$\therefore k_{-ex} = \frac{k_{add}}{2}, \qquad (3.174)$$

Expressing Eqs. (3.173) and (3.174) in terms of exchange constants

$$C_{ex} = C_{-ex} = \frac{k_{add}}{2k_p}. \qquad (3.175)$$

Treating the exchange reaction kinetics as pseudo first order inasmuch as [$P^\bullet$] is constant under stationary state condition, Eq. (3.172) can be written in the form

$$R_{ex} = k_{ex}[P^\bullet][B] = R_{act} = k_{act}[B], \qquad (3.176)$$

where k_{act} is the rate constant of activation of the polymeric RAFT agent (dormant polymer) and given by

$$k_{act} = k_{ex}[P^\bullet]. \qquad (3.177)$$

Substituting $R_p/(k_p[M])$ for [$P^\bullet$] in Eq. (3.177) one gets

$$k_{act} = k_{ex}[P^\bullet] = R_p k_{ex}/(k_p[M]) = R_p C_{ex}/[M]. \qquad (3.178)$$

Similarly, the rate of deactivation of chain radical is given by

$$R_{deact} = R_{-ex} = k_{-ex}[B][P^\bullet] = k_{deact}[P^\bullet], \qquad (3.179)$$

where

$$k_{deact} = k_{-ex}[B] = k_{ex}[B] = k_p C_{ex}[B]. \qquad (3.180)$$

The value of k_{act} for polymeric RAFT agents may be determined in a similar way as for polymer nitroxides (Sec. 3.15.4.4). An alternative method of determination of k_{act} also exists, which is based on a relation between polydispersity and conversion.[273]

Knowing k_{act}, C_{ex} may be evaluated using Eq. (3.178) for which the rate of polymerization must also be known. The value of C_{ex} in turn may be used to evaluate k_{add}

using Eq. (3.175). The values of these parameters for some RAFT agents determined by Fukuda and coworkers are given in Table 3.16, which includes also the corresponding values for polystyrene iodide and poly (methyl methacrylate)-macromonomer. The k_{add} value of 1.9×10^6 for the addition of a PS$^\bullet$ radical to a PS dithiobenzoate RAFT agent is extremely large for a radical addition reaction and is close to the diffusion-limited value.[455] Besides, the C_{ex} values confirm that dithiobenzoates are better RAFT agents than dithioacetates as has been inferred from C_{tr} values of alkyl RAFT agents (Table 3.14). Furthermore, the C_{ex} values of PMMA-macromonomer and PS-I in the polymerizations of MMA and Styrene respectively prove that they are poor DT agents compared to the polymeric RAFT agents presented in the table.

It is instructive to use the C_{ex} values to have some concrete idea regarding the number of monomer molecules added in a chain per cycle of activation-deactivation, on average. Thus, using typical values of $C_{ex} = 500$ and $k_p = 1000\,\text{L mol}^{-1}\text{s}^{-1}$, [M] = 10 mol/L, and [RAFT] = 10^{-2} mol/L, we have the frequency of deactivation $k_{deact} = k_{ex}$ [RAFT] = $5000\,\text{s}^{-1}$. As already discussed in Sec. 3.15.3.1, an activated chain grows for $(k_{deact})^{-1}$ s. The chain adds in this time interval $k_p[\text{M}](k_{deact})^{-1}$ number of monomer molecules, which in the present case would be only two at the start of the polymerization. Thus, an activated chain is deactivated after every second monomer addition, on average. As [M] decreases with increase in conversion, the number of monomer molecules added to a chain per cycle of activation and deactivation also decreases from the above value. It is reactivated once in every $(k_{act})^{-1}$ s, k_{act} being equal to $k_{ex}[\text{P}^\bullet]$, which gives for the given value of R_p in the footnote of Table 3.16, $k_{act} = 0.024\,\text{s}^{-1}$, *i.e.*, it is activated, on average, once in every 41.6 s.

3.15.7.3 *Synthesis of block copolymers*

For successful block copolymerization, the sequence of blocks should be so selected that a preceding block is a better leaving group than an immediately succeeding block. This ensures complete linking of the successive blocks. Thus, a block of poly (alkyl methacrylate) should precede a block of polystyrene or a block of polyacrylate but not the other way round.[470] Nevertheless, the block copolymer will always be contaminated with homopolymers corresponding to each block since new chains are introduced by the initiator during the construction of a block.

3.15.7.4 *Star polymers*

Two approaches may be used. In both, a multi-RAFT agent is used with either a common multivalent Z group or a multivalent R group.[445] Star polymer formed using the former type RAFT agent contains the RAFT functionalities at the core of the star (Scheme 3.13). In

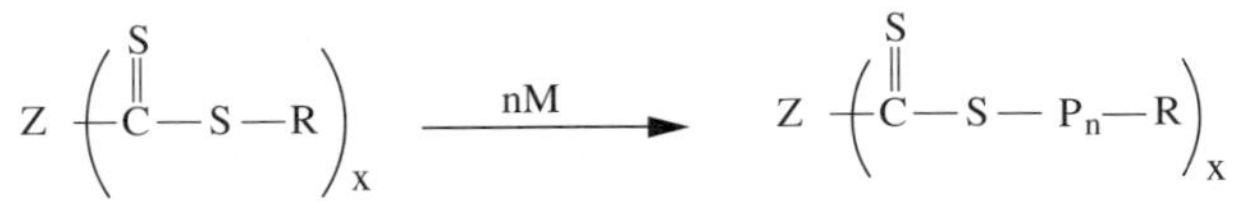

$$Z - \left(\overset{S}{\underset{\|}{C}} - S - R \right)_x \quad \xrightarrow{nM} \quad Z - \left(\overset{S}{\underset{\|}{C}} - S - P_n - R \right)_x$$

Scheme 3.13   Star polymers with RAFT functionalities at the core.

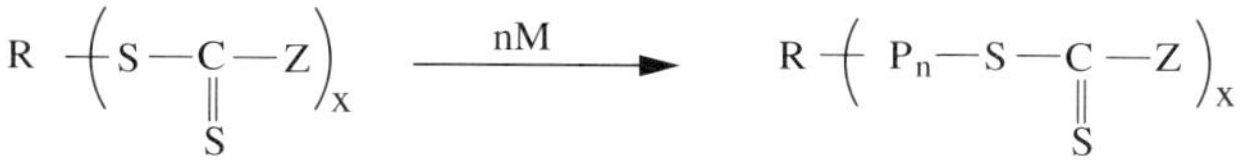

$$R \!-\!\!\left(\!S\!-\!\underset{\underset{S}{\|}}{C}\!-\!Z\right)_{\!x} \xrightarrow{\;nM\;} R\!-\!\!\left(\!P_n\!-\!S\!-\!\underset{\underset{S}{\|}}{C}\!-\!Z\right)_{\!x}$$

Scheme 3.14 Star polymers with RAFT functionalities at the ends of the arms.

contrast, the RAFT functionalities are located at the ends of the arms of the star polymer with the use of the latter type RAFT agent (Scheme 3.14).

References

1. L. Pauling, The Nature of the Chemical Bond, 3rd ed., Cornell University Press, Ithaca, N. Y., p. 261 (1960).
2. T. Alfrey, C.C. Price, *J. Polym. Sci.* (1947) 2, 101.
3. F.R. Mayo, F.M. Lewis, C. Walling, *Discuss. Faraday Soc.* (1947) 2, 285.
4. F.R. Mayo, C. Walling, *Chem. Revs.* (1950) 46, 191.
5. D. Bertin, D. Gigmes, S.R.A. Marque, P. Tordo, *Macromolecules* (2005) 38, 2638.
6. G. Ruchardt, *Angew. Chem. Int. Ed. Engl.* (1970) 9, 830.
7. J.M. Tedder, *Angew. Chem. Int. Ed. Engl.* (1982) 21, 401.
8. B. Giese, *Angew, Chem. Int. Ed. Engl.* (1983) 22, 753.
9. H. Fischer, L, Radom, *Angew. Chem. Int. Ed. Engl.* (2001) 40, 1340.
10. A. Pross, S.S. Shaik, *Acc. Chem. Res.* (1983) 16, 363.
11. P. J. Flory, Principles of Polymer Chemistry, Cornell University Press, Ithaca, New York (1953), Chap. 4.
12. C.H. Bamford, W.G. Barb, A.D. Jenkins, P.F. Onyon, The Kinetics of Vinyl Polymerization by Radical Mechanisms, Butterworths Scientific Pubs., London (1958).
13. R.M. Noyes, *J. Am. Chem. Soc.* (1955) 77, 2042.
14. G. Moad, D.H. Solomon, The Chemistry of Radical Polymerization, 2nd Ed., Elsevier, Amsterdam (2006).
15. J.C. Bevington, *Radical Polymerization*, Academic Press, London (1961).
16. C.H. Bamford, A.D. Jenkins, R. Johnston, *Trans Faraday Soc.* (1959) 55, 1451.
17. W.M. Thomas, E.H. Gleason, J.J. Pellon, *J. Polym. Sci.* (1955) 17, 275.
18. S. Bizilj, D.P. Kelly, A.K. Serelis, D.H. Solomon, K.E. White, Aust. *J. Chem.* (1985) 38, 1657.
19. D.H. Solomon, G. Moad, *Macromol. Symp.* (1987) 10, 109.
20. J.C. Bevington, H.W. Melville, R.P. Taylor, *J. Polym. Sci.* (1954) 12, 449; 1954, 14, 463.
21. J.C. Bevington, S.W. Breuer, T.N. Huckerby, B. J. Hunt, R. Jones, *Eur Polym. J.* (1998) 34, 539.
22. K.C. Berger, *Makromol. Chem.* (1975) 176, 3575.
23. K.C. Berger, G. Meyerhoff, *Makromol Chem.* (1975) 176, 1983.
24. C.A. Barson, J.C. Bevington, B.J. Hunt, *Polymer* (1998) 39, 1345.
25. N.N. Ghosh, B.M. Mandal, *Macromolecules* (1984) 17, 495.
26. J.-M. Bessiere, B. Boutevin, O. Loubet, *Polym. Bull.* (1993) 31, 673.
27. J.C. Bevington, D.E. Eaves, *Trans Faraday Soc.* (1959) 55, 1777.
28. C.H. Bamford, R.W. Dyson, G.C. Eastmond, *Polymer* (1969) 10, 885.
29. C.H. Bamford, E.F.T. White, *Trans Faraday Soc.* (1958) 54, 268.
30. C.H. Bamford, A.D. Jenkins, R. Johnston, *Trans Faraday Soc.* (1959) 55, 179.
31. G. Gleixner, O.F. Olaj, J.W. Breitenbach, *Makromol Chem.* (1979) 180, 2581.
32. M.S. Matheson, E.E. Auer, E.B. Bevilacqua, E.J. Hart, *J. Am. Chem. Soc.* (1949) 71, 497, 2610.

33. M.S. Matheson, E.E. Auer, E.B. Bevilacqua, E.J. Hart, *J. Am. Chem. Soc.* (1951) 73, 1700.
34. M.S. Matheson, E.E. Auer, E.B. Bevilacqua, E.J. Hart, *J. Am. Chem. Soc.* (1951) 73, 5395.
35. N.C. Billingham, A.J. Chapman, A.D. Jenkins, *J. Polym. Sci. Polym. Chem. Ed.* (1980) 18, 827.
36. C.J. Kim, A.E. Hamielec, *Polymer* (1984) 25, 845.
37. F.S. Dainton, M. Tordoff, *Trans. Faraday Soc.* (1957) 53, 499, 677.
38. B. Baysal, A.V. Tobolsky, *J. Polym. Sci.* (1952) 8(5), 529.
39. L.M. Arnett, *J. Am. Chem. Soc.* (1952) 74, 2027.
40. G.V. Schulz, F. Blaschke, *Z. Physik. Chem.* (1942) B51, 75.
41. D.H. Johnson, A.V. Tobolsky, *J. Am. Chem. Soc.* (1952) 74, 938.
42. F.R. Mayo, R.A. Gregg, M.S. Matheson, *J. Am. Chem. Soc.* (1951) 73, 1691.
43. S. Kamenskaya, S. Medvedev, Acta Physicochim (U.R.S.S.) (1940) 13, 565 quoted in Ref. 25.
44. V. Mahedevan, M. Santappa, *Makromol. Chem.* (1955) 16, 119.
45. G.V. Schulz, G. Harborth, *Makromol. Chem.* (1947) 1, 106.
46. C. Walling, L. Heaton, *J. Am. Chem. Soc.* (1965) 87, 38.
47. C. Walling, Free Radicals in Solution (1957), Wiley & Sons. N.Y.
48. K. Nozaki, P.D. Bartlett, *J. Am. Chem. Soc.* (1946) 68, 1686.
49. K. Nozaki, P.B. Bartlett, *J. Am. Chem. Soc.* (1947) 69, 2299.
50. P.D. Bartlett, K. Nozaki, *J. Polym. Sci.* (1948) 3, 216.
51. G.G. Swain, W.H. Stockmeyer, J.T. Clarke, *J. Am. Chem. Soc.* (1950) 72, 5426.
52. R. Hiatt, T. Mill, F.R. Mayo, *J. Org. Chem.* (1968) 33, 1416.
53. J.C. Martin, J.H. Hargis, *J. Am. Chem. Soc.* (1969) 91, 5399.
54. W.A. Pryor, E.H. Morkved, H.T. Bickley, *J. Org. Chem.* (1972) 37, 1999.
55. M. Stickler, E. Dumont, *Makromol. Chem.* (1986) 187, 2663.
56. G. Moad, D.H. Solomon, *The Chemistry of Radical Polymerization*, Elsevier, Amsterdam (2006) Chap. 3.
57. C.E.H. Bawn, D. Verdin, *Trans. Faraday Soc.* (1960) 56, 815.
58. J.C. Bevington, J. Toole, *J. Polym. Sci.* (1958) 28, 413.
59. C.E.H. Bawn, R.G. Halford, *Trans. Faraday Soc.* (1955) 51, 780.
60. Anon "Evaluation of Organic Peroxides from Half-life Data" Technical Bull. Lucidol Division, Pennwalt quoted in Polymer Handbook.
61. I.M. Kothoff, I.K. Miller, *J. Am. Chem. Soc.* (1951) 73, 3055.
62. F.M. Lewis, M.S. Matheson, *J. Am. Chem. Soc.* (1949) 71, 747.
63. C. Decker, *Prog. Polym. Sci.* (1996) 21, 593.
64. G.E. Green, B.P. Stark, S.A. Zahir, *J. Macromol. Sci. Revs. Macromol. Chem.* (1981) C21, 187.
65. S. Maiti, M.K. Saha, S.R. Palit, *Makromol. Chem.* (1969) 127, 224.
66. C.S.H. Chen, *J. Polym. Sci.* (1965) A3, 1127, 1155, 1807.
67. G. Oster, N.-L. Yang, *Chem. Rev.* (1968) 68, 125.
68. W.D. Cook, *Polymer* (1992) 33, 600, 2152.
69. L.E. Mateo, P. Bosch, A.E. Lonzano, *Macromolecules* (1994) 27, 7794.
70. E. Andrzejewska, L.A. Linden, J.F. Rabek, *Makromol. Chem. Phys.* (1998) 199, 441.
71. F. Haber, J.J. Weiss, *Proc. Roy. Soc. (London)* (1934) A147, 332.
72. M.G. Evans, *J. Chem. Soc. (London)* (1947) 266.
73. R.O.C. Norman, in *Chem. Soc. Spl. Publ. — Essays on Free Radical Chemistry; Chem. Soc.*, London (1970) 24, 117.
74. D.A. House, *Chem. Rev.* (1962) 62, 185.
75. E.J. Behrman, J.O. Edwards, *Revs. Inorg. Chem.* (1980) 2, 179.
76. R.G.R. Bacon, *Quart Rev.* (1955) 9, 287.
77. S.R. Palit, T. Guha, R. Das, R.S. Konar, *Encycl. Polym. Sci. Tech.*, H. Mark, N.G. Gaylord, N.M. Bikales, Eds., Vol. 2, p. 299 (1965).
78. A.S. Sarac, *Prog. Polym. Sci.* (1999) 24, 1149.

79. S.R. Palit, B.M. Mandal, *J. Macromol. Sci. Revs. Macromol. Chem.* (1968) C2, 225.
80. B. Vazquez, C. Elvira, J.S. Roman, B. Levenfield, *Polymer* (1997) 38, 4365.
81. I.D. Sideridou, D.S. Achilias, O. Karava, *Macromolecules* (2006) 39, 2072.
82. L. Horner, E. Schwenk, *Angew. Chem.* (1949) 61, 411.
83. L. Horner, *J. Polym. Sci.* (1955) 18, 438.
84. C. Walling, N. Inductor, *J. Am. Chem. Soc.* (1958) 80, 5814.
85. W.A. Pryor, W.H. Hendrickson, *Jr. Tetrahedron Letters* (1983) 24, 1459.
86. T. Sato, S. Kita, T. Otsu, *Makromol. Chem.* (1975) 176, 561.
87. R. Hiatt, K.C. Irwin, C.W. Gould, *J. Org. Chem.* (1968) 33, 1430.
88. J.-F. Lutz, K. Matyjaszewski, *J. Polym. Sci. Polym. Chem.* (2005) 43, 897.
89. T. Sarbu, K.-Y. Lin, J. Ell, D.J. Siegwart, J. Spanswick, K. Matyjaszewski, *Macromolecules* (2004) 37, 3120.
90. S.R. Palit, *Makromol. Chem.* (1959) 36, 89.
91. S.R. Palit, *Makromol. Chem.* (1960) 38, 96.
92. P. Ghosh, P.K. Sengupta, A. Pramanik, *J. Polym. Sci.* (1965) A3, 1725.
93. E. Rizzardo, D.H. Solomon, *J. Macromol. Sci. Chem.* (1979) A13, 997.
94. B.M. Mandal, S.R. Palit, *J. Polym. Sci.*, A-1 (1971) 9, 3301.
95. H.K. Biswas, B.M. Mandal, *Anal. Chem.* (1972) 44, 1636.
96. B.M. Mandal, In Applied Polymer Analysis and Characterization, J. Mitchell, Jr. Ed. Hanser Publishers, Munich-Vienna-N.Y. (1987) Ch. IIB.
97. W.V. Smith, *J. Am. Chem. Soc.* (1949) 71, 4077.
98. G. Ayrey, *Chem. Rev.* (1963) 63, 645.
99. J.C. Bevington, J.R. Ebdon, T.N. Huckerby, N.W.E. Hutton, *Polymer* (1982) 23, 163.
100. G. Moad, D.H. Solomon, S.R. Johns, R.I. Willing, *Macromolecules* (1984) 17, 1094.
101. J.C. Bevington, J.R. Ebdon, T.N. Huckerby, in *NMR Spectroscopy of Polymers*, R.N. Ebbett, Ed; Blackie, London (1993).
102. G. Moad, *Chem. Aust.* (1991) 58, 122.
103. J.C. Bevington, *Trans. Faraday Soc.* (1955) 51, 1392.
104. E. Niki, Y. Kamiya, N. Ohta, *Bull. Chem. Soc. Japan* (1969) 42, 3220.
105. G. Moad, E. Rizzardo, D.H. Solomon, S.R. Johns, R.I. Willing, *Makromol. Chem. Rapid Commun.* (1984) 5, 793.
106. G.S. Hammond, J.N. Sen, C.E. Boozer, *J. Am. Chem. Soc.* (1955) 77, 3244.
107. G.V. Schulz, A. Dinglinger, E. Husemann, *Z. Physik Chem.* (1939) B43, 385.
108. H. Suess, A. Springer, *Z. Physik Chem.* (1937) A181, 81.
109. C. Walling, E.R. Briggs, F.R. Mayo, *J. Am. Chem. Soc.* (1946) 68, 1145.
110. C. Cuthbertson, G. Gee, E.K. Rideal, *Nature* (1937) 140, 889.
111. J.W. Breitenbach, W. Thury, *Experientia* (1947) 3, 281.
112. P.J. Flory, *J. Am. Chem. Soc.* (1937) 59, 241.
113. B.H. Zimm, J.K. Bragg, *J. Polym. Sci.* (1952) 9, 476.
114. F.R. Mayo, *J. Am. Chem. Soc.* (1953) 75, 6133.
115. F.R. Mayo, *J. Am. Chem. Soc.* (1968) 90, 1289.
116. R.R. Hiatt, P.D. Bartlett, *J. Am. Chem. Soc.* (1959) 81, 1149.
117. Y.K. Chang, E. Rizzardo, D.H. Solomon, *J. Am. Chem. Soc.* (1983) 105, 7761.
118. W.C. Buzanowski, J.D. Graham, D.B. Priddy, E. Shero, *Polymer* (1992) 33, 3055.
119. D.B. Priddy, *Adv. Polym. Sci.* (1994) 111, 67.
120. A.N. Hui, A.E. Hamielec, *J. App. Polym. Sci.* (1972) 16, 749.
121. S.R. Palit, S.R. Chatterjee, A.R. Mukherjee, *Encycl. Polym. Sci. Technol.*, H. Mark, N.G. Gaylord, N.M. Bikales Eds. Wiley N.Y., Vol. 3, p. 575 (1965).
122. F.R. Mayo, *J. Am. Chem. Soc.* (1943) 65, 2324.
123. R.A. Gregg, F.R. Mayo, *Disc. Faraday Soc.* (1947) 2, 328.

124. R.A. Gregg, F.R. Mayo, *J. Am. Chem. Soc.* (1948) 70, 2373.
125. R.A. Gregg, F.R. Mayo, *J. Am. Chem. Soc.* (1953) 75, 3530.
126. S. Basu, J.N. Sen, S.R. Palit, *Proc. Roy. Soc. (London)* (1950) A202, 485.
127. C.H. Bamford, E.F.T. White, *Trans. Faraday Soc.* (1956) 52, 716.
128. S.K. Das, S.R. Palit, *Proc. Roy. Soc. (London)* (1954) A226, 82.
129. S.K. Das, S.R. Chatterjee, S.R. Palit, *Proc. Roy. Soc. (London)* (1955) A227, 252.
130. J.T. Clarke, R.O. Howard, W.H. Stockmeyer, *Makromol. Chem.* (1961) 44/46, 427.
131. B.R. Bhattacharyya, U.S. Nandi, *Makromol. Chem.* (1971) 149, 231.
132. G.C. Bhaduri, U.S. Nandi, *Makromol. Chem.* (1969) 128, 183.
133. N. Ya. Kaloforov, E. Borsig, *J. Polym. Sci. Polym. Chem. Ed.* (1973) 11, 2665.
134. R.N. Chadha, J.S. Shukla, G.S. Misra, *Trans. Faraday Soc.* (1957) 53, 240.
135. A.A. Vansheidt, G. Khardi, *Acta Chim. Acad. Sci. Hung. Chem. Abst.* (1960) 54, 6180B.
136. M. Matsumoto, M. Maeda, *J. Polym. Sci.* (1955) 17, 438.
137. D.J. Stein, *Makromol. Chem.* (1964) 76, 170.
138. W.H. Starnes, Jr. et al. *Macromolecules* (1983) 16, 790.
139. S.I. Kuchanov, A.V. Olenin, *Polym. Sci. USSR (Engl. Transl.)* (1973) 15, 2712.
140. P.D. Bartlett, R. Altschul, *J. Am. Chem. Soc.* (1945) 67, 812, 816.
141. K. Nozaki, *Disc. Faraday Soc.* (1947) 2, 337.
142. G. Moad, D.H. Solomon, *The Chemistry of Radical Polymerization*, Elsevier, N. Y. (2006), Chap. 6, p. 279.
143. C. Walling, *J. Am. Chem. Soc.* (1948) 70, 2561.
144. N. Fuhrman, R.B. Mesrobian, *J. Am. Chem. Soc.* (1954) 76, 3281.
145. R.A. Gregg, D.M. Alderman, F.R. Mayo, *J. Am. Chem. Soc.* (1948) 70, 3740.
146. W.V. Smith, *J. Am. Chem. Soc.* (1946) 68, 2059.
147. R.M. Pearson, A.J. Costanza, A.H. Weinstein, *J. Polym. Sci.* (1955) 17, 221.
148. T.A. Fokina et al., *Polym. Sci. USSR* (1967) 8, 2435.
149. J. Chiefari, E. Rizzardo in *Handbook of Radical Polymerization*, T.P. Davis, K. Matyjaszewski Eds. John Wiley & Sons., Hoboken (2002) Chap. 12.
150. G.F. Meijs, E. Rizzardo, *J. Macromol. Sci. Rev. Macromol. Chem. Phys.* (1990) C30, 305.
151. D. Colombani, *Prog. Polym. Sci.* (1999) 24, 425.
152. D. Lim, O. Wichterle, *J. Polym. Sci.* (1958) 29, 579.
153. G.V. Schulz, D.J. Stein, *Makromol. Chem.* (1962) 52, 1.
154. H. Nakamoto, Y. Ogo, S. Imoto, *Makromol. Chem.* (1968) 111, 93.
155. K. Hatada, Y. Terawaki, T. Kitayama, M. Kamachi, M. Tamaki, *Polym. Bull.* (1981) 4, 451.
156. D. Britton, F. Heatley, P.A. Lovell, *Macromolecules* (1998) 31, 2828.
157. T.G. Fox, S. Gratch, *Ann. N.Y. Acad. Sci.* (1953) 57, 367.
158. S. Guha, B. Ray, B.M. Mandal, *J. Polym. Sci. Polym. Chem. Ed.* (2001) 39, 3434.
159. F.A. Bovey, F.C. Schilling, F.L. McCrackin, H.L. Wagner, *Macromolecules* (1976) 9, 76.
160. D.E. Axelson, G.C. Levy, L. Mandelkern, *Macromolecules* (1979) 12, 41.
161. J.C. Randall, J. Macromol, *Sci. Rev. Macromol. Chem. Phys.* (1989) C29, 201.
162. M.J. Roedel, *J. Am. Chem. Soc.* (1953) 75, 6110.
163. W.H. Starnes, Jr., *J. Polym. Sci. Polym. Chem. Ed.* (2005) 43, 2451.
164. Y. Morishima, S.-I. Nozakura, *J. Polym. Sci. Polym. Chem. Ed.* (1976) 14, 1277.
165. M. Asscher, D. Vofsi, *J.C.S.* (1963) 1887, 3921.
166. N.S. Enikolopyan, B.R. Smirnov, G.V. Ponomarev, I.M. Belgovski, *J. Polym. Sci. Polym. Chem. Ed.* (1981) 19, 879.
167. T.P. Davis, D.M. Haddleton, S.N. Richards, *J. Macromol. Sci. Revs. Macromol. Chem. Phys.* (1994) C34, 243.
168. P.A. Clay, R.G. Gilbert, *Macromolecules* (1995) 28, 552.
169. J.P.A. Heuts, T.P. Davis, G.T. Russell, *Macromolecules* (1999) 32, 6019.

170. G. Moad, C.L. Moad, *Macromolecules* (1996) 29, 7727.
171. M.G. Evans, M. Polanyi, *Trans. Faraday Soc.* (1936) 32, 1340; 1938, 34, 11.
172. R.P. Bell, *Proc. Roy. Soc.* (1936) A154, 414.
173. A. Pross, *Theoretical and Physical Principles of Organic Reactivity*, Wiley, New York, 1995.
174. C.H. Bamford, A.D. Jenkins, R. Johnston, *Trans. Faraday Soc.* (1959) 55, 418.
175. G. Henrici-Olive, S. Olive, *Fortschr. Hochpolymer Forsch.* (1961) 2, 496.
176. K. Katagiri, S. Okamura, *J. Polym. Sci.* (1955) 17, 309.
177. C.H. Bamford, A.D. Jenkins, *Trans. Faraday. Soc.* (1963) 59, 530.
178. R.M. Joyce, W.E. Hanford, J. Harman, *J. Am. Chem. Soc.* (1948) 70, 2529.
179. C.M. Starks, *Free Radical Telomerization*, Academic Press, Inc., Orlando, Florida 1974.
180. R. Boutevin, Y. Pietrasanta, M. Taha, *Makromol. Chem.* (1982) 183, 2985.
181. T. Asahara, M. Seno, N. Ohtaur, *Bull. Chem. Soc. Jpn.* (1974) 47, 3142.
182. S. Raynal, J.C. Gautier, M. Gourp, *Eur. Polym. J.* (1979) 15, 317.
183. B. Ameduri, B. Boutevin, *Macromolecules* (1990) 23, 2433.
184. P.D. Bartlett, H. Kwart, *J. Am. Chem. Soc.* (1950) 72, 1051.
185. E.G. Rozantsev, M.D. Gol'dfein, A.V. Trubnikov, *Russ. Chem. Rev. (Engl. Transl.)* (1986) 55, 1070.
186. P.D. Bartlett, G.S. Hammond, H. Kwart, *Disc. Faraday Soc.* (1947) 2, 342.
187. D.J. Hawthorne, D.H. Solomon, *J. Macromol. Sci. Chem.* (1972) 6, 661.
188. G. Moad, D. Shipp, T.A. Smith, D.H. Solomon, *J. Phys. Chem.* (1999) 103, 6580.
189. F.R. Mayo, R.A. Gregg, *J. Am. Chem. Soc.* (1948) 70, 1284.
190. H.W. Melville, W.F. Watson, *Trans. Faraday Soc.* (1948) 44, 886.
191. S.G. Foord, *J. Chem. Soc.* (1940) 48.
192. J.W. Breitenbach, A. Springer, K. Horeischy, Ber. 1938, 71, 1438; 1941, 74, 1386.
193. S.G. Cohen, *J. Am. Chem. Soc.* (1945) 67, 17.
194. C.C. Price, D.H. Read, *J. Polym. Sci.* (1946) 1, 44.
195. G.V. Schulz, *Chem. Ber.* (1947) 80, 232.
196. C.C. Price, *J. Am. Chem. Soc.* (1943) 65, 2380.
197. C.H. Bamford, A.D. Jenkins, R. Johnston, *Trans. Faraday Soc.* (1962) 58, 1212.
198. E.R. Entwistle, *Trans. Faraday Soc.* (1960) 56, 293.
199. W.I. Bengough, W.H. Fairservice, *Trans. Faraday Soc.* (1965) 61, 1206.
200. W.I. Bengough, W.H. Fairservice, *Trans. Faraday Soc.* (1967) 63, 382.
201. W.I. Bengough, T. O'Neil, *Trans Faraday Soc.* (1968) 64, 1014.
202. H. Monteiro, *J. Chim. Phys.*, 192, 59, 9.
203. H. Monteiro, J. Parrod, *Compt. Rend.* (1960) 251, 2026.
204. N.N. Das, M.H. George, *J. Polym. Sci.* (1969) 7A, 269.
205. J.K. Kochi, D.M. Meg, *J. Am. Chem. Soc.* (1965) 87, 522.
206. J.K. Kochi, *Acc. Chem. Res.* (1974) 7, 351; *Rec. Chem. Prog.* (1966) 27, 207.
207. E. Collinson, F.S. Dainton, D.R. Smith, A.J. Trudel, S. Tajuke, *Disc. Faraday Soc.* (1960) 29, 188.
208. B. Atkinson, G.R. Cotten *Trans. Faraday Soc.* (1958) 54, 877.
209. E. Collinson, F.S. Dainton, *Nature* (1956) 177, 1224.
210. F.S. Dainton, P.H. Seaman, *J. Polym. Sci.* (1959) 39, 279.
211. C.E. Barnes, R.M. Elofson, G.D. Jones, *J. Am. Chem. Soc.* (1950) 72, 210.
212. F.A. Bovey, I.M. Kolthoff, *J. Am. Chem. Soc.* (1947) 69, 2113.
213. V.A. Bhanu, K. Kishore, *Chem. Rev.* (1991) 91, 99.
214. C. Decker, A.D. Jenkins, *Macromolecules* (1985) 18, 1241.
215. M. Stickler, In *Comprehensive Polymer Science*, S.G. Eastmond, A. Ledwith, S. Russo, P. Sigwalt Eds. Pergamon, London, Vol. 3, p. 59 (1989).
216. A.M. van Herk, *J. Macromol. Sci. Rev. Macromol. Chem. Phys.* (1997) 37, 633.

217. C.H. Bamford, M.J.S. Dewar, *Proc. Roy. Soc. (London)* (1948) A192, 309; *Disc. Faraday Soc.* (1947) 2, 310.
218. F. Briers, D.L. Chapman, E. Walters, *J. Chem. Soc.* (1926) 562.
219. G.M. Burnett, H.W. Melville, *Proc. Roy. Soc. (London)* (1947) A189, 456.
220. H. Kwart, H.S. Broadbent, P.D. Bartlett, *J. Am. Chem. Soc.* (1950) 72, 1060.
221. B. Yamada, D.G. Westmoreland, S. Kobatake, O'. Konosu, *Prog. Polym. Sci.* (1999) 24, 565.
222. M. Kamachi, *J. Polym. Sci. Polym. Chem. Ed.* (2002) 40, 2699.
223. O.F. Olaj, I. Bitai, F. Hinkelmann, *Makromol. Chem.* (1987) 188, 1689.
224. O.F. Olaj, I. Schnoll-Bitai, *Eur. Polym. J.* (1989) 25, 635.
225. A.P. Aleksandrov, V.N. Genkin, M.S. Kitai, I.M. Smirnova, V.V. Sokolov, *Sov. J. Quant. Electron.* (1977) 5, 547.
226. M. Kamachi, B. Yamada, In *Polymer Handbook*, 4th Ed., J. Bandrup, E.H. Immergut, E.A. Grulke, Eds., John Wiley and Sons, New York (1999) Ch. II, P. 77.
227. M. Buback, R.G. Gilbert, R.A. Hutchinson, B. Klumperman, F-D. Kuchta, B.G. Manders, K.F. O'Driscoll, G.T. Russell, J. Schweer, *Macromol. Chem. Phys.* (1995) 196, 3267.
228. S. Beuermann, M. Buback, T.P. Davis, R.G. Gilbert, R.A. Hutchinson, O.F. Olaj, G.T. Russell, J. Schweer, A.M. van Herk, *Macromol. Chem. Phys.* (1997) 198, 1545.
229. S. Beuermann, M. Buback, T.P. Davis, R.G. Gilbert, R.A. Hutchinson, A. Kajiwara, B. Klumperman, G.T. Russell, *Macromol. Chem. Phys.* (2000) 201, 1355.
230. S. Beuermann, M. Buback, T.P. Davis, N. Garcia, R.G. Gilbert, R.A. Hutchinson *et al.*, *Macromol. Chem. Phys.* (2003) 204, 1338.
231. J.M. Asma, S. Beuermann, M. Buback *et al.*, *Macromol. Chem. Phys.* (2004) 205, 2151.
232. S. Beuerman, M. Bufack, P. Hesse *et al.*, *Pure and Applied Chem.* (2007) 79, 1463.
233. S. Beuermann, M. Buback, *Prog. Polym. Sci.* (2002) 27, 191–254.
234. S. Beuermann, M. Buback, C. Schmaltz, *Macromolecules* (1998) 31, 8069.
235. T.J. Romack, E.E. Maury, J.M. DeSimone, *Macromolecules* (1995) 28, 912.
236. S. Beuermann, D.A. Paqnet Jr. J.H. McMinn, R.A. Hutchinson, *Macromolecules* (1996) 29, 4206.
237. M. Buback, F.-D. Kuchta, *Macromol. Chem. Phys.* (1995) 196, 1887.
238. C. Barner-Kowollik, M. Buback, M. Egorov, T. Fukuda, A. Goto, O.F. Olaj, G.T. Russell, P. Vana, B. Yamada, P.B. Zetterlund, *Prog. Polym. Sci.* (2005) 30, 605.
239. J.B.L. de Kock, A.M. van Herk, A.L. German, *J. Macromol. Sci. Revs. Macromol. Chem.* (2001) C41, 199.
240. M. Buback, H. Hippler, J. Schweer, H.-P. Vogele, *Makromol. Chem. Rapid Commun.* (1986) 7, 261.
241. M. Buback, C. Kowollik, *Macromolecules* (1999) 32, 1445.
242. R. Sack, G.V. Schulz, G. Meyerhoff, *Macromolecules* (1988) 21, 3345.
243. M. Buback, *Makromol. Chem.* (1990) 191, 1575.
244. S.W. Benson, A.M. North, *J. Am. Chem. Soc.* (1962) 84, 935.
245. G.V. Schulz, *Z. Phys. Chem.* (1956) 8, 290.
246. O.F. Olaj, P. Vana, *Macromol. Rapid Commun.* (1998) 19, 433, 533.
247. H.K. Mahabadi, *Macromolecules* (1991) 24, 606.
248. A.A. Gridnev, S.D. Ittel, *Macromolecules* (1996) 29, 5864.
249. G. Moad, E. Rizzardo, D.H. Solmon, A.L.J. Beckwith, *Polym. Bull.* (1992) 29, 647.
250. H. Fischer, L. Radom, *Angew. Chem. Intl. Ed. Eng.* (2001) 40, 1340.
251. R.G.W. Norrish, R.R. Smith, *Nature* (1942) 150, 336.
252. E. Trommsdorff, H. Kohle, P. Lagally, *Macromol. Chem.* (1948) 1, 169.
253. T. Ishige, A.E. Hamielec, *J. Appl. Polym. Sci.* (1973) 17, 1479.
254. C.H. Bamford, W.G. Barb, A.D. Jenkins, P.F. Onyon, "The Kinetics of Vinyl Polymerization by Radical Mechanism", Butterworths, London (1958) Chap. 4.

255. W.H. Carothers, I. Williams, A.M. Collins, J.E. Kirby, *J. Am. Chem. Soc.* (1931) 53, 4203.
256. H. Staudinger, E. Husemann, *Ber Deutsch. Chem. Ges.* (1935) 68, 1618.
257. J.W. Breitenbach, H.F. Kauffmann, G. Zwilling, *Macromol. Chem.* (1978) 180, 2787.
258. A.V. Tobolsky, *J. Am. Chem. Soc.* (1958) 80, 5927.
259. R.H. Gobran, M.B. Berenbaum, A.V. Tobolsky, *J. Polym. Sci.* (1960) 46(148), 431.
260. G.V. Schulz, G. Harborth, *Z. Phys. Chem.* (1939) B43, 25.
261. P.J. Flory, *Principles of Polymer Chemistry*, Cornell University (1953) Chap. 8.
262. S. Penczek, G. Wood, Glossary of terms related to kinetics, thermodynamics and mechanisms of polymerization (IUPAC Recommendation 2007).
263. M.K. Georges, R.P.N. Veregin, P.M. Kazmaier, G.K. Hamer, *Macromolecules* (1993) 26, 2987.
264. B.B. Wayland, G. Poszmik, S.L. Mukherjee, M.J. Fryd, *J. Am. Chem. Soc.* (1994) 116, 7943.
265. M. Koto, M. Kamigaito, M. Sawamoto, T. Higashimura, *Macromolecules* (1995) 28, 1721.
266. J.-S. Wang, K. Matyjaszewski, *J. Am. Chem. Soc.* (1995) 117, 5614.
267. K. Matyjaszewski, J. Xia, *Chem. Rev.* (2001) 101, 2921.
268. M. Kamigaito, T. Ando, M. Sawamoto, *Chem. Rev.* (2001) 101, 3689.
269. S.C. Gaynor, J.-S. Wang, K. Matyjaszewski, *Macromolecules* (1995) 28, 8051.
270. J. Chiefari *et al.*, *Macromolecules* (1998) 31, 5559.
271. K. Matyjaszewski, A.H.E. Muller, *Polym. Prep. (Am. Chem. Soc. Div. Polym. Chem.)* (1997) 38, 6.
272. H. Fischer, *J. Polym. Sci. Polym. Chem.* (1999) 37, 1885.
273. A. Goto, T. Fukuda, *Prog. Polym. Sci.* (2004) 29, 329.
274. A. Goto, T. Fukuda, in *Handbook of Radical Polymerization*, K. Matyjaszewski, T.P. Davis, Eds. Wiley-Inter Science, Chap. 9, p. 407 (2002).
275. K. Matyjaszewski, In *Handbook of Radical Polymerization*, K. Matyjaszewsik, T.P. Davis, Eds. Wiley-Inter Science, Chap. 8, p. 361 (2002).
276. W.A. Braunecker, K. Matyjaszewski, *Prog. Polym. Sci.* (2007) 32, 93.
277. Controlled/Living Radical Polymerization. Progress in ATRP, NMP and RAFT, K. Matyjaszewski Ed. ACS Symp ser. 768, American Chemical Society, Washington DC, 2000.
278. G. Moad, E. Rizzardo, S.H. Thang, *Aust. J. Chem.* (2005) 58, 379.
279. G. Moad, E. Rizzardo, S.H. Thang, *Polymer* (2008) 49, 1079.
280. S. Yamago, K. Iida, J. Yoshida, *J. Am. Chem. Soc.* (2002) 124, 2874, 13666.
281. A. Goto, Y. Kwak, T. Fukuda, S. Yamago, K. Lida, M. Nakajima, J.-I. Yoshida, *J. Am. Chem. Soc.* (2003) 125, 8720.
282. S. Yamago, *et al.*, *J. Am. Chem. Soc.* (2004) 126, 13908.
283. D.F. Grishin, A.A. Moikin, *Vys. Soed Ser A Ser B* (1998) 40, 1266.
284. T. Otsu, M. Yoshida, T. Tazaki, *Makromol. Chem. Rapid Commun.* (1982) 3, 133.
285. T. Otsu, A. Kuriyama, *J. Macromol. Sci. Chem.* (1984) A 21, 961.
286. K. Matyjaszewski, *Macromol Rapid Commun.* (2005) 26, 135.
287. M.J. Parkins, *J. Chem. Soc.* (1964) 5932.
288. E. Daikh, R.G. Finke, *J. Am. Chem. Soc.* (1992) 114, 2939.
289. W. Tang, T. Fukuda, K. Matyjaszewski, *Macromolecules* (2006) 39, 4332.
290. T. Ando, M. Kamigaito, M. Sawamoto, *Tetrahedron* (1997) 53, 15445.
291. P.J. Flory, *J. Am. Chem. Soc.* (1940) 62, 1561.
292. L. Gold, *J. Chem. Phys.* (1958) 28, 91.
293. R. V. Figini, *Makromol. Chem.* (1964) 71, 193; (1967) 107, 170.
294. L.L. Bohm, *Z. Phys. Chem.* (1970) 72, 199; (1974) 88, 297.
295. J.E. Puskas, G. Kaszas, M. Litt, *Macromolecules* (1991) 24, 5278.
296. K. Matyjaszewski, *ACS. Symp. Ser.* (1998) no. 685, K. Matyjaszewski, Ed., American Chemical Society, p. 2; *Macromol. Symp.* (1996) 111, 47.
297. A. Goto, T. Fukuda, *Macromolecules* (1997) 30, 4272.

298. A.H.E. Müller, G. Litvinenko, D. Yan, *Macromolecules* (1995) 28, 7335.
299. D.H. Solomon, E. Rizzardo, P. Caciolli, U.S. Patent, 4, 581, 429 March 27, 1985.
300. D.H. Solomon, *J. Polym. Sci. Polym. Chem.* (2005) 43, 5748.
301. C.J. Hawker, In *Handbook of Radical Polymerization*, K. Matyjaszewski, T.P. Davis, Eds. Wiley-Interscience, Chap. 10, p. 463 (2002).
302. C.J. Hawker, *Acc. Chem. Res.* (1997) 30, 373.
303. D. Benoit, S. Grimaldi, S. Robin, J.-P. Finet, P. Tordo, Y. Gnanou, *J. Am. Chem. Soc.* (2000) 122, 5929.
304. G. S. Ananchenko et al., *J. Polym. Sci. Polym. Chem.* (2002) 40, 3264.
305. A. Goto, T. Fukuda, *Macromol. Chem. Phys.* (2000) 201, 2138.
306. J. Sabek, R. Martschke, H. Fischer, *J. Am. Chem. Soc.* (2001) 123, 2849.
307. S. Marque, H. Fischer, E. Baier, A. Studer, *J. Org. Chem.* (2001) 66, 1146.
308. A. Goto, T. Fukuda, *Macromolecules* (1997) 30, 5183.
309. S. Marque, C.L. Mercier, P. Tordo, H. Fischer, *Macromolecules* (2000) 33, 4403.
310. L. Pauling, *The Nature of the Chemical Bond*, 3rd ed., Cornell University Press, Ithaca, N. Y., pp. 88–105 (1960).
311. A.A. Gridnev, *Macromolecules* (1997) 30, 7651.
312. M.D. Saban, M.K. Georges, R.P.N. Venegin, G.K. Hamer, P.M. Kazmaier, *Macromolecules* (1995) 28, 7032.
313. J.-M. Catala, F. Bubel, S.O. Hammouch, *Macromolecules* (1995) 28, 8441.
314. D. Greszta, K. Matyjaszewski, *Macromolecules* (1996) 29, 5239, 7661.
315. T. Fukuda et al., *Macromolecules* (1996) 29, 6393.
316. M.K. Georges, R.P.N. Veregin, P.M. Kazmaier, G.K. Hamer, M. Saban, *Macromolecules* (1994) 27, 7228.
317. R.P.N. Veregin, P.G. Odell, L.M. Michalak, M.K. George, *Macromolecules* (1996) 29, 4161.
318. P.G. Odell, R.P.N. Veregin, L.M. Michalak, D. Brousmiche, M.K. Georges, *Macromolecules* (1997) 30, 2232.
319. E.E. Malmstrom, C.J. Hawker, R.D. Miller, *Tetrahedron* (1997) 53, 15225.
320. M. Dollin, A.R. Szkurhan, M.K. Georges, *J. Polym. Sci. Polym. Chem.* (2007) 45, 5487.
321. D. Greszta, K. Matyjaszewski, *J. Polym. Sci. Polym. Chem.* (1997) 35, 1857.
322. J.-F. Lutz, P. Lacroix-Desmazes, B. Boutevin, *Macromol. Rapid Commun.* (2001) 22, 189.
323. C. Farcet, M. Lansalot, B. Charleux, R. Pirri, J.-P. Vairon, *Macromolecules* (2000) 33, 8559.
324. N.A. Listigorers, M.K. Georges, P.G. Odell, B. Keoshkerian, *Macromolecules* (1996) 29, 8992.
325. D. Benoit, V. Chaplinski, R. Braslau, C.J. Hawker, *J. Am. Chem. Soc.* (1999) 121, 3904.
326. D. Benoit, E. Harth, P. Fox, R.M. Waymouth, C.J. Hawker, *Macromolecules* (2000) 33, 463.
327. B.B. Wayland, L. Basickes, S.L. Mukerjee, M. Wei, M.L. Fryd, *Macromolecules* (1997) 30, 8109.
328. Z. Lu, M. Fryd, B.B. Wayland, *Macromolecules* (2004) 37, 2686.
329. E. Le Grognec, J. Claverie, R. Poli, *J. Am. Chem. Soc.* (2001) 123, 9513.
330. W.A. Braunecker, Y. Itami, K. Matyjaszewski, *Macromolecules* (2005) 38, 9402.
331. J.P. Claverie, *Res. Descl.* (1998) 416, 1595.
332. A. Debuigne, J.-R. Caille, R. Jerome, *Angew Chem. Intt. Ed.* (2005) 44, 1101.
333. A. Debuigne, J.-R. Caille, C. Detrembleur, R. Jerome, *Angew. Chem. Intl. Ed.* (2005) 44, 3439.
334. A. Debuigne, J.-R. Caille, R. Jerome, *Macromolecules* (2005) 38, 5452.
335. H. Káneyoshi, K. Matyjaszewski, *Macromolecules* (2006) 39, 2757.
336. F. Minisci, *Acc Chem. Res.* (1975) 8, 165.
337. D.P. Curran in *Free Radicals in Synthesis and Biology*, F. Minsci, Ed., Kluwer, Dordrecht, The Netherlands (1989) P. 37.
338. D.P. Curran, *Synthesis* (1988) 489.

339. B. Ameduri and B. Boutevin, *Telomerization in The Encyclopedia of Advanced Materials*, D. Bloor, M.C. Flemings, R.J. Brook, S. Mahajan and R.W. Cahn, Eds. Pergamon Press, Oxford, U.K., p. 2767 (1994).
340. M. Julia, G. Le Thuillier, L. Saussine, *J. Organometal Chem.* (1979) 177, 211.
341. K. Matyjaszewski, J. Xia, *Chem. Rev.* (2001) 101, 2921.
342. M. Kamigaito, T. Ando, M. Sawamoto, *Chem. Rev.* (2001) 101, 3689.
343. V. Cossens, T. Pintauer, K. Matyjaszewski, *Prog. Polym. Sci.* (2001) 26, 337.
344. K. Matyjaszewski, J. Xia in *Handbook of Radical Polymerization*, K. Matyjaszewski, T.P. Davis, Eds., Wiley-Interscience, Chap. 11, p. 523 (2002).
345. N.V. Tsarevsky, K. Matyjaszewski, *Chem. Rev.* (2007) 107, 2270.
346. (a) K. Ohno, A. Goto, T. Fukuda, J. Xia, K. Matyjaszewski, *Macromolecules* (1998) 31, 2699. (b) A. Goto, T. Fukuda, *Macromol. Chem. Rapid Commun.* (1999) 20, 633.
347. T. Ando, M. Kamigaito, M. Sawamoto, *Macromolecules* (2000) 33, 2819.
348. K. Matyjaszewski, J.-L. Wang, T. Grimaud, D.A. Shipp, *Macromolecules* (1998) 31, 1527.
349. (a) H.C. Brown, *Science* (1946) 103, 385. (b) H.C. Brown, H.L. Berneis, *J. Am. Chem. Soc.* (1953) 75, 10.
350. A.K. Nanda, K. Matyjaszewski, *Macromolecules* (2003) 36, 8222.
351. M.B. Gillies, K. Matyjaszewski, P.-O. Norrby, T. Pintauer, R. Poli, P. Richard, *Macromolecules* (2003) 36, 8551.
352. W. Tang, N.V. Tsarevsky, K. Matyjaszewski, *J. Am. Chem. Soc.* (2006) 128, 1598.
353. V. Percec, H.-J. Kim, B. Barboiu, *Macromolecules* (1995) 28, 7970.
354. V. Percec, B. Barboju, H.-J. Kim, *J. Am. Chem. Soc.* (1998) 120, 305.
355. M. Matsuyama, M. Kamigaito, M. Sawamoto, *J. Polym. Sci. A. Polym. Chem.* (1996) 34, 3585.
356. T. Grimaud, K. Matyiaszewski, *Macromolecules* (1997) 30, 2216.
357. D.M. Haddleton, C. Waterson, *Macromolecules* (1999) 32, 8732.
358. T. Nishikawa, M. Kamigaito, M. Sawamoto, *Macromolecules* (1999) 32, 2204.
359. H. Uegaki, Y. Kotani, M. Kamigaito, M. Sawamoto, *Macromolecules* (1997) 30, 2249; (1998) 31, 6756.
360. H. Uegaki, M. Kamigaito, M. Sawamoto, *J. Polym. Sci. Polym. Chem. Ed.* (1999) 37, 3003.
361. D.P. Chatterjee, U. Chatterjee, B.M. Mandal, *J. Polym. Sci. A. Polym. Chem. Ed.* (2004) 42, 4132.
362. J.F. Hester, P. Banerjee, Y.-Y. Won, A. Akhakul, M.H. Acar, A.M. Mayes, *Macromolecules* (2002) 35, 7652.
363. S. Inceoglu, S.C. Olugebefola, M.H. Acar, A.M. Mayes, *Des Monomers Polym.* (2004) 7, 181.
364. Y. Kotani, M. Kamigaito, M. Sawamoto, *Macromolecules* (2000) 336, 6746.
365. K. Endo, A. Yachi, *Polym. Bull.* (2001) 46, 363.
366. D.A. Shipp, J.-L. Wang, K. Matyjaszewski, *Macromolecules* (1998) 31, 8005.
367. K. Matyjaszewski, D.A. Shipp, J.-L. Wang, T. Grimaud, T.E. Patten, *Macromolecules* (1998) 31, 6836.
368. K. Matyjaszewski, D.A. Shipp, G.P. Mcmurtry, S.G. Gaynor, T. Pakula, *J. Polym. Sci. A. Polym. Chem. Ed.* (2000) 38, 2023.
369. D.M. Haddleton, A.M. Heming, D. Kukulj, S.G. Jackson, *Chem. Commun.* (1998) 1719.
370. A. Goto, T. Fukuda, *Macromol. Rapid Commun.* (1999) 20, 633.
371. K. Matyjaszewski, K. Davis, T. E. Patten, M. Wei, *Tetrahedron* (1997) 45, 15321.
372. J.-F. Lutz, K. Matyjaszewski, *Macromol. Chem. Phys.* (2002) 203, 1385.
373. W. Tang, K. Matyjaszewski, *Macromolecules* (2006) 39, 4953.
374. M. Destarac, J.M. Besslere, B. Boutevin, *Macromol. Rapid Commun.* (1997) 18, 967.
375. K. Matyjaszewski, T.E. Pattern, J. Xia, *J. Am. Chem. Soc.* (1997) 119, 674.
376. D.M. Haddleton, C.B. Jasieczek, M.J. Hannon, A.J. Shorter, *Macromolecules* (1997) 30, 2190.
377. J. Xia, K. Matyjaszewski, *Macromolecules* (1997) 30, 7697.

378. J. Xia, S.G. Gaynor, K. Matyjaszewski, *Macromolecules* (1998) 31, 5958.
379. J. Xia, K. Matyjaszewski, *Macromolecules* (1999) 32, 2434.
380. H. Tang et al., *J. Am. Chem. Soc.* (2006) 128, 16277.
381. J. Queffelec, S.G. Gaynor, K. Matyjaszewski, *Macromolecules* (2000) 33, 8629.
382. T.E. Patten, J. Xia, T. Abernathy, K. Matyjaszewski, *Science* (1996) 272, 866.
383. K. Matyjaszewski, B. Gobelt, H.-J. Paik, C.P. Horwitz, *Macromolecules* (2001) 34, 430.
384. T. Ando, M. Kamigaito, M. Sawamoto, *Macromolecules* (2000) 33, 5825.
385. K. Matyjaszewski, Y. Nakagawa, C.B. Jasieczek, *Macromolecules* (1998) 31, 1535.
386. A.K. Nanda, K. Matyjaszewski, *Macromolecules* (2003) 36, 599.
387. A.K. Nanda, K. Matyjaszewski, *Macromolecules* (2003) 36, 1487.
388. S.K. Jewrajka, U. Chatterjee, B.M. Mandal, *Macromolecules* (2004) 37, 4325.
389. U. Chatterjee, S.K. Jewrajka, B.M. Mandal, *Polymer* (2005) 46, 1575.
390. S.B. Lee, A.J. Russell, K. Matyjaszewski, *BioMacromolecules* (2003) 4, 1386.
391. S. McDonald, S.R. Rannard, *Macromolecules* (2001) 34, 8600.
392. K.L. Robinson, M.A. Khan, M.V. de PazBanez, X.S. Wang, S.P. Armes, *Macromolecules* (2001) 34, 3155.
393. K.L. Robinson, M.V. dePazBanez, X.S. Wang, S.P. Armes, *Macromolecules* (2001) 34, 5799.
394. I.Y. Ma, et al., *Macromolecules* (2002) 35, 9306.
395. X-Bories-Azeau, S.P. Armes, *Macromolecules* (2002) 35, 10241.
396. Y. Li, S.P. Armes, X. Jin, S. Zhu, *Macromolecules* (2003) 36, 8268.
397. S.K. Jewrajka, B.M. Mandal, *Macromolecules* (2003) 36, 311.
398. S.K. Jewrajka, B.M. Mandal, *J. Polym. Sci. Polym. Chem.* (2004) 42, 2483.
399. N.V. Tsarevsky, T. Pintauer, K. Matyjaszewski, *Macromolecules* (2004) 37, 9768.
400. W.A. Braunecker, N.V. Tsarevsky, T. Pintauer, R.R. Gil, K. Matyjaszewski, *Macromolecules* (2005) 38, 4081.
401. (a) V. Percec et al. *J. Am. Chem. Soc.* (2006) 128, 14156.
(b) N. H. Nguyen, B.M. Rosen, V. Percec, *J. Polym. Sci. Polym. Chem.* (2010) 48, 1752.
402. (a) J.T. Rademacher, M. Baum, M.E. Pallack, W.J. Brittain, W.J. Simonpick, Jr. *Macromolecules* (2000) 33, 284.
(b) N.V. Tsarevsky, W.A. Braunecker, S.J. Brooks, K. Matyjaszewski, *Macromolecules* (2006) 39, 6817.
403. G. Chambard, B. Klumperman, A.L. German, *ACS Symp. Ser.* (2000) 768, 197.
404. G. Chambard, B. Klumperman, A.L. German, *Macromolecules* (2002) 35, 3420.
405. T. Pintauer, P. Zhou, K. Matyjaszewski, *J. Am. Chem. Soc.* (2002) 124, 8196.
406. C. Yoshikawa, A. Goto, T. Fukuda, *Macromolecules* (2003) 36, 908.
407. K.A. Davis, H.-J. Paik, K. Matyjaszewski, *Macromolecules* (1999) 32, 1767.
408. J.L. Wang, T. Grimaud, K. Matyjaszewski, *Macromolecules* (1997) 30, 6507.
409. (a) T. Pintauer et al., *Macromolecules* (2004) 37, 2679. (b) K. Matyjaszewski et al., *Macromolecules* (2001), 34, 5125.
410. M. Wakioka, K.-Y. Baek, T. Ando, M. Kamigaito, M. Sawamoto, *Macromolecules* (2002) 35, 330.
411. (a) S. Liu, S. Elyashiv, A. Sen, *J. Am. Chem. Soc.* (2001) 123, 12738. (b) S. Elyashiv, N. Greinert, A. Sen, *Macromolecules* (2002) 35, 7521.
412. S. Liu, A. Sen, *J. Polym. Sci. Polym. Chem.* (2004) 42, 6175.
413. R. Venkatesh, B. Klumperman, *Macromolecules* (2004) 37, 1226.
414. A. Ramakrishnan, R. Dhamodharan, *Macromolecules* (2003) 36, 1039.
415. J. Xia, K. Matyjaszewski, *Macromolecules* (1997) 30, 7692.
416. G. Moineau, P. Dubois, R. Jerome, T. Senninger, P. Teyssie, *Macromolecules* (1998) 31, 545.
417. J. Xia, K. Matyjaszewski, *Macromolecules* (1999) 32, 5199.
418. J. Gromoda, K. Matyjaszewski, *Macromolecules* (2001) 34, 7664.

419. M. Li, N.M. Jahed, K. Min, K. Matyjaszewski, *Macromolecules* (2004) 37, 2434.
420. W. Jakubowski, K. Matyjaszewski, *Macromolecules* (2005) 38, 4139.
421. K. Min, H. Gao, K. Matyjaszewski, *J. Am. Chem. Soc.* (2005) 127, 3825.
422. H. Tang, M. Radosz, Y. Shen, *Macromol. Rapid Commun.* (2006) 27, 1127.
423. K. Matyjaszewski, S. Coca, S.G. Gaynor, M. Wei, B.E. Woodworth, *Macromolecules* (1998) 31, 5967.
424. W. Jakubowski, K. Min, K. Matyjaszewski, *Macromolecules* (2006) 39, 39.
425. W. Jakubowski, K. Matyjaszewski, *Angew Chem. Intl. Ed. Eng.* (2006) 45, 4482.
426. H. Dong, W. Tang, K. Matyjaszewski, *Macromolecules* (2007) 40, 2974.
427. K. Min, H. Gao, K. Matyjaszewski, *Macromolecules* (2007) 40, 1789.
428. J. Pietrasik, H. Dong, K. Matyjaszewski, *Macromolecules* (2006) 39, 6384.
429. K. Matyjaszewski *et al.*, *Proc. Natl. Acad. Sci. USA* (2006) 103, 15309.
430. B.W. Mao, L.H. Gan, Y.Y. Gan, *Polymer* (2006) 47, 3017.
431. W. Jakubowski, B. Kirci-Denizli, R.R. Gil, K. Matyjaszewski, *Macromol. Chem. Phys.* (2008) 209, 32.
432. K. Matyjaszewaski, S.M. Jo, H.-J. Paik, D.A. Shipp, *Macromolecules* (1999) 32, 6431.
433. I. Li, B.A. Howell, K. Matyjaszewski, T. Shigemoto, P.B. Smith, D.B. Priddy, *Macromolecules* (1995) 28, 6692.
434. D.A. Shipp, J.-L. Wang, K. Matyjaszewski, *Macromolecules* (1998) 31, 8005.
435. S.C. Hong, T. Pakula, K. Matyjaszewski, *Macromol. Chem. Phys.* (2001) 202, 3392.
436. L. Mueller, W. Jacubowski, W. Tang, K. Matyjaszewski, *Macromolecules* (2007) 40, 6464.
437. C. Burguiere, M.A. Dourges, B. Charleux, J.-P. Vairon, *Macromolecules* (1999) 32, 3883.
438. J. Nicholas, *et al.*, *Macromolecules* (2006) 39, 8274.
439. D.M. Haddleton, C. Waterson, *Macromolecules* (1999) 32, 8732.
440. K. Matyjaszewski, *Polym. Intl.* (2003) 52, 1559.
441. J. Huang, S. Jia, D.J. Siegwart, T. Kowalewski, K. Matyjaszewski, *Macromol. Chem. Phys.* (2006) 207, 801.
442. J. Chiefari, Y.K. Chong, F. Ercole, J. Krstina, J. Jeffery, T.P.T. Le, R.T.A. Mayadunne, G.F. Meijs, C.L. Moad, G. Moad, E. Rizzardo, S.H. Thang, *Macromolecules* (1998) 31, 5559.
443. T.P.T. Le, G. Moad, E. Rizzardo, S.H. Thang, *Int. Pat. Appl.*, WO 9801478, 1998.
444. P. Cacioli, D.G. Hawthorne, R.L. Laslett, E. Rizzardo, D.H. Solomon, *J. Macromol. Sci. Chem.* (1986) 23, 839.
445. G. Moad, E. Rizzardo, S.H. Thang, *Aust. J. Chem.* (2005) 58, 379.
446. G. Moad, E. Rizzardo, S.H. Thang, *Polymer* (2008) 49, 1079.
447. S. Perrier, P. Takolpuckdee, *J. Polym. Sci. Polym. Chem.* (2005) 43, 5347.
448. C. Barner-Kowollik *et al.*, *J. Polym. Sci. Polym. Chem.* (2006) 44, 5809.
449. D.G. Hawthorne, G. Moad, E. Rizzardo, S.H. Thang, *Macromolecules* (1999) 32, 5457.
450. D. Charmot, P. Corpart, H. Adam, S.Z. Zard, T. Biadatti, G. Bouhadir, *Macromol. Symp.* (2000) 150, 23.
451. R.T.A. Mayadunne, E. Rizzardo, J. Chiefari, Y.K. Chong, G. Moad, S.H. Thang, *Macromolecules* (1999) 32, 6977.
452. M. Destarac, D. Charmot, X. Franck, S.Z. Zard, *Macromol. Rapid Commun.* (2000) 21, 1035.
453. J. Chiefari, R.T.A. Mayadunne, C.L. Moad, G. Moad, E. Rizzardo, A. Postma, M.A. Skidmore, S.H. Thang, *Macromolecules* (2003) 36, 2273.
454. Y.K. Chong, J. Krstina, T.P.T. Le, G. Moad, A. Postma, E. Rizzardo, S.H. Thang, *Macromolecules* (2003) 36, 2256.
455. A. Goto, K. Sato, Y. Tsuji, T. Fukuda, G. Moad, E. Rizzardo, S.H. Thang, *Macromolecules* (2001) 34, 402.
456. S. Perrier, C. Barner-Kowollick, J.F. Quinn, P. Vana, T.P. Davis, *Macromolecules* (2002) 35, 8300.

457. M.S. Donovan, A.B. Lowe, B.S. Sumerlin, C.L. McCormick, *Macromolecules* (2002) 35, 4123.
458. G. Moad, R.T.A. Mayadunne, E. Rizzardo, M. Skidmore, S. Thang, *ACS Symp. Ser.* (2003) 854, 520.
459. S. Perrier, C. Barner-Kowollik, J.F. Quinn, P. Vana, T.P. Davis, *Macromolecules* (2002) 35, 8300.
460. R.T.A. Mayadunne, E. Rizzardo, J. Chiefari, J. Krstina, G. Moad, A. Postma, S.H. Thang, *Macromolecules* (2000) 33, 243.
461. Y. Kwak, A. Goto, T. Fukuda, *Macromolecules* (2004) 37, 1219.
462. M.J. Monteiro, H. de Brouwer, *Macromolecules* (2001) 34, 349.
463. C. Barner-Kowollik, J.F. Quinn, D.R. Morsley, T.P. Davis, *J. Polym. Sci. Polym. Chem.* (2001) 39, 1353.
464. G. Moad et al., *Macromol. Symp.* (2002) 182, 65.
465. C.L. Moad, G. Moad, E. Rizzardo, S.H. Thang, *Macromolecules* (1996) 29, 7717.
466. L. Hutson et al., *Macromolecules* (2004) 37, 4441.
467. C. Barner-Kowollik, J.F. Quinn, T.L.U. Nguyen, J.P.A. Heuts, T.P. Davis, *Macromolecules* (2001) 34, 7849.
468. C. Barner-Kowollik, J.F. Quinn, D.R. Morsley, T.P. Davis, *J. Polym. Sci. Polym. Chem.* (2001) 39, 1353.
469. T. Fukuda, A. Goto, Y. Kwak, C. Yoshikawa, Y.-D. Ma, *Macromol. Symp.* (2002) 182, 53.
470. B.Y.K. Chong, T.P.T. Le, G. Moad, E. Rizzardo, S.H. Thang, *Macromolecules* (1999) 32, 2071.

Problems

3.1 What are the typical ranges of R_i and $[P^\bullet]$ in a conventional radical polymerization? Comment on the values.

3.2 The bulk polymerization of a monomer ($[M] = 10\,\text{mol/L}$) reaches 50% conversion in 5 h. Calculate (a) the rate of polymerization in mol $L^{-1}\,s^{-1}$ unit, (b) the percentage conversion in 5 h if the monomer is diluted with an equal amount of solvent assuming that the polymerization is first order in monomer. (Ans: (a) $3.8 \times 10^{-4}\,\text{mol}\,L^{-1}\,s^{-1}$ at start and half as much at 50% conversion, (b) 50%.)

3.3 In a conventional radical polymerization, the monomer half life is 3 h. How long will the next half life be? (Ans: 3 h, assuming initiator decomposition is negligibly small).

3.4 The initiator-derived end group in a chain polymer measures 1 mmol in 100 g of polymer. Calculate the kinetic chain length and $\overline{M}_n$ if disproportionation accounts for 80% of total termination (assuming chain transfer to be absent and monomer molecular weight = 100). (Ans: $\nu = 1000$, $\overline{M}_n = 111100$.)

3.5 Given the following data, calculate $\overline{DP}_n$ of polystyrene. $\nu = 10^4$, $C_M = 5 \times 10^{-5}$, $C_I = 0$, $C_S = 10^{-3}$, $[S]/[M] = 1$ (S = solvent), $k_{td} = 0$. Calculate also the percentages of polymer molecules formed by each of the following reactions: (i) transfer to monomer and solvent each, and (ii) initiation and termination reactions combined. (Ans: 1st part: $\overline{DP}_n = 909$. 2nd part: (i) 4.55% and 90.9%, and (ii) 4.55%).

3.6 The effect of chain transfer is diluted in polymerizations with short kinetic chain lengths — explain.

3.7 Low molecular weight polymers are likely to be less branched than high molecular weight ones. Comment on the statement.

3.8 Chain transfer of acetone in the polymerization of styrene is faster than in the polymerization of MMA, even though the PS radical is less reactive than the PMMA radical. What could be the possible reason?

3.9 In polymerizations initiated by a thermal initiator molecular weight decreases with increase in temperature, whereas in those initiated by a redox initiator it is hardly affected with change in temperature. Explain how this is possible. How would the molecular weight change with increase in temperature in a photopolymerization?

3.10 Polymerization of styrene at 60°C reaches dead end with AIBN as the initiator, but that of methyl methacrylate does not do so. On the other hand, polymerization of both monomers reaches dead end when initiated with a redox initiator. Explain.

3.11 In solution polymerization of allyl acetate, low molecular weight polymer ($\overline{DP}_n \approx 14$) is formed irrespective of initiator and/or monomer concentrations. Explain how this is possible.

3.12 TEMPO mediated NMP of styrene is successfully performed at 120°C, whereas that of methyl acrylate is not — why?

3.13 Some NMP cannot be performed at practically useful rates without conventional initiation. Explain with examples.

3.14 Copper mediated normal ATRP of methyl methacrylate is uncontrolled with Me$_6$TREN as the ligand in the copper complex, whereas ARGET ATRP using the same complex as catalyst is controlled. How is this possible?

3.15 In living polymerization, initiation should be at least as fast as propagation in order that polymer of low polydispersity is formed. In this respect, use of slowly initiating conventional initiator in RAFT as well as in some NMP appears paradoxical. Explain the paradox.

3.16 In block ATR copolymerization of methyl acrylate and methyl methacrylate, which monomer is to be polymerized first and why? (Ans: MMA).

3.17 Referring to problem 3.16 the block sequence to be followed can be reversed by two measures. What are these? Discuss their principles of action.

3.18 NMP of methyl methacrylate is uncontrolled. What could be the reason? How might the polymerization be controlled?

3.19 ATRP and RAFT polymerizations of methyl methacrylate are not satisfactorily controlled with the use of a model initiator and RAFT agent respectively, while the problem does not show up with styrene or methyl acrylate as the monomer. Explain why it is so.

3.20 ATRP of ethylene is not feasible due to an extremely low equilibrium constant. Yet ATR copolymerization of ethylene and methyl acrylate is successful with the usual Cu based catalysts. Explain what makes it possible.

3.21 PMDETA or Me$_6$TREN cannot be used as ligands in copper catalysts in the ATRP of methyl methacrylate in aqueous alcohol media due to disproportionation of the catalysts. In contrast, the problem does not show up with the use of acrylamide as the monomer. How is this possible?

3.22 What are the problems in achieving high molecular weight living polymers by radical polymerization? How can these be solved?

Chapter 4

Anionic Polymerization

Anionic polymerization refers to chain polymerization in which the active center is anionic in character. Depending on the concentration of the active center as well as the dielectric constant, the dipole moment, and the counterion coordinating ability of the solvent and/or any additive present in the polymerization medium various ionic species: ions, ion pairs, and higher order ion aggregates, may exist in equilibrium with each other. Besides, the anion may be carbon-, oxygen-, sulfur-, or nitrogen-centered. Nevertheless, carbanionic polymerization is referred to commonly as anionic polymerization. However, the reactivity of carbanions is much higher than that of the other anions. Hence, carbanionic polymerization has been dealt with exclusively in this chapter. Other anionic polymerizations have been discussed in the chapter on ring-opening polymerization (Chap. 7).

In general, vinyl monomers capable of polymerizing by the anionic mechanism are those with substituents that resonance stabilize the carbanions.[1,2] Examples include nonpolar monomers such as styrene, butadiene, and isoprene, polar monomers such as (meth)acrylates, vinyl pyridines, vinyl ketones, N–alkylacrylamides, N,N–dialkylacrylamides, acrylonitrile, and cyanoacrylates. However, although ethylene does not have an anion-stabilizing substituent, it undergoes polymerization to high molecular weight polymer under suitable conditions with the use of the highly reactive initiator system comprising alkyllithium and tetramethylethylenediamine[1,2] (Sec. 4.1.2.2).

Anionic polymerization is the first chain polymerization, which could be conducted without the occurrence of chain termination and chain transfer. Such polymerization was given the name "living" by its discoverer Michael Szwarc. The discovery made possible for the first time exert immense control on polymerization, molecular weight and its distribution, polymer end groups, and molecular architecture.[3,4] Most of the foundational work on living polymerization were done using styrene, butadiene, isoprene, and methyl methacrylate as monomers.

4.1 Living Anionic Polymerization

4.1.1 *General features*

Chain termination and transfer in living anionic polymerization are prevented by proper choice of solvents and initiators. Hydrocarbons and ethers are the solvents of choice by virtue of their extremely low acidity and high resistance to nucleophilic attack. The former property makes proton transfer to carbanionic species (both free ion and ion pair) unfavorable[3,4]

$$\sim\!\!\sim\!\!\sim \overline{C}\,Mt^+ + RH \xrightarrow{\quad\times\quad} \sim\!\!\sim\!\!\sim C-H + R^-\,Mt^+,$$

where Mt^+ is an alkali ion and RH is a hydrocarbon or an ether solvent, and $\sim C^-Mt^+$ represents a polymer with an ion pair at chain end. The free ion is not shown in the above representation of the absence of proton transfer for the sake of brevity. This practice has been followed also with other reactions throughout this chapter, unless indicated otherwise.

Ethers are, of course, attacked nucleophilically by carbanions

$$\sim\!\!\sim\!\!\sim C^- + R-O-R' \longrightarrow \sim\!\!\sim\!\!\sim CR/R' + {}^-OR'/R.$$

However, the reaction is too slow to affect the living character in the time scale of polymerization even with the strongly nucleophilic polystyryl or polydienyl anion (Sec. 4.1.3.1).

Other solvents such as chlorinated hydrocarbons, esters, and ketones are unsuitable since they are more acidic and susceptible to nucleophilic attack by carbanionic species. Like the polar solvents, the polar monomers and their polymers also undergo nucleophilic attack on polar groups. This makes living anionic polymerization of polar monomers difficult to accomplish (Sec. 4.1.4).

In the first living polymerization described by Szwarc, styrene was polymerized at room temperature in THF. Sodium naphthalene was used as the initiator by virtue of its solubility and ability to initiate polymerization rapidly by electron transfer to monomer, which is important for achieving polymer of narrow MWD (Sec. 4.1.1.1).[3-5] Although polymerizations of nonpolar monomers such as styrene, butadiene, and cyclopentadiene initiated by sodium naphthalene in ethereal solvents were discovered much earlier by Scott, the living nature of these polymerizations was not recognized then.[6]

However, even though no spontaneous termination exists, living anionic polymerization is terminated by adventitious impurities like water, carbon dioxide, and oxygen, as shown in Scheme 4.1.[3]

Hence, the polymerization must be performed using stringently purified monomers, solvents, and reagents in vessels from which air and moisture have been scrupulously excluded. However, purposeful termination ("killing") of the polymerizations with appropriate reagents leads to technologically useful end-functionalized polymers[3-5] (Sec. 4.1.5).

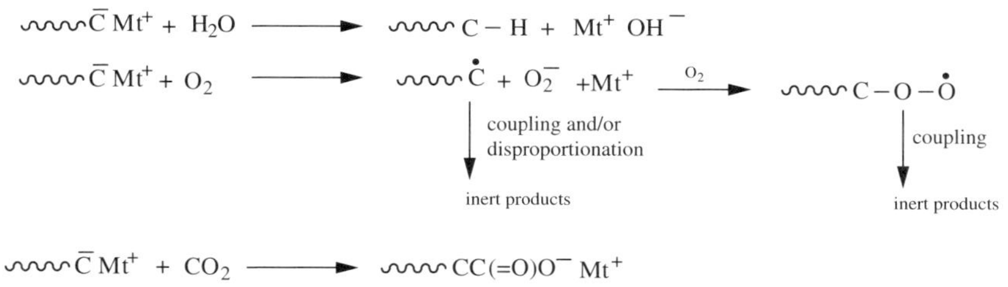

$$\sim\sim\sim\bar{C}\,Mt^+ + H_2O \longrightarrow \sim\sim\sim C-H + Mt^+\,OH^-$$

$$\sim\sim\sim\bar{C}\,Mt^+ + O_2 \longrightarrow \sim\sim\sim\dot{C} + O_2^- + Mt^+ \xrightarrow{O_2} \sim\sim\sim C-O-\dot{O}$$

↓ coupling and/or disproportionation

inert products

↓ coupling

inert products

$$\sim\sim\sim\bar{C}\,Mt^+ + CO_2 \longrightarrow \sim\sim\sim CC(=O)O^-\,Mt^+$$

Scheme 4.1 Adventitious termination in living anionic polymerization.

The propagation reaction may be represented as

$$P_n^* + M \rightarrow P_{n+1}^*, \tag{4.1}$$

where P_n^* represents all types of ionic active centers (free ions, ion pairs, and ion pair aggregates) of chain length n, and M represents the monomer.

Due to the absence of termination, the active center concentration remains constant during polymerization. Accordingly, the polymerization exhibits first order kinetics. The rate of monomer disappearance is given by

$$-\frac{d[M]}{dt} = k_p^{app}[P^*][M], \tag{4.2}$$

where [M] is the monomer concentration, [P*] is the total concentration of living ends (active centers), *i.e.*, $[P^*] = \sum[P_n^*]$, and k_p^{app} is the apparent rate constant of propagation, which takes into account the different rate constants of propagation on different coexisting ionic species (Sec. 4.1.3.1).

Integration of Eq. (4.2) between limits $[M] = [M]_0$ and [M] at time t = 0 and t respectively gives

$$\ln\frac{[M]_0}{[M]} = k_p^{app}[P^*]t. \tag{4.3}$$

The living nature of the polymerization may be verified from the resumption of polymerization that occurs when fresh monomer is added to a living polymer solution after the polymerization of the previously added monomer is completed.[3-5] When the two monomers are different, a block copolymer forms when certain electronegativity requirements for the monomers are met (Sec. 4.1.6). The block copolymer is a di- or a triblock according as the living polymer has respectively one or both ends living.[4] The triblock copolymer may be prepared also by sequential polymerization of the concerned monomers using a monofunctional initiator. Polymers of various topologies and architectures such as linear, graft, star, comb, and dendritic (co)polymers can be prepared using appropriately designed initiators and/or coupling of living chains using suitable coupling agents. The syntheses of some of these (co)polymers are discussed in Secs. 4.1.7 and 4.1.8.

4.1.1.1 *Molecular weight and molecular weight distribution*

The degree of polymerization in a living polymerization is given by Eq. (4.4)

$$\overline{DP}_n = \frac{[M]_0 f_c}{[I]_0}, \qquad (4.4)$$

where $[M]_0$, $[I]_0$, and f_c represent initial monomer concentration, initial initiator concentration, and fractional conversion respectively.

Molecular weight distribution (MWD) should be that of Poisson under the ideal conditions discussed in Chap. 3 in the section on living radical polymerization (Sec. 3.15.3). Broadening of MWD due to: (i) initiation slower than propagation and (ii) involvement of more than one kind of propagating species differing in reactivity and slow exchange between them has been discussed also there. The effect of the latter in living anionic polymerization needs elaboration since the types of propagating species are different from those existing in living radical polymerization.

The propagating ionic species in ionic polymerization may be free ions, ion pairs, and associated ion pairs existing in equilibrium with each other. Besides, the ion pairs may be of two kinds, *viz.*, contact, and solvent-separated (Scheme 4.2). All the species may participate in propagation at widely different rates or some of the species may not be present or reactive at all. Under these circumstances, if the exchange between the various species is not fast in the time scale of propagation, all molecules will not grow at about the same rate resulting in broadened MWD. In the case of slow exchange between the species, MWD may not be even unimodal.[7-11]

$$(P_n^- Mt^+)_2 \rightleftharpoons 2\, P_n^-\, Mt^+ \rightleftharpoons 2\, P_n^-\, |\,S\,|\, Mt^+ \rightleftharpoons 2\, P_n^- + 2\, Mt^+\, (S) \qquad (4.5)$$

| dimeric ion pair | monomeric contact ion pair | monomeric solvent-separated ion pair | free anion |

Scheme 4.2 Various ionic species that may coexist in equilibrium in living anionic polymerization.

Consider the relatively simple case of a polymerization involving dormant dimeric and active monomeric ion pairs existing in dynamic equilibrium, which is encountered, for example, in the living anionic polymerization of methyl methacrylate in ethereal solvents (Sec. 4.1.4.2).[11]

$$(P_n^- Mt^+)_2 \underset{k_A}{\overset{k_D}{\rightleftharpoons}} 2\, P_n^-\, Mt^+$$
$$(P^{\pm})_2 \qquad\qquad\qquad (P^{\pm})$$

The dimeric ion pairs are propagation-inactive, whereas the monomeric ones are active and do all the polymerization. Accordingly, adaptation of the PDI Eq. (3.140) developed in

Chap. 3 to this system gives

$$\text{PDI} = 1 + \frac{k_p^{\pm}[I]_0}{k_A[P^{\pm}]} \left(\frac{2}{f_c} - 1\right), \qquad (4.6)$$

where $k_A[P^{\pm}]$ is the pseudo first order rate constant of association (deactivation) of reactive monomeric ion pairs to dormant dimeric ion pairs, $[P^{\pm}]$ is the total concentration of monomeric ion pairs of all chain lengths, $[I]_0$ is the initial initiator concentration, f_c is the fractional conversion, and $k_p^{\pm}$ is the rate constant of propagation on monomeric ion pairs. The rate constants $k_A[P^{\pm}]$ and $k_p^{\pm}$ are respectively the equivalents of k_{deact} and k_p in Eq. (3.140).

Since $[P^{\pm}] = \alpha[I]_0$, where α is the degree of dissociation, one gets from Eq. (4.6) by substituting $\alpha[I]_0$ for $[P^{\pm}]$

$$\text{PDI} = 1 + \frac{k_p^{\pm}}{\alpha k_A} \left(\frac{2}{f_c} - 1\right). \qquad (4.7)$$

Thus, polymer with narrow MWD is obtained when $k_p^{\pm} \ll \alpha k_A$, *i.e.*, exchange (deactivation) is much faster than propagation.[11] The quantity $\bar{n} = k_p^{\pm}[M]/\alpha k_A c$ gives, on average, the ratio of the frequency of propagation on an ion pair to that of deactivation, c being the active center concentration, *i.e.*, $[I]_0$. With $[M]/c$ typically 200, to start with, and $k_p^{\pm}/\alpha k_A = 0.05$, $\bar{n}$ turns out to be 10. This means that a chain adds 10 monomer molecules in a cycle of activation-deactivation at the start of polymerization, on average. The number gradually decreases with the progress of polymerization as $[M]$ decreases.

PDI equation has been derived also for systems in which the propagating species comprise free ions and ion pairs existing in equilibrium with each other, both being propagation-active, although to vastly different extents.[10]

There are various other potential sources of MWD broadening also. One of these originates from the state of dynamic equilibrium that exists between the propagating chains and the monomer or between the chains themselves. Even though the chains are of similar lengths, as originally produced, slow equilibration takes place in which monomer units from the living ends exchange between chains. However, this process usually takes a very long time to reach the equilibrium distribution.[12] Another involves intermolecular interchange reaction in which polymer molecules exchange sections of chains between themselves. This reaction is however absent in vinyl polymerization but common in ring-opening polymerization (Chap. 7, Sec. 7.1.4). The equilibrium distribution attained in both the two processes is equivalent to the most probable distribution (Chap. 2, Sec. 2.5).[12]

Besides, molecular weight distribution may be broadened also by nonchemical factors. For instance, since carbanionic polymerizations are very fast (completed in seconds), appropriate procedure is required to be used for adequate mixing of the reactants and the solvents so that uniform concentrations of monomer and active centers, and uniform temperature are maintained throughout the polymerizing mass from the very beginning to the end of polymerization.[12]

4.1.1.2 Long term stability of living polymers

Even though living anionic polymerizations are free from transfer and termination, slow side reactions always occur, which impairs the long term stability of the living polymers.[1] The commonly encountered side reactions are: β-hydride elimination, rearrangement, and nucleophilic attack on monomer, polymer, and solvent, as discussed below.[13]

β-Hydride elimination from polystyrene sodium is much faster in THF than in benzene. The reaction converts the living end to a nonliving unsaturated end, the metal being converted to the hydride (reaction 4.8).[14] In addition, the allylic proton in the unsaturated end of the polymer so formed is acidic enough to be transferred easily to a carbanion. Thus, the ultimate result is polystyrene ended with 1,3-diphenylallyl anion, which is too stable to propagate.[13]

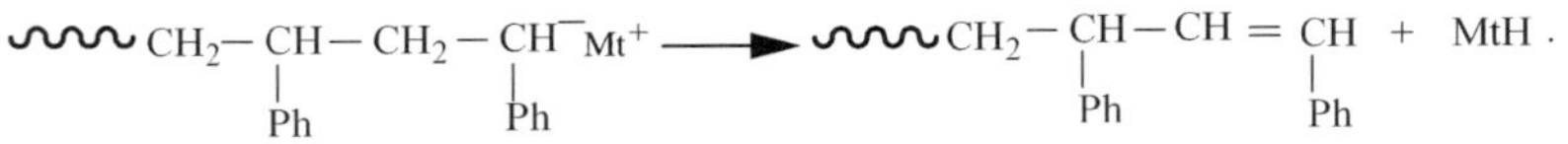

$$\sim\sim\sim CH_2-\underset{Ph}{CH}-CH_2-\underset{Ph}{\overset{-}{CH}}\,Mt^+ \longrightarrow \sim\sim\sim CH_2-\underset{Ph}{CH}-CH=\underset{Ph}{CH} + MtH \quad (4.8)$$

$$\sim\sim\sim CH_2-\underset{Ph}{CH}-CH=\underset{Ph}{CH} + \sim\sim\sim CH_2-\underset{Ph}{\overset{-}{CH}}\,Mt^+ \longrightarrow \sim\sim\sim CH_2-\underset{Ph}{\overset{\cdot\cdot}{C}}\cdots\underset{Ph}{\overset{\cdot\cdot}{CH}}\cdots\underset{}{CH}\,Mt^+ + \sim\sim\sim CH_2\underset{Ph}{CH_2} \quad (4.9)$$

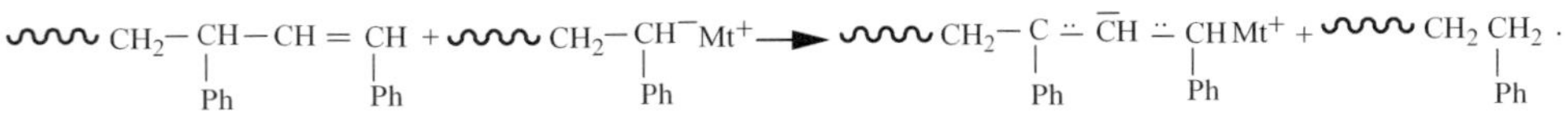

However, with lithium as counterion, living polystyrene and polydienes remain stable for several weeks not only in hexane and cyclohexane but also in benzene.[2] Apparently, β-hydride elimination does not occur in nonpolar solvents when Li^+ is the counter ion.

Furthermore, ethereal solvents are nucleophilically attacked by free carbanion, as already pointed out and shown below

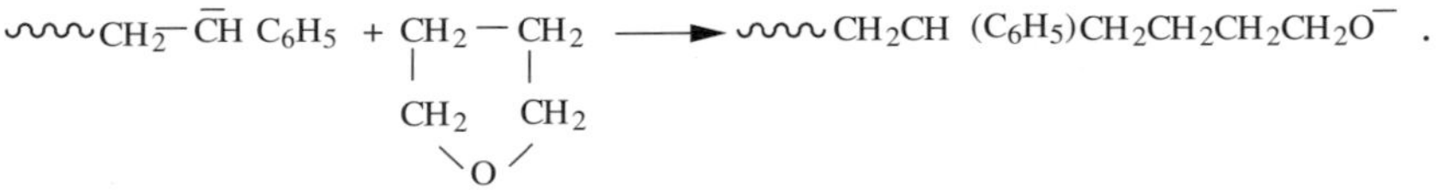

$$\sim\sim\sim CH_2-\overset{-}{C}H\,C_6H_5 + \underset{\underset{O}{CH_2\,\,\,\,CH_2}}{CH_2-CH_2} \longrightarrow \sim\sim\sim CH_2CH\,(C_6H_5)CH_2CH_2CH_2CH_2O^- \quad (4.10)$$

Inasmuch as the proportion of free carbanion increases with dilution, extremely diluted solutions of living polymers are relatively poor in stability. Thus, the decay rate of a 10^{-5} molar solution of living polystyrene (Na^+ counterion) in THF is about a few percent per minute at room temperature.[77]

Ethereal solvents also facilitate carbanion rearrangement in living polydienes as shown with polyisoprenyl anion (reaction 4.11)[15]

$$\sim\sim\sim CH_2-\underset{CH_3}{\overset{|}{C}}=CH-\overset{-}{C}H_2\,Mt^+ \longrightarrow \sim\sim\sim CH_2-\underset{\overset{|}{\overset{-}{C}H_2}\,Mt^+}{C}=CH-CH_3 \quad (4.11)$$

Stability of polar living polymers is particularly poor since both the monomer and the polymer are nucleophilically attacked by living chain ends resulting in terminaton.[16]

4.1.1.3 *Initiators*

Initiation may be effected by either addition of anionic species (both free anions and ion pairs) or electron transfer from alkali and alkaline earth metals, or radical anions to monomer.

4.1.1.3.1 Initiation by nucleophile addition

Lithium alkyls, aryls, and arylalkyls are suitable as initiators of polymerization of nonpolar monomers in hydrocarbon solvents by virtue of their solubility.[2] The high nucleophilicity of the alkyls makes them unstable in ethereal solvents. However, several alkali aryls or arylalkyls such as cumyl, fluorenyl, diphenylmethyl, and 1,1-diphenylhexyl are low enough in nucleophilicity to be stable in these solvents.[17]

Suitable difunctional initiators are exemplified by tetraphenyl disodium butane and (α-methylstyrene) tetramer disodium (or dipotassium) used in polar solvents. Difunctional initiators suitable for use in hydrocarbon solvents are typically organolithium compounds for solubility reasons, *e.g.*, diisopropenylbenzene butyllithium diadduct or bis(α-styryl) phenylene butyllithium diadduct.[17] However, for fast and efficient initiation, the nucleophilicity of the initiator and the electron affinity of the monomer should be matched. Thus, the weakly nucleophilic fluorenyl salt fails to initiate polymerization of styrene but is capable of initiating, albeit somewhat slowly, polymerization of the electrophilic methyl methacrylate;[18,19] but such correlation does not always hold. For example, the *primary* carbanionic salt, benzylalkali, is a poor initiator of styrene polymerization, whereas the less nucleophilic *tertiary* carbanionic salts, *e.g.*, oligo (α-methylstyryl)alkali and cumylalkali, are good. The higher degree of dissociation of the *tertiary* salt makes the concentration of the initiating carbanion much larger than that of the propagating ion. Besides, the smaller K_d of the propagating ion pair exerts a rate enhancing effect on initiation by the initiating ion pair.[20] This factor is very strong in the initiation of the polymerization of methyl methacrylate particularly with Li^+ counterion due to the extrememly low K_d of the propagating ion pair.[20]

Polymerization of electrophilic monomers may be initiated also by nucleophiles of much lower nucleophilicity than carbanions. For example, methyl methacrylate is polymerized by alkoxides in the presence of alcohols to yield very low molecular weight polymers, the alcohol acting as both a solubilizer of alkoxide and a chain transfer agent.[2] When the monomers are strongly electrophilic such as nitroethylene, α-cyanoacrylates, or acrylonitrile, they can be initiated even with uncharged nucleophiles such as trialkylphosphine or pyridine. Remarkably, α-cyanoacrylates are polymerized very fast at room temperature by adventitious water. This property is utilized in *in situ* formation of their polymers, which have excellent adhesive action on a variety of surfaces. The polymers can even join tissues and, accordingly, the precursor monomers are used in suture-less surgery.[21]

4.1.1.3.2 Initiation by electron transfer from alkali or alkaline earth metal

Initiation of polymerization by alkali metals in hydrocarbon solvents is likely to take place through electron transfer from the metal to the adsorbed monomer as shown in Scheme 4.3

for lithium metal and isoprene monomer. The metal has to be present in finely dispersed form so that the consequent large surface brings about faster electron transfer.[22,23]

The primary electron transfer reaction (4.12) shown in Scheme 4.3 results in the formation of a radical anion $M^{\cdot-}$ bound to the metal surface (only one resonance form of $M^{\cdot-}$ is shown to avoid complexity in presentation). However, the radical anion either receives a second electron from the metal surface (reaction 4.13) or undergoes coupling (reaction 4.14). In both cases, a dianion is generated, the two dianions differing only in the number of monomer units in their chains. Propagation follows on both ends of the dianions.

$$Li + CH_2 = C(CH_3) - CH = CH_2 \rightleftharpoons \dot{C}H_2 - C(CH_3) = CH - \overline{C}H_2 \; Li^+ (M^- Li^+), \quad (4.12)$$

$$\dot{M}^- Li^+ + Li \longrightarrow Li^+ \; ^-M^- \; Li^+ \quad (4.13)$$

and / or

$$2 \; \dot{M}^- Li^+ \longrightarrow Li^{+-}M - M^- Li^+ \quad (4.14)$$

Scheme 4.3   Election transfer initiation by lithium.

Nonpolar monomers and hydrocarbon solvents are not capable of solvating the ions. This lack of solvation energy prevents Li^+ ions to overcome the binding energy of the metal lattice and to be extracted from the metal surface into the solution. The growing dianion thus remains attached to the metal. However, as the chains become longer with the progress of polymerization, their desorption from the metal surface may be facilitated due to the larger gain in conformational entropy, which occurs in the process.[24] It becomes easier, lesser is the lattice energy of the alkali metal.

4.1.1.3.3   Initiation by electron transfer from aromatic radical anions

Initiation by direct electron transfer from alkali metal to monomer is relatively slow. This causes broadening of molecular weight distribution of the polymer produced. On the other hand, initiation by electron transfer from radical anion in homogeneous solution is fast enough to eliminate the problem.[3,4]

A radical anion is generated when an alkali metal is contacted with an aromatic or an unsaturated molecule of high electron affinity in polar solvents such as THF, dimethoxyethane, and hexamethylphosphoramide. The reaction involves transfer of the valence electron of the alkali atom to the lowest unoccupied molecular orbital (LUMO) of the electron acceptor.[25] Polynuclear hydrocarbons, such as naphthalene (Np), anthracene, and phenanthrene, ethylenes containing aromatic substituents such as α-methylstyrene, stilbene, and 1,1-diphenylethylene, and aromatic ketones such as benzophenone and fluorenone, can act as electron acceptors.

The exothermic solvation of alkali ion by polar solvent facilitates electron transfer. Thus, reaction (4.15) takes place in THF but not in benzene.

$$\text{Na} + \text{napthalene} \xrightleftharpoons{\text{THF}} \text{Na}^+ (\text{THF}) + (\text{napthalene})^{\cdot -} \quad . \tag{4.15}$$

The position of the equilibrium depends upon the solvation energy of the alkali ion. Thus, the reaction proceeds to ~95% completion in THF but only less than 1% in diethyl ether at room temperature.[18] With an alkaline earth metal as electron donor, the much stronger polar solvent hexamethylphosphoramide is needed.

When the radical anion, represented as $\text{Ar}^{\cdot -}$, is brought into contact with a suitable monomer (M), an electron transfer to the monomer takes place leading to initiation.[3,4,18] The latter becomes visibly evident through an instantaneous change in color. Thus, addition of styrene to a green solution of sodium naphthalene in THF changes the color of the solution instantaneously to cherry red.

$$\underset{\text{(M)}}{\text{Np}^{\cdot -} + \text{C}_6\text{H}_5\text{CH}=\text{CH}_2} \rightleftharpoons \underset{(\text{M}^{\cdot -})}{\text{Np} + (\text{C}_6\text{H}_5\text{CH}=\text{CH}_2)^{\cdot -}} \tag{4.16}$$

$$\underset{}{2\,(\text{C}_6\text{H}_5\text{CH}=\text{CH}_2)^{\cdot -}} \longrightarrow \underset{(^-\text{M}^-)}{\text{C}_6\text{H}_5\bar{\text{C}}\text{HCH}_2\text{CH}_2\bar{\text{C}}\text{HC}_6\text{H}_5} \quad . \tag{4.17}$$

The electron transfer reaction (4.16) is followed immediately by irreversible dimerization of the radical ion (reaction 4.17). The latter makes electron transfer quantitative even when the equilibrium is not favorably disposed to the radical anion side.[18] However, only biphenyl and naphthalene radical anions are effective as initiators due to the comparatively low electron affinity of their precursor hydrocarbons and the absence of side reactions.

A facile reaction competing with dimerization is the addition of monomer to the monomeric radical anion resulting in separation of the radical center from the anion center

$$\text{M}^{\cdot -} + \text{M} \longrightarrow {}^{\cdot}\text{M}-\text{M}^- \quad . \tag{4.18}$$

However, no perceptible radical polymerization occurs before the radicals undergo mutual termination, which is attributed to the extremely large k_t, $ca.,10^{10}$ L mol.$^{-1}$ s^{-1}, of small radicals and one time generation of large quantity of radicals.[18]

The electron affinity of α-methylstyrene and of 1,1-diphenylethylene is high enough to effect direct and efficient electron transfer from alkali metal to them producing radical anions, which dimerize to yield the respective dimeric dianions.[26]

$$\text{Na} + \text{CH}_2 = \text{CPh}_2 \xrightleftharpoons{\text{THF}} (\text{CH}_2 = \text{CPh}_2)^{\cdot -} \text{Na}^+ \longrightarrow \text{Na}^+ \text{Ph}_2 \bar{\text{C}} - \text{CH}_2 - \text{CH}_2 - \bar{\text{C}} \text{Ph}_2 \text{Na}^+ \quad . \tag{4.19}$$

$$\text{K} + \text{CH}_2 = \text{CPhCH}_3 \xrightleftharpoons{\text{THF}} (\text{CH}_2 = \text{CPhCH}_3)^{\cdot -} \text{K}^+ \longrightarrow \text{K}^+ \text{CH}_3\text{Ph} \bar{\text{C}} - \text{CH}_2 - \text{CH}_2 - \bar{\text{C}} \text{Ph} \text{CH}_3 \text{K}^+. \tag{4.20}$$

The dimeric α-methylstyrene dipotassium is reacted with requisite amount of α-methylstyrene to prepare tetrameric α-methylstyrene dipotassium $(K^+PhCH_3\overline{C}-CH_2-C(Ph)(CH_3)-CH_2-)_2$, which is used as an initiator. One α-methylstyrene molecule adds to each end of the dianion in the normal head-to-tail fashion.[27] The first addition of monomer is much faster than further additions, which makes tetramer preparation possible using the requisite amount of α-methylstyrene. Sometimes, oligo (α-methylstyrene) disodium is used instead of the tetramer. However, it can be prepared only at low temperature due to the relatively low ceiling temperature of head-to-tail polymerization.[23]

4.1.1.3.4 Electron transfer initiation by electrochemical means

Initiation by electron transfer to monomer may also take place electrochemically. It takes place directly at the cathode forming a monomer radical anion, which then dimerizes to give a dianion[28]

$$M + e \text{ (cathode)} \longrightarrow M^{\cdot-} . \quad (4.21)$$

$$2M^{\cdot-} \longrightarrow {}^-M-M^- . \quad (4.22)$$

Alternatively, the cation of the electrolyte is reduced at the cathode to metal, which then transfers an electron to monomer. The carbanionic nature of the active center is evident from the observation that electrochemical polymerization of styrene in THF in the presence of an alkali tetraphenylborate or tetraalkylaluminate used as supporting electrolyte proceeds in the cathode compartment exhibiting the characteristic absorption spectrum of polystyrene carbanion in the catholyte.

4.1.2 *Living anionic polymerization of nonpolar monomers in hydrocarbon solvents*

The usually used initiators of polymerization of nonpolar monomers in hydrocarbon solvents are lithium metal and organolithium compounds. One reason for this is the solubility of the organolithium compounds in hydrocarbon solvents. The other is the important microstructure they promote in polydienes (Sec. 4.1.2.4).[29]

Lithium alkyls exist in associated states in hydrocarbon solvents even at very low concentrations. The degree of association depends on the nature of the organic moiety. Thus, whereas n-butyllithium and ethyllithium exist as hexamers,[30,31] s-butyllithium and t-butyllithium exist as tetramers,[32,33] as determined from colligative property measurements in benzene or hexane. These are, however, not ion pair aggregates unlike the aggregated alkyls of other alkali metals. This is due to the covalent nature of C–Li bond, which has, of course, high ionic character.[1] Evidences exist for the nonionic nature of alkyllithiums in the crystalline state, unlike other alkylalkalis. In fact, the crystals of methyl- and ethyllithium comprise aggregates involving multicenter bonds, which typically exist in the crystals or aggregates of electron deficient compounds.[34] However, only the monomeric alkyllithiums are believed to be involved in initiation, although no direct experimental evidence exists in support of their existence.[2,35]

Unlike the alkyllithiums, other organolithiums, such as benzyl-, allyl-, and diphenylmethyl-lithium, in which the organic moiety bearing the negative charge is resonance stabilized, exist as ion pair aggregates in hydrocarbon solvents.[2] The ionized structure of the living ends is evident from the fact that the polymer solutions are colored unlike the alkyllithiums, the absorption spectra being the same as those of the corresponding polymer-lithium ion pairs in the polar solvent THF.

4.1.2.1 *Kinetics*

Initiation mechanism in benzene is different from that in aliphatic or alicyclic hydrocarbons, *e.g.*, hexane or cyclohexane.[36,37] In benzene, it involves the addition of monomeric alkyllithium existing in equilibrium with alkyllithium aggregates to monomer (reactions 4.23 and 4.24), whereas in the nonaromatic solvents it involves the direct addition of alkyllithium aggregates to monomer (reaction 4.25).[37]

$$(R\,Li)_n \;\xrightleftharpoons{K_D}\; n\,R\,Li. \tag{4.23}$$

$$R\,Li + M \;\xrightarrow{k_i}\; R\,M^-\,Li^+. \tag{4.24}$$

$$(R\,Li)_n + M \;\xrightarrow{k_i'}\; (R\,Li)_{n-1}(R\,M^-\,Li^+). \tag{4.25}$$

In general, lower is the degree of aggregation of the initiator higher is the rate of initiation. Thus, initiation is faster with *s*-BuLi than with *n*-BuLi[36–39] (Fig. 4.1). In fact, with the latter, initiation may not be complete even at the completion of polymerization.

Propagation in styrene (S) polymerization following complete consumption of *n*-BuLi initiator in hydrocarbon solvents is 0.5 order in polystyrene lithium (PSLi) over the concentration range of *ca.*, 10^{-5} to 10^{-1} mol/L.[36,38,40] This may be explained assuming dimeric nature of PSLi and very low degree of dissociation of the dimer to monomer. It is also assumed that the monomeric PSLi is propagation-active; the dimeric one is not.

$$(PS^-\,Li^+)_2 \;\xrightleftharpoons{K_D}\; 2\,PS^-\,Li^+. \tag{4.26}$$

$$PS^-\,Li^+ + S \;\xrightarrow{k_p^{\pm}}\; PSS^-\,Li^+. \tag{4.27}$$

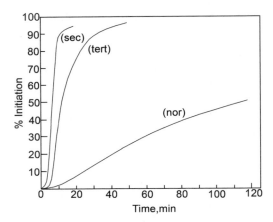

Figure 4.1 Percent initiation vs. time plot in the polymerization of styrene in cyclohexane at 40°C. [t-BuLi] = 1.10 × 10⁻³ mol/L, [styrene] = 0.085 mol/L; [s-BuLi] = 1.19 × 10⁻³ mol/L, [styrene] = 0.093 mol/L; [n-BuLi] = 9.95 × 10⁻⁴ mol/L, [styrene] = 0.260 mol/L. "Reprinted with permission from Ref. 37. Copyright © 1975 American Chemical Society."

From Eq. (4.27), R_p may be expressed as

$$R_p = k_p^{\pm}[PS^-Li^+][S]. \qquad (4.28)$$

The concentration of the monomeric PS^-Li^+ may be related to that of the dimeric one through the equilibrium constant K_D as

$$[PS^-Li^+] = k_D^{1/2}[(PS^-Li^+)_2]^{1/2} \quad (\text{for } K_D \ll 1). \qquad (4.29)$$

Substituting Eq. (4.29) for $[PS^-Li^+]$ in Eq. (4.28) one obtains

$$R_p = K_D^{1/2} k_p^{\pm}[(PS^-Li^+)_2]^{1/2}[S], \qquad (4.30)$$

$$\approx K_D^{1/2} k_p^{\pm} \left(\frac{c}{2}\right)^{1/2}[S], \qquad (4.31)$$

where c is the concentration of the total polystyryllithium in equiv/L, i.e., c = $[PS^-Li^+]$ + $2[(PS^-Li^+)_2] \approx 2[(PS^-Li^+)_2]$ (since degree of dissociation is small), c being also equal to the concentration of the initiator in equiv/L for completed initiation. The polymerizations occur without termination and transfer. Hence, they are living.

In contrast, contradictory reaction orders ranging from 1/2 to 1/6 have been reported for propagation in the polymerizations of isoprene and butadiene in hydrocarbon solvents.[40–45]

Different reaction orders determined by different workers or even by the same workers[43,44] indicate that the systems are too complex to be treated quantitatively through the schemes of initiation and propagation presented above.[46,47] In fact, small angle neutron scattering (SANS) studies showed that the association of the ionic species is not limited to just one type of aggregate. For example, major aggregates found in benzene solutions of oligostyrene (M.W. 2600) with butadienyllithium head group are dimers and star-shaped structures with mean number of arms being as large as 10, the degree of aggregation depending also on the length and the concentration of chains.[48] The complexity in aggregation apart, the assumption of the inactivity of the aggregates may not be valid.[46,47]

4.1.2.2 Effect of additives

Polar additives such as ethers, *tert*-amines, and sulfides decrease or even eliminate aggregation of organolithium compounds effecting increase in rates of initation and propagation.[49–52] These compounds interact specifically by coordinating with lithium. The *tert*-amine, tetramethylethylenediamine, is very effective in this regard. It forms a monomeric chelated complex (**1**), which is soluble in hydrocarbon solvents in all proportions. The complex is a powerful initiator, which is attributed to not only its monomeric nature but also the ionic character of the Li–C bond being increased to that of the corresponding ion pair. Dissociation to free ions may also occur to some extent.[1,53–55]

The effect of additives on initiation is much larger than on propagation, which may be due to the higher degree of aggregation of the organolithium initiator compared to that of the propagating chain end (*vide supra*). Thus, using anisole as the additive, to the extent of five to 40 times the equivalent concentration of *n*-BuLi, initiation of polymerization of styrene or dienes is completed in minutes at 30°C, while propagation remains virtually unaffected.[52] However, different additives affect the polymer microstructure and the chain end stability differently. Thus, whereas the *cis*-1,4 structure in polyisoprene is not formed at all in THF, even when used in small amounts, diphenylether and anisole do not affect the microstructure to any great degree, even when used undiluted.[56] On the other hand, the chain end stability is higher in anisole than in THF.[57,58]

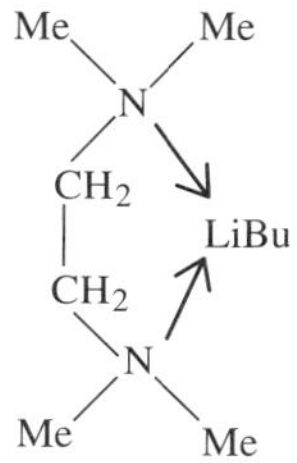

$$
\begin{array}{c}
\text{Me} \diagdown \diagup \text{Me} \\
\text{N} \\
| \\
\text{CH}_2 \\
| \qquad \text{LiBu} \\
\text{CH}_2 \\
| \\
\text{N} \\
\text{Me} \diagup \diagdown \text{Me}
\end{array}
$$

(**1**)

In contrast to the above nonionic Lewis bases, the ionic additives such as lithium alkoxides retard polymerization initiated by organolithium compounds. This is because they are aggregated more strongly than polystyryl- or polydienyllithium propagating species. Mixed complex formed between the propagating species and the more aggregated LiOR increases the state of aggregation of the former. As a result, the reactivity of the propagating species decreases and retardation of polymerization occurs.[59,60]

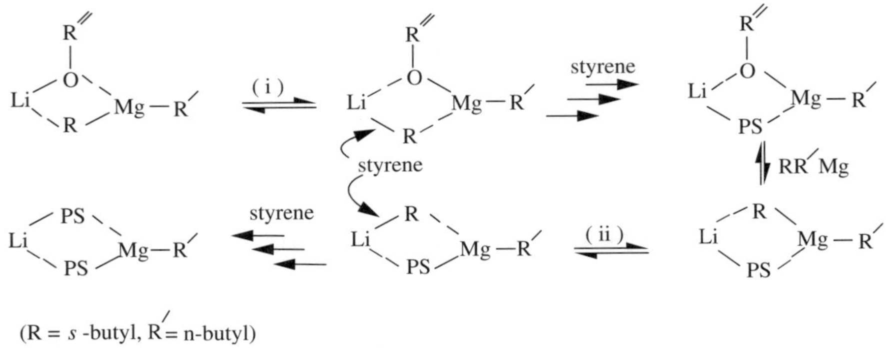

Scheme 4.4 Mechanism of styrene polymerization initiated by "ate" complexes formed between lithium alkoxide and n, sec-dibutyl magnesium.[62]

4.1.2.3 Living anionic polymerization at elevated temperatures — retarded anionic polymerization

Similar retardation of polymerization as effected by lithium alkoxides, described above, also occurs in the presence of aluminum, magnesium, or zinc alkyls, which form mixed complexes with the propagating chain ends.[49,61]

However, "ate" complexes formed between lithium alkoxides and magnesium or aluminum alkyls initiate bulk polymerization of styrene at elevated temperatures ca., $\geq 100°C$, although the individual retarders fail to do so.[62,63] Activation involves ligand exchange in the "ate" complex forming an alkyllithium compound complexed with a new alkoxymetal (Mg or Al)–alkyl derivative (equilibrium (i) in Scheme 4.4). Excess alkyl can effect further exchange giving mixed alkyl complex, which exists in equilibrium with the active form (equilibrium ii).[62] Initiation and subsequent propagation involve addition of monomer to the complexed organolithium.

These polymerizations are of industrial importance since they can be conducted rather easily like radical polymerizations. In addition, the alkoxides may be replaced by the cheaper and readily available hydrides still making efficient initiators.[64]

4.1.2.4 Stereospecificity

Polymerization of isoprene initiated by lithium or alkyllithium initiators in bulk or in hydrocarbon solvents yields cis-1,4-polyisoprene with more than 93% cis content similar to natural rubber.[29] Initiation by other alkali metals such as sodium and potassium gives trans-3,4- and trans-1,4-polyisoprenes.[65–69] Polar compounds bring about changes in microstructure.[56,58,70] For example, 1, 4, adition does not occur in THF.[70]

When butadiene is used in place of isoprene, the microstructure is again largely 1,4, but of mixed cis and trans geometry when the initiator is lithium. In this polymerization too, the proportion of 1,2 polymer increases in the presence of polar substances.[71]

The all *cis*-1, 4 structure of polyisoprene promoted by lithium or its alkyl may be explained based on the coordination of the diene with Li$^+$ ion in the chain end prior to incorporation into the chain.

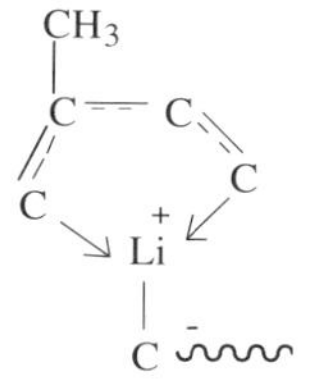

Although butadiene can also coordinate with Li$^+$ ion, the *cis* configuration in the last monomer unit is retained in polyisoprene due to the prevention of rotation by the steric hindrance exerted by the methyl group.[1] A polar solvent such as THF, which strongly coordinates with Li$^+$ ion, precludes coordination with dienes resulting in loss of microstructure control.

In contrast, polymerization of polar monomers in THF occurs stereospecifically. For instance, polymerization of MMA in THF yields *syndio*-rich living polymer.[74,75] Although polymerization in toluene also proceeds stereospecifically giving *isotactic* PMMA in the absence of additive, it is nonliving for reasons discussed later (Sec. 4.1.4). However, the use of triethylaluminum as additive makes it living.

4.1.3 Living Anionic Polymerization of Styrene in Ethereal Solvents

4.1.3.1 Kinetics

The rate of monomer disappearance follows first order kinetics as shown in Fig. 4.2 for living anionic polymerization of styrene in tetrahydrofuran with sodium and cesium as counter ions.[76–80] This proves that the active center concentration remains constant during polymerization, which would be expected if initiation were not slower than propagation and termination were absent.

The first order rate law may be expressed by Eq. (4.32)

$$\ln \frac{[M]_0}{[M]} = k_p^{app}[P^*]t = k_p^{app}ct, \qquad (4.32)$$

where $[P^*]$ is the combined concentration, c, of active centers, which comprise both free ions and ion pairs of all chain lengths, $[M]$ is the monomer concentration at time t, and $[M]_0$ is that at t = 0, and k_p^{app} is the apparent rate constant of propagation. According to Eq. (4.32), k_p^{app} is obtainable by dividing the slope of the first order plot for monomer disappearance by the living end concentration.

It is evident from Fig. 4.2 that polymerizations are very fast, being completed in seconds at 25°C. This is very unlike radical polymerization, which takes hours to complete and that too at elevated temperatures. The difference is partly due to the much higher active center

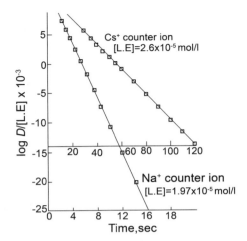

Figure 4.2 First order kinetics plot for monomer disappearance in living anionic polymerization of styrene in THF at 25°C. Log D in the ordinate refers to log $A_t - \log A_\infty$ where A is the absorbance of the polymerizing solution at an absorption band of styrene. [L.E.] is the living end concentration represented by c in Eq. (4.32). "Reprinted with permission from Ref. 76. The absolute rate constants of anionic polymerization by free ions and ion-pairs of living polystyrene. Copyright © 1964 Elsevier."

concentration and partly to the higher k_p^{app} in the anionic polymerization. In the two examples given in Fig. 4.2, the active center concentration is about 100 times as high as that in a typical radical polymerization and the k_p^{app} with Na^+ counter ion (Fig. 4.3) is about 10 times the k_p in radical polymerization (which is 240 at 50°C), whereas with Cs^+ counter ion the two are comparable. On the other hand, at active center concentrations above ca., 10^{-3} mol/L, k_p^{app} with Na^+ counter ion is comparable with k_p in radical polymerization but the concentration factor is overwhelmingly large.

Propagation involving ions and ion pairs may be represented as follows.

$$P_n^- Mt^+ \xrightleftharpoons{K_d} P_n^- + Mt^+, \qquad (4.33)$$

$$P_n^- Mt^+ + M \xrightarrow{k_p^\pm} P_{n+1}^- Mt^+, \qquad (4.34)$$

$$P_n^- + M \xrightarrow{k_p^-} P_{n+1}^-, \qquad (4.35)$$

where P_n^- is the polymeric anion of degree of polymerization n, Mt^+ is the alkali counter anion, $P_n^- Mt^+$ is the polymeric ion pair, M is the monomer, k_p^- and $k_p^\pm$ are the propagation rate constants on free ion and ion pair respectively, and K_d is the dissociation constant.

Based on the propagation reactions (4.34) and (4.35) the rate of propagation R_p may be expressed as

$$R_p = \frac{-d[M]}{dt} = (k_p^\pm [P^- Mt^+] + k_p^- [P^-])[M], \qquad (4.36)$$

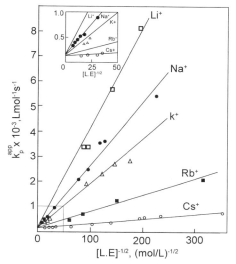

Figure 4.3 Plot of k_p^{app} vs. $[L.E.]^{-1/2}$ in the living anionic polymerization of styrene in THF at 25°C. The alkali counter ions are indicated in the figure. "Reprinted with permission from Ref. 76, The absolute rate constants of anionic polymerization by free ions and ion-pairs of living polystyrene. Copyright © 1964 Elsevier."

where $[P^-Mt^+] = \sum_{n=1}^{\infty}[P_n^-Mt^+]$ and $[P^-] = \sum_{n=1}^{\infty}[P_n^-]$; all other symbols have been defined earlier.

Representing the degree of dissociation of the ionic species as α, Eq. (4.36) transforms to

$$R_p = \frac{-d[M]}{dt} = ((1-\alpha)k_p^{\pm} + \alpha k_p^-)[P^*][M], \quad (4.37)$$

$$= k_p^{app}[P^*][M], \quad (4.38)$$

where $[P^*]$ represents the combined concentration of the ionic species, free ions and ion pairs, and k_p^{app} is the apparent rate constant of propagation defined by

$$k_p^{app} = (1-\alpha)k_p^{\pm} + \alpha k_p^- = k_p^{\pm} + \alpha(k_p^- - k_p^{\pm}). \quad (4.39)$$

Integration of Eq. (4.38) gives Eq. (4.32) presented above and also given earlier as Eq. (4.3). The degree of dissociation α may be related to the dissociation constant K_d of the ion pair-free ion equilibrium represented by Eq. (4.33) as

$$K_d = [P^-][Mt^+]/[P^-Mt^+] \quad (4.40)$$

$$= [P^-]^2/[P^-Mt^+] \quad \text{(since } [P^-] = [Mt^+]\text{)} \quad (4.41)$$

$$= \frac{\alpha^2[P^*]^2}{(1-\alpha)[P^*]} \quad (4.42)$$

$$= \frac{\alpha^2 c}{(1-\alpha)} \quad \text{(where } c = [P^*]\text{)}. \quad (4.43)$$

In a solvent of low dielectric constant such as THF, K_d is very small, and $\alpha \ll 1$ (say, less than 0.05).[76,77] Accordingly, $(1-\alpha) \approx 1$. This approximation reduces Eq. (4.43) to (4.44),

$$K_d = \alpha^2 c, \qquad (4.44)$$

or

$$\alpha = K_d^{1/2} c^{-1/2}. \qquad (4.45)$$

Substituting Eq. (4.45) for α in Eq. (4.39) one obtains[76–79]

$$k_p^{app} = k_p^{\pm} + (k_p^{-} - k_p^{\pm}) K_d^{1/2} / c^{1/2}. \qquad (4.46)$$

Thus, a plot of k_p^{app} vs. the reciprocal square root of active center concentration should be a straight line with intercept $= k_p^{\pm}$ and slope $= (k_p^{-} - k_p^{\pm}) K_d^{1/2}$. Figure 4.3 shows such plots for polymerizations of styrene in THF. Each straight line plot corresponds to an active center having a particular alkali counter ion.[76,77] The values of $k_p^{\pm}$ on ion pairs with different counter ions obtained as the intercepts of the respective straight lines in Fig. 4.3 are given in Table 4.1. With $k_p^{\pm}$ so obtained, k_p^{-} can be determined from the slopes of each line provided the dissociation constant K_d of the corresponding ion pair is also known.

The dissociation constant may be determined from the studies on conductance of living polymer solutions and the knowledge of ion conductance values.[81] Alternatively, it may be determined from the kinetic studies of propagation in the presence of common ion salts, which have much higher dissociation constants than the living ion pairs. A suitable salt is an alkali tetraphenylborate, $MtBPh_4$.[77,80,81] Due to the common ion effect, the dissociation of living end ion pair is depressed such that at concentrations of the added salt above certain threshold $[P^*]$ virtually equals $[P^-Mt^+]$. Under such condition, the degree of dissociation of the living ends may be expressed as

$$\alpha = \frac{[P^-]}{[P^*]} = \frac{[P^-]}{[P^-Mt^+]}. \qquad (4.47)$$

Ideally, activities instead of concentrations should be used. However, Szwarc observed that, in most cases, the activity coefficients are close to unity.[1,81] Also, Schulz et al. calculated the activity coefficients using the Debye-Huckel equation and concluded that the values do not affect the results much.[87]

Substitution from Eq. (4.40) in Eq. (4.47) gives

$$\alpha = K_d / [Mt^+]. \qquad (4.48)$$

Proceeding further by substituting Eq. (4.48) for α in Eq. (4.39) one obtains

$$k_p^{app} = k_p^{\pm} + (k_p^{-} - k_p^{\pm}) K_d / [Mt^+]. \qquad (4.49)$$

Thus, a plot of k_p^{app} vs. $1/[Mt^+]$ should be linear with the slope given by $(k_p^{-} - k_p^{\pm}) K_d$ and the intercept given by $k_p^{\pm}$. Such a plot is shown in Fig. 4.4 in the living polymerization of styrene (Na^+ counter ion) in THF in the presence of sodium tetraphenylborate at concentrations much above the living end concentration.[77,81] The concentration of the alkali

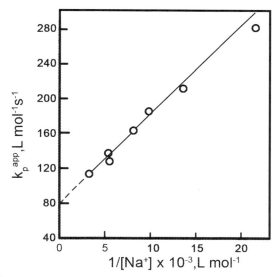

Figure 4.4 Plot of k_p^{app} vs. $1/[Na^+]$ in the living anionic polymerization of styrene in THF at 25°C in the presence of sodium tetraphenylborate. "Reprinted with permission from Ref. 77. Copyright © 1965 American Chemical Society."

ion is calculated from the concentration of the added salt and its dissociation constant. From the value of $(k_p^- - k_p^\pm)K_d$ and that of $(k_p^- - k_p^\pm)K_d^{1/2}$, already obtained as discussed above, K_d is easily determined.

The above treatment holds for polymerizations of styrene in several moderately polar ethereal solvents besides THF, e.g., 3-methyltetrahydrofuran, tetrahydropyran, and oxepane.[76–78] In the more polar dimethoxyethane, however, K_d is relatively high so that the approximation $(1 - \alpha) \approx 1$ is not valid requiring different treatment of the kinetics.[87]

In contrast to the dissociation of the polystyryl ion pair observed in THF and other ethereal solvents mentioned above, no evidence of the dissociation is obtained in dioxane in polymerizations conducted at 25° to 80°C.[82,85] For instance, dilution of living ends in dioxane does not increase the rate of polymerization, i.e., k_p^{app} is independent of living end concentration. The inability of dioxane to effect ion pair dissociation is attributable to its poor alkali ion coordinating power and very low dielectric constant.

The values of $k_p^\pm$ on polystyrene alkali ion pairs in dioxane and THF, and of k_p^- as well as of K_d in the latter, all at 25°C, are given in Table 4.1. The values reveal that k_p^- is about 1000 times larger than $k_p^\pm$. For this reason, the contribution of free ion fraction to propagation may exceed that of ion pair even though the concentration of the free ion is very small. For instance, at a concentration of 10^{-2} mol/L of living ends (Li^+ counter ion), it may be calculated that the free ion fraction would be only 0.0043 but the major amount of polymerization (64%) would be effected by them. The contribution would be larger; smaller is the living end concentration. Thus, at the 10^{-5} mol/L concentration of living ends, the ion fraction would increase to 0.136 and essentially all the polymerizations would be effected

Table 4.1 Rate constants of propagation on polystyrene alkali ion pairs at 25°C*.

	THF ($\epsilon = 7.20$)[a]			Dioxane ($\epsilon = 2.2$)			
	$k_p^{\pm}$, L mol^{-1}s^{-1}			$k_p^{\pm}$, L mol^{-1}s^{-1}			
Counterion	Szwarc[b]	Schulz[c]	$K_d \times 10^{7,b}$ mol/l	Szwarc[d]	Dainton[e]	$E_p^{\pm\,e}$, kJ/mol	$A_p^{\pm\,e}$, l mol^{-1}s^{-1}
Li$^+$	160		1.9	0.9	—	—	—
Na$^+$	80	~200	1.5	3–4	6.5	~40	~10^8
K$^+$	~70		0.7	20	28	~25	~10^6
Rb$^+$	~60		—	21.5	34	~19	~10^5
Cs$^+$	22		0.026	24.5	15	—	—

*$k_p^- = 65\,000$ (ref. 77), $130\,000$ (refs. 80, 87) L mol^{-1} s^{-1} and independent of the ethereal solvent with $E_p^- = $ ~16 kJ/mol, $A_p^- = $ ~10^8 L mol^{-1} s^{-1} (Refs. 77, 87). [a]The Arrhenius plots in THF are not linear (*vide* Sec. 4.1.3.2). [b]Refs. 77, 81; [c]Refs. 80, 87; [d]Ref. 85; [e]Ref. 82.

by ions. These calculations are based on the k_p and K_D values given by Szwarc *et al.* (Table 4.1).

$k_p^{\pm}$ is smaller in dioxane than in THF with all alkali counter ions excepting Cs$^+$. With the latter, $k_p^{\pm}$ is about the same in the two solvents. Furthermore, the trend of variation of $k_p^{\pm}$ with the size of the alkali ion in THF is different from that in dioxane. These results can be explained taking into account the nature of the ion pairs existing in the two solvents.[81]

Earlier, Winstein concluded from the studies of certain solvolysis reactions that ion pairs could be of two types, (i) contact, and (ii) solvent-separated, which exist in equilibrium, with each other. The latter is increasingly favored as the dielectric constant, dipole moment, and ion solvating power of the solvent become larger.[89] The first physical evidence of the existence of the two forms was provided by Hogen-Esch and Smid from the studies of the absorption spectra of fluorenylalkali salts in THF. Separate absorption peaks are observed for the two forms (Fig. 4.5).[90] The proportion of contact ion pair decreases with decrease in temperature until at $-50°$C it becomes negligibly small.

With polystyrene alkali ion pair, the following equilibrium, therefore, is envisaged in THF and similar alkali-ion-solvating solvents.

$$PS^- Mt^+ \underset{}{\overset{K_{c,s}}{\rightleftharpoons}} PS^- \| Mt^+ \quad . \quad (4.50)$$
$$\text{contact ion pair} \qquad \text{solvent-separated ion pair}$$

Propagation on contact ion pair involves separation of the oppositely charged ions in the transition state. The larger is the size of the bare (unsolvated) alkali ion smaller is the energy required to separate the paired ions. Accordingly, the activation energy decreases and the propagation constant increases with the increase in the crystal radius of the alkali ion: Li$^+$ < Na$^+$ < K$^+$ < Rb$^+$ < Cs$^+$. The $k_p^{\pm}$ values in dioxane given in Table 4.1 measured in two different laboratories follow this trend, although the individual values given by the two groups of workers are somewhat different and the value with Cs$^+$ counter ion given by the Dainton group showing deviation from the trend. Nevertheless, it may be inferred that the ion pairs are of the contact type in dioxane. This is in conformity with the weak polar character

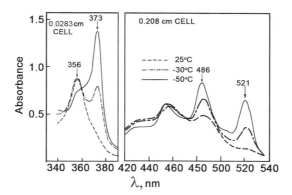

Figure 4.5 Absorption spectra of fluorenylsodium in THF at various temperatures. "Reprinted with permission from Ref. 90. Copyright © 1966 American Chemical Society."

and poor alkali-ion-solvating power of dioxane. However, the effect of decreasing activation energy on the rate constant is largely diminished with the decrease in the frequency factor with the increase in the cation size.

Solvent-separated ion pair is much more reactive than contact ion pair. In fact, its reactivity is comparable with that of the free anion and virtually independent of not only the size of the alkali counterion but also the ethereal solvent. The free anion is not coordinated with the solvent as is evident from the independence of k_p^- on the nature of the ethereal solvent.[87] In the ion pair, too, the anionic part would not be coordinated with the solvent separating it from the cationic part and hence the independence in reactivity.

The $k_p^{\pm}$ values in THF and dioxane with Cs^+ as the counterion are nearly equal. This suggests that this ion pair is of the contact type in THF also. However, we shall see later that a very small fraction of this ion pair exists also in solvent-separated form in THF at lower temperatures.[88] With the other alkali counter ions, $k_p^{\pm}$ is larger in THF than in dioxane (Table 4.1). This suggests that a fraction of the ion pairs with each of these alkali counter ions exists also in the solvent-separated form at 25°C. The fraction is larger smaller is the crystal radius of the alkali ions since the energy of solvation is larger.

The equilibrium (4.50) is exothermic due to the high exothermicity of the alkali ion solvation overcoming the endothermicity of separating the paired ions to accommodate the solvent molecule in between them. Accordingly, the fraction of the more reactive solvent-separated ion pair decreases with the increase in temperature.

4.1.3.2 *Temperature dependence of $k_p^{\pm}$*

Figure 4.6 shows the Arrhenius plots for $k_p^{\pm}$ on ion pairs in THF with sodium and cesium counter ions. The plot with sodium counter ion is wavy with a crest and a trough,[87,88] whereas the one with cesium counter ion is only slightly curved.[88] The results have been attributed to the effects of the variation of the relative population of contact, and solvent-separated ion pairs superposed with the variation of the respective rate constants with temperature.[86–88,91,92]

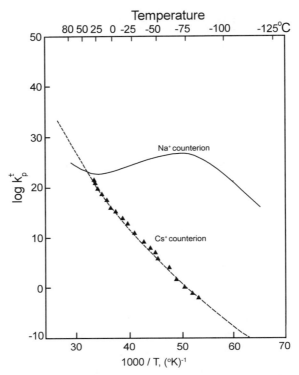

Figure 4.6 Arrhenius plots for the living anionic polymerization of styrene with Na$^+$ and Cs$^+$ counterions in THF. For Cs$^+$ counterion the points are experimental data and the dashed curve is theoretically calculated using Eq. (4.52). "Reprinted with permission from Ref. 87, The influence of polar solvents on ions and ion pairs in the anionic polymerization of styrene and Ref. 88, Anionic polymerization of styrene in tetrahydrofuran (THF) with cumylcesium as initiator. Copyright ©1975 Elsevier Ltd."

Referring to the equilibrium (4.50) between the ion pairs $k_p^\pm$ may be expressed as

$$k_p^\pm = f\, k_{p\,(S)}^\pm + (1-f) k_{p\,(C)}^\pm, \tag{4.51}$$

where $k_{p\,(S)}^\pm$ and $k_{p\,(C)}^\pm$ represent respectively the rate constant of propagation on solvent-separated, and contact ion pair, $k_{p\,(S)}^\pm$ being much larger than $k_{p\,(C)}^\pm$, and f represents the fraction of the solvent-separated ion pair. Inasmuch as equilibrium (4.50) is exothermic, the solvent-separated ion pair concentration and, with it, f increases with decrease in temperature.

Quantitatively speaking, f is related to $K_{c,s}$ as

$$\frac{1}{f} = 1 + \frac{1}{K_{c,s}}.$$

$K_{c,s}$ may be related to $k_{p\,(s)}^\pm$ and $k_{p\,(c)}^\pm$ (with the approximation $k_{p\,(s)}^\pm - k_{p\,(c)}^\pm = k_{p\,(s)}^\pm$) as

$$k_p^\pm = k_{p\,(c)}^\pm + k_{p\,(c)}^\pm K_{c,s}/(1 + K_{c,s}). \tag{4.52}$$

The constants in Eq. (4.52) and their temperature variations have been determined in various ethereal solvents. It may be calculated using these values that $K_{c,s}$ of polystyrene sodium in THF increases from $\sim 10^{-3}$ at 80°C to ~ 100 at -120°C.[87] Therefore, in this temperature range, f increases from ~ 0.001 to ~ 0.99.

Thus, with sodium counter ion, the ion pair is almost exclusively solvent-separated at -120°C. The segment of the Arrhenius line below ca., -75°C has the usual negative slope (positive activation energy). However, the slope gradually decreases (activation energy decreases) as the temperature increases from above -120°C becoming nearly zero at ca., -75°C. The decrease in slope is due to the decrease in f (Eq. 4.51) so that $k_p^{\pm}$ increases less rapidly with temperature than that it would have done if solely solvent-separated ion pair existed. The second term in Eq. (4.51) is too small to prevent the decrease. In the temperature range of -75°C to 25°C, f decreases with increase in temperature so rapidly that $k_p^{\pm}$ decreases in spite of the increase in $k_{p(s)}^{\pm}$ and the increase in the second term. As a result, the net activation energy becomes negative.[86-88,92] Above ca., 25°C, the proportion of the solvent-separated ion pair becomes very small so that the second term in Eq. (4.51) dominates and the activation energy becomes again positive reaching ultimately the value of the propagation on contact ion pair at sufficiently high temperatures.

In contrast, with cesium as the counterion, f is very small so that its increase with decrease in temperature is reflected in the Arrhenius plot only as small curvature, the activation energy remaining positive.[88]

4.1.3.3 Ion association beyond ion-pair

From the studies on the conductivity of tetraalkylammonium salts in moderately polar solvents Fuoss and Kraus concluded that ions might undergo higher order association beyond ion pairs as their concentrations are increased. The immediately higher order of association beyond ion pair is represented by triple ion, which forms by the association of ion pair with free ion as shown in Scheme 4.5.[93]

Formation of triple ions upsets the dissociation equilibrium of ion pairs causing more dissociation of the latter.

With one-ended living anionic polymers in ethereal solvents, triple ion formation may occur at relatively high concentrations. On the other hand, with two-ended living polymers, the existence of triple ion may be exhibited at low bulk concentrations if the chain length is short. In this case, the high local living ends concentration causes intramolecular triple ion formation to take place. Szwarc et al. found evidence of this in α, ω-polystyryl dianion paired with Cs^+ ion, which does not form solvent-separated ion pair in THF.[81,94] The

$$Mt^+ X^- \rightleftharpoons Mt^+ + X^-$$

$$Mt^+ + Mt^+ X^- \rightleftharpoons Mt^+ X^- Mt^+$$

$$X^- + Mt^+ X^- \rightleftharpoons {}^-X Mt^+ X^-$$

Scheme 4.5 Equilibria involving free ion, ion pair, and triple ion.

equillibria involved are shown below. The formation of triple ion (**4**) upsets the dissociation equilibrium of the *bis* ion pair (**2**) causing more dissociation to occur. Accordingly, the Cs^+ ion concentration becomes larger, and the free monoanion (**3**) concentration becomes smaller than that obtains in the absence of triple ion formation. As a result, the rate of polymerization is reduced.

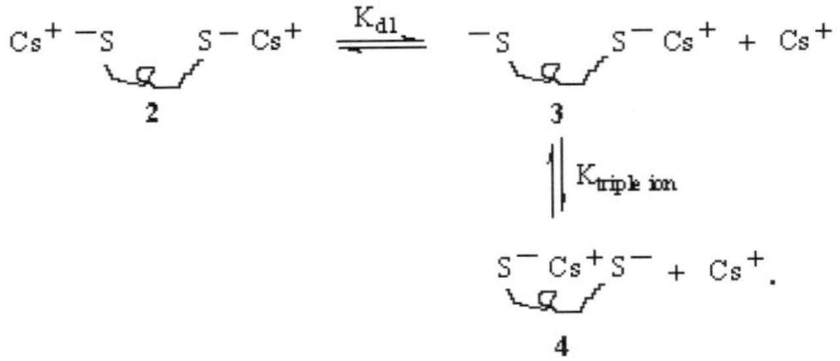

$$Cs^+ {}^-S\underset{2}{\smile} S^- Cs^+ \xrightleftharpoons{K_{d1}} {}^-S\underset{3}{\smile} S^- Cs^+ + Cs^+$$

$$\updownarrow K_{\text{triple ion}}$$

$$S^- Cs^+ S^-\underset{4}{\smile} + Cs^+$$

4.1.4 Living anionic polymerization of polar monomers

4.1.4.1 Side reactions

As already pointed out in Sec. 4.1.1, polar monomers are difficult to polymerize in a living manner by anionic means due to side reactions, which involve nucleophilic attack of initiator and propagating chain end on polar groups in monomer and polymer.

Reactions (4.53) through (4.56) show these side reactions in the polymerization of MMA.[16b,19,95–97] Polymerization in nonpolar solvents presents additional complexities due to very strong association of ion pairs to multiple aggregates and slow exchange between them. This factor makes nonpolar solvents unsuitable for polymerization. However, living polymerization is achieved in these solvents using various additives (*vide infra*).

The methoxide formed in the reactions may also initiate polymerization[2,19] as well as associate with the carbanionic species effecting a change in reactivity.[95] Besides, the vinyl ketones formed in the reactions with monomer may undergo copolymerization.[11]

1. Termination by monomer

a) $\quad R^- Mt^+ + CH_2 = \underset{\underset{O}{\|}}{\overset{\overset{CH_3}{|}}{C}}-C-OCH_3 \longrightarrow CH_2 = \underset{\underset{O}{\|}}{\overset{\overset{CH_3}{|}}{C}}-C-R + CH_3O^- Mt^+.$

(4.53)

b) $\quad \sim\sim\sim C^- Mt^+ + CH_2 = \underset{\underset{CH_3\,O}{|\;\|}}{C-C}-OCH_3 \longrightarrow CH_2 = \underset{\underset{CH_3\,O}{|\;\|}}{C-C}-C\sim\sim\sim + CH_3O^- Mt^+.$

(4.54)

2. Termination by polymer

a) intermolecular

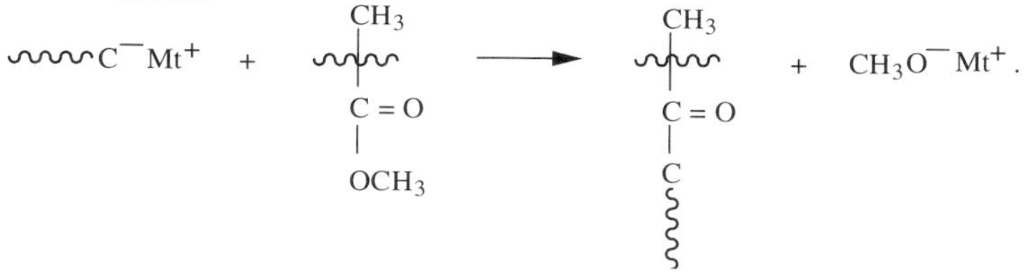

$$\sim\sim\sim C^- Mt^+ + \sim\sim\begin{matrix}CH_3\\|\\\sim\sim\\|\\C=O\\|\\OCH_3\end{matrix} \longrightarrow \sim\sim\begin{matrix}CH_3\\|\\\sim\sim\\|\\C=O\\|\\C\\\sim\sim\end{matrix} + CH_3O^- Mt^+.$$

(4.55)

b) intramolecular (backbiting)

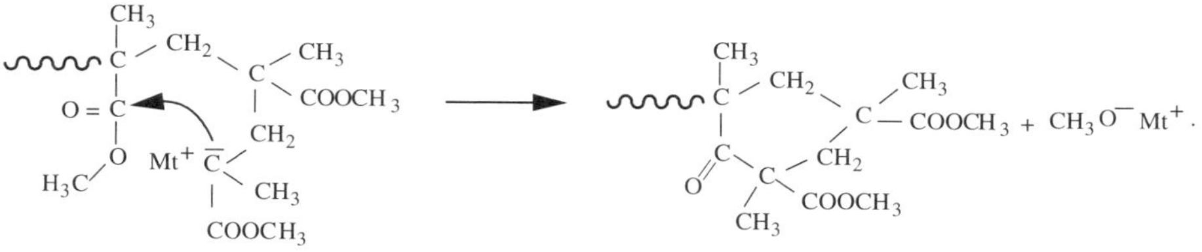

(4.56)

With acrylates as monomers, an additional side reaction may occur, which involves transfer of in-chain tertiary protons to the propagating chain,[11]

$$\sim\sim CH_2-CH\sim\sim + \sim\sim C^- Mt^+ \longrightarrow \sim\sim CH_2-C\sim\sim, Mt^+ + \sim\sim CH.$$
$$\quad\quad |\quad\quad\quad\quad\quad\quad\quad\quad\quad\quad\quad\quad |$$
$$\quad COOR \quad\quad\quad\quad\quad\quad\quad\quad\quad\quad COOR$$

4.1.4.2 Living anionic polymerization of methyl methacrylate in polar solvents in the absence of additives

Termination involving initiator and monomer (reaction 4.53) can be suppressed using a sterically hindered and weakly nucleophilic initiator such as diphenylmethyllithium (DPML),[97] 1,1-diphenylhexyllithium (DPHL),[95,98] or triphenyllithium.[98,100] However, initiators with lesser bulk and stronger nucleophilicity such as cumylalkali and oligo (α-methylstyrene) alkali are also reasonably good.[19,99] Thus, with oligo (α-methyl styrene) monosodium as initiator, polymerization occurs with initiator efficiency of 0.7–0.8.[19] Monomer consumption follows first order kinetics in the temperature range $-50°$ to $-100°C$ indicating that the living ends are stable, at least, in the time scale of polymerization. These results refer to polymerizations, which were conducted in the presence of excess common ion salt in order to prevent side reaction by the more reactive free ion. The active centers are, therefore, exclusively ion pairs, and k_p^{app} is equivalent to $k_p^{\pm}$. This conclusion is also supported by the observation that the Arrhenius plot of $k_p^{\pm}$ is linear with the usual negative slope indicating the existence of only one kind of ion pair.[19,91] This is in sharp contrast with two kinds of ion pair, viz., contact, and solvent-separated, observed in styrene polymerization in THF where the Arrhenius plot is nonlinear (Fig. 4.6).

With increase in living end concentrations, k_p^{app} was found to decrease monotonically suggesting the formation of propagation-inactive (or less active) dimeric ion pair (Eq. 4.57) and possibly higher associates.[101,102] The ion pair with lithium counter ion is more prone to forming such associates. For instance, the reaction order with respect to PMMA-Li in THF at $-65°C$ decreases from 0.75 to 0.58 as the concentration is increased from 1.2×10^{-4} to 2.5×10^{-3} mol/L. This suggests progressive increase in dimerization.[102] The reaction order would have been reduced to 0.5 were the ion pair converted completely to the dimer only and the degree of dissociation of the dimer to the propagation-active monomer were small.

$$2 \, PMMA^- Li^+ \xrightleftharpoons{K_A} \left(PMMA^- Li^+\right)_2 \qquad (4.57)$$
$$P^\pm \qquad\qquad\qquad \left(P^\pm\right)_2$$

When the difunctional initiator oligo (α-methylstyrene) disodium is used, intramolecular dimerization of living ends occurs in short chains.[103,104] The average end-to-end distance in short chain polymer being small, the local concentration of chain ends is high. This facilitates intramolecular dimerization of ion pairs even though their bulk concentrations are too small to effect intermolecular dimerization. The intramolecular formation of triple ions like in polystyrene dicesium, discussed earlier (Sec. 4.1.3.3), is precluded due to the extremely low dissociation constant of PMMA-alkali ion pair (*vide infra*) and excess common ion salt used.

4.1.4.2.1 Nature of ion pairs and their aggregates

The living end in PMMA is not true carbanionic. For instance, ^{13}C NMR studies in THF of methyl-α-lithioisobutyrate (**5**), a small molecular model of PMMA-Li, suggest a planar structure of the anion, which is largely enolate in character. Replacement of Li^+ by other alkali ions probably increases somewhat the carbanionic character. The two methyl groups are nonequivalent in the lithium salt indicating considerable double bond character in the $>C-C(O)$ bond.[105]

$$\begin{array}{c} CH_3 \\ | \\ CH_3-C \quad - \\ \diagup \! \! \diagup \\ ^+M \quad C-OCH_3 \\ \diagdown \! \! \diagdown \\ O \quad - \end{array}$$

(**5**)

Strong bonding between the ester enolate and the alkali ion makes a tight ion pair. However, intramolecular termination (backbiting) observed in MMA polymerization in THF suggests that structure **5** or its peripherally solvated form **6** exists in equilibrium with forms **7** and **8** (Eq. 4.58) in which the metal ion is coordinated intramolecularly with the ester group in the antepenultimate and the penultimate monomer unit respectively.[19,91] Structure **7** facilitates termination. This was evident from the identification of the cyclic β-ketoester, the product of backbiting (Eq. 4.56) in the polymerized mass.[106–109] Intramolecular coordination of the metal ion has also been suggested in the interpretation of NMR spectra of PMMA-Li solution in THF.[110]

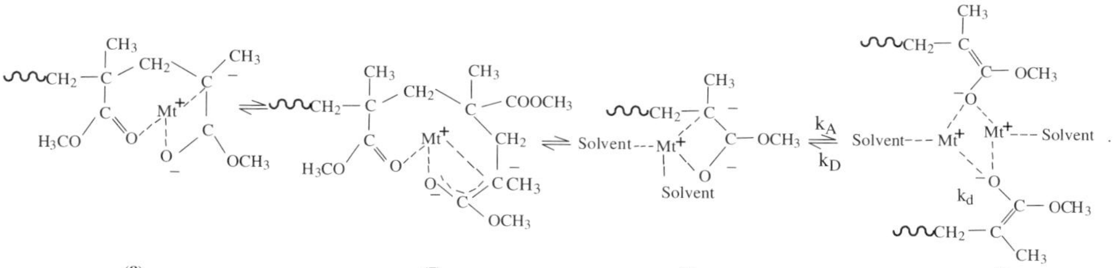

$$\text{(8)} \rightleftharpoons \text{(7)} \rightleftharpoons \text{(6)} \xrightleftharpoons[k_D]{k_A} \text{(9)} \quad (4.58)$$

The positions of the various solvation equilibria in Eq. (4.58) depend on the alkali ion and the coordinating power of the solvent molecules. In solvents of strong coordinating power such as 1,2-dimethoxyethane (DME), structure **6** is believed to be strongly favored due to chelation of Mt^+ by the solvent. Accordingly, polymerization occurs without termination even at 0°C in DME.[111]

4.1.4.2.2 Rate constants of propagation

Further insight into ion pair solvation is obtained from the analysis of the rate constant of propagation on ion pair ($k_p^{\pm}$). The values of $k_p^{\pm}$ on various PMMA-alkali ion pairs are given in Table 4.2. The order of variation of $k_p^{\pm}$ with counterion size in MMA polymerization in THF is completely different from that observed in styrene polymerization. The dimerization tendency of PMMA-alkali ion pairs in THF, as already discussed, suggests that they are not solvent-separated; but the invariance of $k_p^{\pm}$ with counter ion size (from Na^+ to Cs^+) is not supportive of contact ion pair character either (*vide* Sec. 4.1.3.1).

Table 4.2 Rate constant of propagation and Arrhenius parameters in anionic polymerization of methyl methacrylate in THF.[a,b]

Metal ion	Poly(methyl methacrylate)$^-$Mt$^+$		
	$k_p^{\pm}(-40°C)$	E_a^{112}, kJ/mol	log A^{112}
Li$^+$	100[112]	24	7.4
Na$^+$	800[19,91]	18.3	7.0
K$^+$	750[112]	19.3	7.2
Cs$^+$	860[91]	19.5	7.3

[a]All rate constants in $L\,mol^{-1}s^{-1}$. [b]Reprinted from A.H.E. Muller, *Kinetics and Mechanisms in the Anionic Polymerization of Methacrylic Esters*, in, Recent Advances in Anionic Polymerization, T. Hogen-Esch, J. Smid Eds., Elsevier Science Pub. Co., Inc., 1987, p. 205 with permission from Elsevier and author.

The remarkably low value of $k_p^{\pm}$ on PMMA-Li testifies to the predominant enolic character of the chain end. The higher values of $k_p^{\pm}$s on ion pairs with counter ions larger than Li$^+$ are in line with their somewhat greater carbanionic character. The dissociation constant of the ion pair is extremely low ca., $K_d < 10^{-9}$ (with Na$^+$ or Cs$^+$ counter ion at $-78°$C).[113] This is attributed to tight binding of alkali ion with enolate end of the polymer through chelation as shown in structures in Eq. (4.58).[91] Such intramolecular chelation of alkali ion was also considered to be responsible for the low K_d of living poly(vinyl pyridine).[114–116]

Peripheral solvation of the chelated alkali ion in the contact ion pair makes $k_p^{\pm}$ on it larger than that ordinarily expected on the contact ion pair.[91]

4.1.4.3 Ligated living anionic polymerization

Living character of polymerizations of alkyl methacrylates is greatly improved with the addition of various ligands, which form σ, μ, or σ, μ complexes with Li$^+$ ion or the enolate in the living end depending on the ligand type.[11,117–119] The complexes set up new equilibria with the dimeric living end as shown in Scheme 4.6 with LiCl used as the additive.

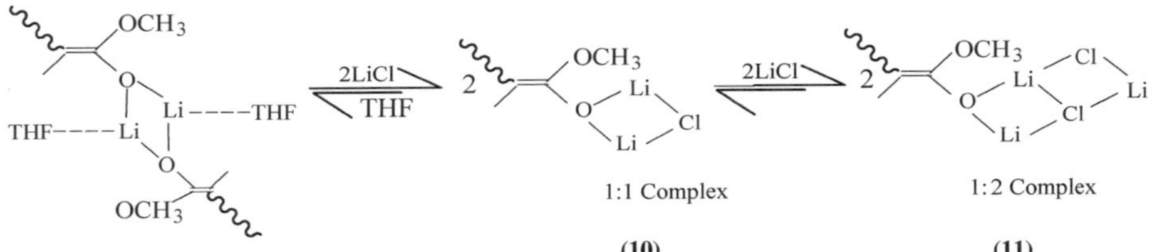

Scheme 4.6 Suggested structures of various species existing in equilibrium in the living anionic polymerization of MMA in THF in the presence of LiCl. "Reprinted with permission from Ref. 11, Anionic vinyl polymerization-50 years after Michael Szwarc. Copyright © 2007 Elsevier Ltd."

Most importantly, the exchange between the propagation-inactive dimeric ion pair (**9**) and the propagation-active monomeric ion pairs (**10**) and (**11**), becomes much faster than propagation resulting in narrow MWD polymer.[11,102,120,121] ^{7}Li NMR spectroscopy of living PMMA-Li in the presence of LiCl confirms complex formation. Thus, the chemical shift of Li with reference to LiCl undergoes a decrease from 0.41 ppm in PMMA-Li to 0.147 ppm in 1:1 PMMA : LiCl.[119]

Kinetic studies reveal that LiCl does not reduce termination, which means that the intramolecularly complexed contact ion pair (**7**) must be one of the species involved in the various equilibria existing in the polymerizing mass apart from the ones shown in Scheme 4.6. In other words, the intramolecularly solvated forms **6**, **7**, and **8** shown in Eq. (4.58) are also involved in the equilibria.

LiCl also improves the living character of the anionic polymerization of *tertiary* acrylates and some functionalized methacrylates, *e.g.*, 2-(dimethylamino)ethyl methacrylate.[122] Initiator efficiency approaches unity and PDI becomes less than 1.1. Figure 4.7 demonstrates the remarkbale narrowing of MWD effected by LiCl in the polymerization of *t*-butyl

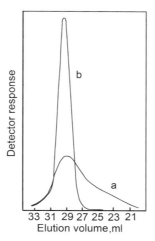

Figure 4.7 GPC traces of poly(*t*-butyl acrylate) prepared by living anionic polymerization of *t*-butyl acrylate in THF at −75°C. Curve a, without LiCl additive; curve b, with LiCl additive (used at a concentration five times that of the living ends). "Reprinted with permission from Ref. 119. Copyright © 1987 American Chemical Society."

acrylate.[99,119] Besides *tertiary* acrylates and various alkyl methacrylates, both 2-vinyl pyridine and 4-vinyl pyridine undergo living anionic polymerization in the presence of LiCl.[123,124]

$LiClO_4$ is more effective than LiCl in improving the living polymerization of MMA. For instance, polymerization is reportedly living even at −20°C in THF in the presence of $LiClO_4$ at $[LiClO_4]/[PMMA-Li] = 10$.[125,126] The initiator efficiency is close to unity and PDI = 1.08. In this system also, like in LiCl assisted polymerizations, no reduction in termination is observed in the kinetic studies of the polymerization.[125] Hence, the narrow MWD is attributable to fast exchange between PMMA-Li dimer and $LiClO_4$ complexed monomeric living ends.

Use of triethylaluminum as the ligand leads to living polymerization even in toluene at −78°C.[127] The ligand also exerts *tacticity* control giving *syndiotactic* polymer. *Isotactic*-rich polymer with broad MWD forms in the absence of the ligand. Use of various coligands with triethylaluminum such as Lewis bases[128,129] and tetraalkylammonium halides[130,131] gives faster polymerization and narrower MWD. When cesium halides are used as coligands, living polymerization of MMA can be performed as well at higher temperatures ca., −20°C.[132] Even polymerization of a primary acrylate, such as *n*-butyl acrylate in which termination is more severe than in MMA polymerization, can be performed livingly below −65°C using cesium fluoride as the coligand. This feat suggests that termination must be largely reduced in this system. This is in sharp contrast to the action of LiCl or $LiClO_4$ ligand, which does not reduce termination, as we have learnt above in this section.

Ligands of the σ, μ type proved to be the most effective additive.[133,134] Thus, with the use of lithium 2-methoxyethoxide as the ligand, extremely fast living polymerization (half life of milliseconds) of *n*-butyl acrylate occurs even in toluene at relatively high temperature, ca., −20°C.[135]

Apart from (meth)acrylates and vinyl pyridines, alkylacrylamides have also been polymerized livingly. For instance, polymerization of N,N-dimethylacrylamide in THF at $-78°C$ using organometallic initiators is living, although, only when Cs^+ is the counter ion. When Cs^+ is replaced with other alkali ions, heterogeneous nonliving polymerization occurs.[136]

On the other hand, the use of Lewis acids such as Et_2Zn, Et_3B, or Et_3Al as additives gives not only living polymerization (with Li^+ or K^+ counter ion) but also influences the *tacticity* of the polymer formed.[137,138] Thus, poly(N,N-diethylacrylamide) formed is highly *isotactic* when LiCl is the additive and an organolithium is the initiator. In contrast, highly *syndiotactic* and *atactic* polymers are obtained using Et_2Zn and Et_3B respectively as additives. Apparently, the coordination of the amidoenolate (**12**) with different Lewis acids controls *tacticity*.[11]

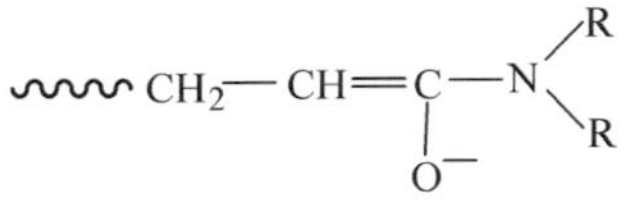

$$\sim\sim\sim CH_2-CH=C-N \begin{smallmatrix} R \\ \\ R \end{smallmatrix}$$
$$\qquad\qquad\qquad | \\ \qquad\qquad\qquad O^-$$

(**12**)

4.1.5 End functionalized polymers

Polymers with designed functional end groups may be prepared by forcible termination ("killing") of living ends with appropriate reagents. For example, carboxyl and hydroxyl groups may be introduced using carbon dioxide[139–141] and ethylene oxide[142] respectively as chain terminators.

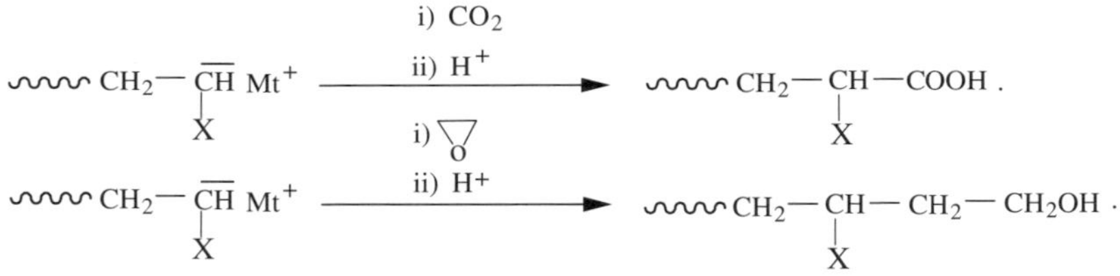

$$\sim\sim\sim CH_2-\overline{CH}\ Mt^+ \quad \xrightarrow[\text{ii) } H^+]{\text{i) } CO_2} \quad \sim\sim\sim CH_2-CH-COOH$$
$$\qquad\qquad | \qquad\qquad\qquad\qquad\qquad\qquad\qquad\qquad | \\ \qquad\qquad X \qquad\qquad\qquad\qquad\qquad\qquad\qquad\qquad X$$

$$\sim\sim\sim CH_2-\overline{CH}\ Mt^+ \quad \xrightarrow[\text{ii) } H^+]{\text{i) } \triangle O} \quad \sim\sim\sim CH_2-CH-CH_2-CH_2OH$$
$$\qquad\qquad | \qquad\qquad\qquad\qquad\qquad\qquad\qquad\qquad\qquad | \\ \qquad\qquad X \qquad\qquad\qquad\qquad\qquad\qquad\qquad\qquad\qquad X$$

Polymers with two living ends give α, ω-difunctional polymers also called "telechelic" polymers. The carbanions should be strongly nucleophilic (*e.g.*, polystyryl and polydienyls) and a polar solvent should be used for quantitative reactions. Reaction with ethylene oxide is required to be performed at low temperature in order to prevent polymerization.[142] Carboxyl groups may be introduced also by using a cyclic anhydride as terminator[143]

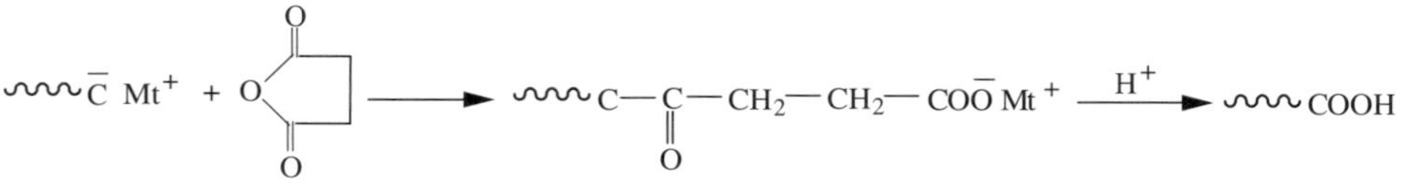

$$\sim\sim\sim \overline{C}\ Mt^+ + \underset{O}{\underset{\|}{O}}\!\!\!\bigcirc\!\!\!\underset{O}{\underset{\|}{O}} \longrightarrow \sim\sim\sim C-C-CH_2-CH_2-CO\overline{O}\ Mt^+ \xrightarrow{H^+} \sim\sim\sim COOH$$

Functionalization by various other groups, e.g., halogen, nitrile, and thiol may be performed using appropriate terminating agents.[144]

Vinyl unsaturation may be introduced to synthesize macromonomers. For instance, poly(alkyl methacrylate) macromonomer is prepared by the reaction of the corresponding living polymer with p-vinylbenzyl chloride (VBC) or methacryloyl chloride.[145,146]

$$\sim\sim CH_2-\underset{COOCH_3}{\overset{CH_3}{\underset{|}{\overset{|}{C^-}}}} Mt^+ \;+\; ClCH_2-\!\!\!\bigcirc\!\!\!-CH=CH_2 \;\longrightarrow\; \sim\sim CH_2-\underset{COOCH_3}{\overset{CH_3}{\underset{|}{\overset{|}{C}}}}-CH_2-\!\!\!\bigcirc\!\!\!-CH=CH_2 \;+\; MtCl\,.$$

$$\sim\sim CH_2-\underset{COOCH_3}{\overset{CH_3}{\underset{|}{\overset{|}{C^-}}}} Mt^+ \;+\; Cl-\underset{O}{\overset{CH_3}{\underset{\|}{\overset{|}{C}}}}-\overset{CH_3}{\underset{|}{C}}=CH_2 \;\longrightarrow\; \sim\sim CH_2-\underset{CH_3OOC}{\overset{CH_3}{\underset{|}{\overset{|}{C}}}}-\underset{O}{\overset{CH_3}{\underset{\|}{\overset{|}{C}}}}-\overset{CH_3}{\underset{|}{C}}=CH_2 \;+\; MtCl\,.$$

A quantitative reaction takes place in spite of the low nucleophilicity of the carbanionic species involved. In fact, with strongly nucleophilic carbanionic species, like polystyryl, or polydienyls, the nucleophilicity needs to be reduced, for example, by reacting them with 1,1-diphenylethylene before they are reacted with VBC or methacryloyl chloride.[145,147,148] Macromonomers are very useful for preparing graft copolymers with nearly uniform graft lengths by way of copolymerizing them with suitable monomers.

4.1.6 Synthesis of block copolymers

The direct method of synthesis of block copolymers involves sequential living polymerization of the concerned monomers in the order of increasing electronegativity. This is because the living polymer prepared from an electronegative monomer may not succeed in initiating the polymerization of an appreciably less electronegative one.[2] For instance, the synthesis of a diblock copolymer of styrene and MMA may be successfully done when the appreciably less electronegative styrene is polymerized first using a monofunctional initiator but not when MMA is polymerized first. However, the synthesis will not be clean inasmuch as the polystyryl anion or the ion pair is too high in nucleophilicity to be incapable of inducing side reaction (Eq. 4.54) in the polymerization of MMA. The problem is overcome by way of reducing nucleophilicity, for example, by adding a few drops of 1, 1-diphenylethylene to the living polystyrene solution. This monomer does not homopolymerize but adds to polystyrene living end giving the desired result.[149]

In contrast, when two monomers differ in electronegativity only to a small degree, e.g., styrene and butadiene, or styrene and isoprene, their diblock copolymers can be synthesized irrespective of the sequence followed.[12] However, polymerizing dienes first in a hydrocarbon solvent gives polystyrene block with broader MWD. This is due to the greater degree of association of polydienyllithiums making initiation of styrene polymerization slower than propagation. The shortcoming in this case can be rectified by adding some polar solvent before polymerization of styrene is undertaken. The polar solvent disintegrates the aggregates of both polydienes and polystyrene living ends forming monomeric species.

This strategy is followed in the synthesis of the thermoplastic elastomers, polystyrene-*b*-polybutadiene-*b*-polystyrene (SBS) or polystyrene-*b*-polyisoprene-*b*-polystyrene (SIS) in hydrocarbon solvents.[150]

An alternative method prepares the diblock copolymer of styrene and diene first using only half the diene required to make the triblock. The diblock is then coupled with dimethyldichlorosilane[2]

$$2PS-b-PBD^-Li^+ + Me_2SiCl_2 \rightarrow PS-b-PBD-Si(Me_2)-PBD-b-PS + 2LiCl.$$

The ABA triblock copolymers may be synthesized also by first synthesizing the middle B block with two living ends using a difunctional initiator and then extending the polymer chain from both ends using monomer A. However, this sequential method does not succeed if the monomer sequence does not observe the increasing electronegativity rule, particularly when the electronegativity difference is quite large. However, the alternative method described above can be used successfully in this case also.

4.1.7 Synthesis of star polymers

Three methods exist for the synthesis of star polymers.

1. Initiation by a multifunctional initiator

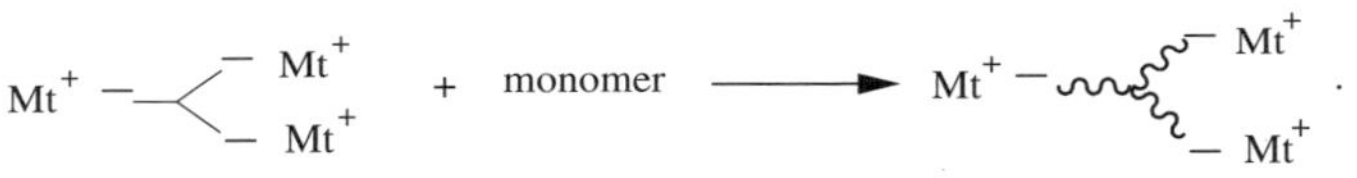

2. Termination by a multifunctional terminator[151–153]

3. Polymerization of a difunctional monomer initiated by a living chain

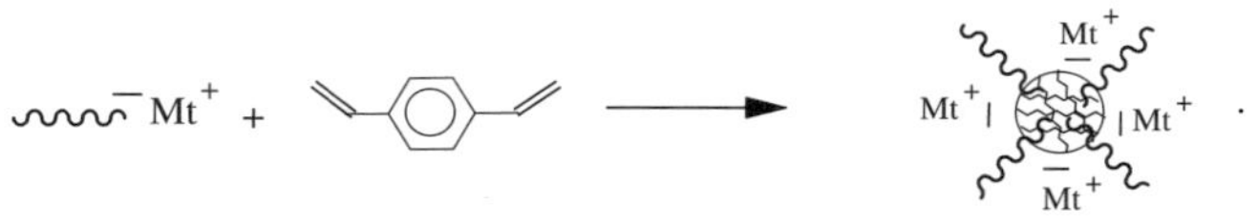

The first method suffers from the solubility problem with most initiators. In addition, due to steric constraints, there may be nonuniformity in arm lengths particularly when the number of arms is large. The solubility problem may be eliminated by appropriate initiator design, an example being the trifunctional initiator **14**, which has three diphenylalkyllithium type initiating centers. The initiator may be prepared by the reaction of *s*-BuLi and 1,3,5-tris (1-phenylethenyl)benzene (**13**), which is incapable of homopolymerization.[154]

Star polymers with a large number of arms may be prepared using a microgel initiator, which may be obtained by polymerizing divinylbenzene at high dilution using *s*-BuLi initiator. The number of arms depends on the mol ratio of *s*-BuLi to divinylbenzene[155–158]

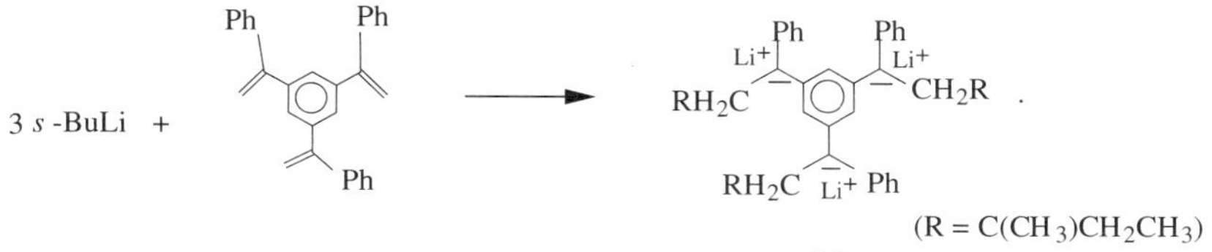

[Structure (13): 1,3,5-tris(1-phenylethenyl)benzene with 3 s-BuLi reacting to give structure (14): trilithiated adduct with RH₂C groups (R = C(CH₃)CH₂CH₃)]

3 s-BuLi + (13) → (14)   (R = C(CH$_3$)CH$_2$CH$_3$)

The second method has proved to be very satisfactory. The terminators usually contain chloro- or bromobenzyl or chlorosilyl functions.[143,151,152] Some of these are compounds **15** through **19**.

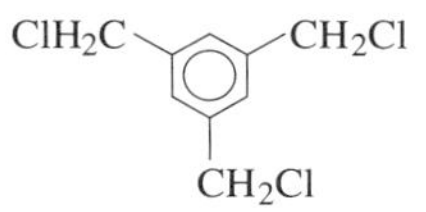

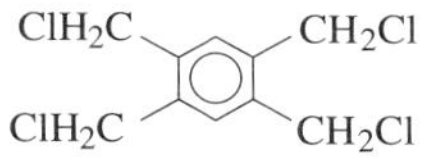

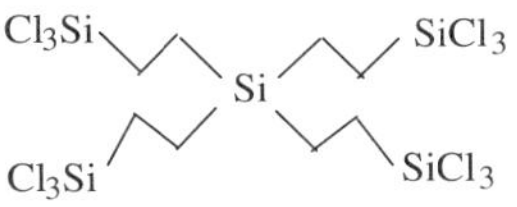

[Structures 15-19: (15) 1,3,5-tris(chloromethyl)benzene; (16) 1,2,4,5-tetrakis(chloromethyl)benzene; (17) SiCl$_4$; (18) Si$_2$Cl$_6$; (19) Si(CH$_2$CH$_2$SiCl$_3$)$_4$]

Termination results in the substitution of polymer chain for halogen atom in the terminator. Chlorosilances are preferable since they terminate cleanly. Separating Si–Cl groups by methylenes as in **19** increases linking efficiency.[153] Dendrimers with chlorosilane groups have been used as terminators to prepare star polymers with more than one hundred number of arms.[153]

Asymmetric star polymers (different arm lengths, different functional end groups of the arms) and topologically asymmetric star-block copolymers as well as miktoarm star polymers (chemically different arms) may also be prepared using suitable synthesis strategy.[153] For example, a living polymer can be terminated using large excess of methyltrichlorosilane. The excess amount is evaporated off and the resulting polymer with methyldichlorosilane end function is used as terminator of other living polymers, which may be the same polymer but of different molecular weights[159] or a different polymer.[160–162] In the former case, star polymer with molecular weight asymmetry and, in the latter case, miktoarm star polymer is obtained.

PS$^-$ Li$^+$ + CH$_3$SiCl$_3$ (excess) → PS Si(CH$_3$)Cl$_2$ $\xrightarrow{\text{PI}^- \text{Li}^+ \text{ (excess)}}$ PS Si(CH$_3$)(PI)$_2$.

In the third method, a difunctional monomer such as divinyl benzene[163–165] or ethylene glycol dimethacrylate[166] is polymerized using a living polymer as initiator to give star polymers with large number of arms. Similarly, reacting the nonhomopolymerizable 1,3,5-tris(1-phenylethenyl) benzene (**13**) with a living polymer gives a three-arm star polymer.[154]

4.1.8 Synthesis of comb polymers

Comb-shaped or comb polymers are graft copolymers with multiple branches attached to the main chain backbone. Three methods are available for their preparation (i) "grafting onto", (ii) "grafting from", and (iii) "grafting through".[153]

In the "grafting onto" method, polymers, which provide graft copolymer backbones, are prepared with pendant electrophilic groups such as anhydrides, esters, pyridine, benzylic halide, *etc.*[167] These polymers are then reacted with living polymers, which get terminated on the electrophilic groups. For instance, polystyrene may be partially chloromethylated[168] or bromomethylated[169] and reacted in THF with a living polymer to yield graft copolymer with branches randomly distributed along the main chain backbone, as shown below for grafting of poly(vinyl pyridine) (PvPy)

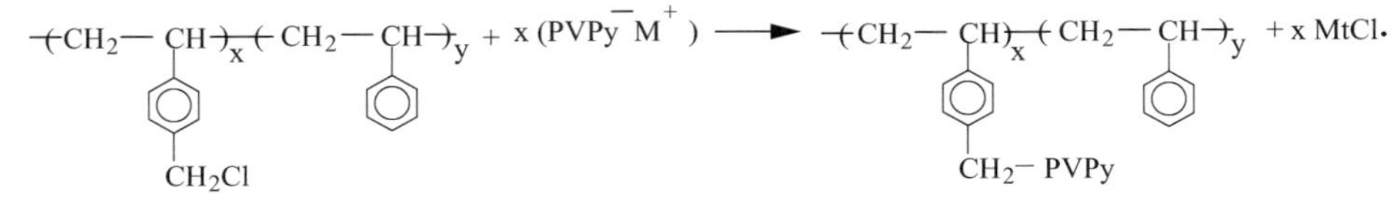

$$\text{-(CH}_2\text{-CH)}_x\text{-(CH}_2\text{-CH)}_y + x\,(\text{PVPy}^- \text{M}^+) \longrightarrow \text{-(CH}_2\text{-CH)}_x\text{-(CH}_2\text{-CH)}_y + x\,\text{MtCl}$$

(with pendant CH_2Cl on left structure and CH_2–PVPy on right structure; phenyl rings on the other unit)

Alternatively, styrene may be copolymerized radically with *p*-vinylbenzyl chloride to prepare the backbone polymer with pendant benzyl chloride groups. Homopolymer of *p*-vinylbenzyl chloride may be used also to prepare very densely branched graft copolymer.

In the "grafting from" method, anionic initiating sites are generated along the main chain backbone and living polymerization of the same or another monomer is carried out. A convenient method of generating anionic initiating sites involves metalation of allylic, benzylic or aromatic C–H bonds, present either in the chain backbone or in pendant groups, by *s*-BuLi complexed with the chelating agent N,N,N′,N′-tetramethylethylenediamine (TMEDA) (Sec. 4.1.2.2), as shown below for grafting from polybutadiene.[154,170,171]

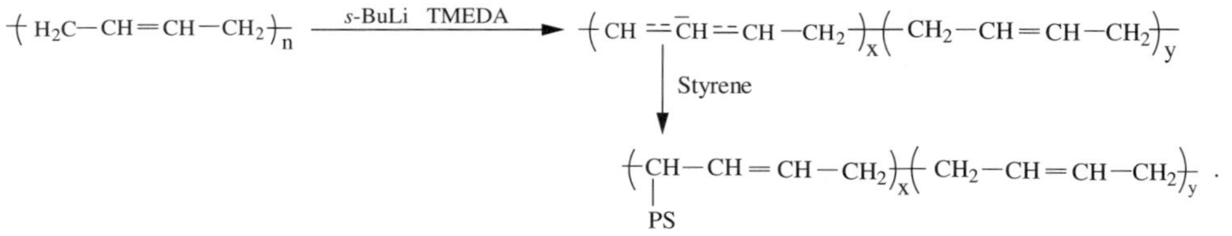

$$\text{-(H}_2\text{C-CH=CH-CH}_2\text{)}_n \xrightarrow{s\text{-BuLi, TMEDA}} \text{-(CH=}\overline{\text{CH}}\text{=CH-CH}_2\text{)}_x\text{-(CH}_2\text{-CH=CH-CH}_2\text{)}_y$$

$$\downarrow \text{Styrene}$$

$$\text{-(CH-CH=CH-CH}_2\text{)}_x\text{-(CH}_2\text{-CH=CH-CH}_2\text{)}_y$$
(with PS branch on the CH)

In the "grafting through" method, a macromonomer, which is a precursor of branches in graft copolymer, is copolymerized with a monomer

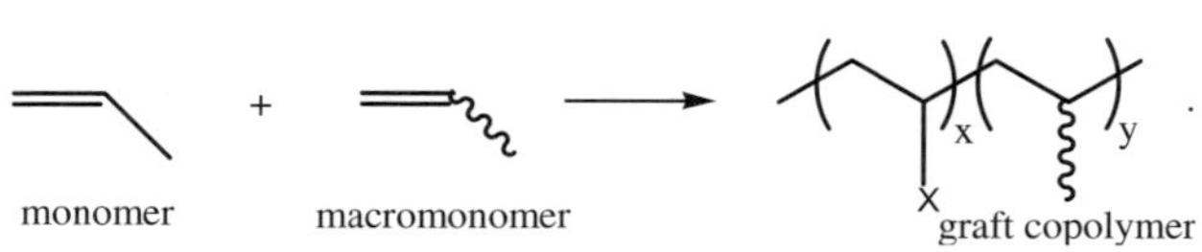

monomer + macromonomer → graft copolymer

Macromonomer synthesis by living anionic polymerization method has been described earlier (Sec. 4.1.5). The branch density can be controlled by the mol ratio of monomer to macromonomer. Homopolymerization of the macromonomer itself gives rise to comb polymer with equally spaced teeth.[172]

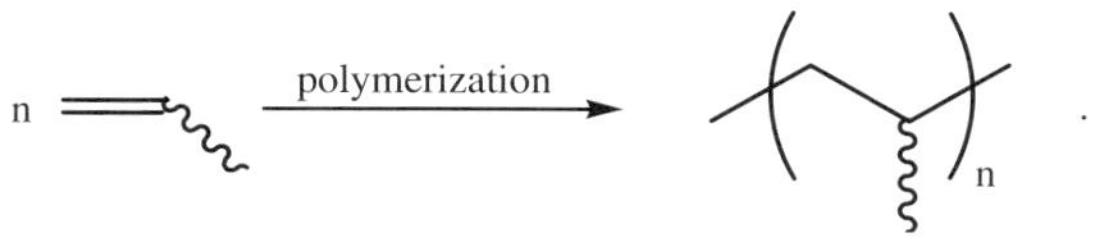

4.1.9 *Group transfer polymerization*

The classical living anionic polymerization (LAP) of alkyl methacrylates is successful only at low temperatures. Industrial scale polymerization, however, requires above-ambient temperature of operation for the process to be cost-effective. Group transfer polymerization (GTP) discovered by Webster in Du Pont met this requirement satisfactorily.[173]

In GTP, a silyl ketene acetal initiator (In) is reversibly activated by a nucleophilic catalyst (Nu$^-$) used in small amount (*ca.*, 0.1 mol% of initiator) for initiation to take place (Scheme 4.7). Fluorides, bifluorides, cyanides, azides, benzoates, and bibenzoates are usually used as nucleophiles in the form of organo-soluble salts containing tris(dimethylamino)sulfonium, trispiperidinosulfonium, or tetrabutylammonium as positive ions. Benzoates and bibenzoates being weaker nucleophiles than fluorides and bifluorides require higher temperature *ca.*, 80°C to perform.[176] THF, acetonitrile, and DMF are usually the solvents used in the polymerization.[173–175]

Propagation also involves reversible activation of silyl ketene acetal at chain end by the nucleophile as shown in Scheme 4.8 for GTP of MMA. The nucleophile attacks the electron deficient Si center in the dormant polymer (P_n) and displaces the polymeric ester enolate (P_n^-), which is propagation-active.[177,178] The mechanism is referred to as 'dissociative'.

The polymer grows intermittently through many cycles of activation–deactivation during the time of polymerization. As discussed in Sec. 4.1.1.1, the exchange between the active (P_n^-) and dormant (P_n) forms of the polymer must be fast in the time scale of propagation in order that a narrow MWD polymer is formed. The difference between LAP and GTP in the polymerization of methyl methacrylate lies in the nature of the dormant species. In LAP it is the dimer of the ester enolate (*vide* Sec. 4.1.4.2), whereas in GTP it is the silyl ketene acetal ended polymer (P_n).

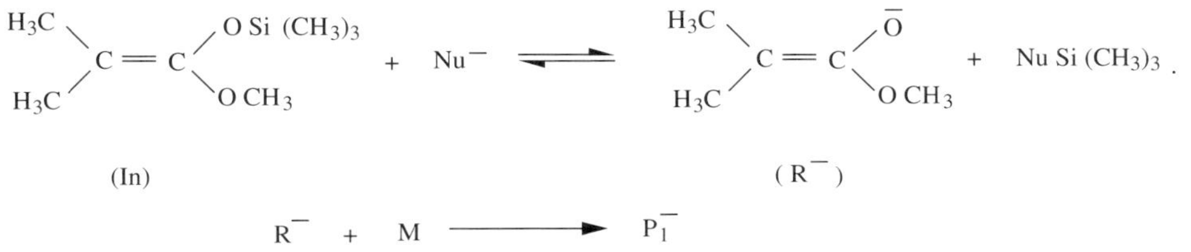

Scheme 4.7 Dissociation mechanism of initiation in GTP.

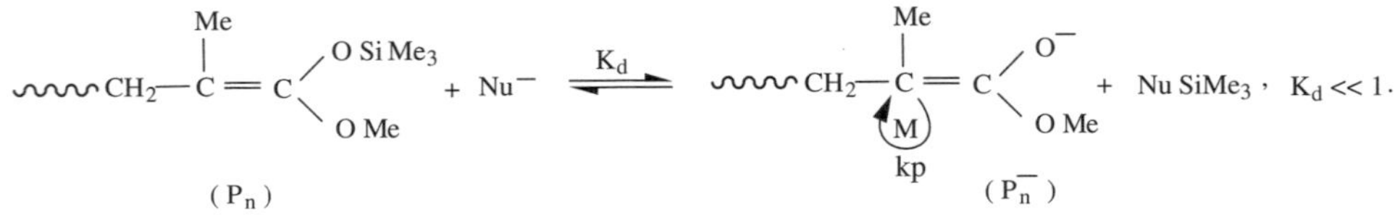

248   *Fundamentals of Polymerization*

$$\text{\textasciitilde CH}_2-\underset{\underset{\text{Me}}{|}}{C}=C\genfrac{}{}{0pt}{}{\diagup \text{O SiMe}_3}{\diagdown \text{O Me}} + \text{Nu}^- \xrightleftharpoons{K_d} \text{\textasciitilde CH}_2-\underset{\underset{\text{Me}}{|}}{C}=C\genfrac{}{}{0pt}{}{\diagup \text{O}^-}{\diagdown \text{O Me}} + \text{Nu SiMe}_3, \; K_d \ll 1.$$

(P_n) (P_n^-) with M, k_p

(M = monomer, k_p = propagation rate constant.)

Scheme 4.8   Dissociative mechanism of reversible deactivation in GTP of MMA.

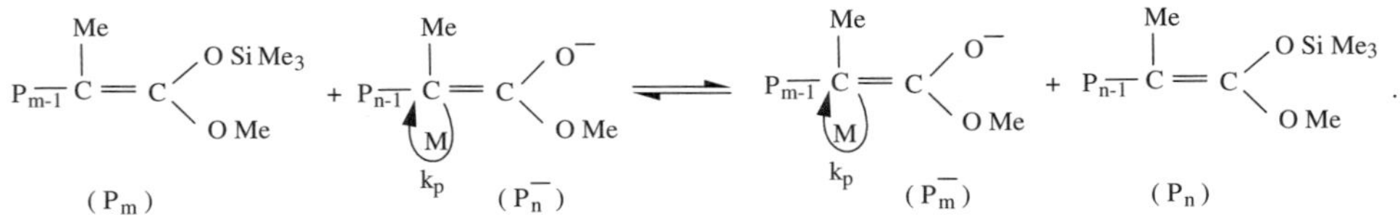

$$P_{m\text{-}1}\underset{\underset{\text{Me}}{|}}{C}=C\genfrac{}{}{0pt}{}{\diagup \text{O SiMe}_3}{\diagdown \text{O Me}} + P_{n\text{-}1}\underset{\underset{\text{Me}}{|}}{C}=C\genfrac{}{}{0pt}{}{\diagup \text{O}^-}{\diagdown \text{O Me}} \rightleftharpoons P_{m\text{-}1}\underset{\underset{\text{Me}}{|}}{C}=C\genfrac{}{}{0pt}{}{\diagup \text{O}^-}{\diagdown \text{O Me}} + P_{n\text{-}1}\underset{\underset{\text{Me}}{|}}{C}=C\genfrac{}{}{0pt}{}{\diagup \text{O SiMe}_3}{\diagdown \text{O Me}}$$

(P_m) (P_n^-) (P_m^-) (P_n)

(M = monomer, k_p = propagation rate constant.)

Scheme 4.9   Degenerative transfer mechanism of exchange in GTP [179].

Too much nucleophile makes the propagation-active enolate concentration too high leading to deterioration of living character.[173] The reversible nature of activation makes possible activate all initiator molecules using only a very small amount of nucleophile.

The exchange between the dormant and the active forms of the propagating species may also take place by degenerative transfer, which has been discussed in Chap. 3 (*vide* Sec. 3.15) in living radical polymerization. In GTP, degenerative transfer may be represented as in Scheme 4.9.[11,179] The ester enolate P_n^- acts as the nucleophile, which abstracts the trimethylsilyl group from the dormant polymer P_m activating the latter to ester enolate P_m^- and converting itself into the dormant polymer P_n.[179] Some nucleophiles may bind strongly to silicon in trimethylsilyl, bifluoride being an example, making activation (Scheme 4.8) irreversible. In such systems, degenerative transfer becomes the only means of exchange.[11]

Besides the mechanism discussed so far, activation of the initiator or the dormant polymer may also take place while the nucleophile is bonded to the silicon atom, as shown in Scheme 4.10. This mechanism of activation is referred to as associative.[180] The pentacoordinate siliconate (**21**) chain end is the active form, which undergoes Michael addition to monomer with concomitant migration of silyl group to the carbonyl oxygen of the newly added monomer.[11]

The dissociative mechanism is supported by the observation of occasional backbiting termination similar to that encountered in classical LAP (reaction 4.56).[181] However, it has been suggested that the mechanism of GTP strongly depends on the nature of the nucleophilic catalyst. As discussed above, a strong silicon binding catalyst, like bifluoride, appears to operate through the irreversible dissociative mechanism, whereas catalysts, like oxyanions, which bind to silicon less strongly, may operate through both associative and dissociative mechanisms.[11]

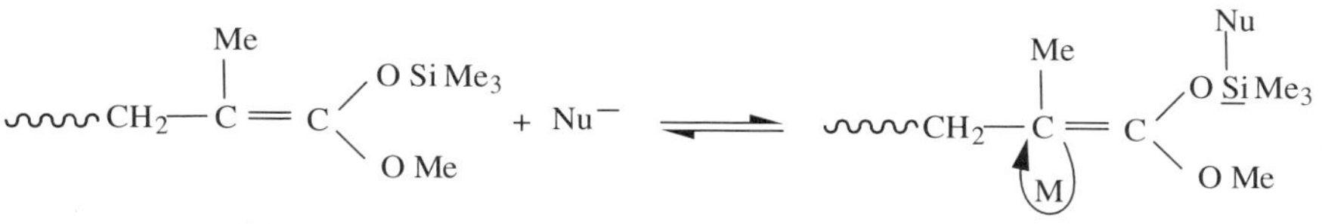

$$\underset{(20)}{\underset{\text{Dormant}}{\sim\sim\text{CH}_2-\underset{\text{Me}}{\overset{\text{Me}}{\text{C}}}=\text{C}\overset{\text{OSiMe}_3}{\underset{\text{OMe}}{\diagdown}}}}+\text{Nu}^-\rightleftharpoons\underset{(21)}{\underset{\text{active}}{\sim\sim\text{CH}_2-\underset{(M)}{\overset{\text{Me}}{\text{C}}}=\text{C}\overset{\overset{\text{Nu}}{|}\text{OSiMe}_3}{\underset{\text{OMe}}{\diagdown}}}}$$

Scheme 4.10 Associative mechanism of reversible activation of dormant polymer molecules.

Apart from methacrylates, several other monomers containing electron deficient double bonds, *e.g.*, acrylates, acrylonitrile, methacrylonitrile, N,N-dimethylacrylamide, and 2-methylene-4-butyrolactone, may also be polymerized. Of the monomers, methacrylates work best. Acrylates polymerize very fast. However, in the polymerization of acrylates, isomerization of the silyl ketene acetal takes place, which moves from terminal to internal positions in the chain (Eq. 4.59). The internal silyl ketene acetal being unreactive due to steric hindrance eventually makes all chains dead.[182]

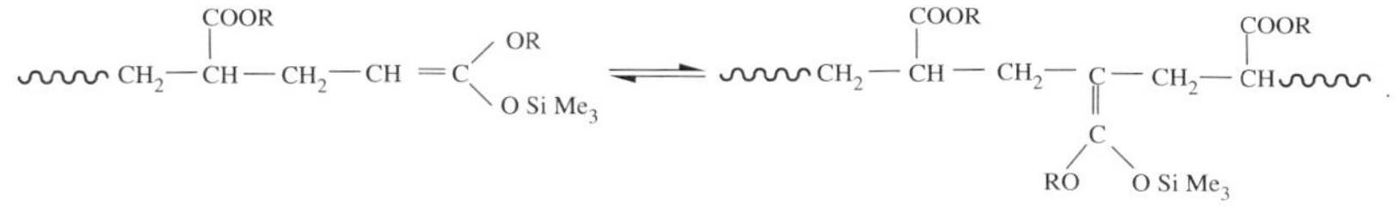

$$\sim\sim\text{CH}_2-\overset{\overset{\text{COOR}}{|}}{\text{CH}}-\text{CH}_2-\text{CH}=\text{C}\overset{\text{OR}}{\underset{\text{OSiMe}_3}{\diagdown}}\rightleftharpoons\sim\sim\text{CH}_2-\overset{\overset{\text{COOR}}{|}}{\text{CH}}-\text{CH}_2-\underset{\overset{\|}{\text{C}}\overset{\diagdown}{\underset{\text{OSiMe}_3}{\text{RO}}}}{\text{C}}-\text{CH}_2-\overset{\overset{\text{COOR}}{|}}{\text{CH}}\sim\sim \quad (4.59)$$

An initiator, which models the propagating chain end, is suitable since rate of initiation becomes comparable with that of propagation leading to formation of narrow MWD polymer. Thus, dimethyl ketene methyl trimethylsilyl acetal (In) in Scheme 4.7 is an efficient initiator for the polymerization of MMA. Similarly, 2-(trimethylsilyl)isobutyronitrile (**22**) is a suitable initiator for the polymerization of methacrylonitrile.

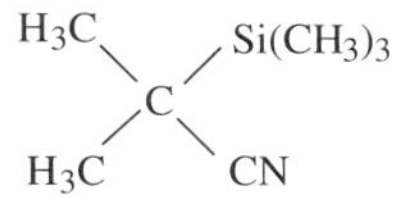

$$\underset{(22)}{\overset{\text{H}_3\text{C}}{\underset{\text{H}_3\text{C}}{\diagdown}}\overset{\diagup\text{Si}(\text{CH}_3)_3}{\underset{\text{CN}}{\text{C}}\diagdown}}\qquad\underset{(23)}{\overset{+}{\text{H}_2\text{C}=\text{C}}\overset{\diagup\text{CH}_3}{\underset{\text{OCH}_3}{\diagdown}}\text{C}=\text{O}:\longrightarrow\text{ZnCl}_2}$$

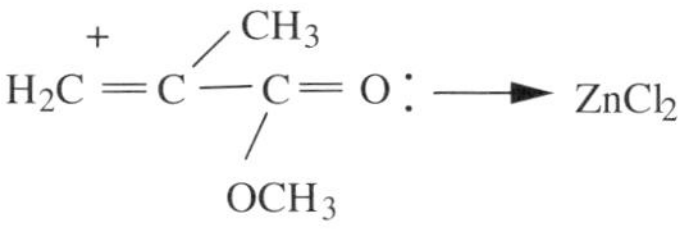

Besides nucleophilic catalysts, electrophilic Lewis acids catalysts, *e.g.*, zinc halides, dialkylaluminum chlorides, and dialkylaluminum oxides, also work but by a different mechanism and at a very high concentration *ca.*, 10 mol percent of initiator.[173,174] Toluene and dichloromethane are usually used as solvents in polymerization. However, the nucleophilic catalysts are preferred since the costly catalyst removal step is avoided, the catalyst amount being too small to affect polymer properties if left in the polymer unremoved.

The electrophilic catalysts increase the electrophilicity of a monomer by coordinating with the polar substituents thereby activating the monomer as shown in **23** with $ZnCl_2$

and MMA. Similar complexes also form with the polar groups in polymer. Hence, a large amount of catalyst is required,[173,176] as already noted. Activation facilitates addition of monomer to the silyl ketene acetal.[173]

Like LAP, GTP has been used to synthesize (co)polymers of various architectures such as block copolymers, star polymers, hyperbranched polymers, networks, *etc*.

Polymerizations are terminated by protic compounds but not by oxygen. Hence, they must be conducted under anhydrous conditions. For the same reason, monomers containing hydroxyl and carboxyl groups are polymerized only after the groups are protected.

4.1.10 *Metal–free living anionic polymerization*

It may be recalled that termination in classical anionic polymerization of alkyl(methyl) acrylates is due to intramolecular nucleophilic attack of ester enolate chain end on ester carbonyl of the antepenultimate monomer unit (Eq. 4.56). This is facilitated by the coordination of the ester carbonyl with the metal counter ion associated with the chain end. Replacement of metal ion with a non-metal ion is expected, therefore, to improve the living character of polymerization. This was indeed found to be true.[183–188] For instance, rapid polymerization of *n*-butyl acrylate, methyl acrylate, acrylonitrile, and methyl methacrylate occurs in THF at room temperature using tetrabutylammonium salts of various nucleophiles yielding polymer with relatively narrow MWD. Since the tetrabutylammonium ion undergoes Hofmann elimination in the presence of bases, only relatively weak bases (nucleophiles) such as malonyl, fluorenyl, or 1,1 diphenylhexyl carbanions, alkyl or aryl thiolates, and isobutyrates need to be used as initiators.[11,188] However, these nucleophiles being weaker than the propagating ester enolate bring about slow initiation — slower than propagation — effecting MWD broadening. Initiation may even be incomplete resulting in low initiator efficiency.[189]

With the use of tetraphenylphosphonium ion (Ph_4P^+) in place of Bu_4N^+ Hofmann elimination is suppressed due to the absence of the reactive alkyl β-hydrogens. Thus, tetraphenylphosphonium triphenylmethyl ($Ph_3\bar{C}\overset{+}{P}Ph_4$) initiated polymerization of MMA in THF proceeds fast at room temperature and is completed in seconds giving quantitative yield of polymer with relatively narrow MWD.[190] Detailed kinetic studies of the polymerization suggested that the active centers (ester enolate and its ion pair) exist in dynamic equilibrium with phosphor ylide dormant species, as shown in Scheme 4.11.[191,192]

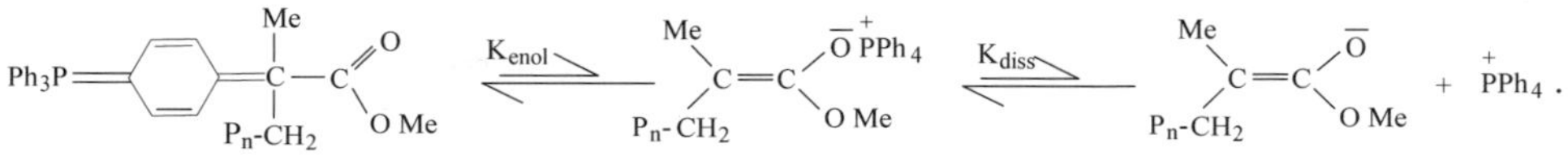

Scheme 4.11 Reversible deactivation in MMA polymerization in presence of PPh_4^+ counter ion.[192]

References

1. M. Szwarc, *Carbanions, Living Polymers and Electron Transfer Processes*, Interscience, New York, Chap. 8, p. 476 (1968).
2. S. Bywater, "Anionic Polymerization" in *Ency. Polym. Sci. & Eng.* H.F. Mark, N.M. Bikales, C.G. Overberger, G. Menges, Eds., Wiley-Interscience, New York, Vol. 2, p. 1 (1985).
3. M. Szwarc, *Nature* (1956), 178, 1168.
4. M. Szwarc, M. Levy, R. Milkovich, *J. Am. Chem. Soc.* (1956), 78, 2656.
5. M. Szwarc, *Carbanions, Living Polymers and Electron Transfer Processes*, Interscience, New York, Chap. 1, p. 1 (1968).
6. N.D. Scott, U.S. Patent 2,181,771 (1939).
7. R.V. Figini, *Makromol. Chem.* (1964), 71, 193.
8. B.D. Coleman, T.G. Fox, *J. Am. Chem. Soc.* (1963), 85, 1241.
9. M. Szwarc, J.J. Hermans, *J. Polym. Sci.* (1964), B2, 815.
10. (a) R.V. Figini, *Makromol. Chem.* (1967), 107, 170.
    (b) R.V. Figini, G. Lohr, G.V. Schulz, *J. Polym. Sci. Polym. Letters* (1965), 3, 985.
11. D. Baskaran, A.H.E. Muller, *Prog. Polym. Sci.* (2007), 32, 173.
12. M. Szwarc, *Carbanions, Living Polymers and Electron Transfer Processes*, Interscience, New York, Chap. 2, p. 27 (1968).
13. M. Szwarc, *Carbanions, Living Polymers and Electron Transfer Processes*, Interscience, New York, Chap. 12, p. 639 (1968).
14. G. Spach, M. Levy, M. Szwarc, *J. Chem. Soc.* (1962), 355.
15. D.N. Bhattacharyya, J. Smid, M. Szwarc, *J. Polym. Sci.* (1965), A3, 3099.
16. (a) M. Szwarc, A. Rembaum, *J. Polym. Sci.* (1956), 22, 190.
    (b) H. Schreiber, *Makromol. Chem.* (1959), 36, 86.
17. P. Rempp, E. Franta, J-E. Herz, *Adv. Polym. Sci.* (1988), 86, 89.
18. M. Szwarc, *Adv. Polym. Sci.* (1983), 49, 1.
19. V. Warzelhan, H. Hocker, G.V. Schulz, *Makromol. Chem.* (1978), 179, 2221.
20. C.J. Chang, T.E. Hogen-Esch, *Tetrahedron Lett.* (1976), 17, 323.
21. H.W. Coover, Jr., J.M. McIntire "2-Cyanoacrylic Ester Polymers" in *Encyl. Polym. Sci. & Engg.*, H.F. Mark, N.M. Bikales, C.G. Overberger, G. Menges, Eds., Wiley-Interscience, New York, Vol. 1, p. 299 (1985).
22. M. Morton, L.J. Felters, *Rubber Chem. Tech.* (1975), 48, 359.
23. M. Szwarc, *Makromol. Chem.* (1960) 35, 132.
24. C.G. Overberger, N. Yamato, *J. Polym. Sci.* (1965), B3, 569; (1964) A4, 3101.
25. B.J. McClelland, *Chem. Rev.* (1964), 64, 301.
26. C.E. Frank *et al. J. Org. Chem.* (1961), 26, 307.
27. D.H. Richards, R.L. Williams, *J. Polym. Sci.* (1973), 11, 89.
28. B.L. Funt, D. Richardson, S.N. Bhadani, *Can. J. Chem.* (1966), 44, 711.
29. F.W. Stavely *et al. Ind. Eng. Chem.* (1956), 48, 778.
30. D. Margerison, J.P. Newport, *Trans. Faraday Soc.* (1963), 59, 2058.
31. G. Wittig, F.J. Meyer and G. Lange, *Ann.* (1951), 571, 167.
32. S. Bywater, D.J. Worsfold, *J. Organomet. Chem.* (1967), 10, 1.
33. M. Weiner, G. Vogel and R. West, *Inorg. Chem.* (1962), 1, 654.
34. F.A. Cotton, G. Wilkinson, C.A. Murillo, M. Bochmann, *Advanced Inorganic Chemistry*, 6th Ed., John Wiley & Sons, New York, p. 106 (1999).
35. K.F. O'Driscoll, A.V. Tobolsky, *J. Polym. Sci.* (1959), 35, 259.
36. F.J. Welch, *J. Am. Chem. Soc.* (1959), 81, 1345.
37. J.E.L. Roovers, S. Bywater, *Macromolecules* (1975), 8, 251.
38. D.J. Worsfold, S. Bywater, *Can. J. Chem.* (1960), 38, 1891; (1964) 42, 2884.

39. H.L. Hsieh, *J. Polym. Sci.* (1965), A3, 163.
40. M. Morton, L.J. Fetters, E.E. Bostick, *J. Polym. Sci.* (1963), C1, 311.
41. Yu L. Spirin, A.R. Grantmakher and S.S. Medvedev, *Dokl. Akad. Nauk SSSR* (1962), 146, 368.
42. Yu L. Spirin, D.K. Polyakov, A.R. Grantmakher and S.S. Medvedev, *Dokl. Akad. Nauk SSSR* (1961), 139, 899.
43. A.F. Johnson, D.J. Worsfold, *J. Polym. Sci. A* (1965), 3, 449.
44. D.J. Worsfold, S. Bywater, *Macromolecules* (1972), 5, 393.
45. Yu L. Spirin et al. *J. Polym. Sci.* (1962), 58, 1181.
46. R.N. Young, L.J. Fetters, J.S. Huang and R. Krishnamoorti, *Polym. Int.* (1994), 33, 217.
47. L.J. Fetters et al. *Macromol. Symp.* (1997), 121, 1.
48. J. Stellbrink et al. *Macromolecules* (1999), 32, 5321.
49. F.J. Welch, *J. Am. Chem. Soc.* (1960), 82, 6000.
50. S. Bywater, D.J. Worsfold, *Can. J. Chem.* (1962), 40, 1564.
51. E.N. Kropacheva, B.A. Dolgoplosk and E.M. Kuznetsova, *Dokl. Akad. Nauk SSSR* (1960), 130, 1253.
52. G.M. Burnett, R.N. Young, *Eur. Polym. J.* (1966), 2, 329.
53. G.C. Eberhardt, W.A. Butte, *J. Org. Chem.* (1964), 29, 2928.
54. A.W. Langer, Trans. N.Y. *Acad. Sci.* (1965), 27, 41.
55. A.V. Tobosky, C.E. Rogers, *J. Polym. Sci.* (1959), 38, 205.
56. A.V. Tobolsky, C.E. Rogers, *J. Polym. Sci.* (1959), 40, 73.
57. S. Bywater, A.F. Johnson, D.J. Worsfold, *Can. J. Chem.* (1964), 42, 1255.
58. A. Gourdenne, P. Sigwalt, *Eur. Polym. J.* (1967), 3, 481.
59. J.E.L. Roovers, S. Bywater, *Trans. Faraday Soc.* (1966), 62, 1876.
60. A. Guyot, J. Vialle, *J. Macromol. Sci.* (1970), A4, 107.
61. H.L. Hsieh, I.W. Huang, *Macromolecules* (1986), 19, 299.
62. S. Menoret et al. *Macromolecules* (2002), 35, 4584.
63. D.B. Patterson, A.F. Halasa, *Macromolecules* (1991), 24, 1583.
64. S. Carlotti et al. *Makromol. Chem. Rapid Commun.* (2006), 27, 905.
65. H. Morita, A.V. Tobolsky, *J. Am. Chem. Soc.* (1957), 79, 5853.
66. A.V. Tobolsky, R.J. Boudreau, *J. Polym. Sci.* (1961), 51, S53.
67. R.S. Stearns, L.E. Forman, *J. Polym. Sci.* (1959), 41, 381.
68. F.C. Foster, J.L. Binder, *Adv. Chem. Ser.* (1957), No. 17, 7.
69. D.J. Worsfold, S. Bywater, *Can. J. Chem.* (1964), 42, 2884.
70. H. Hsieh, D.J. Kelley, A.V. Tobolsky, *J. Polym. Sci.* (1957), 26, 240.
71. A.A. Korotkov, *Angew, Chem.* (1958), 70, 85.
72. B.D. Coleman, T.G. Fox, *J. Am. Chem. Soc.* (1963), 85, 1241; *J. Chem. Phys.* (1963), 38, 1065.
73. T.G. Fox et al., *J. Am. Chem. Soc.* (1958), 80, 1768.
74. G. Lohr, G.V. Schulz, *Makromol. Chem.* (1973), 172, 137; *Eur. Polym. J.* (1974), 10, 121.
75. A.H.E. Muller, H. Hocker, G.V. Schulz, *Macromolecules* (1977), 10, 1086.
76. D.N. Bhattacharyya, C.L. Lee, J. Smid, M. Szwarc, *Polymer* (1964), 5, 54.
77. D.N. Bhattacharyya, C.L. Lee, J. Smid, M. Szwarc, *J. Phys. Chem.* (1965), 69, 612.
78. H. Hostalka, R.V. Figini, G.V. Schulz, *Makromol. Chem.* (1964), 71, 198.
79. H. Hostalka, G.V. Schulz, *Z. Phys. Chem. N.F.* (1965), 45, 286.
80. H. Hostalka, G.V. Schulz, *Polym. Letters* (1965), 3, 175.
81. M. Szwarc, *Pure & App. Chem.* (1966), 12, 127.
82. F.S. Dainton et al. *Makromol. Chem.* (1965), 89, 257.
83. L.L. Bohm, G.V. Schulz, *Makromol. Chem.* (1972), 153, 5.
84. G. Lohr, S. Bywater, *Can. J. Chem.* (1970), 48, 2031.
85. D.N. Bhattacharyya, J. Smid, M. Szwarc, *J. Phys. Chem.* (1965), 69, 624.
86. L.L. Bohm et al. *Adv. Polym. Sci.* (1972), 9, 1.

87. B.J. Schmitt, G.V. Schulz, *Eur. Polym. J.* (1975), 11, 119.
88. G. Lohr, G.V. Schulz, *Eur. Polym. J.* (1975), 11, 259.
89. S. Winstein, G.C. Robinson, *J. Am. Chem. Soc.* (1958), 80, 169.
90. T.E. Hogen-Esch, J. Smid, *J. Am. Chem. Soc.* (1966), 88, 307.
91. R. Kraft et al., *Macromolecules* (1978), 11, 1093.
92. T. Shimomura, K.J. Tolle, J. Smid, M. Szwarc, *J. Am. Chem. Soc.* (1967), 89, 796.
93. R.M. Fuoss, C.A. Kraus, *J. Am. Chem. Soc.* (1933), 55, 2387.
94. D.N. Bhattacharyya, J. Smid, M. Szwarc, *J. Am. Chem. Soc.* (1964), 86, 5024.
95. D.M. Wiles, S. Bywater, *Trans. Faraday Soc.* (1965), 61, 150.
96. R.K. Graham, D.L. Dunkelberger, N.W. Goode, *J. Am. Chem. Soc.* (1960), 82, 400.
97. D.M. Wiles, S. Bywater, *J. Phys. Chem.* (1964), 68, 1983.
98. T. Ishizone, K. Yoshimura, A. Hirao, S. Nakahama, *Macromolecules* (1998), 31, 8706.
99. S.K. Varshney et al. *Macromolecules* (1990), 23, 2618.
100. K. Ishiza, K. Mitsutani, T. Fukutomi, *J. Polym. Sci. Polym. Lett. Ed.* (1987), 25, 287.
101. C.B. Tsvetanov, A.H.E. Muller, G.V. Schulz, *Macromolecules* (1985), 18, 863.
102. D. Kunkel, A.H.E. Muller, L. Lochmann, M. Janata, *Macromol. Symp.* (1992), 60, 315.
103. V. Warzelhan, G.V. Schultz, *Makromol. Chem.* (1976), 177, 2185.
104. V. Warzelhan, H. Hocker, G.V. Schultz, *Makromol. Chem.* (1980), 181, 149.
105. L. Vancea, S. Bywater, *Macromolecules* (1981), 14, 1321.
106. W.E. Goode, F.H. Owens, W.L. Myers, *J. Polym. Sci.* (1960), 47, 75.
107. L. Lochmann, M. Rodova, J. Petranek, D. Lim, *J. Polym. Sci. Polym. Chem.* (1974), 12, 2295.
108. M. Janata, L. Lochmann, A.H.E. Muller, *Makromol. Chem.* (1990), 191, 2253.
109. A.H.E. Muller, L. Lochmann, J. Trekoval, *Makromol. Chem.* (1986), 187, 1473.
110. W. Fowells, C. Schuerch, F.A. Bovey, F. Hood, *J. Am. Chem. Soc.* (1967), 89, 1396.
111. R. Kraft, A.H.E. Muller, H. Hocker, G.V. Schulz, *Makromol. Chem. Rapid Commun.* (1980), 1, 363.
112. H. Jeuck, A.H.E. Muller, *Makromol. Chem. Rapid Commun.* (1982), 3, 121.
113. G.Lohr, G.V. Schulz, *Makromol. Chem.* (1973), 172, 137: *Eur. Polym. J.* (1974), 10, 121.
114. M. Fisher, M. Szwarc, *Macromolecules* (1970), 3, 23.
115. D. Honnore, J.C. Favier, P. Sigwalt, M. Fontanille, *Eur. Polym. J.* (1974), 10, 425.
116. M. Tardi, P. Sigwalt, *Eur. Polym. J.* (1973), 9, 1369.
117. J.S. Wang, R. Jerome, P. Teyssie, *J. Phys. Org. Chem.* (1995), 8, 208.
118. D. Baskaran, *Prog. Polym. Sci.* (2003), 28, 521.
119. R. Fayt, R. Forte, R. Jacobs, R. Jerome, T. Ouhadi, P. Teyssie, S.K. Varshney, *Macromolecules* (1987), 20, 1442.
120. M. Janata, L. Lochmann, A.H.E. Muller, *Makromol. Chem.* (1993), 194, 625.
121. M. Janata, L. Lochmann, P. Vlcek, J. Dybal, A.H.E. Muller, *Makromol. Chem.* (1992), 193, 101.
122. S. Creutz, Ph. Teyssie, R. Jerome, *Macromolecules* (1997), 30, 6.
123. J.W. Klein, J.P. Lamps, Y. Gnanou, P. Rempp, *Polymer* (1991), 32, 2287.
124. S.K. Varshney, X.F. Zhong, A. Eisenberg, *Macromolecules* (1993), 26, 701.
125. D. Baskaran, S. Sivaram, *Macromolecules* (1997), 30, 1550.
126. D. Baskaran, A.H.E. Muller, S. Sivaram, *Macromolecules* (1999), 32, 1356.
127. T. Kitayama et al. *Polym. Bull.* (1988), 20, 505.
128. H. Schlaad et al., *Macromolecules* (1998), 31, 573.
129. H. Schlaad, A.H.E. Muller, *Macromol. Symp.* (1996), 107, 163.
130. H. Schlaad, B. Schmitt, A.H.E. Muller, *Angew. Chem. Intl. Ed.* (1998), 37, 1389.
131. H. Schlaad, A.H.E. Muller, *Macromolecules* (1998), 31, 7127.
132. B. Schmitt, W. Stauf, A.H.E. Muller, *Macromolecules* (2001), 34, 1551.
133. P. Bayard et al. *Polym. Bull.* (1994), 32, 381.
134. J-S. Wang, R. Jerome, P. Bayard, P. Teyssie, *Macromolecules* (1994), 27, 4908.

135. A. Maurer *et al. Polym. Prep. (Am. Chem. Soc., Div. Polym. Chem.)* (1997), 38(1), 467.
136. X. Xie, T.E. Hogen-Esch, *Macromolecules* (1996), 29, 1746.
137. M. Kobayashi, S. Okuyama, T. Ishizone, S. Nakahama, *Macromolecules* (1999), 32, 6466.
138. M. Kobayashi, T. Ishizone, S. Nakahama, *Macromolecules* (2000), 33, 4411.
139. J. Trotman, M. Szwarc, *Makromol. Chem.* (1960), 37, 39.
140. D.P. Wyman, V.R. Allen, T. Altares, *J. Polym. Sci.* (1964), 2A, 4545.
141. R.P. Quirk, W.C. Chen, *Makromol. Chem.* (1982), 183, 2071.
142. D.H. Richards, M. Szwarc, *Trans. Faraday Soc.* (1959), 55, 164.
143. P. Rempp, E. Franta, J.E. Herz, *Adv. Polym. Sci.* (1988), 86, 1471.
144. C.A. Uraneck, J.N. Short, R.P. Zelinsky, U.S. Pat. 3, (1966), 135, 716.
145. P. Rempp, P. Lutz, P. Masson, E. Franta (1984), *Makromol. Chem. Suppl.* (1984), 8, 3.
146. B.C. Anderson *et al. Macromolecules* (1981), 14, 1599.
147. G. Schulz, R. Milkovitch, *J. App. Polym. Sci.* (1982), 27, 4773.
148. R. Asami, M. Takaki, H. Hanahata, *Macromolecules* (1983), 16, 628.
149. D. Freyss, P. Rempp, H. Benoit, *J. Polym. Sci. Polym. Lett.* (1964), 2, 217.
150. J.T. Bailey, E.T. Bishop, W.R. Hendricks, G. Holden, R.N. Legge, *Rubber Age* (1966), 98(10), 69.
151. M. Morton *et al. J. Polym. Sci.* (1962), 57, 471.
152. T. Orofino, F. Wenger, *J. Phys. Chem.* (1963), 67, 566.
153. N. Hadjichristidis, M. Pitsikalis, S. Pispas, H. Iatrou, *Chem. Rev.* (2001), 101, 3747.
154. R.P. Quirk, Y. Tsai, *Macromolecules* (1998), 31, 8016.
155. H. Eschwey, M. Hallensleben, W. Burchard, *Makromol. Chem.* (1973), 173, 235.
156. W. Burchard, H. Eschwey, *Polymer* (1975), 16, 180.
157. P. Lutz, P. Rempp, *Makromol. Chem.* (1988), 189, 1051.
158. C. Tsitsilianis *et al. Macromolecules* (1991), 24, 5897.
159. R.W. Pennisi, L.J. Fetters, *Macromolecules* (1988), 21, 1094.
160. J.W. Mays, *Polym. Bull* (1990), 23, 247.
161. H. Iatrou, E. Siakali-Kioulafa, N. Hadjichristidis, J. Roovers, J.W. Mays, *J. Polym. Sci. Polym. Phys.* (1995), 33, 1925.
162. A. Avgeropoulos, N. Hadjichristidis, *J. Polym. Sci. Polym. Chem.* (1997), 35, 813.
163. J.W. Mays, N. Hadjichristidis, L.J. Fetters, *Polymer* (1988), 29, 680.
164. D.J. Worsfold, J.G. Zilliox, P. Rempp, *Can. J. Chem.* (1969), 47, 3379.
165. L.-K. Bi, L.J. Fetters, *Macromolecules* (1976), 9, 632.
166. V. Efstratiadis *et al. Polym. Intl.* (1994), 33, 171.
167. M. Pitsikalis, M. Pispas, J.W. Mays, N. Hadjichristidis, *Adv. Polym. Sci.* (1998), 1351.
168. J. Selb, Y. Gallot, Polymer, (1979), 20, 1259.
169. M. Pitsikalis *et al. J. Polym. Sci. Polym. Chem.* (1999), 37, 4337.
170. J. Falk, D.J. Hoeg, R. Schlott, J.F. Pendelton, *Rubber Chem. Technol.* (1973), 46, 1004.
171. N. Hadjichristidis, J. Roovers, *J. Polym. Sci.* (1978), A16, 851.
172. Y. Ederle, F. Isel, S. Grutke, P. Lutz, *Macromol. Symp.* (1998), 132, 197.
173. O.W. Webster, *J. Polym. Sci. Polym. Chem.* (2000), 38, 2855.
174. O.W. Webster *et al. J. Am. Chem. Soc.* (1983), 105, 5706.
175. D.Y. Sogah, W.R. Hertler, O.W. Webster, G.M. Cohen, *Macromolecules* (1987), 20, 1473.
176. I.B. Dicker *et al.*, *Macromolecules* (1990), 23, 4034.
177. R.P. Quirk, G.P. Bidinger, *Polym. Bull.* (1989), 22, 63.
178. R.P. Quirk, J. Ren, *Macromolecules* (1992), 25, 6612.
179. A.H.E. Muller, G. Litivenko, D. Yan, *Macromolecules* (1996), 29, 2346.
180. A.H.E. Muller, G. Litivenko, D. Yan, *Macromolecules* (1995), 26, 2339.
181. W.J. Brittain, I.B. Dicker, *Macromolecules* (1989), 22, 1054.
182. O.W. Webster, "Group Transfer Polymerization" in *Encycl. Polym. Sci. & Eng.*, Vol. 7, p. 580 (1983).

183. M.T. Reez, T. Knauf, U. Minet, C. Bingel, *Angew. Chem. Int. Ed.* (1988), 27, 1371.
184. M.T. Reez, *Angew. Chem. Int. Ed.* (1988), 27, 994.
185. S. Sivaram et al. *Macromolecules* (1991), 24, 1698.
186. S. Sivaram et al. *Polym. Bull.* (1991), 25, 77.
187. D.J.A. Raj, P.P. Wadgaonkar, S. Sivaram, *Macromolecules* (1992), 25, 2774.
188. F. Bandermann et al. *Macromol. Chem. Phys.* (1995), 196, 2335.
189. D. Baskaran et al. *Macromolecules* (1999), 32, 2865.
190. A.P. Zagala, T.E. Hogen-Esch, *Macromolecules* (1996), 29, 3038.
191. D. Baskaran, A.H.E. Muller, *Macromolecules* (1997), 30, 1869.
192. D. Baskaran et al. *Macromolecules* (1997), 30, 6695.

Problems

4.1 Living anionic polymerization can be conducted in ethereal solvents but not in chlorinated hydrocarbons but the reverse is true for living cationic polymerization. Explain.

4.2 How would you establish the absence of transfer reaction in a living anionic polymerization?

4.3 Arrange the following monomers in the order of increasing electronegativity: butadiene, methyl methacrylate, and styrene. Explain the order. If you are to make a ternary triblock copolymer using the sequential living anionic polymerization of the three monomers, what sequence would you follow and why?

4.4 Negative activation energy is encountered in living anionic polymerization of styrene in THF in certain temperature region when sodium is the counterion. Why the same phenomenon is not observed when Cs^+ is the counterion or styrene is replaced with methyl methacrylate?

4.5 Polystyrene lithium is more reactive than other polystyrene alkalis in THF but the opposite is true when polystyrene is replaced with poly(methyl methacrylate). Explain.

4.6 When LiCl is used as an additive in the anionic polymerization of methacrylates, the MWD becomes much narrower. How does this happen?

4.7 If you are to prepare a diblock copolymer of styrene and methyl methacrylate what monomer sequence would you use in the preparation and why?

4.8 How would you reduce side reaction in the initiation step when polystyrene lithium is to be used to initiate the polymerization of MMA?

4.9 Can you prepare hydroxy monotelechelic poly(methyl methacrylate) using ethylene oxide as the terminator of monoended living PMMA anion? Explain.

4.10 Styrene can be polymerized livingly by living anionic polymerization but not by group transfer polymerization. Justify the statement.

4.11 Discuss what is common between the principles underlying GTP and LAP of methyl methacrylate.

Chapter 5

Coordination Polymerization

Coordination polymerization is a chain polymerization in which both the monomer and the active center are coordinated to the polymerization catalyst prior to the incorporation of the monomer in the polymer chain. The polymerization exhibits various degrees of stereochemical control on the structure of the polymer depending on the nature of the catalyst and reaction conditions.

Ethylene, α-olefins (the linear ones), and conjugated dienes are the most important monomers that are polymerized by this route. Of these monomers, the α-olefins are not polymerizable to form high polymer by any other method. The field made a start when Ziegler discovered in 1953 that ethylene is easily polymerized in a hydrocarbon solvent to high molecular weight polymer at normal pressure and moderate temperature using triethylaluminum or diethylaluminum chloride in conjunction with titanium tetrachloride as catalysts.[1,2] In the following year, Natta discovered stereochemical control exerted by the catalysts on the structure of polypropylene. The discoveries not only gave a new dimension to polymer science but also revolutionized the polyolefin industry. Ziegler and Natta were honored with a Nobel Prize in 1963 and the catalysts were named after them.

5.1 Ziegler–Natta Catalysts

The first generation Ziegler–Natta catalysts are comprised of two components: (i) a Group IV to VIII transition metal compound, which may be a halide, a subhalide, an oxyhalide, an alkoxide, a β-diketonate, a cyclopentadienyl dihalide, and so on, and (ii) a Group I to IV base metal alkyl or hydride. The transition metal compound is essentially the precatalyst, which is alkylated, and thus activated, by the base metal alkyl to form the real catalyst. In the literature, however, the two component mixture is usually referred to as the catalyst or the catalyst system, although, sometimes the base metal alkyl is referred to as the cocatalyst and the transition metal compound as the catalyst. The higher generations Ti-based catalysts supported on activated magnesium chloride contain a third and a fourth component, which are often Lewis bases, added to increase the activity and the stereospecificity. These additives also influence the molecular weight of the polymer to be formed. The first catalysts discovered, as already introduced, are $AlEt_3 + TiCl_4$ and $AlEt_2Cl + TiCl_4$. The two components of the catalysts on mixing together in a hydrocarbon solvent react to yield the real catalyst, which is insoluble in the medium leading to heterogeneous catalysis.[1,2] The active sites of the catalysts are not identical. This was immediately evident when the Natta group obtained a mixture of both crystalline (*isotactic*) and amorphous (*atactic*) polypropylene in the polymerization of propylene.[3-6]

The real catalyst from aluminum triethyl and titanium tetrachloride is formed in reactions (5.1) through (5.4).[7,8] At temperatures above *ca.*, $-30°C$ and at low Al to Ti ratios $TiCl_4$ is reduced to $TiCl_3$, which precipitates as crystalline solid (indicated by the subscript 's'), and $AlEt_3$ is converted to $AlEt_2Cl$ liberating an ethyl radical. The latter undergoes disproportionation and coupling reactions forming ethylene and butane respectively. Further reaction of $TiCl_3$ crystals with $AlEt_2Cl$ occurs resulting in the alkylation of Ti on the crystal surface (reaction 5.4). The alkylated $TiCl_3$ crystals act as the real catalysts. In fact, mixtures of separately prepared crystalline $TiCl_3$ and $AlEt_2Cl$ make very effective catalysts. Polymerization takes place by insertion of monomer into Ti-C bond.

$$TiCl_4 + Al(C_2H_5)_3 \longrightarrow TiCl_{3(s)} + Al(C_2H_5)_2Cl + \overset{\bullet}{C}_2H_5. \qquad (5.1)$$

$$TiCl_4 + Al(C_2H_5)_2Cl \longrightarrow TiCl_{3(s)} + Al(C_2H_5)Cl_2 + \overset{\bullet}{C}_2H_5. \qquad (5.2)$$

$$2 \overset{\bullet}{C}_2H_5 \begin{cases} \longrightarrow C_4H_{10} \quad \text{(combination product).} \\ \longrightarrow C_2H_4 + H_2 \quad \text{(disproportionation products).} \end{cases} \qquad (5.3)$$

$$TiCl_{3(s)} + Al(C_2H_5)_2Cl \longrightarrow Ti(C_2H_5)Cl_{2(s)} + Al(C_2H_5)Cl_2. \qquad (5.4)$$

5.1.1 *Isotactic polypropylene*

The isospecificity of the catalysts comprising variously prepared $TiCl_3$ and soluble $AtEt_2Cl$ depends upon the crystal modification of $TiCl_3$. Thus, the violet colored α-, γ-, and δ-$TiCl_3$ all having a layer lattice make highly isospecific catalysts,[9-13] the γ and δ types being more active than the α, whereas the β-$TiCl_3$, which has a linear (chain like) structure exhibits low isospecificity.[14] It, however, makes an effective catalyst for the polymerization of isoprene to *cis*-1, 4-polyisoprene. Reaction of $TiCl_4$ and AlR_3 or AlR_2Cl yields β-$TiCl_3$, which may be converted to the γ form by heating at 100–200°C and to the δ form on treatment with $TiCl_4$ at moderate temperatures *ca.*, <100°C.[15] The α form may be prepared by the reduction of $TiCl_4$ with hydrogen or aluminum. Reduction by aluminum concurrently yields $AlCl_3$. The δ form may be obtained by prolonged grinding of the α- or γ- $TiCl_3$.

Figure 5.1 shows the stereochemical model of the structural layer, which constitutes the layer lattices of α-, γ-, and δ- $TiCl_3$, as given by Natta *et al.*[9] The structural layer is constituted of a layer of Ti atoms sandwitched between two close-packed Cl atom layers. The

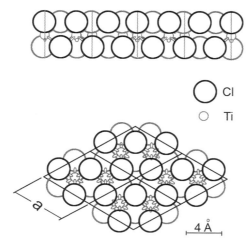

Figure 5.1 Stereochemical model of the structural layer which characterizes the layer lattice structures of α-, γ-, and δ-modifications of TiCl$_3$. "Reprinted with permission from Ref. 9. Copyright ©1961 John Wiley & Sons."

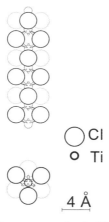

Figure 5.2 Stereochemical model of the linear macromolecule (TiCl$_3$) which represents the structural unit in fiber shaped lattice of β-TiCl$_3$. "Reprinted with permission from Ref. 9. Copyright ©1961 John Wiley & Sons."

Ti atoms are located in the octahedral interstices of the Cl atom packing, which is hexagonal in α-TiCl$_3$, cubic in γ-TiCl$_3$, and irregular in δ-TiCl$_3$. The last one may be described as a statistical "average" of the modes of packings in the other two forms.[9]

The stereochemical model of the structural unit in fiber shaped β-TiCl$_3$ lattice is shown in Fig. 5.2. The structural unit represents a linear polymer of TiCl$_3$, which has the chemical structure (**1**) shown in Fig. 5.3.[9] The non terminal Ti atoms are octahedrally coordinated with six Cl atoms as in other forms.

The activity of these first generation catalysts is rather low giving yields of about 1 kg polypropylene/g catalyst.[20] With such low catalyst productivity the polymer contains too much metal, which makes it unsuitable for many applications or processing without

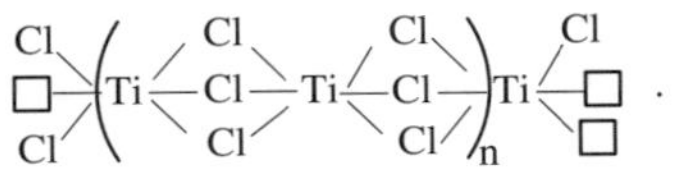

$$\square\!\!\!\!>\!\!Ti\!\!<\!\!\genfrac{}{}{0pt}{}{Cl}{Cl}\!\!\left(\genfrac{}{}{0pt}{}{Cl}{Cl}\!\!>\!\!Ti\!\!<\!\!\genfrac{}{}{0pt}{}{Cl}{Cl}\!\!>\!\!Ti\!\!<\!\!\genfrac{}{}{0pt}{}{Cl}{Cl}\right)_n\!\!Ti\!\!<\!\!\genfrac{}{}{0pt}{}{Cl}{\square}$$

(1)

Figure 5.3 Chemical structure of a linear polymer of $TiCl_3$. A square box at the terminals of the chain represents a vacant coordination site.

impairing polymer properties. Therefore, the metal has to be removed by chemical treatment, a process called "deashing".[15] Furthermore, with many of these catalysts, the *atactic* content in the polymer is too high, which must be reduced for achieving good mechanical properties. This is done by removing the *atactic* polymer using a solvent extraction process. However, both the two purification processes increase the cost of the polymer.

An increase in activity was achieved with the so called "second generation catalysts," which may be prepared by reacting $TiCl_4$ with $AlEt_2Cl$ followed by treating the product with ether and then with $TiCl_4$.[17] The ether treatment removes the aluminum chloride that forms concurrently with β-$TiCl_3$ during the reduction of $TiCl_4$ with $AlEt_2Cl$. As a result, more titanium atom sites become available for reaction and the activity is increased. The subsequent $TiCl_4$ treatment brings about transformation of $TiCl_3$ from β- to the more active and isospecific δ-modification, as stated earlier.

Further increase in activity and isospecificity was achieved with the use of activated $MgCl_2$–supported catalysts,[18] which are treated with suitable electron donor compounds (Lewis bases) in order to increase isospecificity by preventing undesirable coordination of propylene with coordinately unsaturated Mg atoms (*vide infra*).[19–26]

Magnesium chloride has a layer–lattice structure like α-, γ-, and δ- $TiCl_3$.[18] The ionic radii of Mg^{+2} and Ti^{+3} ions are very close to each other, being 0.65 Å and 0.68 Å respectively,[18] so that $TiCl_3$ is strongly adsorbed on $MgCl_2$ when prepared in the latter's presence. In addition, activated $MgCl_2$ is nanocrystalline providing large surface for the adsorption of $TiCl_3$. It may be prepared by physical methods such as ball-milling $MgCl_2$ in the presence of ethyl benzoate[15,24] or by chemical methods such as reacting a magnesium alkyl with a chlorinating agent or with $TiCl_4$.[15]

Treatment of the support with electron donors follows various protocols. One of these uses two different electron donors at two different stages of the supporting process. In the first stage, an electron donor (designated "internal") is used during the treatment of the support with $TiCl_4$. In the second stage, another electron donor (designated "external") is used during the alkylation of the supported $TiCl_4$ with AlR_3 so that the coordination sites of magnesium falling vacant due to any internal donors lost in this stage is filled up again (Sec. 5.1.6.1).[20,25]

However, the internal donor loss may be prevented and the need for external donors obviated using suitable bidentate donors, such as some 2,2-disubstituted 1,3-dimethoxypropanes, which is done in the fifth generation catalysts.[26] The higher generations catalysts have an order of magnitude higher activity than the second generation catalysts (Table 5.1). Titanium constitutes 0.5–3% of the catalysts.

Table 5.1 Activity of typical catalysts of various generations in the polymerization of propylene.

Generation of catalyst	Catalyst components (typical examples)	Polypropylene yield, kg/g catalyst
1st	$TiCl_3(AlCl_3)/AlEt_2Cl$	$1^{c,20}$
2nd	$TiCl_4/AlEt_2Cl/ether/TiCl_4$	$3-5^{c,20}$
3rd	$MgCl_2/TiCl_4/^a$ethyl benzoate/$AlR_3/^b$aromatic ester[20]	$15-30^{d,15}$
4th	$MgCl_2/TiCl_4/^a$phthalate ester/$Al\,R_3/^b$alkoxy silane[25]	$30-80^{d,15}$
5th	$MgCl_2/TiCl_4/$diether$^a/Al\,R_3^{26}$	$8-160^{d,15}$

^{a}internal donor; ^{b}external donor; ^{c}polymerization in hexane at 70°C, 0.7 MPa, 4h; ^{d}polymerization of liquid propylene in presence of hydrogen at 70°C, for 1–2 h.

Catalysts have also been developed for the *isotactic* polymerization of higher α-olefins. However, the catalytic activity decreases with increase in the size of the α-substituent in the monomer.[15]

5.1.2 Syndiotactic polypropylene

Syndiotactic polymerization of propylene was achieved by Natta and coworkers using soluble catalyst systems comprising vanadium tetrachloride or vanadium (acetyl acetonate)$_3$, AlR_2Cl, and a Lewis base such as anisole at low temperatures, *e.g.*, below $-70°C$. Replacement of AlR_2Cl with AlR_3 results in stereoirregular polypropylene. α-Olefins higher than propylene do not yield *syndiotactic* polymers.[27–29] Vanadium is reduced to +3 oxidation state when VCl_4 is treated with aluminum alkyls/alkyl halides, AlR_3 or AlR_2Cl, forming $VRCl_2$ which should be catalytically active.[30] It has been suggested that $VRCl_2$ does not exist freely. Either it forms binuclear complexes with itself or with the aluminum alkyls.[31] A pentacoordinated vanadium (**2**) has been proposed to be involved in *syndiotactic* polymerization of propylene.[31] The square box shown in the structure represents a vacant co-ordination site to be filled by propylene before its insertion into the V–R bond. However, the active form of the catalyst accounts for only a small fraction of the total amount of vanadium used in the system.[30] The regio- and stereoregularity of the polymer is poor. The microstructure comprises *syndiotactic* blocks interposed by short atactic blocks.[31]

(**2**)

5.1.3 *Polyethylene and ethylene copolymers*

Polyethylene produced by Ziegler catalysts is essentially linear and is known as high density polyethylene (Chap. 1, Sec. 1.7.1). The productivity of the original Ziegler catalysts is low. The present day highly active catalysts used in HDPE production are based on $TiCl_4$ supported on activated magnesium chloride in combination with $AlEt_3$ or $AlEt_2Cl$ used as cocatalysts. However, not all the HDPEs are produced in the world using these catalysts. A large fraction is produced using Phillips catalysts, which are CrO_3 supported on SiO_2–Al_2O_3. These and other supported transition metal oxide catalysts were discovered in Phillips and Standard Oil laboratories respectively at about the same time as the Ziegler catalyst was.

Apart from HDPE, the various LLDPEs, which are copolymers of ethylene and small amounts (*ca.*, 10%) of α-olefins, are commercially produced using the supported Ziegler catalysts. These catalysts account for about 90% of the world production; the metallocene catalysts account for the rest 10%.[15] The Phillips and Standard Oil catalysts are unsuitable.

In contrast to the LLDPEs, the rubbery ethylene-propylene copolymers containing 35 to 65 mol% of either monomer or the corresponding EPDM rubbers are produced using the soluble vanadium based catalysts as are used to produce *syndiotactic* polypropylene but at a higher temperature *ca.*, 0°C.[32–34]

5.1.4 *Polymerization of conjugated dienes*

All the four regular stereoisomers of polybutadiene, *viz.*, *cis*-1,4-, *trans*-1,4-, *isotactic*-1,2-, and *syndiotactic*-1,2-, are obtained using various soluble catalysts.[3] Natta attributed this to the coordination of both double bonds of the diene to the transition metal of the catalyst. This effects a particular orientation of the inserting monomer with respect to the growing chain depending on the metal and the ligands in the catalytically active transition metal complex.[3]

However, heterogeneous catalysts like $TiI_4/Al(i-C_4H_9)_3$ (Al/Ti = 3 to 8)[35,36] and $CoCl_2/AlR_2Cl$[37] were developed to produce *cis*-1,4-polybutadiene.[38] The two catalysts differ in the type of their metal–carbon bonds. The bond is σ for titanium (**3**) and π-allyl for cobalt[7] (**4**).

$$Ti-CH_2-CH=CH-CH_2R$$

σ Mt—C bond

π- allyl Mt—C bond

3 **4**

cis-1,4-polyisoprene is produced using $TiCl_4/Al(i-C_4H_9)_3$/ether as the catalyst system.[39,40] As discussed earlier, β-$TiCl_3$ is the precatalyst formed *in situ*. Also, as previously discussed, β-$TiCl_3$ has a fiber type crystal structure. Since the structural unit of the fiber is a $TiCl_3$ macromolecule (Fig. 5.3), it follows that for electrical neutrality, the terminal Ti atoms of the macromolecule are coordinated with only four, or five Cl atoms instead of six for the non terminal Ti atoms, as shown in structure **1**. Thus, in the lateral surface of the crystal the Cl atom vacancies exist at the fiber ends. The four coordinated terminal Ti atoms become active sites following alkylation by the cocatalyst, which replaces a loosely held chlorine by an alkyl group (Sec. 5.1.6.1). The two vacant coordination sites are filled by the diene.[15]

5.1.5 *Degree of stereoregularity*

The degree of stereoregularity of a polymer may be determined by various methods such as solvent extraction, infra red-, ^{1}H and ^{13}C NMR spectroscopy, and measurement of crystallinity (fraction of crystalline polymer) by x-ray diffraction. In the solvent extraction method, the stereoirregular fraction is extracted from the whole polymer. The weight fraction of the insoluble polymer provides a measure of the degree of stereoregularity. For example, the *atactic* fraction in polypropylene prepared by using an isospecific catalyst is removed by extraction with boiling heptane. The weight fraction of the heptane insoluble fraction gives the degree of stereoregularity, which in this case is called "*isotacticity* index" (I.I.). Alternatively, the polymer is dissolved in boiling xylene and the *isotactic* fraction is crystallized out at room temperature (*ca.*, 20°C). The weight fraction of the polymer insoluble in xylene at room temperature, called "xylene index" (X.I.), is also used to express the degree of stereoregularity. ^{13}C NMR spectroscopic measurements have revealed that the heptane or xylene insoluble fractions contain mmmm pentads in excess of 90%.[42,43] However, *isotactic* polypropylene of very low molecular weights also are soluble in heptane and xylene.

^{13}C NMR spectroscopy is widely used to determine the degree of stereoregularity. It is common practice to express the degree of stereoregularity (*e.g.*, syndiotactic or isotactic) in terms of the percentage of *isotactic* and *syndiotactic* diads, triads, *etc.*, in the polymer. The stereochemistry of the diads and the triads is represented by the following symbols: *isotactic* (*meso*) diads (m), *syndiotactic* (racemic) diads (r), *isotactic* triads (mm or i), *syndiotactic* triads (rr or s), and *heterotactic* triads (mr or h). The diads and triads are shown in Fig. 5.4 using Fischer projection formulas rotated through 90°.

Determination of polymer microstructure by high resolution ^{13}C NMR spectroscopy has revealed that the use of Lewis bases as catalyst components in higher generations catalysts affects not only the proportion of *isotactic* and *atactic* fractions but also their microstructures. Thus, the *isotactic* fraction of polypropylene prepared by using the 4th generation catalysts contains a low (1 to 2 wt %) amount of *syndiotactic* blocks. In fact, the polymers have been shown to be stereoblocks in which very long highly *isotactic* blocks are interposed with short

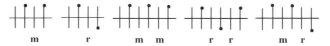

Figure 5.4 Various diads and triads are shown in rotated Fischer projection formulas of portions of polypropylene chains. A vertical line without dots represents H−C−H and the one with a dot represents CH$_3$−C−H with the dot representing a CH$_3$ group. The horizontal line represents the main chain.

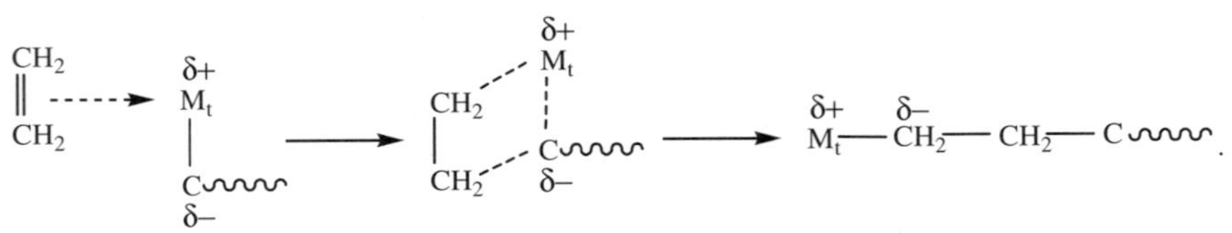

Figure 5.5 Basic mechanism of insertion of olefin into M_t–C bond.

isotactoid (weakly *isotactic*, m-diad content in the range of 70 to 90%), and *syndiotactoid* (*syndiotactic* r-diad content in the range of 80 to 90%) or syndiotactic blocks.[41–45]

5.1.6 Mechanism

Natta proposed a coordinated anionic mechanism, which involves coordination of monomer to transition metal in an active site of the catalyst followed by insertion of the coordinated monomer into the highly polarized transition metal-carbon bond, which is formed *in situ* through the alkylation of the transition metal in the precatalyst by the base metal alkyl cocatalyst, as discussed earlier.[46–49] In support of the perceived insertion into the M_t–C bond rather than into the base metal-carbon bond, M_b–C, it may be cited that some organometallic catalysts containing only titanium as metal atom also exhibit isospecificity.[50–52]

However, insertion may occur in either of the two regiochemical modes shown in Scheme 5.1 with propylene monomer. The mode of insertion depends on the nature of the transition metal as well as that of the carbon, *i.e.*, *primary* or *secondary*, in the M_t–C bond. Insertion of propylene into titanium–carbon bond occurs overwhelmingly in the 1,2 mode (95 to 99% depending on the nature of the catalyst).[14,53–55] The regiospecificity of the *atactic* polymer is only little lower than that of the *isotactic* polymer. In contrast, insertion into vanadium–carbon bond shows high preference for the 2,1 mode when the carbon is *secondary*, and only a low preference for the 1,2 mode when the carbon is *primary*.[31,56–63] As a result, long sequences of 2,1 insertions alternate with short sequences of 1,2 insertions. Overall, the 2,1 insertions comprise ca., 85% of the total.[53] It has also been established that 2,1 insertions into vanadium–carbon bond in the soluble vanadium based catalysts are syndiotactic, whereas 1,2 insertions are *atactic*.[63–65] Thus, the regioblocks of *syndiotactic* propylene are also stereoblocks comprising long *syndiotactic* blocks interposed by short *atactic* blocks.

The 2,1 insertions yield tail-to-head linkages between monomers (the unsubstituted end of the monomer being tail and the substituted end being head), whereas 1,2 insertions yield head-to-tail linkages. A regioblock structure is essentially a multiblock of head-to-tail (H–T) and tail-to-head (T–H) polymers.

A change over from 2,1 to 1,2 insertion yields a head-to-head linkage, whereas that from 1,2 to 2,1 yields a tail-to-tail linkage. As shown in the regiostructure in Fig. 5.6, at the point of change from H–T to T–H one T–T linkage occurs. Such linkages have been shown by numbers 1 and 3 on the chain, whereas a change of mode of addition from T–H to H–T gives a head-to-head (H–H) linkage between blocks identified by the number 2 on

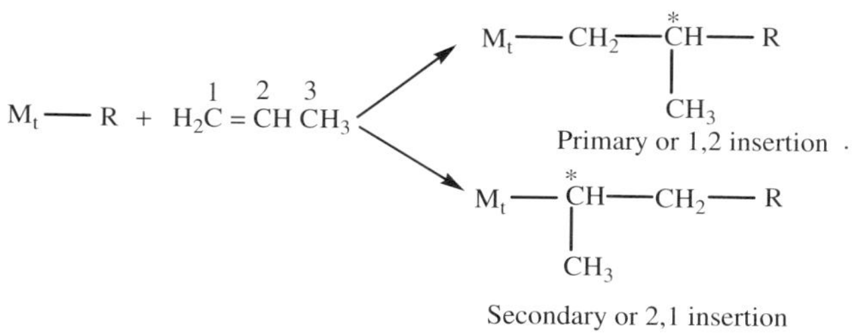

Coordination Polymerization 267

$$M_t—R + H_2C=CH\ CH_3 \begin{matrix} \nearrow \\ \searrow \end{matrix} \begin{matrix} M_t—CH_2—\overset{*}{C}H—R \\ | \\ CH_3 \\ \text{Primary or 1,2 insertion} \\ \\ M_t—\overset{*}{C}H—CH_2—R \\ | \\ CH_3 \\ \text{Secondary or 2,1 insertion} \end{matrix}$$

Scheme 5.1 Regiochemical modes of insertion of propylene into transition metal carbon bond (M_t–R representing a transition metal alkyl with other ligands of M_t omitted to avoid complexity.)

$$M_t—H—T—H—T\overset{1}{—}T—H—T—H—T—H\overset{2}{—}H—T—H—T—H—T\overset{3}{—}T—H—T—H—T—H$$

Figure 5.6 A regioblock chain growing from the left to the right by monomer insertion into M_t–C bond.

the chain. Higher is the concentration of H–H and T–T linkages lower is the regioregularity of the polymer. The H–H units may be estimated from the ^{13}C NMR spectra of the methyl region in polypropylene. In order to increase the accuracy of the measurement, the intensity of the resonance is magnified using propylene with ^{13}C enriched carbon in the methyl group to the extent of ca., 30%.[60] The T–T linkage may be estimated by infra red spectroscopy[58] or making use of methylene carbon resonance in ^{13}C NMR spectroscopy.[62]

5.1.6.1 *Isotactic propagation*

The absolute configuration of the chiral carbon in each newly added monomer unit (Scheme 5.1) depends on not only the regiochemistry but also the stereochemistry of monomer insertion. As discussed above, the monomer insertion into the Ti–C bond is regiochemically almost completely 1,2. The stereochemistry of insertion is always *cis* irrespective of the type of propagation, *isospecific* or *syndiospecific*.[66–68] However, it depends on the prochiral face of the olefin involved in the reaction. The difference in the stereochemistry of *cis* 1,2- insertion of α-olefin into the Ti–C bond is shown in Fig. 5.7 for the involvement of the two different prochiral faces of the olefin in the reaction.

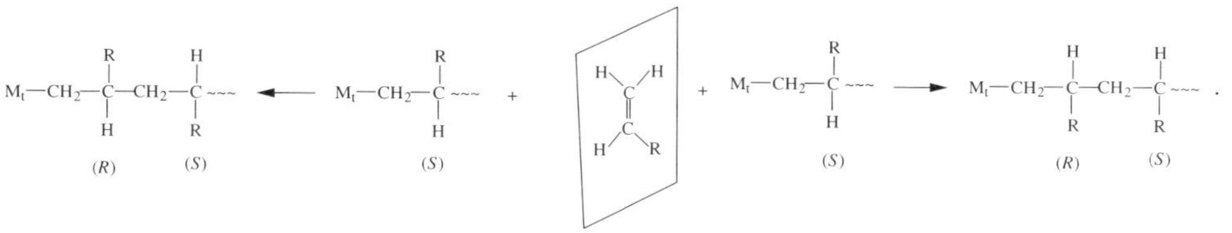

Figure 5.7 The difference in the stereochemistry of *cis* 1,2 insertion of α-olefin into the Mt–C bond due to the involvement of the two different prochiral faces of the olefin. "Reproduced with permission from Ref. 53, Stereospecific polymerization of propylene: An outlook 25 years after its discovery. Copyright Wiley–VCH Verlag GmbH & Co. KGaA."

Thus, for isospecfic propagation, the active center (**5**) must discriminate between the two prochiral faces of the α-olefin so that the same prochiral face is attacked in each insertion step.[53,69] For this to happen, it must have center(s) of chirality. Looking at the structure **5** of the active center the β carbon atom (with respect to the metal) in the chain head is chiral. Besides, the metal atom itself can be a center of chirality.[69] However, it should maintain its absolute configuration during the insertion process. This it can do by remaining bound to the solid catalyst surface.[53]

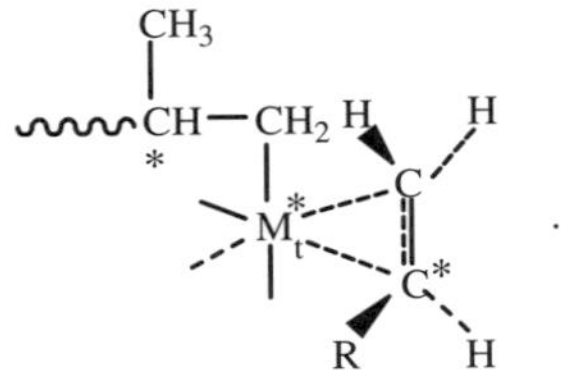

(5)

Direct evidence demonstrating the chirality of the transition metal atoms in isospecific catalysts was obtained from the observation that the heterogeneous isospecific catalyst $TiCl_3/AlR_3$ polymerizes a *racemic* α-olefin to the corresponding *racemic* polymer. Thus, the chiral metal, over all, is racemic.[70] In contrast, the homogeneous aspecific catalyst $Ti(CH_2C_6H_5)_4/Al(CH_2C_6H_5)_3$ produces a copolymer of the two enantiomers of the *racemic* monomer. which cannot be resolved by elution chromatography using an optically active stationary phase.[53,71] On the other hand, there are evidences which suggest that the chirality of the β-carbon atom at the chain head is not essential for isospecific propagation.[65,72]

There exist two views on the active center, monometallic and bimetallic, as shown by structures **6** and **7** respectively.[7,73,74]

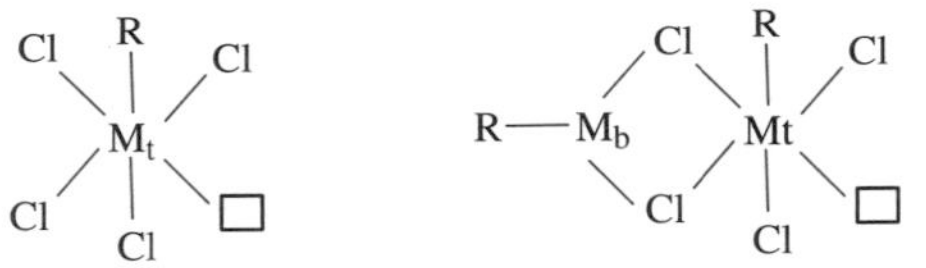

Monometallic　　　　　　　Bimetallic
(6)　　　　　　　　　　　　(7)

A square box in the structures represents a vacant coordination site; M_t is the transition metal and M_b is the base metal. For isospecific propagation, the view of monometallic active center has gained general acceptance. This is discussed below first for the unsupported catalysts and then for the supported catalysts.

Referring back to the layer lattice structure of α-$TiCl_3$ (Fig. 5.1) discussed earlier, the electrical neutrality of the crystal requires the existence of chlorine atom vacancies on the lateral surface of the crystal, on the edges of the chlorine atom planes. No such vacancies exist in the basal planes of the crystal.[16,75–77] Titanium atoms situated inside the lattice are octahedrally coordinated with six chlorine atoms, which are shared with other titanium

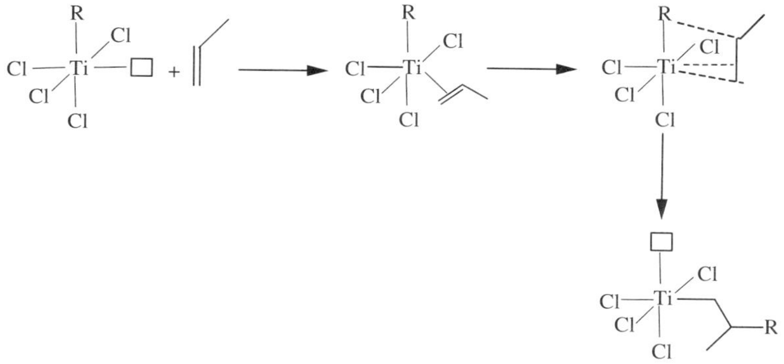

Scheme 5.2 The monometallic mechanism of monomer insertion into the transition metal–carbon bond. "Reprinted with permission from Ref. 73, The formation of isotactic polypropylene under the influence of Ziegler–Natta catalysts. Copyright © 1960 Elsevier Science Ltd, and from Ref. 81, Ziegler–Natta catalysts III. Stereospecific polymerization of propene with the catalyst system $TiCl_3$-$AlEt_3$. Copyright © 1964 Elsevier Science (USA)."

atoms (Fig. 5.1), but, practically all those present on the edges of the titanium atom planes have one coordination site vacant each. However, not all the chlorines coordinated to each of these titanium atoms are equivalent. One of these is loosely held being bound to a single Ti atom. These titanium atoms have the potential to form active sites following monoalkylation by the cocatalyst, which replaces the loosely held chlorine.[76–80] Thus, only a very small fraction of titanium atoms constitute active sites. An olefin fills the vacant coordination site forming a π bond with a vacant d orbital of titanium.[73]

As for the mechanism of insertion, the one proposed by Cossee[73] and later refined by Arlman and Cossee[81] gained general acceptance. The mechanism (Scheme 5.2) proposes a concerted four center reaction involving titanium and carbon atoms of the Ti–C bond in the titanium alkyl (polymer) and the two π bonded carbon atoms of the olefin. A 4-membered ring transition state is formed, which is followed by migration of the alkyl (polymer) to one end of the olefin. The π bonding between titanium and the olefin effects weakening of the labile Ti–C bond, which facilitates the migration of the alkyl (polymer) group.[73]

The steric interaction between the approaching olefin and the ligands surrounding the metal forces the monomer to assume a particular configuration. However, the interchange of the position of the vacant d orbital and the metal orbital involved in bonding with the alkyl (polymer), which occurs following alkyl migration, may not present identical steric environment for the next monomer insertion. This may cause a less regio- and stereoselective insertion. Thus to achieve high isospecificity, the polymer chain has to migrate to the original position so that the monomer occupies the same vacant site before each insertion step.

As for the activity of the supported catalysts, it may be recalled that magnesium chloride also has a layer–lattice structure like α-, γ-, and δ- $TiCl_3$.[18] The magnesium ions present on the edges of the magnesium chloride lamellae are coordinately unsaturated. The number of coordination sites filled may be four or five as against six for the bulk magnesium ions. This follows from a consideration parallel to that made earlier in this section in the discussion of the lamellar lattice structure of $TiCl_3$. The magnesium ions with four coordination sites filled are more acidic than those are with the five sites filled.[15] It has been suggested that the internal donor coordinates with the more acidic magnesium, while TiCl$_4$, which exists in the dimeric Ti$_2$Cl$_8$ form, coordinates with the less acidic magnesium.[15,82] The absence of coordination of the internal donor ethyl acetate with TiCl$_4$ was confirmed from analytical studies.[83] On treatment with AlEt$_3$, the supported dimeric titanium tetrachloride is reduced to Ti$_2$Cl$_6$, which is subsequently ethylated giving rise to chiral isospecific active sites.[82]

Some of the internal donors may be lost during the treatment with AlEt$_3$. The external donor is used to make up the loss and keep all coordination sites of magnesium filled lest such sites become additionally available for monomer coordination and reduce isospecificity. The formation of stereoblocks is attributable to the rapid isomerization of the catalytic sites leading to variation in steric hindrance near the active sites.[15]

Further insight into the mechanism came from the studies on the stereospecificity of the first monomer (propylene) insertion into the titanium–carbon bond in which the carbon is enriched with ^{13}C. Two successive 1,2 insertions of the monomer can give rise to two stereoisomers at the chain head. The stereoisomers are shown in Scheme 5.3 with structures drawn according to the rotated Fischer's projection formula as *erythro* and *threo* isomers.[84] For stereospecific insertion, the *erythro* product should be formed exclusively. The enriched carbon resonance in ^{13}C NMR spectrum appears at two positions due to the different chemical shifts for the two configurations. The ratio of the resonance intensity of the *erythro* to that of the *threo* isomer gives the degree of stereospecificity of initiation.[84]

Such studies have revealed that insertion of propylene into Ti–CH$_3$ bond in the catalyst system δ TiCl$_3$/Al(CH$_3$)$_3$ is stereoirregular, whereas that into Ti–CH$_2$CH$_3$ bond in δ TiCl$_3$/Al(C$_2$H$_5$)$_3$ is stereoregular, although the degree of the latter is less than that of the average stereoregularity of the polymer. In fact, the stereospecificity of insertion into the Ti–alkyl bond increases with the size of the alkyl group in the order: methyl < ethyl < isopropyl.[85] Obviously, the nonbonded interaction between the methyl group of an

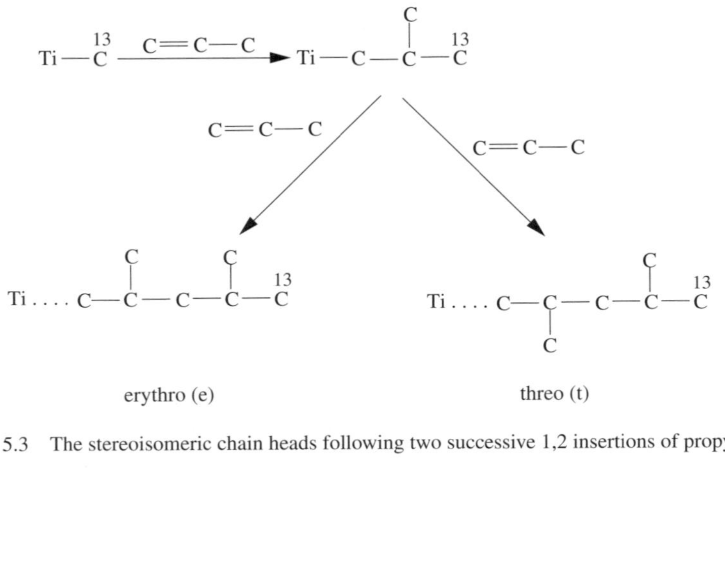

Scheme 5.3 The stereoisomeric chain heads following two successive 1,2 insertions of propylene.[85]

Table 5.2 Average *isotacticity* of polypropylene and stereoregularity of the first inserted propylene into Ti-CH$_2$CH$_3$ bond using activated-MgCl$_2$-supported catalysts with or without Lewis bases.[a]

Catalyst	Ti%	Y[b]	isotactic fraction[c]		
			wt%	(mm)	e/t
(a) MgCl$_2$–TiCl$_4$	2.4	55	29	0.91	2.2
(b) MgCl$_2$–EB–TiCl$_4$	1.4	44	70	0.94	2.8
(c) MgCl$_2$–BEHP–TiCl$_4$	3.0	55	66	0.94	3.4

[a]Cocatalyst, Al (^{13}CH$_2$CH$_3$)$_3$–Zn(CH$_2$CH$_3$)$_2$: EB, ethyl benzoate; BEHP, *bis*(2-ethylhexyl) phthalate; [b]yield in g polymer/g catalyst.h; [c]wt%, heptane insoluble fraction in weight%; (mm), mol fraction of isotactic triads by NMR; e/t, intensity ratio (integrated peak area) of resonances related to the isotactic (e) and syndiotactic (t) placement of the first propene unit (see text). "Reprinted with permission from Ref. 84. Copyright © 1990 American Chemical Society."

incoming monomer with the β carbon (with respect to titanium) in the Ti–alkyl (polymer) plays a crucial role in the stereospecificity of propylene insertion.[84]

Similar studies on the stereoregularity of the first monomer insertion into Ti–CH$_2$CH$_3$ bond in activated MgCl$_2$–supported catalysts have shown that it increases in the presence of Lewis bases as does the average isotacticity of the polymer (Table 5.2).[84]

The nonbonded interaction of the monomer with the halide ligands is also important for *isotactic* placement of the monomer. This is evident from the fact that the first monomer insertion into Ti–CH$_3$ formed in the catalyst system δTiCl$_3$/Al(CH$_3$)$_2$I is stereoregular,[86] although, as described above, it is not so when the cocatalyst is changed to Al(CH$_3$)$_3$. Indeed, Natta *et al.* observed that the isospecificity of the δ TiCl$_3$/Al(CH$_3$)$_2$I catalyst system is very high.[87] This result has been attributed to the exchange of some Cl ligands bonded to Ti by I and the higher steric interaction of I than Cl forcing *isotactic* placement of the monomer unit in the growing chain.[86]

5.1.6.2 Syndiotactic propagation

It has been discussed earlier (Scheme 5.1) that *syndio*specific propagation is associated with *secondary* insertion of propylene on *secondary* vanadium alkyl bond.

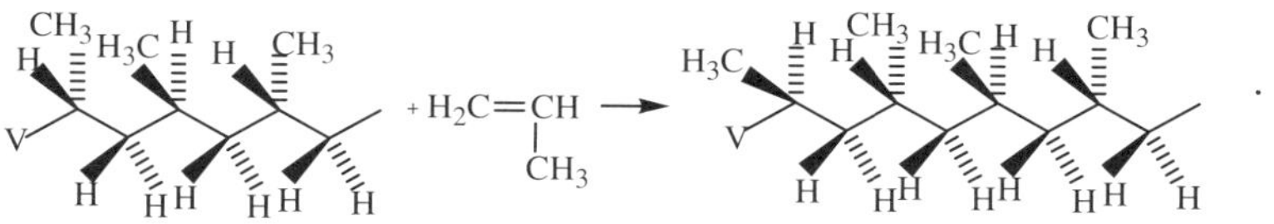

The *syndio*specificity originates from the chirality of the *secondary* carbon of the last added monomer unit attached to vanadium. Steric interaction between complexing propylene and vanadium bound *secondary* alkyl group of the last added propylene forces *syndiotactic* placement. However, for the mechanism to succeed, rotation around the vanadium *secondary* carbon bond must be absent. The unusual resistance of the bond towards hydrolytic attack by methanol proves its stability,[88] which has been cited to indicate configurational stability of the *secondary* carbon.[53] Besides, the low temperature of polymerization helps to impart the needed stability. The inability of higher α-olefins to polymerize in *syndiotactic* mode is attributable to high steric hindrance imposed by monomers for *secondary* insertion.[53]

The lack of *syndio*specificity of the VCl_4–AlR_3 catalysts as opposed to the existence of the property in the VCl_4–AlR_2Cl catalysts has been attributed to the inability of the former to make long-sequences of *secondary* insertions of propylene.[60] Thus, the difference between the two catalysts lies in the difference in their relative rates of *secondary vs. primary* insertion. The relative rate for *secondary* insertion is much greater with VCl_4–AlR_2Cl than with VCl_4–AlR_3 catalyst.

5.1.7 *Polymerization of acetylene*

Acetylene was polymerized in hexane by Natta and coworkers using $AlEt_3$/$Ti(OPr)_4$ as the catalyst. Polymerization at elevated temperatures gave the *trans* polymer as red powder, whereas at low temperatures the *cis* polymer was obtained as black powder. However, the polymers are insoluble and sensitive towards atmospheric oxidation. Accordingly, the materials did not generate much interest. However, nine years later, in 1967, Shirakawa and coworkers discovered a method of preparing polyacetylene in the form of shining films using a catalyst similar to that used by Natta *et al. viz.*, $AlEt_3$/$Ti(OBu)_4$.[89] By a fortuitous error, they used nearly thousand times greater amount of catalyst than is usually used. Another key factor was to let the acetylene react with the concentrated catalyst solution on the surface avoiding bubbling acetylene into the solution or mechanically stirring the solution. Polymerization at 150°C gives the silvery *trans* polymer film, whereas polymerization at low temperature gives the copper colored *cis* polymer film. It is now history that ten years later such films helped usher in the remarkable new field of polymer electronics (*vide* Chap. 1, Sec. 1.7.3) and H. Shirakawa shared the Nobel prize in chemistry, 2000 with A. G. MacDiarmid and physicist A. J. Heeger.

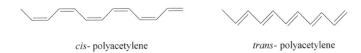

cis- polyacetylene   trans- polyacetylene

5.1.8 *Chain transfer and regulation of molecular weight*

The Ziegler–Natta polymerization is generally free from chain termination but not from chain transfer.[3,90] The latter occurs to monomer, base metal alkyl, and purposefully added transfer agents such as metal alkyl or hydrogen leading to decrease in molecular weight.[91,92]

Thermal cleavage and monomer transfer[92,93]

Two types of thermal cleavage may occur in propylene polymerization, as shown below.

$$M_t-CH_2-\underset{\underset{CH_3}{|}}{CH}-P_n \xrightarrow{\text{below 60°C}} M_t-H + H_2C=\underset{\underset{CH_3}{|}}{C}-P_n. \tag{5.5}$$

$$M_t-H + CH_2=CHCH_3 \longrightarrow M_t-CH_2-CH_2-CH_3. \tag{5.6}$$

$$M_t-CH_2-\underset{\underset{CH_3}{|}}{CH}-P_n + CH_2=CHCH_3 \xrightarrow{> 80\ °C} M_t-CH_2-CH=CH_2 + P_n-\underset{\underset{CH_3}{|}}{CH}-CH_3. \tag{5.7}$$

The β-hydride elimination shown in reaction (5.5) may proceed without assistance of the monomer.[91,92] However, the metal hydride produced may subsequently react with the monomer. The two steps together represent monomer transfer.[92] However, reaction (5.5) is insignificant at low temperature with Ti-based catalysts. Reaction (5.7) also represents a monomer transfer although the transferred monomer retains the double bond.

Alkyl transfer[91]

This type of transfer is represented by reaction (5.8)

$$M_t-P_n + M_b-R \longrightarrow M_t-R + M_b-P_n, \tag{5.8}$$

where M_t is the transition metal and M_b is the base metal. Zinc diethyl has a high chain transfer constant and is often chosen as a transfer agent to regulate the molecular weight.[7]

Transfer with hydrogen

Hydrogen is also used as a transfer agent to regulate the molecular weight.[7] The reactions involved may be represented as follows:

$$M_t-P_n + H_2 \longrightarrow M_t-H + P_n-H. \tag{5.9}$$

$$M_t-H + CH_2=CHCH_3 \longrightarrow M_t-CH_2-CH_2-CH_3. \tag{5.10}$$

5.1.9 Branching in polyethylene

Ideally, according to the mechanism of polymerization so far discussed, the polyethylene prepared by using Z-N catalysts should be linear. In reality, some short branches are formed in the polymer (ca., 3 per 1000 carbon atoms). This has been attributed to copolymerization of ethylene with small amount of 1-butene and other higher α-olefins, which may be present as impurities in the monomer and/or may be generated in the system during polymerization through some metal ions impurities.[7] For example, at a nickel impurity centre the polymerization of ethylene is limited to dimerization yielding 1-butene.[94]

5.2 Metallocene Catalysts

Metallocene catalysts, like the original Ziegler–Natta catalysts, are comprised of two components, a Group IV metallocene, *e.g.*, zirconocene or titanocene being one, and methylaluminoxane (MAO) being the other. The metallocene is actually the precatalyst, which is activated by MAO to form the real catalyst. These are soluble catalysts and the active sites in the catalytic species are all of the same type. Hence, they are referred to as single-site catalysts. These catalysts produce polyolefins with lower polydispersities and more uniform distribution of comonomers in copolymers of ethylene and α-olefins than are obtained using Ziegler–Natta catalysts. Besides, they are much more active. The polymers exhibit better film clarity and tensile strength, and contain lower amounts of oligomers.

Metallocene catalysts were discovered by Kaminsky and coworkers about 25 years after the discovery of the classical Ziegler catalysts.[95,96] MAO is prepared by the controlled hydrolysis of trimethylaluminum.[97,98] It has the basic unit

$$H_3C\!-\!Al(CH_3)\!-\!O\!-\!Al(CH_3)\!-\!O\!-\!Al(CH_3)\!-\!O\!-\!Al(CH_3)_2$$

These units associate to satisfy the unsaturated aluminum atoms through coordination in the best way possible. Three to four such units form a cage structure, which is believed to be its active form.[99,100] In analogy to methyl siloxane the compound is so named. The corresponding ethyl aluminoxane is not very active. However, MAO is required to be used in very high excess, *e.g.*, MAO: metallocene = 5000: 1 (mol/mol).

Zirconocene/MAO catalysts exhibit very high activity. For example, biscyclopentadienyl–zirconium dichloride and MAO reportedly produced nearly 40 metric tons of polyethylene (molecular weight 78 000) per g zirconium in 1 h at 95°C and 8 atm ethylene pressure.[101] This sort of activity is ten to hundred times larger than that observed with the classical Ziegler catalysts.[98] Additionally, the molecular weight distribution is narrower, $M_w/M_n \approx 2$. Oligomers are formed only in traces. The last two features testify to the single-site nature of the catalyst.

LLDPE is produced with similar efficiency as HDPE by copolymerizing ethylene with 1-butene and/or 1-hexene. The copolymers have higher tensile property and other mechanical properties due to the much lower amount of oligomers formed compared to that produced with the Ziegler–Natta catalysts.

Metallocene catalysts also proved to be much more active than the vanadium-based Ziegler–Natta catalysts in the copolymerization of ethylene and cycloolefins, such as cyclobutene, cyclopentene, and norbornene, without ring-opening of the cyclic olefins taking place.[102]

For *isotactic* propagation of polypropylene chain, a chiral metallocene is required to be used. However, the mobility of the ligands bonded around the metal center must be either prevented or rendered slower than propylene insertion. This is achieved in *ansa* metallocenes.[98] In such a metallocene, the two cyclopentadiene ligands sandwitching the metal are substituted to make them rigid. Furthermore, they are linked with a $-CH_2-CH_2-$ bridge. The cyclopentadiene ligands are thus, so to say, strapped (*ansa*) around the metal. The *ansa* zirconocene, ethyl(tetrahydroindenyl)$_2$zirconium dichloride, [Et(THind)$_2$ZrCl$_2$] (**8**), has three stereoisomers, *R*, *S*, and *meso*. The *meso* isomer produces the *atactic* polypropylene, whereas the *R* or *S* isomer produces the *isotactic* polymer.[98] Suitable substituents in the indenyl rings and bridges linking the latter have led to catalysts, which produce polypropylenes with very high *isotacticity ca.*, 99%.[103]

For *syndiotactic* propagation, the catalyst should have C_s symmetry, an example being isopropyl(cyclopentadienyl-1-fluorenyl)zirconium dichloride (**9**).[104] It has a bent sandwich structure, which provides two different bonding positions for the inserting monomer. *Syndiotacticity* is believed to originate from active site isomerization with each monomer addition.

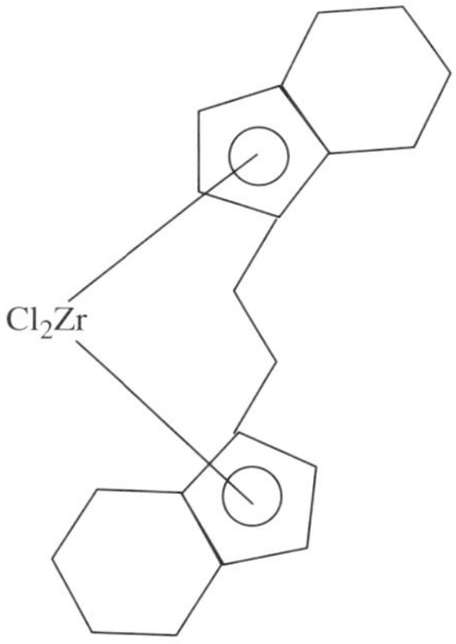

(**8**) 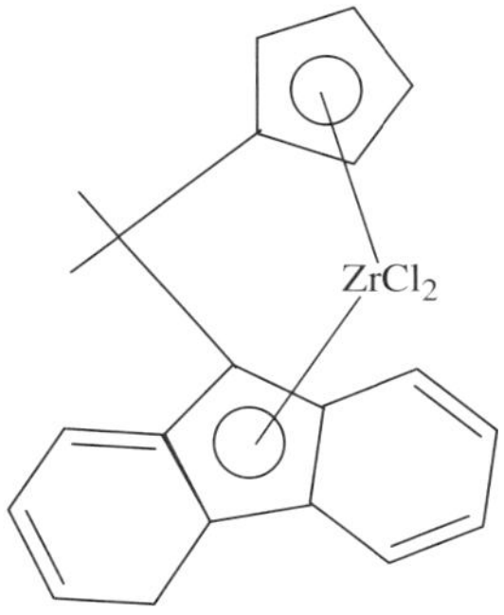   (**9**)

5.2.1 *Mechanism of activation*

The mechanism is not clearly established. In particular, the need for high excess of MAO and its role during polymerization are not well understood.[95] The plausible mechanism is shown in Scheme 5.4. Metallocene and MAO form a complex, which transforms to the monomethylated metallocene.[98] The latter reacts with more MAO to give the catalytically active cationic metal species.[105–108] All these reactions are reversible.

Besides MAO, weakly coordinating cocatalysts such as tetraphenylborate, carborane, and perfluorophenylborate also act as activators.[107,109,110] As regards the requirement of a large excess of MAO, one view attributes it to the additional role of MAO as a scavenger of impurities. However, Kaminsky's group proposed deactivation and reactivation reactions

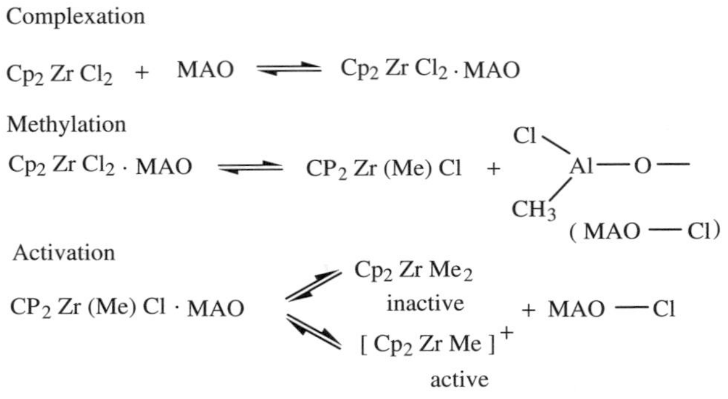

Complexation

$$Cp_2ZrCl_2 + MAO \rightleftharpoons Cp_2ZrCl_2 \cdot MAO$$

Methylation

$$Cp_2ZrCl_2 \cdot MAO \rightleftharpoons Cp_2Zr(Me)Cl + \underset{CH_3}{\overset{Cl}{>}}Al-O- \quad (MAO-Cl)$$

Activation

$$CP_2Zr(Me)Cl \cdot MAO \rightleftharpoons \begin{matrix} Cp_2ZrMe_2 \\ \text{inactive} \\ [Cp_2ZrMe]^+ \\ \text{active} \end{matrix} + MAO-Cl$$

Scheme 5.4 Mechanism of activation of metallocene precatalyst. "Reprinted with permission from Ref. 98. Copyright © 2004 John Wiley & Sons."

(5.11) and (5.12) respectively to explain the need for a large excess of MAO.[111] Propagation proceeds with the coordination of olefin with the active cationic metal species prior to insertion into the M_t–C bond.

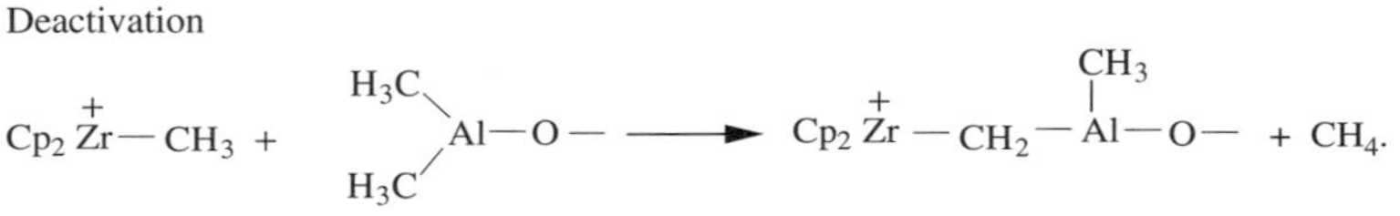

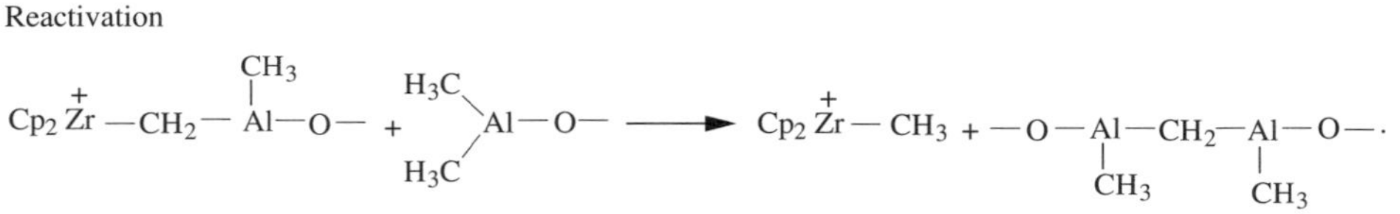

Deactivation

$$Cp_2\overset{+}{Zr}-CH_3 + \underset{H_3C}{\overset{H_3C}{>}}Al-O- \longrightarrow Cp_2\overset{+}{Zr}-CH_2-\underset{|}{\overset{CH_3}{Al}}-O- + CH_4. \quad (5.11)$$

Reactivation

$$Cp_2\overset{+}{Zr}-CH_2-\underset{|}{\overset{CH_3}{Al}}-O- + \underset{H_3C}{\overset{H_3C}{>}}Al-O- \longrightarrow Cp_2\overset{+}{Zr}-CH_3 + -O-\underset{|}{\overset{}{Al}}-CH_2-\underset{|}{\overset{}{Al}}-O-. \quad (5.12)$$
$$ CH_3 \quad\quad CH_3$$

5.3 Late Transition Metal Catalysts

Late transition metal catalysts are new additions to the list of olefin polymerization catalysts. They hold promise of producing polyethylene or poly(α-olefin)s with novel microstructure leading to novel properties and applications. Besides, due to the low oxophilicity they have much greater functional group tolerance than early transition metal ones have and, accordingly, are suitable for the copolymerization of ethylene or α-olefins and polar comonomers.

α-Diimine complexes of Pd(II) and Ni(II) and pyridyl bis-imine complexes of cobalt and iron proved effective catalysts in the polymerization of ethylene, α-olefins, and cyclic olefins or copolymerization of olefins and polar monomers.[112–118] The α-diimine complex of Pd is represented by the chemical structure **10**.[115,118] The catalyst activity may be tuned by appropriately changing the substituents.

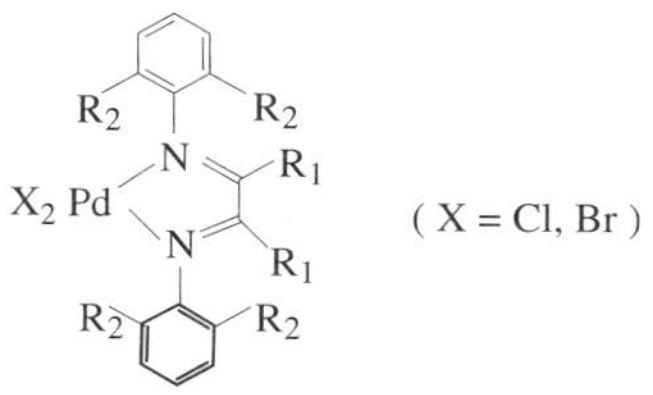

Coordination Polymerization 277

(10) (X = Cl, Br)

The suggested mechanism of polymerization of ethylene based on low temperature NMR spectroscopic studies is shown in Scheme 5.5.[115] The catalyst complex (**11**) is essentially a precatalyst. It is activated by MAO or a modified MAO (MMAO) to form the catalyst.[117] MMAO is prepared by the controlled hydrolysis of trimethylaluminum and triisobutylaluminum. As with the metallocene catalysts discussed in the preceding section, a large excess of MAO/MMAO is required to be used. MAO or MMAO monomethylates the metal and yields a coordinatively unsaturated cationic metal complex (**12**), which is catalytically active. The olefin coordinates with the catalyst prior to its insertion into the metal-methyl bond. The insertion gives **14**. Continued repetition of these two steps leads to linear polymer.

Higher molecular weight polyethylene is obtained with Pd catalysts rather than with Ni. On the other hand, Pd catalysts give highly branched polymer the density of which may be as low as 0.85.[115] However, unlike in Ziegler–Natta polymerization, the branching does not involve generation of α-olefins followed by their copolymerization with ethylene. A "chain walking" mechanism has been proposed for branch formation as shown also in Scheme 5.5. The cationic metal alkyl species **15** exists in equilibrium with **16** in which the β hydrogen atom in the alkyl group is simultaneously bonded to carbon and the

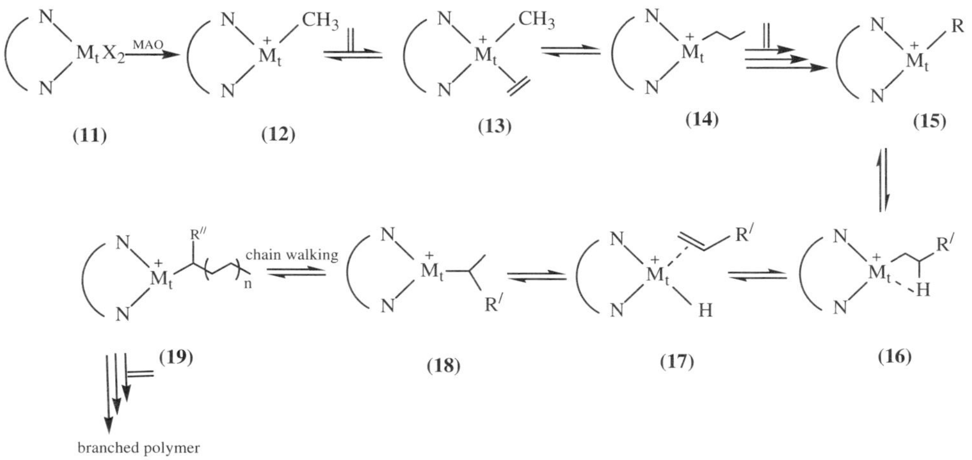

Scheme 5.5 Mechanism for the polymerization of ethylene using Pd or Ni α-diimine complexes activated by MAO.[115]

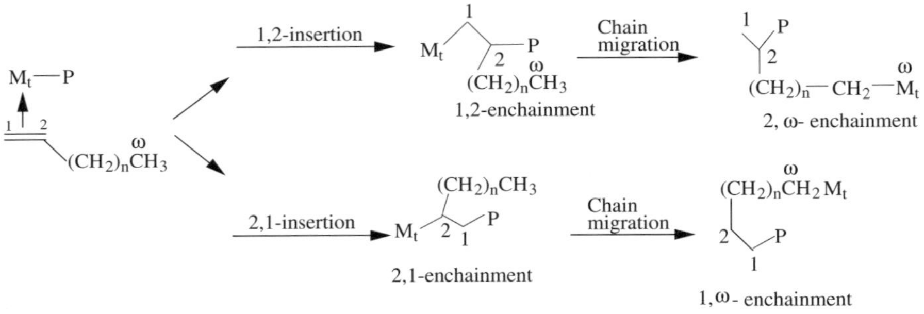

Scheme 5.6 Pathways to 1,2-, 1,ω-, 2,1- and 1,ω-enchainment of α-olefins. Mt–P represents the metal catalyst bonded to polymer.[115]

metal (β-agostic interaction). As a result, the C—H bond is activated facilitating β-hydride elimination forming **17**. Readdition of the hydride to the alkene gives **18**. A series of the above reactions on the same chain leads to migration of the metal down the chain to various positions depending on the series lengths forming alkyl branches longer than methyl as shown in **19**. Frequent repetition of the above events during chain growth gives rise to highly branched polymer without the kind of long branches that exists in LDPE.[115]

With α-olefins, the polyolefin microstructure is substantially different from those obtained with the Ziegler–Natta or the metallocene catalysts. With propylene, for example, the polymer obtained is essentially an ethylene-propylene copolymer.[115] This result can be explained using the chain walking mechanism in the insertion of α-olefins, which can give rise to both 1,ω- and 2, ω-enchainment (Scheme 5.6). The former arises from 2,1-insertion followed by chain migration of the metal to the ω carbon atom. This accounts for the presence of ethylene residues in the polymer.

Copolymerization of ethylene and α-olefins with functional monomers, such as acrylates, acrylic acid, methyl vinyl ketone, CO, crotonaldehyde, *etc.*, has been performed successfully.[115]

The other group of late metal complexes which showed promise are the pyridyl bis-imine complexes of cobalt and iron.[116–119] One of the most active iron based catalyst has bulky aryl substituents in the imine groups as shown in complex **20**.

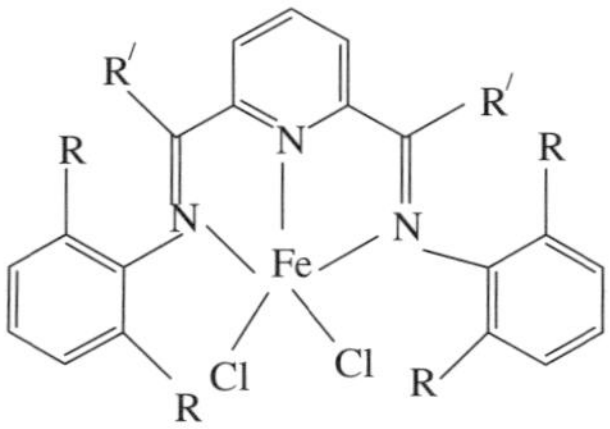

(**20**)

By using different R and R' groups the activity of the catalyst and the molecular weight of the polymer can be further tuned. A five-coordinate pseudo-square-pyramidal conformation for the complex has been suggested from crystallographic studies. The aryl groups on the imino nitrogens stand nearly perpendicular to the ligand co-ordination plane.[115] Bis-chelate formation is prevented by the bulky substituents on the imino nitrogens resulting in high activity for ethylene polymerization.[116,120] The polyethylene obtained is strictly linear and of high molecular weight in contrast to the highly branched polyethylene obtained with Pd diimine catalysts discussed above. Obviously, the "chain walking" does not take place with these catalysts.[115] The late metal catalysts, however, have not yet found commercial use in polyolefin production. Many of these catalysts initially exhibit very high activity, which goes on decreasing during the course of polymerization.[115]

5.4 Living Polymerization of Alkenes

As has been discussed in Sec. 5.1.8, Ziegler–Natta polymerizations are generally free from termination. However, transfer reactions to monomer and metal alkyls do occur (Eqs. 5.5 to 5.8). In order to achieve living polymerization, these reactions should be absent. This may be achieved by lowering the temperature of polymerization. In addition, for the sake of uniform growth of all the polymer molecules, polymerization is required to be carried out in the homogeneous phase using soluble catalysts.

Natta and coworkers provided the first indication of living polymerization of propylene conducted at $-78°C$ using the soluble catalyst $VCl_4 + AlEt_2Cl$.[28,29] The molecular weight of the *syndio*-rich polypropylene increased linearly with time over a period of 25 h. Later, Doi *et al.* found the molecular weight polydispersity to be between 1.4 and 1.9, which is rather high for a living polymer. They, however, succeeded to obtain *syndio*-rich polypropylene of low PDI (PDI = $1.07 - 1.18$) replacing VCl_4 with $V(acetylacetonate)_3$.[121–123] The molecular weight of the polymer increased linearly with time over 15 h. The living character is appreciably lost by working at temperatures, which are even slightly greater than $-65°C$. However, only one polymer chain forms per twenty-five vanadium atoms. The situation improves to yield one polymer chain per vanadium atom by replacing acetylacetonato with 2-methyl-1,3-butanedionato ligand.

Living polymerization by metallocene catalysts has also been achieved. Monomer transfer is prevented by resorting to low reaction temperature, while alkyl transfer is eliminated by using boron-based activators.[124,125] Various other catalysts including the late transition metal ones and activators have yielded living polymerizations. However, because of the low turn over (mols of monomer polymerized per mol of catalyst) the living co-ordination polymerization is yet to be adopted by the industry.[126]

References

1. K. Ziegler, E. Holzkamp, H. Breil, H. Martin, *Angew Chem.* (1955), 67, 541.
2. K. Ziegler, Nobel Lecture Dec., 12, 1963, in, Nobel Lectures — Chemistry, 1963–70, Elsevier, Amsterdam p. 6, (1972); *Rubber Chem. Technol.* (1965), 38, p. xxiii.
3. G. Natta, Nobel Lecture Dec., 12, 1963, in, Nobel Lectures — Chemistry, 1963–1970, Elsevier, Amsterdam, p. 27, (1972).
4. G. Natta, *Angew Chem.* (1955), 67, 430.

5. G. Natta et al., *J. Am. Chem. Soc.* (1955), 77, 1708.
6. G. Natta, *J. Polym. Sci.* (1955), 16, 143.
7. J. Boor, Jr. *Ziegler-Natta Catalysts and Polymerizations*, Academic Press, New York (1979).
8. W. Cooper, *Stereospecific Polymerization*, in, Progress in High Polymers Vol. 1, J.C. Robb and F.W. Peaker Eds., Heywood & Co., London, p. 281, (1961).
9. G. Natta, P. Corradini, G. Allegra, *J. Polym. Sci.* (1961), 51, 399.
10. G. Natta, *J. Polym. Sci.* (1959), 34, 531.
11. G. Natta, I. Pasquon, A. Zambelli, G. Gatti, *J. Polym. Sci.* (1961), 51, 387.
12. G. Natta, I. Pasquon, *Adv. Catal.* (1959), 9, 1.
13. G. Natta, *Chim. Ind. (Milan)* (1960), 42, 1207.
14. G. Natta et al., *Chim. Ind. (Milan)* (1956), 38, 124.
15. J.C. Chadwick, *Ziegler-Natta Catalysts*, in, *Encyc. Polym. Sci. & Tech.*, 3rd ed., H. Mark Ed., Wiley-Interscience, Hoboken, New Jersey, vol. 8, p. 517, (2003).
16. E.J. Arlman, *J. Catal.* (1964), 3, 89.
17. J.P. Hermans, P. Henrioulle (1972), US Pat 4210738 (to Solvay).
18. R.N. Haward, A.N. Roper, K.L. Fletcher, *Polymer* (1973), 13, 365.
19. P. Galli, L. Luciani, G. Geechin, *Angew. Makromol. Chem.* (1981), 94, 63.
20. P.C. Barbe, G. Geechin, L. Noristi, *Adv. Polym. Sci.* (1987), 81, 1.
21. E.P. Moore, *Polypropylene Handbook, Polymerization, Characterization, Properties, Processing, Applications*, Hanser Publishers, Munich (1996).
22. G.A. Razuvaev, K.S. Minsker, G.T. Fedoseeva, V.K. Bykhovskii, *Polym. Sci. USSR* (1961), 2, 299.
23. R.L. McConnel et al., *J. Polym. Sci. A* (1965), 3, 2135.
24. B.L. Goodall, S. van der Vens, *Polypropylene and other Polyolefins. Polymerization and Characterization*, Elsevier, Amsterdam, Ch. 1, (1990).
25. S. Parodi et al. (1981) Eur. Pat. 45977 (to Montedison).
26. E. Albizzati et al., *Macromol. Symp.* (1995), 89, 73.
27. G. Natta, L. Porri, G. Zanini, L. Fiore, *Chim. Ind. (Milan)* (1959), 41, 526.
28. G. Natta, I. Pasquon, A. Zambelli, *J. Am. Chem. Soc.* (1962), 84, 1488.
29. A. Zambelli, G. Natta, I. Pasquon, *J. Polym. Sci.* (1963), C4, 411.
30. G. Natta et al., *Makormol. Chem.* (1965), 81, 161.
31. A. Zambelli, G. Allegra, *Macromolecules* (1980), 13, 42.
32. G. Natta et al., *J. Polym. Sci.* (1961), 51, 411.
33. G. Natta, *Rubber Plastics Age* (1957), 38, 495.
34. G. Natta, G. Crespi, *Rubber Age* (1960), 87, 459.
35. W.M. Saltman, T.H. Link, *Ind. Eng. Chem. Prod. Res. Dev.* (1964), 3, 199.
36. W. Copper, in, *The Stereo Rubbers*, W.M. Saltman Ed., John Wiley & Sons, Inc., N.Y., p. 24, (1977).
37. M. Gippin, *Rubber Chem. Technol.* (1962), 35, 1066.
38. M. Kerns, S. Henning, M. Rachita, in *Encyc. Polym. Sci. & Tech.*, 3rd Ed., H. Mark Ed., Wiley-Interscience, Hoboken, New Jersey, Vol. 5, p. 317, (2003).
39. S.E. Horne et al., *Ind. Eng. Chem.* (1956), 48, 784.
40. E. Schoenberg, H.A. Marsh, S.J. Walters, W.M. Saltman, *Rubber Chem. Technol.* (1979), 52, 526.
41. V. Busico et al., *Macromolecules* (1999), 32, 4173.
42. J.C. Randall, *Macromolecules* (1997), 30, 803.
43. V. Busico, R. Cipullo, P. Corradini, R. De Biasio, *Macromol. Chem. Phys.* (1995), 196, 491.
44. R. Paukkeri, T. Vaananen, A. Lehtinen, *Polymer* (1993), 34, 2488.
45. R. Paukkeri, E. Iiskola, A. Lehtinen, H. Salminen, *Polymer* (1994), 35, 2636.
46. G. Natta, *Angew. Chem.* (1956), 68, 393.

47. G. Natta, *Experientia Suppl.* (1957), 7, 21.
48. G. Natta, F. Danusso, D. Sianesi, *Makromol. Chem.* (1959), 30, 238.
49. W.L. Carrick, *J. Am. Chem. Soc.* (1958), 80, 6455.
50. G. Natta, P. Pino, G. Mazzanti, R. Lanzo, *Chim. Ind. (Milan)* (1957), 39, 1032.
51. C. Beerman, H. Bestian, *Angew. Chem.* (1959), 71, 618.
52. N. Yermakov, V. Zakharov, *Adv. Catal.* (1975), 24, 173.
53. P. Pino, R. Mulhauft, *Angew. Chem. Intl. Ed.* (1980), 19, 857.
54. Y. Takegami, T. Suzuki, *Bull. Chem. Soc. Japan* (1969), 42, 848.
55. A. Zambelli, P. Locatelli, E. Rigamonti, *Macromolecules* (1979), 12, 156.
56. A. Zambelli, C. Wolfsgruber, G. Zannoni, F.A. Bovey, *Macromolecules* (1974), 7, 750.
57. F.A. Bovey, M.C. Sacchi, A. Zambelli, *Macromolecules* (1974), 7, 752.
58. A. Zambelli, C. Tosi, C. Sacchi, *Macromolecules* (1972), 5, 649.
59. A. Zambelli, G. Bajo, E. Rigamonti, *Makromol. Chem.* (1978), 179, 1249.
60. A. Zambelli, P. Locatelli, E. Rigamonti, *Macromolecules* (1979), 12, 156.
61. A. Zambelli, C. Tosi, *Adv. Polym. Sci.* (1974), 15, 31.
62. T. Asakura *et al.*, *Makromol. Chem.* (1977), 178, 791.
63. Y. Doi, T. Asakura, *Makromol. Chem.* (1975), 176, 507.
64. A. Zambelli *et al.*, *Macromolecules* (1971), 4, 475.
65. C. Wolfsgruber, G. Zannoni, E. Rigamonti, A. Zambelli, *Makromol. Chem.* (1975), 176, 2765.
66. G. Natta, M. Farina, M. Peraldo, *Chim. Ind. (Milan)* (1960), 42, 255.
67. T. Miyazawa, T. Ideguchi, *J. Polym. Sci. B* (1963), 1, 389.
68. A. Zambelli, M. Giongo, G. Natta, *Makromol. Chem.* (1968), 112, 183.
69. G. Natta, *J. Inorg. and Nucl. Chem.* (1958), 8, 589.
70. P. Pino, *Adv. Polym. Sci.* (1965), 4, 393.
71. P. Pino *et al.*, in, *Coordination Polymerization*, J.C.W. Chien Ed., Academic Press, New York, P. 25, (1975).
72. F. Ciardelli, P. Locatelli, M. Marcetti, A. Zambelli, *Makromol. Chem.* (1974), 175, 923.
73. P. Cossee, *Tetrahedron Letters* (1960), no. 17, 12.
74. G. Natta, G. Mazzanti, *Tetrahedron* (1960), 8, 86.
75. G. Natta, I. Pasquon, *Adv. Catal.* (1959), 9, 1.
76. E.J. Arlman, *J. Catal.* (1964), 3, 89.
77. H. Hargitay, L. Rodriguez, M. Miotto, *J. Polym. Sci.* (1959), 35, 559.
78. J. Boor, Jr., *J. Polym. Sci. C* (1963), 1, 257.
79. L.A.M. Rodriguez, H.M. van Looy, *J. Polym. Sci. A-1* (1966), 4, 1971.
80. L.A.M. Rodriguez, H.M. van Looy, *J. Polym. Sci. A-1* (1966), 4, 1951.
81. E.G. Arlman, P. Cossee, *J. Catal.* (1964), 3, 99.
82. V. Busico *et al.*, *Makromol. Chem.* (1985), 186, 1279.
83. M. Terano, T. Kataoka, T. Keii, *Makromol. Chem.* (1987), 188, 1477.
84. M.C. Sacchi, C. Shan, P. Locatelli, I. Tritto, *Macromolecules* (1990), 23, 383.
85. A. Zambelli, M.C. Sacchi, P. Locatelli, G. Zannoni, *Macromolecules* (1982), 15, 211.
86. A. Zambelli *et al.*, *Macromolecules* (1980), 13, 798.
87. G. Natta, I. Pasquon, A. Zambelli, G. Gatti, *J. Polym. Sci.* (1961), 51, 387.
88. A. Zambelli, C. Sacchi, *Makromol. Chem.* (1974), 175, 2213.
89. T. Ito, H. Shirakawa, S. Ikeda, *J. Polym. Sci. Polym. Chem.* (1975), 13, 1943; (1974), 12, 11.
90. G. Natta, I. Pasquon, E. Giachetti, *Angew. Chem.* (1957), 69, 213.
91. G. Natta, I. Pasquon, E. Giachetti, *Chim. Ind. (Milan)* (1958), 40, 97.
92. G. Natta, I. Pasquon, *Adv. Catal.* (1959), 11, 1.
93. G. Bier, *Makromol. Chem.* (1964), 70, 44.
94. K. Ziegler, E. Holzekamp, H. Breil, H. Martin, *Angew. Chem.* (1955), 67, 541.
95. H. Sinn, W. Kaminsky, H.J. Vollmer, R. Woldt, DE. Patent, 3007725, 29 Feb., (1980).

96. H. Sinn, W. Kaminsky, H.J. Vollmer, *Angew. Chem. Int. Ed.* (1980), 19, 390.
97. W. Kaminsky, M. Miri, H. Sinn, R. Woldt, *Makromol. Chem. Rapid Commun.* (2000), 21, 1333.
98. W. Kaminsky, *J. Polym. Sci. Polym. Chem. Ed.* (2004), 42, 3911.
99. H. Sinn, *Macromol. Symp.* (1995), 97, 27.
100. M. Ystenes et al., *J. Polym. Sci. Polym. Chem.* (2000), 38, 3106.
101. W. Kaminsky, H. Sinn, IUPAC 28th Macromolecular Symposium 12 July, Amherst, USA, Proceedings p. 247, (1982).
102. W. Kaminsky, R. Spiehl, *Makromol. Chem.* (1989), 190, 515.
103. W. Spaleck et al., *Makromol. Symp.* (1995), 89, 237.
104. J.A. Ewen, R.L. Jones, A. Razavi, J.P. Ferrara, *J. Am. Chem. Soc.* (1988), 110, 6255.
105. A.K. Shilova, F.S. Dyachkovskii, A.E. Shilov, *J. Polym. Sci. Part C* (1967), 16, 2333.
106. J.J. Eisch, S.I. Pombrick, G.X. Jheng, *Organometallics* (1993), 12, 3856.
107. R.F. Jordan, W.E. Dasher, S.F. Echols, *J. Am. Chem. Soc.* (1986), 108, 1718.
108. M. Bochmann, L.M. Wilson, *J. Chem. Soc. Chem. Commun.* (1986), 1610.
109. P. Gassman, M.R. Callstrom, *J. Am. Chem. Soc.* (1987), 109, 7875.
110. X. Yang, C.L. Stern, T. Marks, *J. Organometallics* (1991), 10, 840.
111. W. Kaminsky, R. Steiger, *Polyhedron* (1988), 7, 2375.
112. L.K. Johnson, C.M. Killian, M. Brookhart, *J. Am. Chem. Soc.* (1995), 117, 6414.
113. C.M. Killian, D.J. Tempel, L.K. Johnson, M.J. Brookhart, *J. Am. Chem. Soc.* (1976), 118, 11664.
114. A.S. Abu-Surrah, B. Rieger, *Angew. Chem. Int. Ed.* (1996), 35, 2475.
115. S.D. Ittel, L.K. Johnson, M. Brookhart, *Chem. Rev.* (2000), 100, 1169.
116. B.L. Small, M. Brookhart, A.M.A. Bennett, *J. Am. Chem. Soc.* (1998), 120, 4049.
117. G.J.P. Britovsek, V.C. Gibson, D.F. Wass, *Angew. Chem. Int. Ed.* (1999), 38, 428.
118. G.J.P. Britovsek et al., *Chem. Commun.* (1998), 849.
119. V.C. Gibson, D.F. Wass, *Chem. Br.* (1999), 7, 20.
120. V.C. Gibson, S.K. Spitzmesser, *Chem. Rev.* (2003), 103, 283.
121. Y. Doi, M. Takada, T. Keii, *Bull. Chem. Soc. Jpn.* (1979), 52, 1802.
122. Y. Doi, S. Ueki, T. Keii, *Macromolecules* (1979), 12, 814.
123. Y. Doi, S. Ueki, T. Keii, *Macromol. Chem. Phys.* (1979), 180, 1359.
124. J. Sassmannshausen, M.E. Bochmann, J. Rosch, D. Lilge, *J. Organomet. Chem.* (1997), 548, 23.
125. Y. Fukui, M. Murata, K. Soga, *Macromol. Rapid Commun.* (1999), 20, 637.
126. J.D. Gregory et al., *Prog. Polym. Sci.* (2007), 32, 30.

Problems

5.1 What are the components of Ziegler–Natta catalysts? What are their functions?

5.2 Would the initiation of propylene polymerization be stereoregular using the catalyst system $\sigma\text{-TiCl}_3 + \text{Al(CH}_3)_3$? Would the stereoregularity of propagation be similar with that of initiation? Explain your answer.

5.3 Discuss the choice of $MgCl_2$ as catalyst support. Why is activation needed? How is the activation done? How much is the amount of Ti reduced due to supporting?

5.4 Would the isospecificity of the titanium based catalysts decrease or increase if some of the Cl atoms in $TiCl_3$ are replaced by iodine? Explain your answer.

5.5 Why the use of low temperature is important in the *syndiotactic* polymerization of propylene using soluble vanadium based catalysts?

5.6 Metallocene catalysts produce polymers with narrower MWD and lesser oligomers content than Z–N catalysts do. What makes this change possible?

5.7 Which catalysts are used to produce dendritic polyethylene? Which step in the mechanism is responsible for such branching? Is the density of the polymer higher or lower than LDPE? Explain your answers.

5.8 What are the structural features that make possible for a metallocene catalyst to produce *isotactic* propylene?

5.9 How will you explain the formation of short chain branches in polyethylene prepared by Ziegler-Natta catalysts?

5.10 How was the chirality of the catalytic centers in Ti-based isospecific catalysts proved?

Chapter 6

Cationic Polymerization

The chain carrier in cationic polymerization is an electrophile, such as a carbenium ion, an oxonium ion, a sulfonium ion, an iminium ion, an oxazolinium ion, a siloxonium ion, *etc*. Accordingly, a wide variety of nucleophilic monomers, belonging to the families of olefins and heterocyclics, is polymerized by this route. Of the cations mentioned, the carbenium ion is extremely high in reactivity. Accordingly, the carbenium ionic (referred to hereafter as carbocationic) polymerization is treated exclusively in this chapter. Other cationic polymerizations are discussed in the chapter on ring-opening polymerization. Similar separate treatment was given to carbanionic polymerization earlier in this book by virtue of the high reactivity of carbanions *vis-à-vis* other anions.

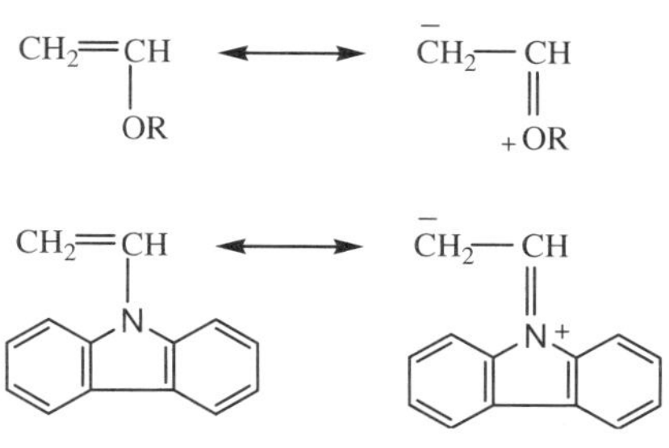

For a vinyl monomer to be capable of undergoing cationic polymerization, it should have an electron-rich double bond. This may occur through either the inductive (+I) effect of electron-releasing substituent(s) or the resonance interaction of the double bond with the substituent(s). The resonance effect is particularly strong so that vinyl ethers and N-vinylcarbazole are highly nucleophilic and very reactive in carbocationic polymerization.

6.1 The Nucleophilicity and Electrophilicity Scales

Rate constants of reactions of various kinds of electrophiles, *e.g.*, carbocations, metal-π complexes, and diazonium ions, with π-, σ-, and n-nucleophiles have been expressed quantitatively as

$$\log k(20°C) = s(N + E), \tag{6.1}$$

Table 6.1 The values of N, E, and s parameters of some vinyl monomers and their carbocations[1-7,a]

Monomer	N	s	E
N-vinylcarbazole	5.02	0.94	2.41
Vinyl ethyl ether	3.92	0.90	–
p-Methoxystyrene	3.30	–	4.7
α-Methylstyrene	2.35	1.0	–
P-Methylstyrene	(1.70)	(1.00)	–
Isobutylene	1.11	0.98	8.9
Isoprene	1.10	0.98	8.3
Styrene	0.78	0.95	9.5
2,4,6-Trimethylstyrene	0.68	1.09	6.04
P-Chlorostyrene	(0.21)	(1.00)	–
Butadiene	−0.87	(1.00)	–
Chloroprene	(−1.59)	–	–

[a]The values in parentheses are approximate ones.

where k is the rate constant, E is the electrophilicity parameter, N is the nucleophilicity parameter, and s is a nucleophile-dependent slope parameter.[1,2]

The N and s parameters of many nucleophiles and E parameters of many electrophiles have been determined from a correlation analysis of the rate constants, which are not diffusion-limited, according to Eq. (6.1) arbitrarily setting E of $(p\text{-MeOC}_6\text{H}_4)_2\text{C}^+\text{H}$ as zero and s of 2-methyl-1-pentene as one.[1,2] The N and s parameters of some cationically polymerizable vinyl monomers and E parameters of some of the corresponding carbocations are given in Table 6.1.[1-7] The N and E parameters change in mutually reverse direction with change in monomer (exceptions existing). This is because the substituent that increases the electron density at the β carbon atom in the monomer decreases the positive charge density at the α carbon atom of the corresponding carbocation.

6.2 Bronsted Acids as Initiators

Bronsted acids (HA) can initiate polymerization of vinyl monomers. The reactions involved in the polymerization may be represented as in Scheme 6.1. For the sake of simplicity, the active center is shown in the scheme as ion pair instead of free ion, although both may be present in a real polymerization. In addition, atoms or groups attached to the two unsaturated carbon atoms in the monomer are omitted for the same reason.

$$\text{H}^+\text{A}^- + \text{C=C} \longrightarrow \text{H}-\text{C}-\text{C}^+\text{A}^- \qquad \text{Initiation} \qquad (6.2)$$

$$\text{H}-\text{C}-\text{C}^+\text{A}^- + n(\text{C=C}) \longrightarrow \text{H}-(\text{C}-\text{C})_n-\text{C}-\text{C}^+\text{A}^- \qquad \text{Propagation} \qquad (6.3)$$

$$\text{H}-(\text{C}-\text{C})_n-\text{C}-\text{C}^+\text{A}^- + \text{C=C} \longrightarrow \text{H}-(\text{C}-\text{C})_n-\text{C}=\text{C} + \text{H}-\text{C}-\text{C}^+\text{A}^- \qquad \text{Monomer transfer} \qquad (6.4)$$

$$\text{H}-(\text{C}-\text{C})_n-\text{C}-\text{C}^+\text{A}^- \longrightarrow \text{H}-(\text{C}-\text{C})_n-\text{C}-\text{C}-\text{A} \qquad \text{Termination} \qquad (6.5)$$

Scheme 6.1 Reactions involved in Bronsted acid catalyzed carbocationic polymerization.

In the scheme, termination is shown to occur through ion pair collapse forming a covalent bond C–A at the chain end. Monomer transfer is the most important transfer reaction. It is bimolecular with the monomer directly participating in the transfer reaction in which a β proton of the propagating carbocation is transferred to the monomer. However, monomer transfer may also be preceded by a rate–controlling unimolecular step in which the β proton is eliminated to the counter anion and subsequently taken up by the monomer (*vide infra*). This process is called spontaneous chain transfer. With aromatic olefins, another mechanism of monomer transfer may be predominant. This involves the intramolecular or intermolecular attack of the aromatic nucleus in the polymer by the carbocation resulting in indanyl ring formation with the concomitant release of a proton, which is taken up by the monomer.[8–10] The intramolecular attack gives phenyl substituted indanyl group ended polymer

$$\sim\sim\sim CH_2-CH-CH_2-\overset{+}{C}H \quad \xrightarrow[\text{Styrene}]{-H^+} \quad \sim\sim\sim CH_2-CH-CH_2 + CH_3-\overset{+}{C}H \cdot \qquad (6.6)$$

With HCl as the initiator, polymerization of isobutylene does not occur. The highly nucleophilic conjugate base of the acid, the Cl$^-$ ion, causes termination immediately after initiation, the carbocation being highly electrophilic.[11] With H_2SO_4 and H_3PO_4 as initiators having less nucleophilic conjugate bases, HSO_4^- and $H_2PO_4^-$ respectively, dimers or trimers of isobutylene are formed.[12,13] In contrast, N-vinylcarbazole undergoes rapid polymerization even with HCl as the initiator. The low electrophilicity of the carbocation in this case slows down termination.

Polar solvents help in the ionization of the acid and in the dissociation of the ion pair. Hence, carbocationic polymerization is facilitated in their presence. However, solvents should be nonnucleophilic and, accordingly, chlorohydrocarbon solvents, such as methyl chloride, methylene chloride, and 1,2-dichloroethane, are used. Thus, styrene undergoes rapid polymerization at low temperatures in these solvents with strong acid initiators, *e.g.*, $HClO_4$, H_2SO_4, or CF_3SO_3H.[8,9,15–29] The pattern of polymerization depends upon temperature.[17–21]

Thus, with $HClO_4$ as the initiator, polymerization is completed in a few seconds at a very low temperature, *viz.*, $-97°C$, giving limited yield, which is proportional to the initial concentration of the acid. It can be revived by warming to higher temperatures.[19] Thus, in the temperature range of -78 to $-60°C$ two stages of polymerization are seen: a flash polymerization (Stage I) followed by a slower polymerization (Stage II), which exhibits first order kinetics.[21] The first order rate constant is proportional to the initial acid concentration. Above $-30°C$ (-30 to $25°C$) Stage 1 is not observed.[17,18,22–24] The results of kinetic studies of polymerization at some representative temperatures are shown in Fig. 6.1.

For polymerizations with completed initiation,

$$\overline{M}_n = \frac{f_c[M]_0 M_m}{n[I]_0}, \qquad (6.7)$$

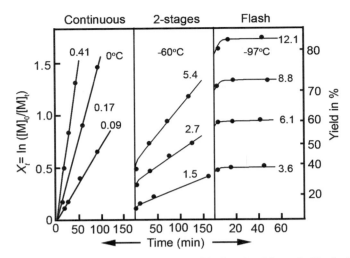

Figure 6.1 Percent yield vs. time (middle and right sections of the figure) and first order kinetic plots (left section) for the disappearance of monomer in the perchloric acid (I) initiated polymerization of styrene in CH_2Cl_2 at various temperatures. $[M]_0 = 0.435$ mol L^{-1} and $[I]_0$ in mmol L^{-1} indicated in the figures. "Reprinted with permission of Wiley-VCH and the deceased author's wife from Ref. 21. Styrene polymerization in methylene chloride by perchloric acid. Copyright © Wiley-VCH Verlag GmbH & Co. KGaA."

where $[M]_0$ and $[I]_0$ are the initial concentrations of monomer and initiator respectively, M_m is the molecular weight of monomer, f_c is the fractional yield, and n is the number of polymer molecules formed per molecule of initiator. In the absence of transfer, n = 1, while in the presence of it, n > 1. Molecular weight measurements revealed that n is close to unity up to about $-71°C$ but more than unity at higher temperatures (Fig. 6.2).

The negative deviation of $[I]_0 M_n$ from the ideal value represented by the dotted line in Fig. 6.2 (Eq. 6.7, n = 1) observed at higher temperatures is attributable to chain transfer reaction. Addition of common anion salt Bu_4NClO_4 brings about reduction in both rate of polymerization and rate of chain transfer (not shown).[20,21] This indicates that both propagation and transfer are facilitated by free ion. Occurrence of termination leads to perchlorate ester as end group.

The nature of the active species in perchloric acid initiated polymerization had been debated for a long time.[30–36] Gandini and Plesch could not detect ionic conductivity of the polymerizing solution.[22–24] They also failed to observe optical absorption expected of the polystyrene carbocation (free or ion paired). These observations led them to postulate that propagation involves covalently bonded perchlorate ester and proceeds through a 6-membered cyclic transition state (**1**). This hypothesis referred to as pseudocationic polymerization was in vogue for a long time. It was discarded eventually.[33–38] In fact, transient optical absorption, tentatively attributed to the free polystyrene cation, and electrical conductivity of the polymerizing solution were observed in stopped-flow studies of Stage 1 polymerizations of styrene and its derivatives initiated by perchloric acid.[25,26]

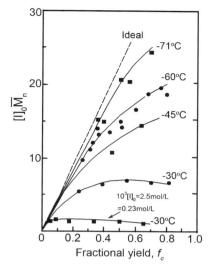

Figure 6.2 Deviation of molecular weight from the ideal value (see text) at temperatures of −71°C and above. Initiator (perchloric acid) concentration is 2.5 mmol/L for all curves except for the bottom one where it is 0.23 mmol/L. "Reprinted with permission of Wiley-VCH and the deceased author's wife from Ref. 21. Styrene polymerization in methylene chloride by perchloric acid. Copyright © Wiley-VCH Verlag GmbH & Co. KGaA."

However, the concentration of the ionic species passes through a maximum of ∼1% of the acid concentration.

(1)

The ionization of the polymer perchlorate (P_nA) was suggested to result from its interaction with the acid HA[21]

$$P_nA + HA \rightleftharpoons P_n^+ \; HA_2^- \rightleftharpoons P_n^+ + HA_2^-,$$

where Pn^+ and $Pn^+HA_2^-$ represent the free ion and the ion pair respectively.

6.3 Lewis Acids as Coinitiators

The first important use of strong Lewis acids (Friedel Crafts halides) was made in the polymerization of isobutylene leading to the commercial production of polyisobutylene and butyl rubber. In 1931 Otto and Müller–Cunradi reported that isobutylene is rapidly polymerized by BF_3 at −100°C in methyl chloride diluent to yield high molecular weight polyisobutylene.[39] Subsequently, Thomas and Sparks discovered butyl rubber in 1937 by copolymerizing isobutylene and butadiene and, later, isoprene at $-100°C$ using $AlCl_3$ as the catalyst and methyl chloride as the diluent.[40–42]

The mechanism of initiation by strong Lewis acid depends on the dryness of the polymerization system and the nature of the acid. Polyani, Evans, Plesch, and coworkers showed that stringently purified and dried isobutylene does not polymerize when brought into contact with BF_3 or $TiCl_4$ but does so at a very fast rate when a trace of water is introduced into the reaction vessel.[43–46] Water and other protogens were therefore considered as cocatalysts. However, the terms 'cocatalyst' and 'catalyst' are of older usage. They have been replaced appropriately with 'initiator' and 'coinitiator' respectively.[10] It was assumed that the coordination complex formed from the Lewis acid and water undergoes ionization in the presence of a monomer and/or a polar solvent leading to initiation as shown in reaction (6.8) for the initiation of isobutylene with the $BF_3.H_2O$ complex. The Lewis acid may also form a π complex with the monomer as shown in reaction (6.9) with $TiCl_4$ and isobutylene, which interacts with H_2O leading to initiation.

$$F_3B \leftarrow OH_2 + H_2C = C(CH_3)_2 \rightarrow (H_3C)_3C^+[BF_3OH]^-. \tag{6.8}$$

$$H_2O + TiCl_4.H_2C = C(CH_3)_2 \rightarrow (H_3C)_2C^+[TiCl_4OH]^-. \tag{6.9}$$

Solvation of ion by solvent provides the necessary driving force for ionization in the presence of a polar solvent. There exists kinetic evidence for complexation of carbocation with monomer and/or solvent.[47]

However, subsequent researchers claimed that some Lewis acids, such as $TiCl_4$, $AlBr_3$, etc., are also capable of initiating isobutylene on their own.[48–51] An unambiguous proof is, of course, difficult to obtain. Some supporting evidences are nevertheless available. For example, the reaction between $AlCl_3$ and the nonpolymerizable monomer, 1,1-diphenylethylene (1,1-DPE), in CH_2Cl_2 under high purity condition at $\leq -30°C$ generates the stable carbocation $Ph_2C^+CH_2-$ with a stoichiometry of one carbocation for two $AlCl_3$ molecules.[52] The course of reaction has been postulated to be as follows:

$$AlCl_3 + CH_2 = CPh_2 \rightarrow Cl_2Al - CH_2 - C(Ph)_2 - Cl \xrightarrow{AlCl_3}$$
$$Cl_2Al - CH_2 - \overset{+}{C}(Ph)_2[AlCl_4]^-.$$

6.3.1 *Controlled initiation*

Due to the presence of adventitious protogens, e.g., H_2O, in polymerization systems it is not possible to control initiation using strong Lewis acids (Friedel Crafts halides) as coinitiators. However, Kennedy and Baldwin discovered that certain dialkylaluminum halides, such as $AlEt_2Cl$, and trialkylaluminums, such as $AtEt_3$, readily coinitiate polymerization of isobutylene with an organic halide having a labile C-Cl bond used as initiator.[53,54] For example, with *t*-butyl chloride as the initiator, the reactions involved in the initiation process may be represented as shown in Scheme 6.2.

The ions shown in the reactions are not necessarily free. They may be paired as well depending on their concentrations, the medium, and the presence or the absence of salts

Ion generation

$$Al(C_2H_5)_3 + (CH_3)_3CCl \longrightarrow (CH_3)_3C^+ + [Al(C_2H_5)_3Cl]^- \quad (6.10)$$

Cationation

$$(CH_3)_3C^+ + CH_2=C\begin{subarray}{l}CH_3\\CH_3\end{subarray} \longrightarrow (CH_3)_3C-CH_2-C^+\begin{subarray}{l}CH_3\\CH_3\end{subarray} \quad (6.11)$$

Scheme 6.2 Controlled initiation using organohalide initiator.

with common anion. Trialkylaluminums are preferred to dialkylaluminum halides because they are weaker Lewis acids and not able to generate ions from adventitious protogens. Therefore, the head group (α end group) in the polymer is derived from the added initiator. In contrast, with dialkylaluminum chloride as the coinitiator partial initiation occurs with the adventitious protogens as initiators. Other metal chlorides or bromides, which are relatively weak Lewis acids, $e.g.$, $TiCl_4$, $SnCl_4$, BCl_3, and $SnBr_4$, are also suitable as coinitiators for controlled initiation. Organic chlorides such as t-chlorides, p-substituted benzyl chlorides, and substituted alkyl or allyl chlorides are suitable as initiators. The carbenium ions generated from these initiators are resonance stabilized and, accordingly, the C-Cl bonds in the initiators are liable to undergo heterolytic cleavage.

6.3.2 Controlled termination

Weak Lewis acids coinitiators also provide control on termination, which takes place by the transfer of a ligand anion from the counter anion to the carbocation as shown in reaction (6.12)

$$\sim\sim\sim C^+ [AlR_3Cl] \longrightarrow \sim\sim\sim C-R + AlR_2Cl. \quad (6.12)$$

When R is larger than methyl, termination also occurs partly by hydride ion transferred from β C–H in the R group as shown in reaction (6.13) for R = ethyl[55]

$$\sim\sim\sim C^+ [Al(C_2H_5)_3Cl]^- \longrightarrow \sim\sim\sim C-H + Al(C_2H_5)_2Cl + C_2H_4. \quad (6.13)$$

However, using the appropriate alkylaluminum coinitiator it is possible to functionalize the polymer end by the desired R group. Thus, polyisobutylenes with functional end groups such as vinyl,[56] phenyl,[57] and cyclopentadienyl[58] have been prepared by this method.

6.3.3 Chain transfer

It has been discussed in Sec. 6.2 that the most important chain transfer reaction in carbocationic polymerization is the monomer transfer, which is of two kinds: (i) direct transfer, which is first order in monomer, and (ii) indirect or spontaneous transfer, which is zero order in monomer. Both kinds may or may not work simultaneously in a polymerization.

The contributions of these reactions to the number average degree of polymerization ($\overline{DP}_n$) in nonliving polymerization may be determined using the Mayo equation.

The $\overline{DP}_n$ equation

The $\overline{DP}_n$ equation (also called the Mayo equation) may be derived by an approach analogous to that used in the case of free radical polymerization assuming a stationary state for the concentrations of the active centers, which may be both free ion and ion pair existing in equilibrium with each other.[59]

The following set of propagation, transfer, and termination reactions involving both free ion and ion pair as active centers are considered.

Initiation

$$\text{In} + \text{Coin} \longrightarrow R^+ + X^- \text{ and } R^+X^- \xrightarrow{M} P_1^+ + X^- \text{ and } P_1^+X^-. \quad (6.14)$$

Propagation

$$P_n^+ + M \xrightarrow{k_p^+} P_{n+1}^+, \qquad R_p^+ = k_p^+[P^+][M], \quad (6.15)$$

$$P_n^+X^- + M \xrightarrow{k_p^\pm} P_{n+1}^+X^-, \qquad R_p^\pm = k_p^\pm[P^+X^-][M], \quad (6.16)$$

$$\text{where } [P^+] = \sum_{n=1}^{\infty}[P_n^+] \text{ and } [P^+X^-] = \sum_{n=1}^{\infty}[P_n^+X^-]. \quad (6.17)$$

Transfer to monomer

i) direct

$$P_n^+ + M \xrightarrow{k_{tr,M}^+} P_n^= + P_1^+, \qquad R_{tr,M}^+ = k_{tr,M}^+[P^+][M]. \quad (6.18)$$

$$P_n^+X^- + M \xrightarrow{k_{tr,M}^\pm} P_n^= + P_1^+X^-, \qquad R_{tr,M}^\pm = k_{tr,M}^\pm[P^+X^-][M]. \quad (6.19)$$

ii) indirect or spontaneous

$$P_n^+X^- \xrightarrow{k_{tr}^\pm} P_n^= + HX,$$
$$HX + M \xrightarrow{fast} P_1^+X^-. \qquad R_{tr}^\pm = k_{tr}^\pm[P^+X^-]. \quad (6.20)$$

Termination

$$P_n^+ + X^- \xrightarrow{k_t^+} P_n, \qquad R_t^+ = k_t^+[P^+]^2. \quad (6.21)$$

$$P_n^+X^- \xrightarrow{k_t^\pm} P_n, \qquad R_t^\pm = k_t^\pm[P^+X^-]. \quad (6.22)$$

Scheme 6.3 Reaction elements in carbocationic polymerization.

In Scheme 6.3, M is the monomer, P_n is the polymer with $\overline{DP}_n = n$, $P_n^=$ is the dead polymer with terminal unsaturation, P_n^+ and $P_n^+X^-$ are the free ion and ion pair chain carrier respectively. Spontaneous elimination of proton from the free ion should be very slow requiring high activation energy and hence not included in the scheme. In addition, transfer reactions with solvent and polymer have not been included in view of the fact that monomer transfer is the predominant transfer reaction. The rate equations for the various reactions are also shown in the right side of the scheme. The square term in the rate equation for termination of the free ion arises due to the bimolecular character of reaction (6.21) and $[X^-]$ being equal to $[P^+]$. The concentration terms for free ion and ion pair in the various rate equations represent the total concentrations of the respective species of all sizes.

Cationic Polymerization

Assuming that a stationary state condition prevails

$$\overline{DP}_n = \frac{R_p}{R_t + R_{tr,M} + R_{tr}} = \frac{R_p^+ + R_p^\pm}{R_t^+ + R_t^\pm + R_{tr,M}^+ + R_{tr,M}^\pm + R_{tr}^\pm}. \quad (6.23)$$

Inverting Eq. (6.23) and subsequently substituting the rate terms from the respective equations given in Scheme 6.3 one obtains

$$\frac{1}{\overline{DP}_n} = \frac{k_{tr,M}^+[P^+] + k_{tr,M}^\pm[P^+X^-]}{k_p^+[P^+] + k_p^\pm[P^+X^-]} + \frac{k_t^+[P^+]^2 + (k_t^\pm + k_{tr}^\pm)[P^+X^-]}{k_p^+[P^+] + k_p^\pm[P^+X^-]} \cdot \frac{1}{[M]}. \quad (6.24)$$

A more amenable form of the equation may be obtained by writing the concentrations of free ion and ion pair in terms of the total concentration of chain carrier [P*], and the degree of dissociation α as follows.[56]

Consider the equilibrium

$$P^+X^- \underset{}{\overset{K_D}{\rightleftharpoons}} P^+ + X^-. \quad (6.25)$$

It follows that

$$K_D = [P^+]^2 / [P^+X^-] = \alpha^2 [P^*] / (1 - \alpha), \quad (6.26)$$

where $[P^*] = [P^+] + [P^+X^-]$ and α is the degree of dissociation.

Using Eq. (6.26) to eliminate the ionic species concentrations in Eq. (6.24) one gets

$$\frac{1}{\overline{DP}_n} = \frac{\alpha k_{tr,M}^+ + (1-\alpha)k_{tr,M}^\pm}{\alpha k_p^+ + (1-\alpha)k_p^\pm} + \frac{(1-\alpha)k_t^+ K_D + (1-\alpha)(k_t^\pm + k_{tr}^\pm)}{\alpha k_p^+ + (1-\alpha)k_p^\pm} \cdot \frac{1}{[M]}. \quad (6.27)$$

The equation is reduced to simpler forms for $\alpha = 0$ and $\alpha = 1$.

For $\alpha = 0$, the chain carriers are exclusively ion pairs, and $K_D = 0$. In this case, Eq. (6.27) reduces to

$$\frac{1}{\overline{DP}_n} = \frac{k_{tr,M}^\pm}{k_p^\pm} + \frac{(k_t^\pm + k_{tr}^\pm)}{k_p^\pm [M]}. \quad (6.28)$$

For $\alpha = 1$, the chain carriers are exclusively free ions. In this case, Eq. (6.27) reduces to (6.29) by substituting Eq. (6.26) for K_D and $[P^+]$ for $[P^*]$.

$$\frac{1}{\overline{DP}_n} = \frac{k_{tr,M}^+}{k_p^+} + \frac{k_t^+[P^+]}{k_p^+[M]}. \quad (6.29)$$

However, since both free ion and ion pair would be present in a real system, Eq. (6.27) should be used to determine monomer transfer constant for which $1/\overline{DP}_n$ is to be plotted against $1/[M]$ (Mayo Plot). However, the plot may be misleading if α changes with monomer concentration. Besides, polymerization should be restricted to low conversion, as discussed in Chap. 3. When this is not possible, the appropriate integrated but complex equation is to be used.[60]

Many organic chlorides initiated and weak Lewis acids coinitiated polymerizations of isobutylene give linear plots of $1/\overline{DP}_n$ vs. $1/[M]$, the lines passing through the origin.

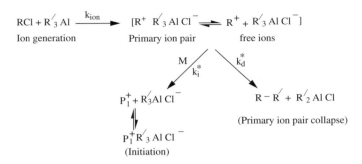

Scheme 6.4 Initiation and side reactions at the initiation stage.

Obviously, for such systems the monomer transfer is insignificant.[56,58,61] However, the analysis is valid only if α does not change with monomer concentration, as argued above.[56] This occurs only when the chain carrier concentration does not change with monomer concentration, i.e., the rate of initiation R_i is independent of [M]. Consider the competing reactions, viz., the initiation, and the primary ion pair collapse following ion generation from the initiator and the coinitiator, as shown in Scheme 6.4.[5]

In the scheme, k_i^* and k_d^* represent the composite rate constants (involving both free ion and ion pair) for initiation and collapse of primary ion pair respectively. The latter brings about wastage of the initiator and consequently a lowering of the initiator efficiency, f. For R_i to be independent of [M], the rate of primary ion pair collapse must be negligible. This makes f equal to unity. In quantitative terms, f is defined by Eq. (6.30).

$$f = k_i^*[M]/(k_i^*[M] + k_d^*). \tag{6.30}$$

The equation reduces to

$$f = 1, \quad \text{if } k_i^*[M] \rangle\rangle k_d^*, \tag{6.31}$$

and

$$f = k_i^*[M]/k_d^*, \quad \text{if } k_i^*[M] \langle\langle k_d^*. \tag{6.32}$$

When f is unity, R_i equals the rate of ionization of the initiator.

The above reasoning was used to explain the difference in the nature of the Mayo plots obtained in the trivinylaluminum (V_3Al) coinitiated polymerization of isobutylene using two different initiators, 3-chloro-1-butene ($3-Cl-1-C_4^=$) and t-butyl chloride (t-BuCl). As Fig. 6.3 shows, the Mayo plot is almost parallel to the 1/[M] axis when ($3-Cl-1-C_4^=$) is the initiator, which indicates that monomer transfer is predominant. In contrast, with t-BuCl as the initiator, the Mayo plot leaves only a small intercept on the $1/\overline{DP}_n$ axis indicating that monomer transfer occurs to only a small extent.[56]

This contradictory result is reconciled with the fact that f is much less than unity with 3-chloro-1-butene initiator but close to unity with t-butyl chloride initiator at relatively low

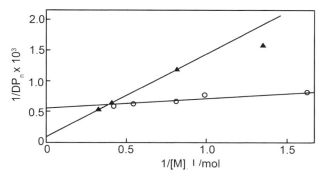

Figure 6.3 Mayo plots in the polymerization of isobutylene using 3-chloro-1-butene and t-butyl chloride as initiators and trivinylaluminum as coinitiator. (o) [3-Cl-1-$C_4^=$], $-72°C$; (▲) [t-BuCl], $-70°C$. Medium: CH_3Cl or CH_2Cl_2 (30% by volume) respectively, rest pentane + isobutylene. "Reprinted with permission from Ref. 56. Copyright © 1985 John Wiley & Sons, Inc."

Table 6.2 Initiator efficiency (f) in isobutylene polymerization coinitiated by trivinylaluminum.[a]

Initiator[b]	Concn. (mol/L × 10^3)	Al/Cl	Temp. (°C)	Yield (g)	(%)	$\overline{M}_n \times 10^{-2}$	f
3-Cl-1-$C_4^=$	2.50	10	−31	0.151	18.6	500	0.24
3-Cl-1-$C_4^=$	2.50	10	−40	0.273	33.1	698	0.31
3-Cl-1-$C_4^=$	2.50	10	−50	0.483	57.7	933	0.41
3-Cl-1-$C_4^=$	6.25	4	−50	0.747	89.2	761	0.31
t-BuCl	2.50	10	−30	0.398	48.9	392	0.81
t-BuCl	2.50	10	−40	0.545	66.1	403	1.10

[a] Recipe: i-C_4H_8 = 1.25 mL; CH_3Cl = 3.25 mL; n-pentane = 0.5 mL; [V_3Al] = 2.5×10^{-2} mol/L.
[b] 3-Cl-1-$C_4^=$ = 3-Chloro-1-butene, t-BuCl = t-butyl chloride. "Reprinted with permission from Ref. 56. Copyright © 1985 John Wiley & Sons, Inc."

temperature (Table 6.2). Hence, the application of the Mayo equation would be misleading in polymerization initiated by the former.

The difference in f for the two initiators may be attributed to the expectedly much lower reactivity of the allylic carbenium ion (**2**).

$$CH_3-\overset{\delta+}{CH}\overset{}{=\!=}CH\overset{}{=\!=}\overset{\delta+}{CH_2} \qquad\qquad CH_3-\underset{CH_3}{\overset{CH_3}{\underset{|}{\overset{|}{C}}}}+$$

(2)   (3)

Some evidence in support may be cited. The two ions **2** and **3** may be considered as models of polybutadiene$^+$ (PBD$^+$) and polyisobutylene$^+$ (PIB$^+$) cations respectively. The literature reports that the rate constant of the addition of PIB$^+$ to isobutylene (k_{11}) is about five orders of magnitude larger than that of the addition of a polyisoprene$^+$ (PiPr$^+$) cation (k_{21}) at $0°C$.[62]

$$\text{\small{CH}_2-\overset{\overset{\text{CH}_3}{|}}{\underset{\underset{\text{CH}_3}{|}}{C}}^+ + CH_2=(CH_3)_2 \xrightarrow{k_{11}} CH_2-\overset{\overset{\text{CH}_3}{|}}{\underset{\underset{\text{CH}_3}{|}}{C}}-CH_2-\overset{\overset{\text{CH}_3}{|}}{\underset{\underset{\text{CH}_3}{|}}{C}}^+}$$

$k_{11} = 1.5 \times 10^8$ L / mol. s at 0°C

$$CH_2 \overset{\delta+|}{\underset{}{-}}\overset{CH_3}{C} \rlap{=}{=} CH \rlap{=}{=} \overset{\delta+}{CH_2} + CH_2 = (CH_3)_2 \xrightarrow{k_{21}} CH_2 - \overset{\overset{\text{CH}_3}{|}}{C} = CH - CH_2 - CH_2 - \overset{\overset{\text{CH}_3}{|}}{\underset{\underset{\text{CH}_3}{|}}{C}}^+$$

$k_{21} = 3 \times 10^3$ L / mol. s at 0°C

Significantly, Mayr equation (Eq. 6.1) predicts only four times higher reactivity inasmuch as the electrophilicity of the two ions are not very different (Table 6.1). Nevertheless, the experimentally determined difference in reactivity is so great that one would expect the PIB$^+$ to be more reactive than PBD$^+$ also, even though the latter is more reactive than PiPr$^+$ in not having an electron-releasing methyl group at the allyl position.

6.3.4 *Molecular weight dependence on temperature of polymerization*

Molecular weight decreases with increase in temperature. From the $\overline{DP}_n$ equation (6.27), it would be an extremely formidable task relating quantitatively $\overline{DP}_n$ with temperature by way of expressing all the rate coefficients in their Arrhenius forms.

Nevertheless, Flory used the molecular weight-temperature data of Thomas et al.[42] in the polymerization of isobutylene induced by BF$_3$ and showed that the plot of log $\overline{M}_v$ vs. 1/T is linear.[63] From the slope of the straight line Flory calculated the activation energy ($E_{\overline{M}_v}$ = -19.3 kJ/mol). Subsequently, Kennedy and coworkers found that the Arrhenius plots for molecular weights ($\overline{M}_v$) fall into two distinct groups: (1) the high activation energy group ($E_{\overline{M}_v}$ = ~ -27.5 kJ/mol), and (2) the low activation energy group ($E_{\overline{M}_v}$ = ~ -4.9 kJ/mol).[64–66] The former group is represented by strong Lewis acid coinitiated polymerizations or the radiation induced polymerizations. The latter group is represented by weak Lewis acid coinitiated polymerizations.

Representative Arrhenius plots for the two groups in polymerizations of isobutylene are shown in Figs. 6.4 and 6.5 respectively.[66,56] The difference in activation energies was attributed to different chain breaking reactions being predominantly operative in the two groups. As discussed earlier (Sec. 6.3.2), the alkylaluminum coinitiated polymerizations are termination dominated,[56–58] whereas the strong Lewis acids coinitiated polymerizations are monomer-transfer dominated.

6.3.5 *Reversible termination*

Kennedy and coworkers discovered that polymerization of isobutylene induced by the H$_2$O/BCl$_3$ initiating system results in a chlorine-terminated polymer.[67] The initiating capability of the polymer was demonstrated by successfully using it as an initiator

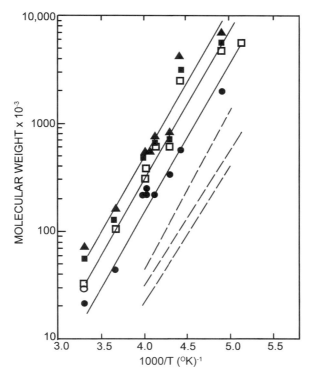

Figure 6.4 Arrhenius plots for molecular weight in the polymerization of isobutylene: initiation by γ radiation (full lines, the three lines are for M_w, M_v, and M_n); coinitiation by different strong Lewis acids (dotted lines, M_v). "Reprinted with permission from Ref. 66(a). Copyright © 1972 John Wiley & Sons Inc. and from Ref. 66(b), Fundamental studies on cationic polymerization IV-Homo and copolymerizations with various catalysts. Copyright © 1965 Elsevier Science Ltd."

in the polymerization of styrene with Et_2AlCl as the coinitiator. The product was a block copolymer, viz., polyisobutylene-b-polystyrene.[68] Subsequently, the appropriate combination of initiator and Lewis acid was developed achieving polymerization with reversible termination[69]

$$P_n^+ \; MtX_{n+1}^- \underset{k_i}{\overset{k_{-i}}{\rightleftarrows}} P_n-X \; + \; MtX_n ,$$

where MtX_n is a relatively weak Lewis acid, e.g., $TiCl_4$, $SnCl_4$, $SnBr_4$, BCl_3, $ZnCl_2$, I_2, etc., and the counteranion is high in nucleophilicity. The latter property plays a key role in eliminating monomer transfer as we have already seen in weak Lewis acid coinitiated polymerization of isobutylene.

Reversible termination is a key concept, which led to the discovery of living carbocationic polymerization and subsequently living radical polymerization[70] (Chap. 3, Sec. 3.15).

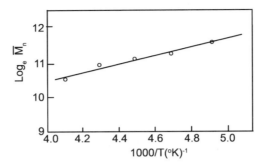

Figure 6.5 Arrhenius plot for molecular weight in a weak Lewis acid coinitiated polymerization of isobutylene; medium, CH_2Cl_2(30% by volume), rest pentane + isobutylene; initiator = t-butyl chloride (2.5 mmol L^{-1}), coinitiator = trivinylaluminun (25 mmol L^{-1}). "Reprinted with permission from Ref. 56. Copyright © 1985 John Wiley & Sons Inc."

6.3.6 Living carbocationic polymerization

The reaction sequences in living carbocationic polymerization involving reversible termination are shown in Scheme 6.5. In the scheme, RX is an organic halide initiator with a labile C-X bond and MtX_n is a relatively weak Lewis acid. The rate constants of the various reactions are written over the arrows between the reactants and products; those involved in initiation have been identified with the superscript 0. The ionic species has been shown as ion pair since living cationic polymerization is best achieved with it rather than with free ion (*vide infra*).

As discussed in the case of living radical polymerization (Sec. 3.15.3), the following conditions apart from the absence of transfer and termination are required to be fulfilled in

Initiation

$$R-X + MtX_n \underset{k^0_{-i}}{\overset{k^0_i}{\rightleftharpoons}} R^+ MtX^-_{n+1} \qquad K^0_i = k^0_i / k^0_{-i} \qquad (6.33)$$

$$R^+ MtX^-_{n+1} + M \xrightarrow{k'_i} P^+_1 MtX^-_{n+1} \qquad (6.34)$$

Propagation

$$P^+_1 MtX^-_{n+1} + M \xrightarrow{k^\pm_p} P^+_2 MtX^-_{n+1}$$
$$P^+_2 MtX^-_{n+1} + M \xrightarrow{k^\pm_p} P^+_3 MtX^-_{n+1} \text{ and so on} \qquad (6.35)$$

Reversble Termination

$$P^+_n MtX^-_{n+1} \underset{k_i}{\overset{k_{-i}}{\rightleftharpoons}} P_n-X + MtX_n, \qquad K_i = k_i / k_{-i} \qquad (6.36)$$

Scheme 6.5 Component reactions in living cationic polymerization.

order to achieve polymer with low PDI:

(i) Initiation should not be slower than propagation.
(ii) The deactivation equilibrium must be dynamic and the exchange between active and dormant states must be much faster than propagation so that a chain undergoes a large number of activation–deactivation cycles during the whole course of polymerization.
(iii) In addition, adventitious initiation notoriously associated with carbocationic polymerization should be absent or reduced to an insignificantly low level.

6.3.6.1 Selection of initiator

Two steps are involved in the initiation process (Scheme 6.5); ionization of the initiator with the help of the metal halide Lewis acid followed by the addition of the ionic species to the double bond in the monomer. The magnitude of the ionization constant depends on the strength of the R-X bond and that of the Lewis acid, the polarity of the medium, and the temperature. Weaker is the R-X bond or stronger is the Lewis acid larger is the K_i^0.

The literature reported values of the ionization constant (K_i^0), the rate constants of ionization (k_i^0), and the reverse reaction (k_{-i}^0) for selected systems are compiled in Table 6.3. It will be evident from the data in the table that K_i^0 increases with the increase in the strength of the Lewis acid: $SbF_5 > TiCl_4 > BCl_3$. Thus, cumyl chloride is almost fully ionized by SbF_5 in CH_2Cl_2 at $-70°C$[71] but only weakly ionized by the weak Lewis acid BCl_3 ($\sim 0.1\%$ at $-65°C$ with [cumyl chloride]$_0$ and [BCl_3] at 0.011 and 0.53 mol L^{-1} respectively[72]).

The ionization equilibrium is exothermic due to the negative enthalpy of solvation of the ionic species produced. Hence, K_i^0 increases with decrease in temperature. Also, it decreases with decrease in solvent polarity such that ionization does not occur in nonpolar solvents such as hydrocarbons. Hence, nonnucleophilic polar solvents, viz., CH_3Cl, CH_2Cl_2, or mixtures of these solvents with hydrocarbons, are used as polymerization media. However, with vinyl ether monomers, this demand on solvent polarity does not exist, polymerization being effected even in the nonpolar solvent toluene.[83,84] In this case, the solvation of the carbocation with the alkoxy groups in the monomer and in the polymer provides the necessary driving force. In case of the polymeric cation, an autosolvation with the alkoxy group in the penultimate unit predominates.[73,84] Even the much weaker Lewis acids, $ZnCl_2$ and I_2, act as effective coinitiators in the polymerizations of these monomers.

For initiation to be not slower than propagation, Eq. (6.37) must be satisfied.

$$K_i^0 k_i' \geq K_i k_p^{\pm}. \tag{6.37}$$

A standard initiator is usually a small molecular model of the polymeric halide for which $K_i^0 \approx K_i$ and $k_i' > k_p$, as in living radical polymerization (Secs. 3.15.4 and 3.15.6.1). The hydrohalogenated monomer is often used as the model initiator, e.g., 1-phenylethyl chloride for styrene, and (2,4,6-trimethylphenyl)-1-ethyl chloride for 2,4,6-trimethylstyrene.[85–87] However, this principle is not universally applicable. For example, t-butyl chloride is unsuitable for initiation of isobutylene polymerization.[61,88,89] This is attributable to the B-strain effect,[90–92] as discussed earlier in respect of the ATRP of methyl methacrylate using

Table 6.3 K_i, k_i and k_{-i} values for some monomer hydrochlorides and polymeric chlorides.

Halide[a]	Lewis acid	Solvent	T (°C)	K_i^0 or K_i Lmol^{-1}	k_i^0 or k_i Lmol^{-1}s^{-1}	k_{-i}^0 or k_{-i} s^{-1}	Ref.
Monomer hydrochlorides				K_i^0	k_i^0	k_{-i}^0	
HIBVECl	BCl$_3$	CH$_2$Cl$_2$	−78	80 ± 5	2.7×10^5	3×10^3	73
CumCl	TiCl$_4$	CH$_2$Cl$_2$	−77	27	−	−	74
(HαMStCl)	BCl$_3$	CH$_2$Cl$_2$	−65	5×10^{-4}	$\sim 10^{-1}$	< 160	72
	SbF$_5$	CH$_2$Cl$_2$	−62	−	$> 2.5 \times 10^5$	−	47
H(αMSt)$_2$Cl	BCl$_3$	CH$_2$Cl$_2$	−65	0.5	−	−	72
H(TMe)StCl	TiCl$_4$	CH$_2$Cl$_2$	−68	24	−	−	75
	BCl$_3$	CH$_2$Cl$_2$	−70	2.3×10^{-3}	3.2	1.4×10^3	76
	BCl$_3$	CH$_2$Cl$_2$	−40	7×10^{-4}	2.7	3.8×10^3	77
HIndCl	SbF$_5$	CH$_2$Cl$_2$	−68	2.6×10^3	$> 10^5$	80	78, 79
	TiCl$_4$	CH$_2$Cl$_2$	−65	13	$\geq 2.4 \times 10^3$	180	78
	SnCl$_4$	CH$_2$Cl$_2$	−40	1.5×10^{-2}	7.9	520	80
Polymeric chlorides				K_i	k_i	k_{-i}	
PIBCl	TiCl$_4$	Hex/MeCl (60/40)	−80	2.5×10^{-8}	0.58	2.3×10^7	81
PStCl	TiCl$_4$	MeChx/MeCl (60/40)	−80	4.4×10^{-9}	0.14	4.7×10^7	5
P(MeOSt)Cl	SnBr$_4$	CH$_2$Cl$_2$	−60	3.1×10^{-2}	−	−	83

[a] HIBVECl = 1 : 1 adduct of isobutyl vinyl ether and HCl, CumCl = Cumyl chloride, H(αMSt)$_2$Cl = 1:1 adduct of the dimer of α-methylstyrene and HCl, H(TMe)StCl = 1:1 adduct of 2,4,6-trimethylstyrene and HCl, HIndCl = 1-chloroindane. For the polymeric chlorides, PIB = polyisobutylene, PSt = polystyrene, and P(MeOSt) = poly(p-methoxystyrene). "Reprinted with permission from Ref. 47, Carbocationic polymerization: Mechanisms and kinetics of propagation reactions. Copyright © 2005 Elsevier Ltd."

model initiator (Chap. 3, Sec. 3.15.6.1). The steric strain released due to rehybridization from sp^3 to sp^2 configuration of the terminal carbon during ionization is larger for the polyisobutylene halide than for the t-butyl halide.[93–95]

The effect is decreased when the initiator is, 2-chloro-2,4,4-trimethylpentane (TMPCl), which is the adduct of HCl and the dimer of isobutylene.[88,89] Still further decrease occurs with the increase in the number of isobutylene units in hydrochlorinated isobutylene (IB) n-mers, H-(IB)$_n$-Cl. Accordingly, the higher mers are better initiators than TMPCl.[93–95] Similarly, the ionization constant of the hydrochlorinated dimer of α-methylstyrene with BCl$_3$ as the coinitiator is 1000 times larger than that of cumyl chloride (hydrochlorinated α-methylstyrene) in CH$_2$Cl$_2$ at −65°C (Table 6.3). Hence, the former is a good initiator for the living polymerization of α-methylstyrene, whereas the latter is not.[9,72]

Greater is the resonance stabilization of a carbocation, greater is the ionization of its precursor halide, but lesser is its rate of addition to a monomer. Thus, the trityl salts, Ph$_3$C$^+$SbCl$_6^-$ and Ph$_3$C$^+$SbF$_6^-$, although fully ionized, are very slow initiators. In addition, steric hindrance exerted by the three phenyl groups in the trityl cation makes initiation further slower.[106] Nevertheless, with the very reactive monomers, such as vinyl ethers[97] and N-vinylcarbazole,[98] they effect generally complete initiation.

In contrast to trityl chloride, cumyl chloride is a very good initiator for a large number of monomers because cumyl cation is not as unreactive as the trityl is, but its resonance stabilization is large enough for K_i^0 to be adequately large. In fact, with cumyl chloride as initiator and $TiCl_4$ as coinitiator, a rapid consumption of monomer occurs in the beginning of polymerization of isobutylene,[99] styrene,[100] or indene.[101] This has been attributed to much larger K_i^0 with cumyl chloride than K_i with these polymeric chlorides. This causes a large increase in ionic species concentration in the beginning of polymerization resulting in rapid consumption of monomer. The ionic species concentration is reduced and stabilized to a lower value as the growing chains undergo the first ion-pair collapse.

Compared to *tert*-alkyl chlorides and cumyl chloride, the corresponding alkanoates (mostly acetate)[102,103] or the methyl ethers[104,105] are more efficient initiators. For example, 2-chloro-2,4,4-trimethylpentane (TMPCl)/BCl_3 is a relatively slow initiating system but not 2,4,4-trimethyl-2-pentyl acetate (TMPOAc)/BCl_3.[102,103] TMPOAc ionizes at a much faster rate than TMPCl, acetate being a better leaving group than chloride, which results in a larger K_i^0.[106]

$$TMPOAc + BCl_3 \xrightleftharpoons{K_i^0} TMP^+AcOBCl_3^- \xrightarrow{nM} P_nCl + BCl_2OAc$$

$$P_n-Cl + BCl_3 \xrightleftharpoons{K_i} P_n^+ BCl_4^- \searrow TMPCl + BCl_2OAc$$

Equation (6.37) is satisfied and initiation is completed early in the polymerization, K_i^0 being greater than K_i. The polymer has only the chloride end group due to the fast replacement of acetate by chloride, which is more nucleophilic. The acetoxydichloroborane that forms in the exchange is too weak a Lewis acid to ionize TMPCl or P_n-Cl. Hence, an excess BCl_3 is required, which is a stronger Lewis acid.[106,107]

6.3.6.2 *Effect of added nucleophiles, proton traps, and common anion salts*

Faster initiation than propagation and reversible termination proved to be not good enough for obtaining living polymers with low polydispersity. For example, PDI >5 was reported for polystyrene prepared with cumyl acetate/BCl_3 initiator system in CH_3Cl diluent at $-30°C$,[108] and under the same conditions, polyisobutylene with PDI $\simeq 1.5$ was prepared.[102,103] However, addition of nucleophiles to the extent of ~ 10 mol% of the Lewis acid helps to narrow the molecular weight distribution. Examples of these nucleophiles include dimethylacetamide, dimethylsulfoxide, and ethyl acetate in the polymerizations of alkenes and arenes;[88,89,103,109,110] dioxane,[111a] ethyl acetate,[111b] and alkyl sulfides[112] in the polymerizations of vinyl ethers.

Regarding the role of the nucleophile, the Kennedy school held the view that it forms donor acceptor complex with the carbocation, as a result of which the reactivity of the latter is reduced. The carbocation thus made less reactive was believed to be less prone to chain transfer, and a narrower MWD is the result. The nucleophile, in effect, reduces the charge of the carbocation and consequently the acidity of the β proton, suppressing thereby chain transfer. The reactivity reduction is apparent from the decrease in the rate of polymerization that occurs.

The Higashimura school also suggested that the nucleophile reduces the reactivity of the carbocation, which is required for the living polymerization to take place when the counteranion is only weakly nucleophilic (*e.g.*, $Et_2AlClOH^-$). The added nucleophile may form either onium ion by reacting with the carbocation or solvate the latter (reaction 6.38).[111,114] Either way, the reactivity of the carbocation is reduced.

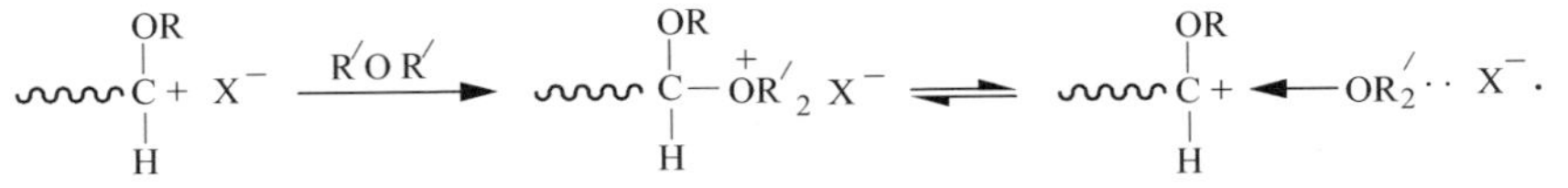

$$\underset{\text{ion pair}}{\overset{\overset{\displaystyle OR}{|}}{\underset{\underset{\displaystyle H}{|}}{\sim\sim\sim C}}+ X^-} \xrightarrow{R'OR'} \underset{\text{onium ion}}{\overset{\overset{\displaystyle OR}{|}}{\underset{\underset{\displaystyle H}{|}}{\sim\sim\sim \overset{+}{C}-OR'_2}}\,X^-} \rightleftharpoons \underset{\text{solvated ion}}{\overset{\overset{\displaystyle OR}{|}}{\underset{\underset{\displaystyle H}{|}}{\sim\sim\sim C+}}\leftarrow OR'_2 \cdots X^-} \qquad (6.38)$$

Reduction in carbocation reactivity may be effected also by suitable nucleophilic counteranion. In fact, the first successful living cationic polymerization was performed with isobutyl vinyl ether using a protonic acid initiator [HB (B = halogen, $CF_2CO_2^-$, *etc.*)], and a weak Lewis acid coinitiator (I_2, ZnX_2, *etc.*).[115,116] The real initiator is, of course, the adduct (**4**) of the protonic acid and the monomer as shown in Scheme 6.6. The Lewis acid ionizes the adduct to generate the initiating carbocation (**5**) stabilized by the nucleophilic binary counter anion, $B^{\delta-}\ldots MtX_n$, effecting reduction in carbocation reactivity.[115]

Besides nucleophiles, a hindered Lewis base such as 2,6-di-*t*-butylpyridine (D*t*BP), which is nonnucleophilic, also reduces PDI.[117–120] The base is believed to be incapable of forming complexes with Lewis acids, but it accepts proton.[119] When present in the polymerization system, it traps the proton eliminated in the unimolecular rate-controlling step of the spontaneous transfer reaction (Scheme 6.3) thereby preventing this type of monomer transfer. The net result is, of course, irreversible termination. However, at sufficiently low temperatures *ca.*, $-80°C$, proton elimination does not occur. The mechanism of action of D*t*BP in bringing about living polymerization has been attributed to that of the common anion salts it forms with the protic impurities generated in the reactions of Lewis acids with adventitious protogens.[119] These salts suppress the dissociation of ion pair by the common ion effect. In this respect, they act in a way similar to the added common ion salts discussed below.[121–125]

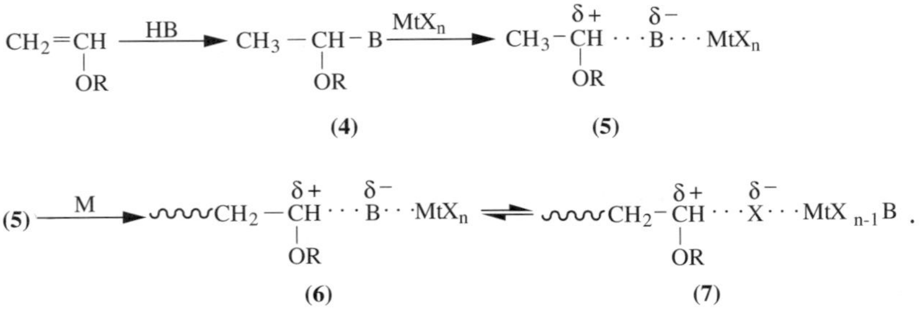

$$CH_2=\underset{\underset{\displaystyle OR}{|}}{CH} \xrightarrow{HB} CH_3-\underset{\underset{\displaystyle OR}{|}}{CH}-B \xrightarrow{MtX_n} CH_3-\overset{\delta+}{\underset{\underset{\displaystyle OR}{|}}{CH}}\cdots\overset{\delta-}{B}\cdots MtX_n$$

$$\qquad\qquad\qquad\qquad (4) \qquad\qquad\qquad (5)$$

$$(5)\xrightarrow{M}\sim\sim\sim CH_2-\overset{\delta+}{\underset{\underset{\displaystyle OR}{|}}{CH}}\cdots\overset{\delta-}{B}\cdots MtX_n \rightleftharpoons \sim\sim\sim CH_2-\overset{\delta+}{\underset{\underset{\displaystyle OR}{|}}{CH}}\cdots\overset{\delta-}{X}\cdots MtX_{n-1}B$$

$$\qquad\qquad\qquad (6) \qquad\qquad\qquad\qquad\qquad (7)$$

Scheme 6.6 Mechanism of living polymerization of alkyl vinyl ethers.[115,116]

However, Webster, Szwarc, Matyjaszewski, and Sigwalt favored a different explanation for the nucleophile effect, which is in keeping with the principles governing molecular weight distribution in reversible termination, as discussed in Chapters 3 and 4.[106,112,126,127] In contrast to the low reactivity assumed for the complexes of carbocations with nucleophiles by the Kennedy and Higashimura schools, these researchers considered the complexes dormant species existing in dynamic equilibrium with the active ionic species. Therefore, they effectively act in the same way as do the dormant polymeric chlorides, which form through reversible termination. The equilibria involved are[106]

$$P-Nu^+ \; MtCl_{n+1}^- \xrightleftharpoons{K_{nu}} P^+ \; MtCl_{n+1}^- + Nu \rightleftharpoons PCl + MtCl_n + Nu,$$

$$\text{dormant} \qquad\qquad \text{active} \qquad\qquad \text{dormant}$$

where P^+ is a carbocation (from initiator or polymer), and Nu is a nucleophile. Apart from the counteranion, the nucleophile also reversibly deactivates the propagating ion pair but at a faster rate. Thus, the deactivation frequency increases, and consequently, the number of exchanges between the dormant species and the active ionic species increases for a given DP and more is the number of exchanges lower is the PDI (Sec. 3.15.3.1, Chap. 3).

The common anion salt formed by complexation of metal halide Lewis acid with halide ion of added quaternary ammonium or phosphonium halide reduces the free ion concentration through suppression of the ion pair dissociation. The life time of free ion is much greater than that of ion pair since the former has to pass through the ion pair stage by way of associating with the oppositely charged ion before being deactivated.[106] The rate of association would be quite slow in view of the very low concentrations of the ions. Due to the greater life time the probability of undergoing the chain transfer reaction is greater for the free ion leading to the decrease in livingness.[106]

6.3.6.3 *Test of livingness*

The living nature of carbocationic polymerizations of several monomers was ascertained by various research groups (Table 6.4) from the following results.

(i) Monomer disappearance follows first order kinetics.
(ii) $\overline{M}_n$ increases linearly with conversion.
(iii) Initiator efficiency is close to unity.
(iv) The PDI is low *ca.*, ~1.1.
(v) Terminal unsaturation in polymer is absent and, in case of polystyrene, phenyl substituted indanyl end group (reaction 6.6) is also absent.
(vi) Indanylation of an aromatic initiator, such as cumyl chloride or acetate, when one such is used, also does not occur. In fact, it is completely suppressed in polymerizations conducted in relatively nonpolar medium such as the 1:1 (v/v) mixture of CH_2Cl_2 and n-hexane at temperatures $\leq -40°C$.[128]
(vii) Block copolymers are obtained using the living polymers as macroinitiators in the polymerizations of suitable monomers.

Table 6.4 Selected examples of living carbocationic polymerizations.

Monomer	Temperature °C	Solvent[a]	Initiator[b]	Coinitiator	Additive[c]	Ref.
Isobutylene	−80	Hex/MeCl (60/40)	TMPOAC, DiCumoMe	$TiCl_4$	Nil	133
	−80	Hex/MeCl (60/40)	CumoMe, DiCumoMe, TMPOMe	$TiCl_4$	DtBP	117
	−80	Hex/MeCl (60/40)	TMPCl	$TiCl_4$	DMA,DMSO	88,89
	−80	Hex/MeCl (60/40)	TMPCl	$TiCl_4$	DtBP	120
	−80	Hex/MeCl (60/40)	t-Bu-m-DCC	$TiCl_4$	DMP	134
	−80	Hex/CH_2Cl_2	TMPCl	$TiCl_4$	Bu_4NCl	122
Styrene	−15	CH_2Cl_2	HStX (X=Cl,Br)	$SnCl_4$	Bu_4NX (X=Cl,Br,I)	85
	−40 to −78	CH_2Cl_2	HStCl	$TiCl_3(Oi\text{-}Pr)$	Bu_4NCl	86
	−15	CH_2Cl_2	HStCl	$SnCl_4$	DtBP	5
	−50 to −80	MeCHX/MeCl (60/40)	Hp-MeStCl	$TiCl_4$	DtBP	5
	−80	MeCHX/MeCl (60/40)	TMPCl	$TiCl_4$	DMA+DtBP	110
	−80	HeX/MeCl (30/70)	TMPCl	$TiCl_4$	Bu_4NCl	135
p-Methoxy–styrene	30	H_2O	Hp-OMeStCl	$Yb(Otf)_3$	—	137
	−15 to 25	CH_2Cl_2	HI	I_2,ZnI_2	Bu_4NX (X=Cl,Br,I)	138
		Toluene	HI	ZnI_2	—	139
	−20 to −60	CH_2Cl_2	Hp-OMeStCl	$SnBr_4$	DtBP	82
p-Cl-styrene	−15 to 25	CH_2Cl_2	H StCl	$SnCl_4$	Bu_4NCl	140
	−80	MeCHX/MeCl (60/40)	Hp-ClStCl, Hp-MeStCl	$TiCl_4$	DtBP	141
	−80	MeCHX/MeCl (60/40)	TMPCl	$TiCl_4$	DMA + DtBP	135, 142
2,4,6-Trimethyl-styrene	−70	CH_2Cl_2	HTMeStCl	BCl_3	DtBP	77
	−20 to −70	CH_2Cl_2	HTMeStCl	BCl_3	DtBP	76
Indene	−70	CH_2Cl_2,CH_2Cl_2/hex (60/40)	CumOMe	$TiCl_4$	—	143
	−75	CH_2Cl_2	CumCl	$TiCl_4$	DMSO	144
	−40	CH_2Cl_2	CumCl	$SnCl_4$	DtBMP	101
	−67	CH_2Cl_2	CumCl	$TiCl_4$	—	129
Alkyl vinyl-ethers	−5 to −35	nonpolar	HI	I_2	—	146
	<0	CH_2Cl_2	HI	I_2	—	147
Butyl vinyl-ether	−15	Hexane	HI	I_2	—	116A, 146

(*Continued*)

Table 6.4 (Continued)

Monomer	Temperature °C	Solvent[a]	Initiator[b]	Coinitiator	Additive[c]	Ref.
Isobutyl vinyl ether	−40 to 25 −40	Toluene, CH_2Cl_2	HI	ZnX_2 (X=Cl,Br,I) or SnX_2 (X=Cl,I)	—	145, 148
	0	Toluene	RCOOH	$ZnCl_2$	—	148
	−40	Toluene	HB	$ZnCl_2$	—	115
	−78	CH_2Cl_2	HCl	$SnCl_4$	Bu_4NX, Bu_4PX	116
p-Methyl styrene	−30	MeCHX/MeCl (50/50)	TMPCl	BCl_3	Bu_4NCl + DtBP	135
	<0	Toluene, CH_2Cl_2	HI	ZnX_2	Bu_4NCl	136
	−15 to −70	CH_2Cl_2	Hp-MeStCl, HTMeStCl	$SnCl_4$	DtBP	131

[a]Hex = n-hexane, MeCHX = methylcyclohexane, solvent mixtures in v/v ratios given in parentheses; [b]CumCl = cumyl chloride, CumOAc = cumyl acetate, CumOMe = cumyl methyl ether, DiCumOMe = dicumyl methyl ether, TMPCl = 2-chloro-2,4,4-trimethylpentane, t-Bu-m-DCC = 5-t-butyl-1,3-bis(2-chloro-2-propyl)benzene, HStCl = 1-chloro-1-phenylethane, Hp-MeStCl = 1-chloro-1-(p-methylphenyl)ethane, Hp-OMeStCl = 1-chloro-1-(p-methoxyphenyl)ethane, Hp-ClStCl = 1-chloro-1-(p-chlorophenyl)ethane, HTMeStCl = 1-chloro-1-(2,4,6-trimethylphenyl)ethane, HB = HI, HCl or RCOOH (R = CF_3, CCl_3, $CHCl_2$, CH_2Cl, CH_3); [c]DMA = dimethylacetamide, DMSO = dimethyl sulfoxide, DtBP = di-t-butylpyridine, DMP = 2,4-dimethylpyridine, DtBPMP = 2,6-di-t-butyl-4-methylpyridine.

6.3.6.4 Kinetics

As in the case of living anionic polymerization, the instantaneous rate of monomer disappearance is given by

$$-\frac{d[M]}{dt} = k_p^{app}[P^*][M], \qquad (6.39)$$

where $k_p^{app} = \alpha k_p^+ + (1 - \alpha)k_p^{\pm}$ and $[P^*]$ = combined concentration of active centers comprising free ions and ion pairs of all sizes. In a living polymerization, $[P^*]$ remains constant throughout the time of polymerization. Integration of Eq. (6.39) provides

$$\ln \frac{[M]_0}{[M]} = k_p^{app}[P^*]t = k_{app}t, \qquad (6.40)$$

where $[M]_0$ and $[M]$ are the monomer concentrations at time zero and t respectively.

Thus, a plot of ln $[M]_0/[M]$ vs. time should be linear, the line passing through the origin. From the slope of the line, k_{app} may be determined. In many systems, polymerizations are fast so that rate measurements are possible only by using methods such as adiabatic calorimetry,[97,129] on-line ATR-FTIR spectroscopy,[119] stopped-flow UV-visible spectroscopy,[25-27] etc. Figure 6.6 reproduces the kinetic plots in the relatively slow polymerizations of isobutylene in the mixed solvent, hexane/CH_3Cl (60/40 v/v), at −25°, −40°, and −60°C using TMPCl/$TiCl_4$ as the initiating system and DtBP as the proton trap.[120] At −60°C, the plot is linear indicating early completion of initiation and absence

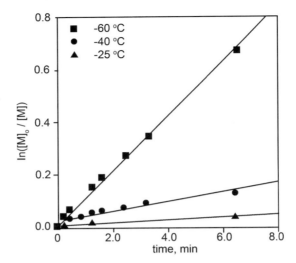

Figure 6.6 First order plots for monomer disappearance in living polymerizations of isobutylene in hexane/methyl chloride solvent mixture (60/40 v/v) at various temperatures indicated in the figure. Recipe: $[TMPCl]_0 = 2$ mmol L^{-1}, $[TiCl_4] = 36$ mmol L^{-1}, [2,6,di-t-butylpyridene] = 3 mmol L^{-1} and [isobutylene]$_0$ = 1.54 mol L^{-1}. "Reprinted with permission from Ref. 120. Copyright © 1998 American Chemical Society."

of irreversible termination. In contrast, at higher temperatures the plots yield lines, which are initially bent, indicating the presence of irreversible termination.

Monomer consumption follows first order kinetics also in the living polymerizations of many other monomers.[85,96,130–132]

6.3.6.5 Molecular weight and molecular weight distribution

Chain transfer brings about reduction in molecular weights. However, higher target molecular weights are more sensitive to the occurrence of chain transfer since the percent decrease in molecular weight for the same rate of transfer is larger longer is the kinetic chain length. The other reason why polymerization with high target molecular weights should be performed to detect the presence of chain transfer is that the reduced molecular weight is more accurately determined.[106] In the absence of chain transfer, the evolution of molecular weight with conversion should be linear (*vide* Eq. 6.7, with n = 1).

Figure 6.7 shows the molecular weight evolution in the polymerization of isobutylene in the presence or absence of the electron donor ethyl acetate with the target $\overline{M}_n$ of 45 000 and the monomer added in several portions.[89] Very good agreements between theory and experiment resulted with or without ethyl acetate.

As for the molecular weight distribution, the theoretical equation (6.41) relating PDI with conversion has been derived with solely ion pair as active center and relatively slow exchange between the dormant and the active species

$$\text{PDI} = 1 + \frac{1}{\beta}\left(\frac{2}{f_c} - 1\right), \quad \text{for } \overline{DP}_n \gg 1, \tag{6.41}$$

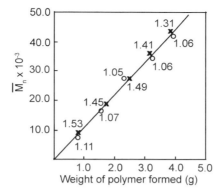

Figure 6.7 Molecular weight increase of polyisobutylene prepared by using CumCl/TiCl$_4$ initiator system with monomer added in increments (5 × 1 mL). The points are experimental values (x, without using any electron donor; o in the presence of ethyl acetate as electron donor); the numbers against each point is the PDI. The line represents the theoretical molecular weights. "Reprinted (in part) with permission from Ref. 89. Copyright © 1990 American Chemical Society."

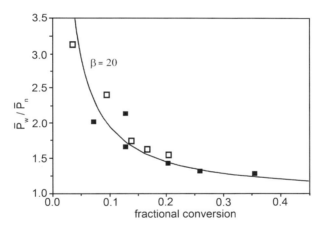

Figure 6.8 Variation of polydispersity index with monomer conversion in cationic polymerization of isobutylene in methylene chloride at −40°C using 2,4,4-trimethyl-2-pentyl chloride as initiator and BCl$_3$ as coinitiator in the presence of the proton trap di-t-butylphenol. Data (Ref. 151) plotted according to Eq. (6.41) with $\beta = 20$.[150] "Reprinted with permission from Ref. 150. Copyright © 1996 American Chemical Society."

where $\beta = k_{-i}/(k_p^{\pm})[I]_0$ and f_c is the fractional conversion.[150] The equation follows from the general PDI equation (Eq. 3.140) derived in Chap. 3 (Sec. 3.15.3.1) for living polymerization working on the principle of reversible deactivation by substituting k_{-i} for k_{deact} and $k_p^{\pm}$ for k_p. According to Eq. (6.41), PDI decreases from a markedly high value at very low conversions initially rapidly and then slowly to the final value (PDI $= 1 + 1/\beta$) at complete conversion.

Figure 6.8 reproduces a fitting[150] of the published data[151] to the equation with $\beta = 20$ and $[I]_0 = 1.16 \times 10^{-2}$ mol L^{-1}. However, there are systems in which PDI remains low with a value ≤1.1 throughout the period of polymerization conducted in the presence of a

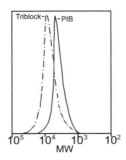

Figure 6.9 GPC traces of α,ω-dichloropolyisobutylene and triblock copolymer polystyrene-b-polyisobutylene-b-polystyrene. "Reprinted with permission from Ref. 152. Copyright © 1991 John Wiley & Sons, Inc."

nucleophile, but not otherwise (Fig. 6.7).[89] This result confirms that the exchange becomes still further faster than propagation in the presence of the nucleophile.

6.3.6.6 Block copolymerization

Living carbocationic polymerization induced by organic halide/Lewis acid initiating system yields polymer capped with halogen atom (Cl, Br, I). The polymer may be used, therefore, as a macroinitiator for the polymerization of other monomers to synthesize block copolymers. For example, synthesis of polystyrene-b-polyisobutylene-b-polystyrene, a commercially important triblock copolymer, was achieved by Kennedy et al. using this method.[152]

The synthesis was done in two stages. In the first stage, α,ω-dichloropolyisobutylene (Cl-PIB-Cl) was synthesized by the polymerization of isobutylene in hexane/CH_3Cl (60/40 v/v) at −80°C using the difunctional initiator 1, 4-di-(2-methoxy-2-propyl) benzene (dicumyl methyl ether) and the coinitiator $TiCl_4$. In the second stage, the difunctional Cl-PIB-Cl was used as the macroinitiator in the living polymerization of styrene using the same solvent, temperature, and coinitiator as were used in the first stage but in the additional presence of a proton trap (2,6-di-t-butylpyridine) and an electron donor (dimethylacetamide). The gpc traces of the macroinitiator and the triblock copolymer are reproduced in Fig. 6.9. As evidenced from the figure, molecular weight increases following block copolymerization. The MWD of the block copolymer is reasonably narrow and monomodal indicating complete utilization of the macroinitiator.[152] The triblock copolymer so prepared exhibits the properties of a thermoplastic elastomer at polystyrene content of 26 to 30%.

6.4 End Functionalized Polymers

The subject is discussed in respect of polyisobutylene.

6.4.1 Living polymerization method

The living carbocationic polymerization of isobutylene yields polymer with t-halogen terminus. Thus, using cumyl methyl ether (**8**), p-dicumyl methyl ether (**9**), tricumyl methyl ether (**10**), and 3,3′,5,5′-tetra(2-acetoxyisopropyl)biphenyl (**11**)[153] as the initiators and BCl_3

as the coinitiator, mono-, di-, tri, and tetra-*t*-chloro telechelic polyisobutylenes respectively were prepared.

Faster initiation effected by the initiator with acetate or methyl ether as the leaving group and excess BCl_3 as the coinitiator has been discussed earlier (Sec. 6.3.6.1). The functionality loss due to indanylation of the initiators is prevented by conducting polymerization at relatively nonpolar media and at temperatures $\geq -40°C^{128}$ (Sec. 6.3.6.3).

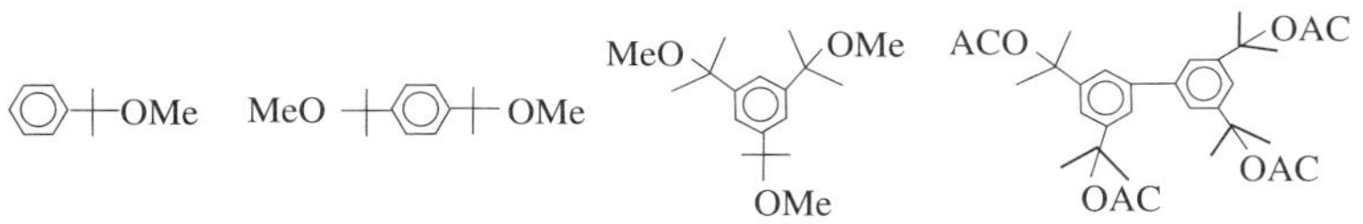

(8)   (9)   (10)   (11)

Dehydrochlorination of the first generation chlorotelechelics by potassium *t*-butoxide yields the corresponding second generation telechelics with isopropylidine end groups.[154]

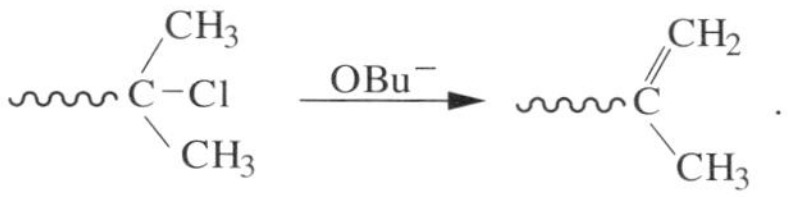

These in turn may be converted by suitable chemistry to a variety of third generation telechelics with terminal functional groups such as hydroxy, aldehyde, epoxy, sulfoxy, chlorosilyl, and phenolic.[155,156] A fourth generation telechelics may be obtained in turn from these polymers. For example, methacrylate-telechelic polyisobutylenes may be prepared from the hydroxy telechelics by treating the latter with methacryloyl chloride in the presence of triethylamine.[157]

Living polymers may also be "killed" by suitable reagents to obtain some other telechelics. Thus, methallyltrimethylsilane terminates the living polymerization of isobutylene producing polyisobutylene with methallyl end group.[158]

$$P_n^+ \; MtCl_{n+1}^- + CH_2 = \underset{CH_3}{\underset{|}{C}} - CH_2 - SiMe_3 \longrightarrow P_n - CH_2 - \underset{CH_3}{\underset{|}{C}} = CH_2 .$$

6.4.2 Inifer method

The primary or first generation chlorotelechelics may be prepared also by an alternative route. It involves slow initiation, reversible termination, and facile transfer to initiator. The initiator transfer agent is called "inifer", which is the short form of "initiator transfer".[159] A suitable inifer for isobutylene polymerization contains cumyl chloride moieties. This is because the larger resonance stabilization of the cumyl cation facilitates the transfer of the chloride ion to the polyisobutylene cation. However, in order to act as an inifer, initiation should be slower than propagation so that the initiator becomes available during the course of polymerization for transfer to take place. This is achieved with the use of the weak Lewis acid coinitiator BCl_3. The mechanism is shown in Scheme 6.7.

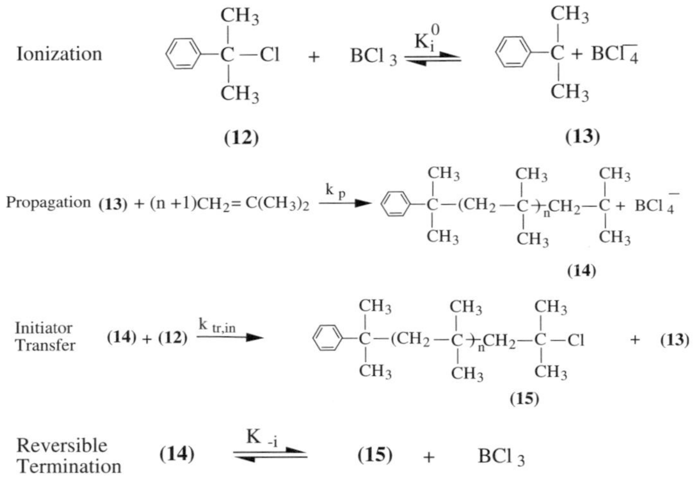

Scheme 6.7 Reactions scheme for chlorotelechelic polymer synthesis using inifer technique.

Thus, the 'inifer' method gives rise to t-Cl ended polymer and the molecular weight of the polymer is controlled by the rate of inifer reaction so that the number of polymer molecules produced by initiator transfer is much greater than that started by initiation.

Di- and tri- chloro telechelics were prepared by this method using 1,4-bis(2-chloro-2-propyl)benzene (p-dicumyl chloride) and 1,3,5-tris(2-chloro-2-propyl) benzene (tricumyl chloride) respectively as inifers.[159,160]

6.5 Photoinitiated Cationic Polymerization

Cationic photoinitiators used in industrial photocuring are onium salts, which generate protonic acid on photolysis in the presence of a hydrogen donor.[161,162] The protonic acid thus generated initiates polymerization of cationically polymerizable monomers. For example, the triarylsulfonium salt $Ar_3S^+MtF_6^-$ ($MtF_6^- = BF_4^-$, PF_6^-, or SbF_6^-) undergoes a homolytic C-S bond cleavage to form radicals upon irradiation with light of 190–365 nm wave lengths. The carbon radical and the sulfenium cation radical thus produced abstract hydrogen atom from the hydrogen donor substrate (RH) producing the strong protonic acid $HMtF_6$ and the substrate radical (R•)

$$Ar_3\overset{+}{S}\ PF_6^- \xrightarrow{h\nu} Ar_2\overset{+}{S}\ PF_6^- + \overset{\cdot}{Ar} \xrightarrow{2RH} Ar_2S + Ar + HPF_6 + 2\overset{\cdot}{R}.$$

Triarylsulfonium salts and, to a lesser extent, diaryliodonium salts are the preferred photoinitiators since they have the desired combination of properties, $e.g.$, high thermal stability and high quantum yield.[162]

The photodecomposition of the initiator occurs very fast so that only a very short irradiation time (<1 s) is required for photocuring.[162] The post irradiation cationic polymerization is also fast and completed in few seconds. Polyfunctional vinyl ethers and oxiranes are usually the monomers used[161,163–16]

6.6 Propagation Rate Constants

It has been experimentally established that when the counteranion is derived from a Lewis acid, $k_p^+ \approx k_p^\pm$ (*vide infra*). Accordingly, $k_p^{app} = k_p^+ = k_p^\pm = k_p$ (say). We shall, therefore, use the symbol k_p to address the rate constant of propagation in Lewis acid coinitiated polymerization, unless indicated otherwise.

Two methods have been used to determine k_p. In one of these, known as the ionic species concentration (ISC) method, the apparent rate constant of propagation ($k_{app} = k_p[P^*]$) is determined from the kinetic studies of polymerization (Sec. 6.3.6.4). The concentration of the active ionic species ($[P^*]$) is also measured by an appropriate method leading to the determination of k_p from k_{app}. In the other method, either the competition between the monomer (rate constant k_p) and an added stronger π-nucleophile (rate constant k_e) for reaction with the ionic species or the equilibria involved in capping the ionic species by 1,1-diarylethylenes are studied. These studies lead to the determination of k_p/k_e from which k_p is evaluated with the value of k_e estimated to be equal to that of a diffusion-limited reaction. This is known as the diffusion clock (DC) method. However, with carbocations which are either sterically hindered (*e.g.*, 2,4,6-trimethylstyrene cation) or resonance stabilized (*e.g.*, *p*-methoxystyrene) k_e is not limited by diffusion. In such cases, a weaker π-nucleophile is used in the competition and k_e is separately determined in independent experiments by measuring the rate of the reaction of the carbocation with the nucleophile.[76,82] The k_p values determined for various monomers are compiled in Table 6.5. The diffusion clock method of determining k_p based on competition experiment is outlined below.

6.6.1 *The diffusion clock method*

6.6.1.1 *Competition experiment (method 1)*

This method of determining k_p was pioneered by Mayr and Roth[170] who polymerized isobutylene in CH_2Cl_2 at $-78°C$ using the hydrochlorinated dimer (HDIBCl or TMPCl) or trimer (HTIBCl) of isobutylene as the initiator (In) and $AlCl_3$ or $TiCl_4$ as the coinitiator in the presence of allyltrimethylsilane (ATMS) or methallyltrimethylsilane (MATMS). The competition between the latter and the monomer may be represented as in Scheme 6.8.

All chains are terminated with the methallyl group. Measurement of the concentration of the first terminated product, $TMP-CH_2C(CH_3)=CH_2$, by gas chromatography led to the evaluation of k_e/k_p using Eq. (6.42) provided all the initiator is used up before the polymerization stops.

$$\frac{k_e}{k_p} = \frac{[TMP - CH_2C(CH_3)=CH_2]}{[In]_0 - [TMP - CH_2C(CH_3)=CH_2]}, \quad (6.42)$$

where $[In]_0$ is the initial concentration of the initiator TMPCl. With k_e/k_p thus measured, k_p is calculated using $k_e = 3 \times 10^9$ L mol^{-1} s^{-1} (diffusion-limited value).

Table 6.5 Propagation rate constants in cationic polymerization.[180]

Monomer	Solvent[a]	Initiator[b]	Temperature °C	Method[c]	k_p L mol^{-1}s^{-1}	$k_p^{\pm}$ L mol^{-1}s^{-1}	k_p^{+} L mol^{-1}s^{-1}	Ref.
Isobutylene	Nil	γ-ray	0 to −78	Kinetics	—	—	1.5×10^8	166
	CH$_2$Cl$_2$	γ-ray	−78	Kinetics	—	—	1.8×10^8	167
	MeCl/Hex (40/60)	RCl/LA	−80	DC	—	$(4.2-4.7) \times 10^8$	$(6.2-6.5) \times 10^8$	168, 169
Styrene	Hex,CH$_2$Cl$_2$	RCl/LA	−80	DC	—	—	$(6 \pm 2) \times 10^8$	170
	CH$_2$Cl$_2$	HClO$_4$	−80	ISC	5×10^4	2×10^3	2×10^4	21, 26
	CH$_2$Cl$_2$	CF$_3$SO$_3$H	−1	ISC	10^3	—	—	29
			−62	ISC	—	—	—	27, 28
	CH$_2$Cl$_2$	γ-ray	30	Kinetics	—	—	—	171
		γ-ray	15		—	—	3.5×10^6	172
p-Chlorostyrene	CH$_2$Cl$_2$	RCl/LA	−50 to −80	DC	—	2×10^9	—	5
	MeCl/MeCHX (40/60)	RCl/LA	−80	DC	—	3×10^9	—	141
p-Methylstyrene	MeCl/MeCHX (40/60)	RCl/LA	−15 to −70	DC	—	1×10^9	—	131
p-Methoxystyrene	CH$_2$Cl$_2$	CF$_3$SO$_3$H	30	ISC	1.3×10^5	—	—	173
		Trityl salt	−15	ISC	7.4×10^4	—	—	174
		RCl/LA	−60	ISC + Competition (NDC)	—	3.8×10^4	—	82
2,4,6-Trimethyl-styrene	CH$_2$Cl$_2$	RCl/LA	−70	ISC	—	1.4×10^4	—	76,77
	CH$_2$Cl$_2$	RCl/LA	−70	Competition (NDC)	—	8.3×10^4	—	76
Indene	CH$_2$Cl$_2$	RCl/LA	−40	ISC	—	10^5	—	101
Isobutyl vinyl ether	CH$_2$Cl$_2$	γ-ray	0	Kinetics	—	—	4×10^4	175, 176
Isobutyl vinyl ether	CH$_2$Cl$_2$	Trityl salts	0	ISC	6.8×10^3	—	—	97a
Ethyl vinyl ether	CH$_2$Cl$_2$	Trityl salts	0	ISC	7×10^3	—	—	97b
N-Vinylcarbazole		—	−40	Kinetics	$(1-2.4) \times 10^4$	—	—	177
			0	Kinetics	$(2.2-4.6) \times 10^5$	—	—	178
			20	Kinetics	9.5×10^5	—	—	179

[a]MeCl = methyl chloride, Hex = hexane, MeCHX = methylcyclohexane, mixed solvent compositions in v/v; [b]RCl = organic chloride, LA = Lewis acid; [c]Kinetics = using kinetic equation for R_p of nonliving polymerization, ISC = ionic species concentration, DC = diffusion clock, competition (NDC) = competition reaction with a nucleophile at a rate lower than the diffusion-limited.

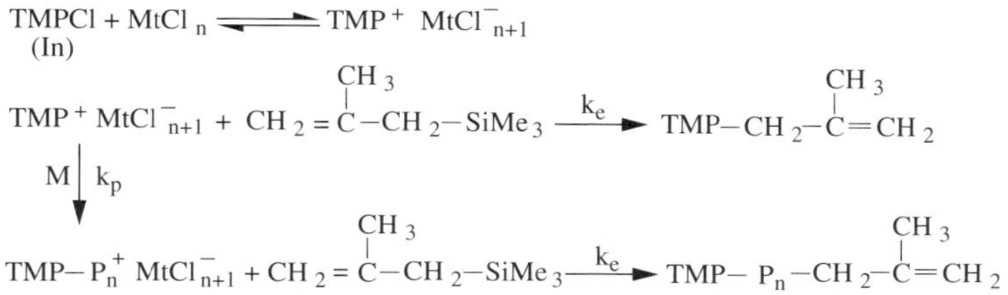

$$\text{TMPCl} + \text{MtCl}_n \rightleftharpoons \text{TMP}^+ \text{MtCl}_{n+1}^-$$
(In)

$$\text{TMP}^+ \text{MtCl}_{n+1}^- + \text{CH}_2=\underset{\underset{\text{CH}_3}{|}}{\text{C}}-\text{CH}_2-\text{SiMe}_3 \xrightarrow{k_e} \text{TMP}-\text{CH}_2-\underset{\underset{\text{CH}_3}{|}}{\text{C}}=\text{CH}_2$$

$$\text{M} \downarrow k_p$$

$$\text{TMP}-\text{P}_n^+ \text{MtCl}_{n+1}^- + \text{CH}_2=\underset{\underset{\text{CH}_3}{|}}{\text{C}}-\text{CH}_2-\text{SiMe}_3 \xrightarrow{k_e} \text{TMP}-\text{P}_n-\text{CH}_2-\underset{\underset{\text{CH}_3}{|}}{\text{C}}=\text{CH}_2$$

Scheme 6.8 Competition reaction between methallyltrimethylsilane and isobutylene (M) for the diisobutylene cation (TMP+).[170]

6.6.1.2 Competition experiment (method 2)

This method is based on the measurement of the yield or the molecular weight of the polymer produced in the competition experiment. The rates of consumption of the monomer and the nucleophile in a living polymerization with fast and complete initiation may be expressed as follows.

$$-\frac{d[M]}{dt} = k_p[P^*][M], \qquad (6.43)$$

and

$$-\frac{d[Nu]}{dt} = k_e[P^*][Nu]. \qquad (6.44)$$

Integration of Eqs. (6.43) and (6.44) gives respectively

$$\ln \frac{[M]_0}{[M]_t} = k_p \int_0^t [P^*]dt, \qquad (6.45)$$

and

$$\ln \frac{[Nu]_0}{[Nu]_t} = k_e \int_0^t [P^*]dt. \qquad (6.46)$$

Dividing Eq. (6.45) by (6.46) one gets for $t = \infty$ (completion of reaction)

$$\frac{k_p}{k_e} = \frac{\ln([M]_0/[M]_\infty)}{\ln([Nu]_0/[Nu]_\infty)} = \frac{\ln(1-f_{c,\infty})}{\ln(1-[In]_0/[Nu]_0)}, \qquad (6.47)$$

where $f_{c,\infty}$ is the final fractional conversion and $[In]_0$ is the initial concentration of the initiator, and $[Nu]_\infty = [Nu]_0 - [In]_0$ since all chains are terminated by the nucleophile for $[Nu]_0 \gg [In]_0$.

Alternatively, $f_{c,\infty}$ in Eq. (6.47) may be substituted using the DP equation

$$\overline{DP}_{n,\infty} = f_{c,\infty}[M]_0/[In]_0, \qquad (6.48)$$

where $\overline{DP}_{n,\infty}$ is the degree of polymerization at the end of the polymerization. Substituting from Eq. (6.48) for $f_{c,\infty}$ in Eq. (6.47) gives

$$\frac{k_p}{k_e} = \frac{\ln(1 - \overline{DP}_{n,\infty}[In]_0/[M]_0)}{\ln(1 - [In]_0/[Nu]_0)}. \qquad (6.49)$$

Both Eqs. (6.47) and (6.49) have been used to determine k_p. The values are in close agreement with each other. If the polymerization is conducted in the presence of a proton trap, the measured rate constant of propagation is $k_p^{\pm}$ inasmuch as the ionic species is solely ion pair, whereas in the absence of a proton trap or of a common anion salt the ionic species is solely free ion and the measured rate constant is k_p^{+}.[169]

6.6.2 Values of rate constants of propagation

The k_p values for several monomers are given in Table 6.5. For four monomers, *viz.*, isobutylene, styrene, *p*-methoxystyrene, and 2,4,6-trimethylstyrene, more than one method have been used. Checks for the agreement between the methods for the values are desirable for reliability. With isobutylene, both the DC method applied to chemically induced polymerization and an indirect ISC method applied to the radiation induced polymerization give the same order of $k_p \approx 10^8$ to 10^9 L mol^{-1} s^{-1}. In respect of the ISC method, a lower value (0.9×10^4 L mol^{-1} s^{-1}) previously calculated by Plesch[167a] using the data of Hayashi, Okamura and coworkers[180] was revised by Williams[167b] to 1.8×10^8 L mol^{-1} s^{-1}. With styrene, the k_p determined by the DC method is 5 to 6 orders of magnitude larger than the value determined by the ISC method. The lower value obtained by the ISC method may be due to an overestimate of the PS cation concentration.[47]

It is also observed that k_p^{+} for polyisobutylene cation is of the same magnitude as $k_p^{\pm}$ with $Ti_2Cl_9^{-}$ as the counteranion. This is in sharp contrast with the large difference observed between k_p^{-} and $k_p^{\pm}$ in the anionic polymerization of styrene. For example, in THF, $k_p^{-} = 65\,000$ and $k_p^{\pm} = 22$ (with Cs^+ counter ion) at 25°C, both values in L mol^{-1}s^{-1} unit.[181] This difference in behavior between the cationic and the anionic polymerizations is attributable to the interionic distance in ion pair in the case of cationic polymerization being much larger due to the much larger size of the counterions. For instance, the crystal radii of $SbCl_6^{-}$ and $CF_3SO_3^{-}$ ions are 3 and 2.96 Å respectively, whereas those of Na^+ and Cs^+ ions are 0.97 and 1.67 Å respectively.[47]

References

1. H. Mayr, M. Patz, *Angew Chem. Int. Ed. Engl.* (1994), 33, 938.
2. H. Mayr, B. Kempf, A.R. Ofial, *Acc. Chem. Res.* (2003), 36, 66.
3. H. Mayr, Rate constants and reactivity ratios in carbocationic polymerizations. Ionic polymerization and related processing, NATO science series E, Vol. 359 Dordecht: Kluwer, p. 99 (1999).
4. H. Schimmel, A.R. Ofial, H. Mayr, *Macromolecules* (2002), 35, 5454.

5. P. De, R. Faust, H. Schimmel, A.R. Ofial, H. Mayr, *Macromolecules* (2004), 37, 4422.
6. H. Mayr, M. Patz, *Macromol. Symp.* (1996), 107, 99.
7. H. Mayr, A.R. Ofial, H. Schimmel, *Macromolecules* (2005), 38, 33.
8. D.C. Pepper, P.J. Reilly, *Proc. Chem. Soc.* (1961), 460.
9. S. Bywater, D.J. Worsfold, *Can. J. Chem.* (1966), 44, 1671.
10. J.P. Kennedy, Cationic Polymerization of Olefins: A Critical Inventory, Wiley-Interscience, 1975.
11. J.P. Kennedy, E. Marechal, Carbocationic Polymerization, Wiley Interscience, N.Y, USA; Krieger, p. 85 (1982).
12. A.V. Buterlov, *Ann. Chem.* (1877), 189, 47.
13. A. Guterbok in Polyisobutylene, Springer, Berlin (1959).
14. G.D. James, *In the Chemistry of Cationic Polymerization*, P.H. Plesch Ed., Pergamon Press, Oxford, p. 542 (1963).
15. D.C. Pepper, *Quart. Rev.* (1954), 8, 88.
16. M.J. Hayes, D.C. Pepper, *Proc. Roy Soc.* (1961), A263, 63.
17. D.C. Pepper, P.J. Reilly, *J. Polym. Sci.* (1962), 58, 639.
18. D.C. Pepper, P.J. Reilly, *Proc. Roy Soc.* (1966), A291, 41.
19. L.E. Darcy, W.P. Millrine, D.C. Pepper, *Chem. Commun.* (1968), 1441.
20. D.C. Pepper, *Makromol. Chem.* (1974), 175, 1077.
21. D.C. Pepper, Macromol. *Chem. Phys.* (1995), 196(3), 963.
22. A. Gandini, P.H. Plesch, *Proc. Chem. Soc.* (1964), 113, 264.
23. A. Gandini, P.H. Plesch, *Proc. Chem. Soc.* (1964), 240.
24. A. Gandini, P.H. Plesch, *J. Chem. Soc.* (1965), 4826.
25. M. DeSorgo, D.C. Pepper, M. Szwarc, *Chem. Commun.* (1973), 419.
26. J.P. Lorimer, D.C. Pepper, *Proc. Roy. Soc. A* (1976), 351, 551.
27. J.P. Vairon, A. Rives, C. Bunel, *Makromol. Chem. Macromol. Symp.* (1992), 60, 97.
28. J.P.Vairon, B. Charleux, M. Moreau, Stopped-flow technique and cationic polymerization kinetics. Ionic polymerizations and related processes of NATO science series E, Dordecht: Kluwer, Vol. 359, p. 177 (1999).
29. T. Kunitake, K. Takarabe, *Macromolecules* (1979), 12, 1061.
30. P.H. Plesch, *Makromol. Chem. Macromol. Symp.* (1988), 13/14, 375, 393.
31. K. Matyjaszewski, *Makromol. Chem. Macromol. Symp.* (1988), 13/14, 389.
32. K. Matyjaszewski, P. Sigwalt, *Makromol. Chem.* (1986), 187, 2299.
33. M. Szwarc, *Macromolecules* (1995), 28, 7309.
34. M. Szwarc, *Macromolecules* (1995), 28, 7312.
35. K. Matyjaszewski, C.-H. Lin, *J. Polym. Sci. A* (1988), 26, 3031.
36. K. Matyjaszewski, *Makromol. Chem. Macromol. Symp.* (1988), 13/14, 433.
37. T. Higashimura, M. Sawamoto, *Macromolecules* (1992), 25, 2587.
38. T. Higashimura, M. Sawamoto, *Macromolecules* (1995), 28, 3747.
39. M. Otto, M. Muller-Cunradi, U.S. Pat. 2, 203, 873 to I.G. Ferbenindustrie, July 26 (1931).
40. J.P. Kennedy, Ref. 10, p. 10.
41. R.M. Thomas, R.W. Sparks, U.S. Pat. 2, 356, 128 to Standard Oil Development Co. (1944).
42. R.M. Thomas, W.J. Sparks, P.K. Frolich, M. Otto, M. Muller-Cunradi, *J. Am. Chem. Soc.* (1940), 62, 276.
43. A.G. Evans, D. Holden, P.H. Plesch, M. Polyani, H.A. Skinner, W.A. Weinberger, *Nature* (1946), 157, 102.
44. A.G. Evans, G.W. Meadows, M. Polyani, *Nature* (1946), 158, 94.
45. A.G. Evans, M. Polyani, *J. Chem. Soc.* (1947), 252.
46. P.H. Plesch, M. Polyani, H.A. Skinner, *J. Chem. Soc.* (1947), 257.
47. P. Sigwalt, M. Moreau, *Prog. Polym. Sci.* (2006), 31, 44.

48. H. Cheradame, P. Sigwalt, *Compt. Rend.* (1964), 259, 4273.
49. H. Cheradame, P. Sigwalt, *Bull Soc. Chim. France* (1970), 3, 843.
50. N.A. Ghanem, M. Marek, *Eur. Polym. J.* (1972), 8, 999.
51. M. Chmelir, M. Marek, O. Wichterle, *J. Polym. Sci. C* (1967), 16, 833.
52. M. Masure, G. Sauvet, P. Sigwalt, *J. Polym. Sci. Polym. Chem.* (1978), 16, 3065.
53. J.P. Kennedy, Belgian Pat, 663, 319 (April 30, 1965).
54. J.P. Kennedy, F.P. Baldwin, Belgian Pat. 663, 320 (April 30, 1965).
55. J.P. Kennedy, S. Rengachary, *Fortschr. Hochpolym. Forsch.* (1974), 14, 1.
56. B.M. Mandal, J.P. Kennedy, *J. Polym. Sci. Polym. Chem.* Ed. (1978), 16(4), 833.
57. J.P. Kennedy, D. Chung, *J. Polym. Sci. Polym. Chem.* Ed. (1981), 19, 2729.
58. P. Kennedy,K.F. Castner, *J. Polym. Sci. Polym. Chem.* Ed. (1979), 17, 2055.
59. P.H. Plesch, Progress in High Polymers Vol. 2, J.C. Robb and W.H. Peaker, Eds., Heywood Books, London, p. 137 (1968).
60. Y. Sakurada, T. Higashimura, S. Okamura, *J. Polym. Sci.* (1958), 33, 496.
61. J.P. Kennedy, S.Y. Huang, S.C. Feinberg, *J. Polym. Sci. Polym. Chem.* Ed. (1977), 15, 2801.
62. F. Williams, A. Shinkawa, J.P. Kennedy, *J. Polym. Sci. Poly. Symp.* Ed. (1976), 56, 421.
63. P.J. Flory in Principles of Polymer Chemistry, Cornell University Press, Ithaca, New York, p. 218.
64. J.P. Kennedy, J.K. Gillham, *Adv. Polym. Sci.* (1972), 10, 1.
65. Ref. 10, p. 125, 131, 135 (for values of E)
66. (a) J.P. Kennedy, A. Shinkawa, F. Williams, *J. Polym. Sci., A-1* (1971), 9(6), 1551
    (b) J.P. Kennedy, R.G. Squires, Polymer (1965), 6 (11), 579.
67. J.P. Kennedy, S.C. Feinberg, S.Y. Huang, *J. Polym. Sci. Polym. Chem.* Ed. (1977), 15, 2869.
68. J.P. Kennedy, S.Y. Huang, S.C. Feinberg, *J. Polym. Sci. Polym. Chem.* Ed. (1978), 16, 243.
69. R. Faust, A. Fehervari, J.P. Kennedy, *J. Macromol. Sci. Chem.* (1982–83), A18, 1209.
70. J.P. Kennedy, *J. Polym. Sci. Polym. Chem.* Ed. (1999), 37, 2285.
71. K. Matyjaszewski, P. Sigwalt, *Macromolecules* (1987), 20, 2679.
72. R. Russell, M. Moreau, B. Charleux, J.P. Vairon, K. Matyjaszewski, *Macromolecules* (1998), 31, 3775.
73. K. Matyjaszewski, M.Teodorescu, C-H. Li n, *Macromol. Chem. Phys.* (1995), 196, 2149.
74. M. Moreau, *Intl. Symp. Ionic Polym. IP'97* (1997), p. 135.
75. M. Moreau, unpublished result quoted in P. Sigwalt, M. Moreau, *Prog. Polym. Sci.* (2006), 31, 44.
76. P. De, L. Sipos, R. Faust, M. Moreau, B. Charleux, J.P. Vairon, *Macromolecules* (2005), 38, 41.
77. J.P. Vairon, M. Moreau, B. Charleux, A. Cretol, R. Faust, *Macromol. Chem. Macromol. Symp.* (2002), 183, 43.
78. M. Givechi, Ph.D. Thesis, Paris 1999 quoted in P. Sigwalt, M. Moreau, *Prog. Polym. Sci.* (2006), 31, 44.
79. M. Givechi, A. Polton, M. Tardi, M. Moreau, P. Sigwalt, J.P. Vairon, *Macromol. Chem. Macromol. Symp.* (2000), 157, 77.
80. A. Polton, M. Moreau, P. Sigwalt quoted in P. Sigwalt, M. Moreau, *Prog. Polym. Sci.* (2006), 31, 44.
81. L. Sipos, P. De, R. Faust, *Macromolecules* (2003), 36, 8282.
82. P. De, R. Faust, *Macromolecules* (2004), 37, 7930.
83. W.C. Hsieh, A. Diffeux, D.R. Squire, V. Stannet, *Polymer* (1982), 23, 427.
84. A. Deffieux, W.C. Hsieh, D.R. Squire, V. Stannet, *Polymer* (1981), 22, 1575.
85. T. Higashimura, Y. Ishihama, M. Sawamoto, *Macromolecules* (1993), 26, 744.
86. T. Hasebe, M. Kamigaito, M. Sawamoto, *Macromolecules* (1996), 29, 6100.
87. P. De, R. Faust, H. Schimmel, A.R. Ofial, H. Mayr, *Macromolecules* (2004), 37, 4422.
88. G. Kaszas, J.E. Puskas, C.C. Chen, J.P. Kennedy, *Polym. Bull* (1988), 20, 413.

89. G. Kaszas, J.E. Puskas, C.C. Chem, J.P. Kennedy, *Macromolecules* (1990), 23, 3909.
90. H.C. Brown, R.S. Fletcher, *J. Am. Chem. Soc.* (1949), 71, 1845.
91. H.C. Brown, A. Stern, *J. Am. Chem. Soc.* (1950), 72, 5068.
92. H.C. Brown, H.L. Berneis, *J. Am. Chem. Soc.* (1953), 75, 10.
93. H. Mayr, M. Roth, R. Faust, *Macromolecules* (1996), 29, 6110.
94. M. Roth, M. Patz, H. Freter, H. Mayr, *Macromolecules* (1997), 30, 722.
95. H. Schlaad, Y. Kwon, R. Faust, H. Mayr, *Macromolecules* (2000), 33, 743.
96. Zs. Fodor, R. Faust, *J. Macromol. Sci.* (1998), A35, 375.
97. (a) F. Subira, G. Sauvet, J.P. Vairon, P. Sigwalt, *J. Polym. Sci. Polym. Symp.* (1976), 56, 221.
    (b) F. Subira, J.P. Vairon, P. Sigwalt, *Macromolecules* (1988), 21, 2339.
98. G. Sauvet, J.P. Vairon, P. Sigwalt, *C.R. Acad. Sci. Paris* (1967), 265, 1090.
99. R.F. Storey, A.B. Donnalley, *Macromolecules* (1999), 32, 7003.
100. R.F. Storey, Q.A. Thomas, *Macromolecules* (2003), 36, 5065.
101. M. Givehchi, M. Tardi, A. Polton, P. Sigwalt, *Macromolecules* (2000), 33, 9512.
102. R. Faust, J.P. Kennedy, *Polym. Bull* (1986), 15, 317.
103. R. Faust, J.P. Kennedy, *J. Polym. Sci. Polym. Chem.* (1987), A25, 1847.
104. (a) M.K. Misra, J.P. Kennedy, *Polym. Bull.* (1987), 17, 7.
     (b) M.K. Misra, B. Wang, J.P. Kennedy, *Polym. Bull.* (1987), 17, 307.
105. M.K. Misra, J.P. Kennedy, *J. Macromol. Sci. Chem.* (1987), A24, 933.
106. K. Matyjaszewski, P. Sigwalt, *Polym. Intl.* (1994), 35, 1.
107. K. Matyjaszeewski, C-H. Lin, *J. Polym. Sci. Polym. Chem.* Ed. (1991), 29, 1439.
108. R. Faust, J.P. Kennedy, *Polym. Bull.* (1988), 19, 21.
109. G. Kaszas, J.E. Puskas, C.C. Chen, J.P. Kennedy, *J. Macromol. Sci. Chem.* (1989), A26, 1099.
110. G. Kaszas, J.E. Puskas, C.C. Chen, J.P. Kennedy, W.G. Hager, *J. Polym. Sci. Polym. Chem.* Ed. (1991), 29, 421.
111. (a) T. Higashimura, Y. Kishimoto, S. Aoshima, *Polym. Bull.* (1987), 18, 111.
     (b) Y. Kishimoto, S. Aoshima, T. Higashimura, *Macromolecules* (1989), 22, 3877.
112. C.G. Cho, B.A. Feit, O.W. Webster, *Macromolecules* (1990), 23, 1918.
113. M. Sawamoto, *Prog. Polym. Sci.* (1991), 16, 111.
114. T. Higashimura, M. Sawamoto, S. Aoshima, Y. Kishimoto, E. Takeuchi in Frontiers of Macromolecular Science, T. Saegusa, T. Higashimura and A. Abe Eds., Blackwell, Oxford, Eng., p. 67 (1989).
115. M. Kamigato, M. Sawamoto, T. Higashimura, *Macromolecules* (1992), 25, 2587.
116. (a) M. Miyamoto, M. Sawamoto, T. Higashimura, *Macromolecules* (1984), 17, 265.
     (b) T. Higashimura, M. Miyamoto, M. Sawamoto, *Macromolecules* (1985), 18, 611.
117. M. Gyor, H-C. Wang, R. Faust, *J. Macromol. Sci. Chem.* (1992), A29, 639.
118. Zs. Fodor, M. Gyor, H.-C. Wang, R. Faust, *J. Macromol. Sci. Chem.* (1993), A30, 349.
119. R.F. Storey, C.L. Curry, L.K. Hendry, *Macromolecules* (2001), 34, 5416.
120. Zs. Fodor, Y.C. Bae, R. Faust, *Macromolecules* (1998), 31, 4439.
121. H. Katayama, M. Kamigaito, M. Sawamoto, T. Higashimura, *Macromolecules* (1995), 28, 3747.
122. T. Pernecker, J.P. Kennedy, B. Ivan, *Macromolecules* (1992), 25, 1642.
123. T. Pernecker, J.P. Kennedy, *Polym. Bull.* (1992), 29, 27.
124. T. Pernecker, T. Kelen, J.P. Kennedy, *J. Macromol. Sci. Chem.* (1993), A30, 399.
125. Gy. Deak, M. Zsuga, T. Kelen, *Polym. Bull.* (1992), 29, 239.
126. M. Szwarc, *Makromol. Chem. Rapid Commun.* (1992), 13, 141.
127. K. Matyjaszewski, *Macromolecules* (1993), 26, 1787.
128. V.S.C. Chang, J.P. Kennedy, B. Ivan, *Polum. Bull.* (1980), 3, 339.
129. L. Thomas, A. Polton, M. Tardi, P. Sigwalt, *Macromolecules*, (1995), 28, 2105.
130. P. Sigwalt, *Macromol. Symp.* (1998), 132, 127.
131. P. De, R. Faust, *Macromolecules* (2005), 38, 5498.

132. M. Ouchi, M. Sueoka, M. Kamigaito, M. Sawamoto, *J. Polym. Sci. Polym. Chem.* Ed. (2001), A39, 1067.
133. G. Kaszas, J.E. Puskas, J.P. Kennedy, *Polym. Bull.* (1987), 18, 123.
134. R.F. Storey, K.R. Choate, Jr. *Macromolecules* (1997), 30, 4799.
135. A. Nagy, I. Majoros, J.P. Kennedy, *J. Polym. Sci. Polym. Chem.* Ed. (1997), A35, 3341.
136. K. Kojima, M. Sawamoto, T. Higashimura, *J. Polym. Sci. Polym. Chem.* Ed. (1990), A28, 3007.
137. K. Satoh, M. Kamigaito, M. Sawamoto, *Macromolecules* (1999), 32, 3829.
138. K. Kojima, M. Sawamoto, T. Higashimura, *Macromolecules* (1990), 23, 948.
139. T. Higashimura, M. Mitsuhashi, M. Sawamoto, *Macromolecules* (1979), 12, 178.
140. S. Kanaoka, Y. Eika, M. Sawamoto, T. Higashimura, *Macromolecules* (1996), 20, 1778.
141. P. De, R. Faust, *Macromolecules* (2004), 37, 9290.
142. J.P. Kennedy, J. Kurian, *Macromolecules* (1990), 23, 3736.
143. L. Thomas, A. Polton, M. Tardi, P. Sigwalt, *Macromolecules* (1992), 25, 5886.
144. L. Thomas, M. Tardi, A. Polton, P. Sigwalt, *Macromolecules* (1993), 26, 4075.
145. M. Miyamoto, C. Okamoto, T. Higashimura, *Macromolecules* (1987), 20, 2693.
146. M. Miyamoto, M. Sawamoto, T. Higashimura, *Macromolecules* (1984), 17, 2228.
147. T. Enoki, M. Sawamoto, T. Higashimura, *J. Polym. Sci. Polym. Chem.* Ed. (1986), 24, 2261.
148. K. Kojima, M. Sawamoto, T. Higashimura, *Macromolecules* (1989), 22, 1552.
149. M. Kamigaito, M. Sawamoto, T. Higashimura, *Macromolecules* (1991), 24, 3988.
150. A.H.E. Muller, G. Litivenko, D. Yan, *Macromolecules* (1996), 29, 2339.
151. L. Balogh, R. Faust, *Polym. Bull.* (1992), 28, 367.
152. G. Kaszas, J.E. Puskas, C.C. Chen, J.P. Kennedy, W.G. Hager, *J. Polym. Sci. Polym. Chem.* Ed. (1991), 29(3), 427.
153. K.J. Huang, M. Zsuga, J.P. Kennedy, *Polym. Bull.* (1988), 19, 43.
154. J.P. Kennedy, V.S.C. Chung, R.A. Smith, B. Ivan, *Polym. Bull.* (1979), 1, 575.
155. B. Ivan, J.P. Kennedy, V.S.C. Chung, *J. Polym. Sci. Polym. Chem.* Ed. (1980), 18, 3177.
156. J.P. Kennedy, B. Ivan, In Designed Polymers by Carbocationic Macromolecular Engineering. Theory and Practice, Hanser Publishers: Munich, p. 173 (1992).
157. J.P. Kennedy, M. Hiza, *J. Polym. Sci. Polym. Chem.* Ed. (1983), 21, 1033.
158. L. Wilezek, J.P. Kennedy, *J. Polym. Sci. Polym. Chem.* (1987), A25, 3255.
159. J.P. Kennedy, R.A. Smith, *J. Polym. Sci. Polym. Chem.* Ed. (1980), 18, 1523.
160. J.P. Kennedy, L.R. Ross, J.E. Lackey, O. Nuyken, *Polym. Bull.* (1981), 4, 167.
161. J.V. Crivello, *Adv. Polym. Sci.* (1984), 62, 2.
162. J.V. Crivello, J.H.W. Lam, *J. Polym. Sci. Polym. Chem.* Ed. (1979), 17, 977.
163. C. Decker, *Prog. Polym. Sci.* (1996), 21, 593.
164. C. Decker, K. Mousa, *J. Polym. Sci. Polym. Chem.* Ed. (1990), 28, 3429.
165. S.C. Lapin, Radiation Curing Science and Technology, S.P. Pappas Ed., Plenum Press, p. 241 (1992).
166. R.B. Taylor, F. Williams, *J. Am. Chem. Soc.* (1969), 91, 3728.
167. (a) P.H. Plesch, *Phil. Trans. Roy. Soc. Lond. Ser A.* (1993), 342(1666), 469.
    (b) F. Williams, *Macromolecules* (2005), 38, 206.
168. H. Schlaad, Y. Kwon, L. Sipos, R. Faust, B. Charleux, *Macromolecules* (2000), 33, 8225.
169. P. De, R. Faust, *Macromolecules* (2005), 38, 9897.
170. M. Roth, H. Mayr, *Macromolecules* (1996), 29, 6104.
171. K. Ueno, F. Williams, K. Hayashi, S. Okamura, *Trans. Faraday Soc.* (1967), 63, 1478.
172. F. Williams, Ka. Hayashi, K. Ueno, Ko. Hayashi, S.Okamura, *Trans. Faraday Soc.* (1967), 63, 1501.
173. M. Sawamoto, T. Higashimura, *Macromolecules* (1979), 12, 581.
174. R. Cotrel, G. Sauvet, J.-P. Vairon, P. Sigwalt, *Macromolecules* (1976), 9, 931.
175. Ka, Hayashi, Ko. Hayashi, S. Okamura, *J. Polym. Sci. A1* (1971), 9, 2305.

176. A.M. Goineau, J. Kohler, V. Stannet, *J. Macromol. Sci. Chem.* (1977), A11, 99.
177. J.M. Rooney, *Makromol. Chem.* (1978), 179, 165.
178. P.M. Bowyer, A. Ledwith, D.C. Sherrington, *Polymer* (1971), 12, 509.
179. M. Rodriguez, L.M. Leon, *Eur. Polym. J.* (1983), 19, 585.
180. K. Ueno, H. Yamaoka, K. Hayashi, S. Okamura, *Int. J. Appl. Radiat. Isot.* (1966), 17, 595.
181. M. Szwarc, *Pure & App. Chem.* (1966), 12, 127.

Problems

6.1 An essential requirement for using the Mayo plot as a diagnostic tool for detecting monomer transfer in cationic polymerization is that the initiator efficiency should be unity. Justify the statement.

6.2 How would you know whether transfer reaction is absent or not in a living cationic polymerization?

6.3 *t*-Butyl chloride (the unimer chloride corresponding to isobutylene) fails to initiate polymerization of the monomer with BCl_3 as the coinitiator but not 2,4,4-trimethyl-2-pentyl chloride (the dimer chloride), which succeeds. What could be the reason?

6.4 Even though 2,4,4–trimethyl-2-pentyl chloride (TMPCL) is able to initiate polymerization of isobutylene with BCl_3 as the coinitiator, the polymerization is not living in the true sense. Explain.

6.5 Referring to problem 6.4, a change of coinitiator from BCl_3 to $TiCl_4$ gives living polymerization, although the MWD of the polymer is relatively broad. What could be the reason? How could the MWD be made narrower?

6.6 In comparison to TMPCl, the corresponding acetate is a much faster initiator. What makes the difference?

6.7 In the inifer method of preparing the chloride-terminated polyisobutylene, BCl_3 is the coinitiator of choice rather than $TiCl_4$. What could be the reason?

6.8 A linear $\overline{M}_n$ vs. conversion plot as a test of livingness may hide the occurrence of transfer reaction when the target molecular weight is low. Justify the statement.

6.9 Refer to Table 6.1. Determine the sequence of monomers to be polymerized in the cationic diblock copolymerization of the following pairs of monomers
 (i) isobutyene + styrene (ii) styrene and isobutyl vinyl ether, and (iii) styrene + *p*-methoxystyrene. Justify your answer.

6.10 Explain how does a nucleophile, a proton trap agent, or a common anion salt each helps in obtaining narrow distribution polymer in cationic living polymerization.

Chapter 7

Ring-Opening Polymerization and Ring-Opening Metathesis Polymerization

A wide variety of cyclic compounds each containing at least one heteroatom or an unsaturation center in the ring undergoes ring-opening polymerization (ROP), as shown schematically in reaction (7.1)

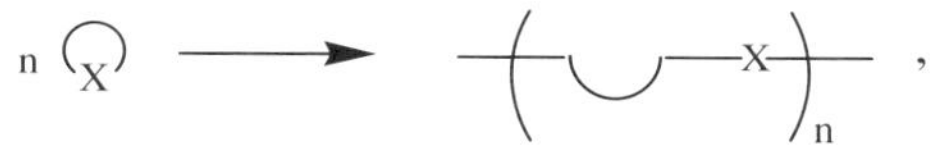

$$n \underset{X}{\bigcirc} \longrightarrow {\Big(}\!\!\smile\!\!-X{\Big)}_n , \qquad (7.1)$$

where X is a heteroatom or a double bond.

The mechanism of ring opening, however, is different for the two groups of monomers. Whereas a heteroatom-carbon bond is opened in the heterocycles through a nucleophilic reaction, the double bond is opened through an olefin metathesis reaction in the cycloalkenes. The polymerization of the latter group of monomers is specifically referred to as ring-opening metathesis polymerization (ROMP). The difference in mechanism notwithstanding, many of the polymerizations of both groups of monomers are living in nature exhibiting chain growth without termination and irreversible chain transfer.

The polymer produced by ROP structurally resembles a condensation polymer. Indeed, some of these polymers are also produced by step polymerization (nylon 6 and silicone polymers being typical examples). However, ROP has several superior features. Step polymerization yields polymer of only moderate molecular weight and that too only at very high extents of reaction. For example, the $\overline{DP}_n$ reaches only 200 at an inordinately high conversion of 99.5% that requires long time to reach. Besides, an extremely high degree of monomer purity, accurate stoichiometric balance between the monomers, prevention of loss of monomers and absence of side reactions are absolute necessities. In contrast, ROP easily yields high molecular weight polymer at appreciably lower conversion and lesser time, and under fewer demanding conditions. Besides, many of these polymerizations, as pointed out above, are living and, accordingly, have the attractive features that go with it. Several polymers produced by ROP have found important biomedical applications besides various other applications: poly (ethylene oxide) and its block copolymers, polyglycolide, polylactides, polyanhydrides and their block copolymers, homopolypeptides and block copolypeptides, silicones, and polyphosphazenes being some examples.

7.1 General Features

7.1.1 Polymerizability

For the conversion of a liquid monomer (l) to its condensed amorphous polymer (c) to take place, the free energy change $\Delta G^0_{l,c}$ must be negative. The latter is related to the entropy and enthalpy changes by the equation

$$\Delta G^0_{l,c} = \Delta H^0_{l,c} - T\Delta S^0_{l,c}.$$

For a discussion on thermodynamics of ROP, cycloalkanes are chosen as model monomers although they themselves fail to polymerize for non-thermodynamic reasons (*vide infra*).[1,2] Table 7.1 gives the semiemperically calculated values of $\Delta G^0_{l,c}$, $\Delta H^0_{l,c}$, and $\Delta S^0_{l,c}$ at 25°C for the hypothetical conversions of liquid cycloalkanes to linear polymers. Inasmuch as no change in the number and type of bonds occurs in ROP, $\Delta H^0_{l,c}$ is primarily determined by the ring strain. On the other hand, $\Delta S^0_{l,c}$ is primarily determined by the loss in translational entropy of the monomer and the gain in conformational entropy caused by ring opening.[1,2] The translational entropy loss remains approximately unchanged with change in ring size, whereas the conformational entropy gain increases with increase in ring size. The sum of the two opposing effects is reflected in the $\Delta S^0_{l,c}$ value given in Table 7.1 for each of several cycloalkanes. The value monotonically (becomes less negative) increases with ring size.[2] The loss in rotational entropy is considered to be balanced by the gain in internal rotation and vibrational entropy of the polymer.[3]

Table 7.1 Thermodynamic parameters for hypothetical polymerization of cycloalkanes at 25°C.[a]

Compound	$\Delta H^0_{l,c}$ kJ/mol	$\Delta S^0_{l,c}$ J(°K)$^{-1}$mol^{-1}	$\Delta G^0_{l,c}$ kJ/mol
Cyclopropane	−113	−69	−92.5
Cyclobutane	−105	−55.2	−89.9
Cyclopentane	−21.7	−42.7	−9.2
Cyclohexane	+2.93	−11.5	+5.8
Cycloheptane	−21.7	−16[b]	−16.7
Cyclooctane	−34.7	−3[b]	−33.6
Methylcyclopropane	−105	−84.5	−79.9
Methylcyclobutane	−100	−72	−78.6
Methylcyclopentane	−17.15	−64	+2.1
Methylcyclohexane	+9.2	−31.8	+18.8
1,1-Dimethylcyclopropane	−97.5	−93.3	−69.4
1,1-Dimethylcyclobutane	−93.3	−75.3	−67
1,1-Dimethylcyclopentane	−13.4	−65.7	+6.27
1,1-Dimethylcyclohexane	+7.5	−35.6	+18

[a] Refs. 1 and 2. Reproduced by permission of the Royal Society of Chemistry.
[b] H.L. Finke *et al.*, *J. Am. Chem. Soc.*, 1956, 78, 5469.

However, thermodynamic feasibility is not sufficient in itself to make polymerization a reality. A mechanism of polymerization must also be available.[1,2] Thus, although $\Delta G^\circ_{l,c}$ of cycloalkanes except cyclohexane is negative, the compounds are not polymerizable due to the lack of a mechanism. In contrast, cycloalkenes or heterocyclic monomers with negative $\Delta G^\circ_{l,c}$ are polymerizable since appropriate mechanisms exist.

Referring to Table 7.1, $\Delta H^\circ_{l,c}$ of 3- and 4-membered cyclic (small ring) monomers is highly negative ca., -100 kJ/mol, which is ascribable to high angle strain. It far outweighs the effect of the unfavorable entropy change, so that polymerization is practically irreversible. On the other hand, $\Delta G^\circ_{l,c}$ of 5-, 6-, or 7-membered cyclic (common ring) monomers is small so that a change in the nature of the ring or the presence of substituents may alter its sign.[1,2] Accordingly, the feasibility of polymerization of such monomers exhibits critical dependence on the nature of the ring and the presence of the substituents. The former is evident from the experimental observations recorded in Table 7.2 and the latter from the $\Delta G^\circ_{l,c}$ data for cyclopentane and its mono- and 1,1-dimethyl derivatives given in Table 7.1. Substitution makes polymerization less exothermic since the resulting steric interference is more in the polymer chain than in the ring monomer. It also causes larger decrease in the entropy of polymerization since substituents reduce freedom of internal rotation in the polymer chain but have little effect on the analogous freedom in the ring monomer, which is small in the first place.[4]

Besides, a change in solvent may be enough to alter polymerizability. However, because of the small $\Delta G^\circ_{l,c}$ in the polymerization of common ring monomers an equilibrium polymerization occurs (Chap. 1, Sec. 1.8).[1,2]

Table 7.2 Polymerizability of some unsubstituted common ring monomers.[a]

| | Polymerizability[b] | | |
| | Ring size | | |
Monomer	5	6	7
Cycloalkenes	+	−	+
Cyclic ethers	+	−	+
Cyclic acetals (one formal group)	+	−	+
(two formal groups)			+
(three formal groups)		+	
Lactones	−	+	+
Cyclic carbonates	−	+	+
Cyclic anhydrides	−	−	+
Lactams	+	+	+
Cyclic amines	−	−	+
Cyclic sulfides	−	−	
Cyclodimethylsiloxanes		+	

[a]Smaller or larger ring monomers of each family are polymerizable; [b]+ sign ≡ polymerizable; − sign ≡ nonpolymerizable; No sign ≡ either the monomer does not exist or polymerizability is not reported in the literature.

As the ring size increases from 7- to 8-membered, $\Delta H^o_{1,c}$ becomes more negative due to the increase in ring strain (*vide* Table 2.1 in Chap. 2), whereas $\Delta S^o_{1,c}$ becomes less negative due to the larger gain in conformational entropy. As a result, $\Delta G^o_{1,c}$ becomes more negative and the monomer undergoes irreversible polymerization to high conversion.

The thermodynamic parameters for polymerizations of cycloalkanes with ring size larger than eight-membered in the medium rings family (8 to 11 ring atoms) are not available. However, the larger ring strain (*vide* Chap. 2, Table 2.1) and the increasingly larger gain in conformational entropy with increasing ring size in this family suggest that the polymerizability would even be greater. In large rings (12 or more ring atoms), the ring strain is virtually nonexistent. As a result, $\Delta H^o_{1,c}$ of unsubstituted large ring monomers would be vanishingly small. On the other hand, $\Delta S^o_{1,c}$ would be positive being dominated by the conformational entropy gain. Hence, polymerization of large ring monomers would be entropy-driven.[1,2]

The trend of polymerizability is different for cyclodimethylsiloxanes $(Si(CH_3)_2O)_n$, symbolically represented as D_n, with structures shown for n = 3 and 4. The Si–O bond length (=1.63 Å) and the bond angles in linear siloxanes, Si–O–Si = 143° and O–Si–O = 110°,[5] are such that the 6-membered D_3 is more strained than the higher members of the D_n family are. However, all members of the D_n family (n ≥ 3) are polymerizable. The enthalpy of polymerization of D_3 is small negative (−15 kJ/mol), whereas that of larger rings, D_4 to D_5, is nearly zero and the driving force for their polymerization is a small positive ΔS, which increases from 3.5 to 5.8 J/mol. SiO group. °K, as the ring size increases from D_4 to D_7 and above.[6,7]

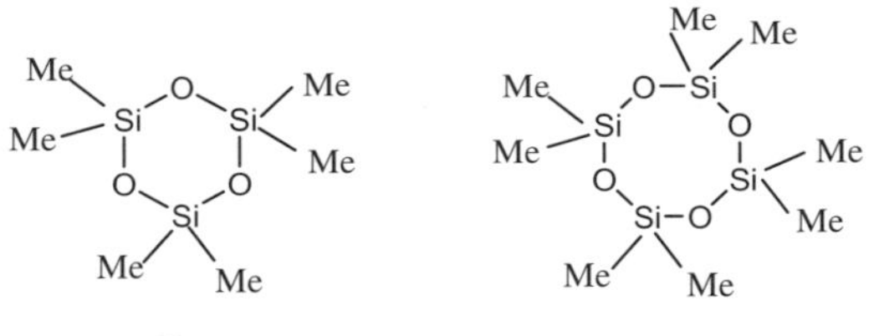

D_3

D_4

7.1.2 Ring-chain equilibrium

One common feature of ROP is that apart from monomeric rings that may exist in equilibrium with polymer (equilibrium polymerization), r-mer rings (r > 1) are formed in various proportions depending on the type of monomer polymerized and the initiator used. However, at the thermodynamic equilibrium, the concentrations of rings and chains are independent of the initiator type. Ring formation takes place through the mechanism of intramolecular chain transfer to polymer called backbiting (Sec. 7.1.4). Jacobson–Stockmayer theory of ring-chain equilibrium discussed in Chap. 2 (Sec. 2.2) predicts a constant equilibrium concentration of large rings (Eq. 2.11) provided the chains existing in equilibrium are long.[10] The weight fraction of rings, however, decreases with increase for monomer polymerized.

The theory also predicts a critical monomer concentration below which the products will be all rings (Eq. 2.15).

7.1.3 *The nature and reactivity of propagating species*

Polymerization of heterocyclic monomers typically takes place anionically and/or cationically with propagating species containing charged heteroatom, *e.g.*, alkoxide, carboxylate, sulfide, silanoate, *etc.*, in anionic ROP, and cyclic *tertiary* oxonium, sulfonium, iminium, siloxonium, and so on, in cationic ROP. These anionic and cationic species are much less reactive than carbanionic and carbenium ionic species respectively, as would be evident from the k_p values given in Table 7.3.

The mechanism of polymerization involves nucleophilic substitution. In cationic ROP, the heteroatom in monomer acts as a nucleophile. It attacks an α carbon atom of the propagating cationic species resulting in chain extension with ring opening

where X is a heteroatom.

Table 7.3 Propagation rate constants involving various ionic propagating species.

Monomer	Ion type	Solvent	Temp.°C	k_p L mol^{-1}s^{-1}	Ref.
		Cationic polymerization $k_p^+ (\simeq k_p^\pm)$			
Styrene	C$^+$	CH$_2$Cl$_2$	−50 to −80	2×10^9	Chap. 6
Isobutylene	C$^+$	CH$_3$Cl/hexane (40/60 v/v)	−80	$\sim 5 \times 10^8$	Chap. 6
Tetrahydrofuran	O$^+$	CH$_3$NO$_2$	25	0.02	35
Oxepane	O$^+$	C$_6$H$_5$NO$_2$	25	0.0005	36
		CH$_2$Cl$_2$	25	0.0006	36
3,3-Dimethylthietane	S$^+$	CH$_2$Cl$_2$	25	0.0093	155
2-Methyloxazoline	N–C$^+$=O	C$_6$H$_5$NO$_2$	25	0.00003	144
		Anionic Polymerization	Temp.°C	k_p^-	$k_p^\pm$	
Styrene	C$^-$	THF	25	65000	22 (Cs$^+$)	Chap. 4
Ethylene oxide	O$^-$	THF	70	–	3.50 (Cs$^+$)	26, 27
			25	–	0.25 (Cs$^+$)	
			70	–	0.94 (K$^+$)	26
ε-Caprolactone	O$^-$	THF	20	–	≥1.70^a (Na$^+$)	92
β-Propiolactone (1 mol/L)	COO$^-$	CH$_2$Cl$_2$	25	0.22	1.7×10^{-3} (K$^+$,DBC^b)	72
Hexamethylcyclotrisiloxane (D$_3$)	Si(Me)$_2$O$^-$	THF	22	–	0.098 (Li$^+$)	27

^{a}Value of k_p^{app}. ^{b}DBC = dibenzo-18-crown-6.

In anionic ROP, the picture is reversed. The nucleophilic propagating anionic species attacks an α carbon atom in the monomer resulting in chain extension with ring opening

$$\sim\sim\sim X^- + X\overset{CH_2}{\underset{CH_2}{\diamond}} \longrightarrow \sim\sim\sim X-CH_2\,CH_2-X^-$$

The reactivity of the cationic species decreases with decrease in electrophilicity

$$C^+ \gg O^+ > S^+ > N^+,$$

and that of the anionic species decreases with decrease in nucleophilicity[8]

$$C^- \gg S^- > O^- > COO^-.$$

A further mechanism of polymerization involves charged monomer rather than charged propagating chain end. This is referred to as "activated monomer" (AM) mechanism[9] (protic acid activated cationic polymerization of cyclic ethers in alcoholic media, polymerization of N-carboxyanhydrides by nonnucleophilic bases, and anionic polymerization of lactams being some examples) (*vide infra*).

7.1.4 Backbiting and intermolecular interchange

The heteroatoms in the polymer chain compete effectively with the monomer for reaction with the propagating chain end. This reaction, referred to variously as intramolecular interchange, intramolecular chain transfer, or backbiting, results in cyclics formation. The reactions are represented schematically by Eqs. (7.2) and (7.3) with cationic and anionic active centers respectively.

Backbiting:

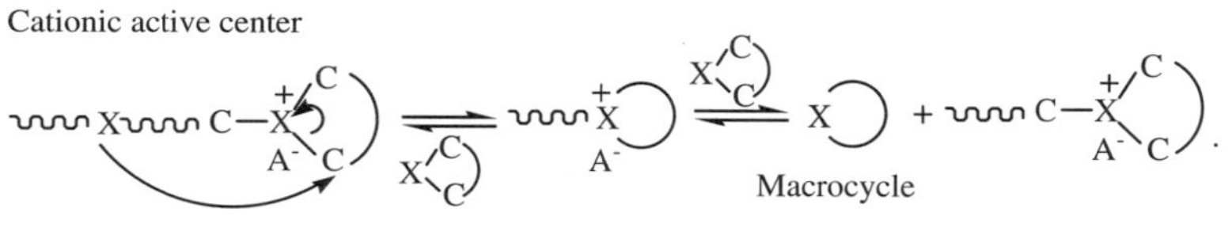

Cationic active center

$$\sim\sim X\sim\sim C-\overset{+}{X}\overset{C}{\underset{A^-\,C}{\diamond}} \rightleftharpoons \sim\sim\overset{+}{X}\,X\overset{C}{\underset{C}{\diamond}} \rightleftharpoons X\bigcirc + \sim\sim C-\overset{+}{X}\overset{C}{\underset{A^-\,C}{\diamond}} \qquad (7.2)$$

Macrocycle

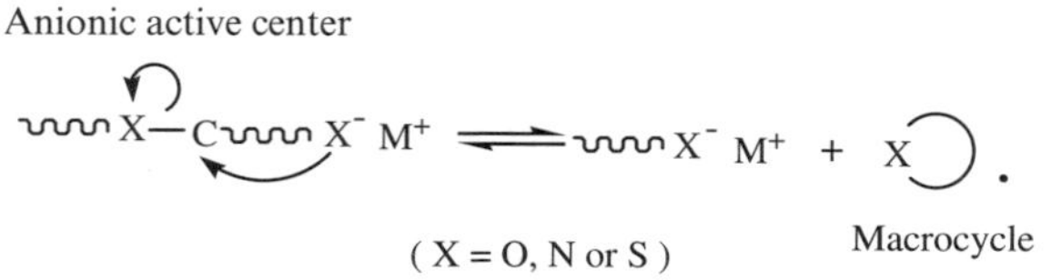

Anionic active center

$$\sim\sim X-C\sim\sim X^-\,M^+ \rightleftharpoons \sim\sim X^-\,M^+ + X\bigcirc. \qquad (7.3)$$

(X = O, N or S)   Macrocycle

Although the reactions are reversible, they may not reach thermodynamic equilibrium by the time polymerization is completed. As a result, cyclics may either exceed or fall short of their equilibrium concentrations. However, the equilibrium concentrations may reach eventually if the system is equilibrated for long enough time. This practice is useful in the former case for reducing cyclics concentration. A typical example is the protonic acid catalyzed ROP of cyclodimethylsiloxanes[11] (*vide* Sec. 7.10). The proportion of higher cyclics in an equilibrated system may be theoretically estimated using Jacobson–Stockmayer equation discussed in Chap. 2 (Eq. 2.13).[10] On the other hand, when backbiting is much slower than propagation, the proportion of cyclics may be minimized by discontinuing polymerization under monomer-depleted conditions. This is because the rate of propagation decreases with the progress of polymerization due to monomer depletion, which is first order in monomer, whereas backbiting being independent of monomer concentration continues unabated.

However, appropriate polymerization mechanism may make the chain end unreactive towards backbiting, the activated monomer mechanism of polymerization of oxiranes in the presence of alcohols being an example (*vide* Sec. 7.2.3). Alternatively, appropriate initiator may reduce the chain end reactivity such that backbiting is reduced to a greater extent than propagation, covalent metal alkoxide initiated polymerization of cyclic esters being an example (*vide* Sec. 7.4.2).

In contrast, intermolecular interchange, which is equivalent to intermolecular chain transfer to polymer, causes shuffling of sections of chains between chains, as shown in reactions (7.4) and (7.5) for cationic and anionic ROP respectively. As a result, the MWD is broadened but the number average molecular weight remains unchanged since no change in the number of molecules occurs.

Intermolecular interchange:

(a) Cationic active center

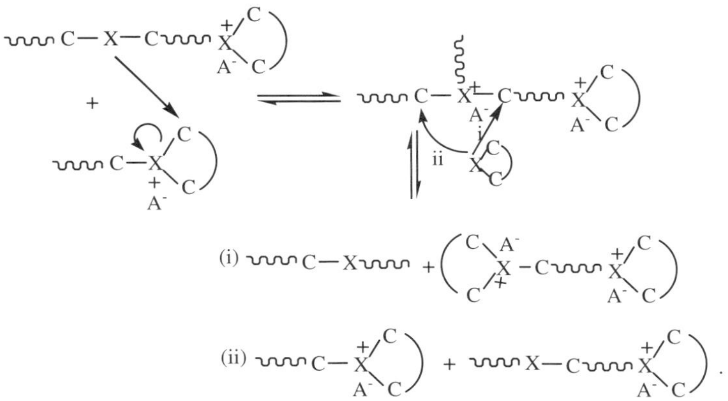

$$ (7.4) $$

(b) Anionic active center

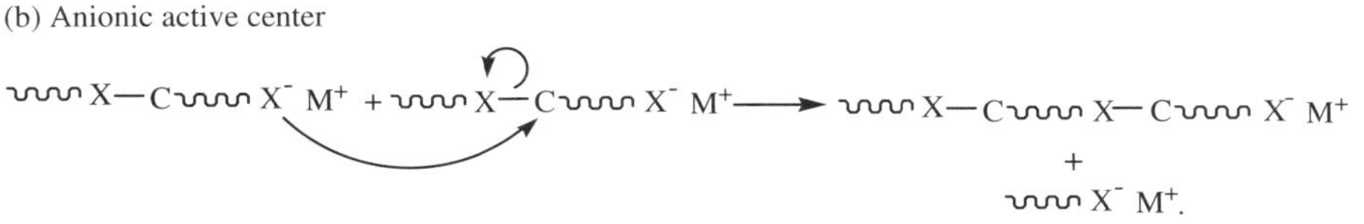

$$ (7.5) $$

As discussed in Chap. 2 (Sec. 2.3), the establishment of equilibrium in the intermolecular interchanges results in the most probable MWD ($\overline{M}_W / \overline{M}_n = 2$). Accordingly, PDI increases

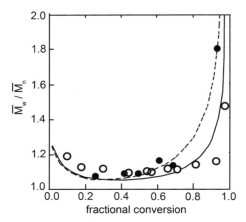

Figure 7.1 Evolution of polydispersity index with conversion due to intermolecular interchange occurring in the living coordinated ROP of L,L-lactide using tin dibutoxide as initiator. Conditions of polymerization: solvent = THF, $[M]_0 = 1$ mol/L, $[Initiator]_0$ (in mmol/L) and temperature = 5, 80°C (●); 8.9, 20°C (○). Points experimental, the dotted and solid curves are theoretically computed for the chain transfer constant $k_p/k_{tr} = 125$ and 200 respectively. "Reprinted with permission from Ref. 13. Copyright © 2000 American Chemical Society."

from near unity toward 2 as polymerization is continued in monomer-depleted conditions due to the long time thus being available for interchange to advance toward equilibrium. An example is shown in Fig. 7.1, which refers to the living coordinated anionic polymerization of L,L-lactide.[12,13] The mechanism is analogous to that shown in reaction (7.5).

7.1.5 Bridged cyclic monomers

The simplest bridged cyclic monomers are the bicyclic monomers, which have three atoms common to the two rings as in norbornene (bicyclo [2.2.1] hept-2-ene, **1**) and 7-oxabicyclo [2.2.1] heptane (**2**). In these monomers, there are three bridges joining the two bridgeheads. Of these three, the one containing the heteroatom or the C=C bond opens for polymerization. Thus, norbornene polymerizes to form linear polymer with a cyclopentane ring in the repeating unit,[14] whereas 7-oxabicyclo [2.2.1] heptane polymerizes to give linear polymer with a cyclohexane ring in the repeating unit.[15]

7.2 Cyclic Ethers

Cyclic ethers are polymerized commonly using cationic initiators. Backbiting and intermolecular transetherification proceed concurrently with propagation in the cationic ROP of lower cyclic ethers. However, these side reactions can be nearly eliminated by conducting acid catalyzed polymerization in the presence of hydroxylic solvents, *viz.*, water and alcohols (*vide* activated monomer mechanism of polymerization).

Among the important commercial polyethers are poly(ethylene oxide) (PEO) as well as amphiphilic block copolymers with PEO as hydrophilic block(s), and polytetrahydrofuran. PEO and its amphiphilic block copolymers have found many biomedical applications besides other applications. Polytetrahydrofuran is used in the synthesis of polyurethanes- and polyesters-based thermoplastic elastomers in which it provides the rubbery soft block (Chap. 2) by virtue of low T_g ($-86°C$) and low T_m ($43°C$).

7.2.1 *Anionic polymerization*

Only the three-membered cyclic ethers, *i.e.*, the epoxides, are polymerizable by anionic means. Alkali metal hydroxides, oxides, amides, alkyls, aryls, and napthalenides are used as initiators.[16–18] Initiation as well as propagation involves nucleophilic attack of the anionic species (both ion and ion pair) on carbon atom bonded with oxygen in monomer as shown in reaction (7.6) for the propagation on ion pair.

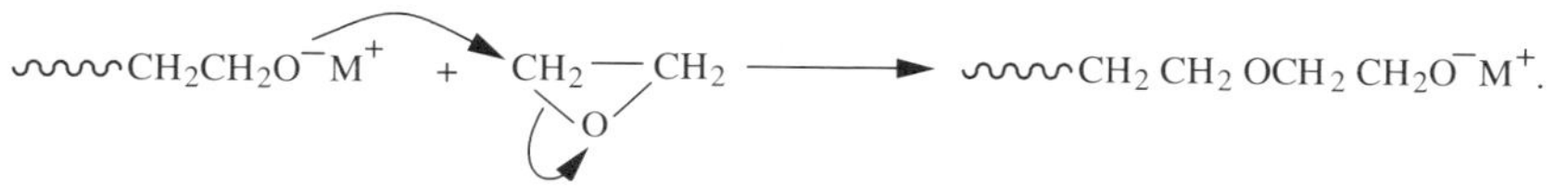

$$\sim\!\!\sim\!\!CH_2CH_2O^-M^+ + \underset{O}{CH_2\!-\!CH_2} \longrightarrow \sim\!\!\sim\!\!CH_2CH_2OCH_2CH_2O^-M^+ \qquad (7.6)$$

Propagation on ion is not shown. This practice is followed throughout this chapter, unless indicated otherwise. Polymerization occurs livingly in an aprotic solvent such as tetrahydrofuran (THF). However, living character is not destroyed in the presence of protic solvents, *viz.*, alcohols, which effect degenerative transfer (*vide* Chap. 3, Sec. 3.15).[19,20] Of course, the degree of polymerization is reduced, the number of polymer molecules being increased by the number of alcohol molecules added.

$$R\!+\!OCH_2CH_2\!+\!_x O^-M^+ + ROH \rightleftharpoons R\!+\!OCH_2CH_2\!+\!_x OH + RO^-M^+.$$

$$R\!+\!OCH_2CH_2\!+\!_y O^-M^+ + R\!+\!OCH_2CH_2\!+\!_x OH \rightleftharpoons R\!+\!OCH_2CH_2\!+\!_y OH + R\!+\!OCH_2CH_2\!+\!_x O^-M^+.$$

In contrast to polymerization of ethylene oxide, that of propylene oxide (PO) proceeds with extensive chain transfer to monomer (reaction 7.7).[21] It involves proton transfer from the methyl group in the monomer resulting in ring opening and strain release, which drives the reaction forward. Similar transfer also occurs from epoxides substituted with alkyl groups larger than methyl.[12,21]

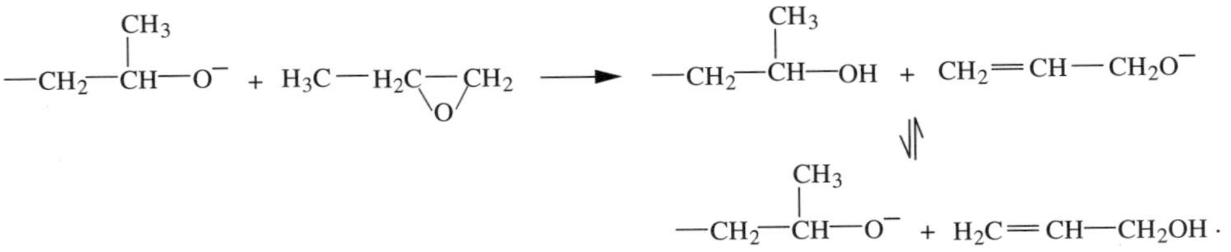

$$-CH_2-\underset{\underset{CH_3}{|}}{CH}-O^- + H_3C-H_2C-\overset{\overset{}{\diagdown}}{\underset{\underset{O}{\diagup}}{C}}-CH_2 \longrightarrow -CH_2-\underset{\underset{CH_3}{|}}{CH}-OH + CH_2=CH-CH_2O^-$$

$$\Updownarrow$$

$$-CH_2-\underset{\underset{CH_3}{|}}{CH}-O^- + H_2C=CH-CH_2OH \cdot$$

(7.7)

Because of the prevalence of the monomer transfer, only low molecular weight polymers are formed.[22] Nevertheless, anionic polymerization of PO is practiced in industry for the production of oligomeric block copolymers with poly(ethylene oxide).[12]

However, monomer transfer is eliminated in polymerizations initiated by the coordinated anionic initiator aluminum porphyrin (**3**).[23] Living polymerization is effected (reaction 7.8) in the presence of alcohol, which acts as a degenerative chain transfer agent like it does in the anionic polymerization discussed above.

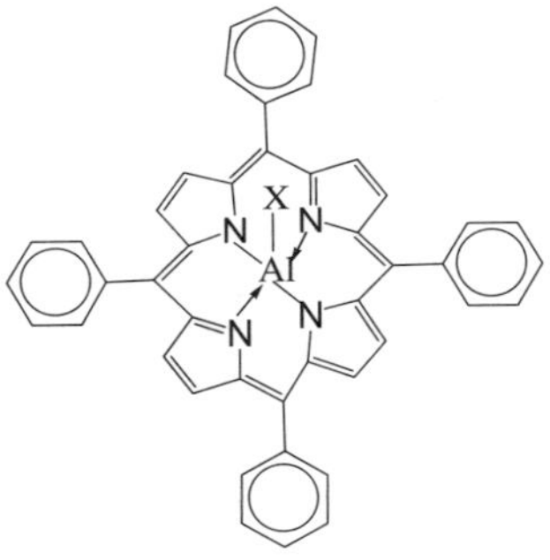

[Aluminum porphyrin structure (**3**)], X = Cl, OR

$$\text{(Al)}-X + x\ \underset{\underset{O}{\diagdown\diagup}}{C-C}^R \longrightarrow \text{(Al)}\!\!-\!\!\{O-\underset{\underset{R}{|}}{C}-C\}_x\!X \xrightleftharpoons{R'OH} \text{(Al)}\!\!-\!\!OR' + H\{O-\underset{\underset{R}{|}}{C}-C\}_x\!X$$

$$\downarrow \text{monomer}$$

$$\text{(Al)}\!\!-\!\!\{O-\underset{\underset{R}{|}}{C}-C\}_y\!OR'$$

$$\Updownarrow R'OH$$

$$H\{O-\underset{\underset{R}{|}}{C}-C\}_y\!OR' + \text{(Al)}\!-\!OR'.$$

(3)

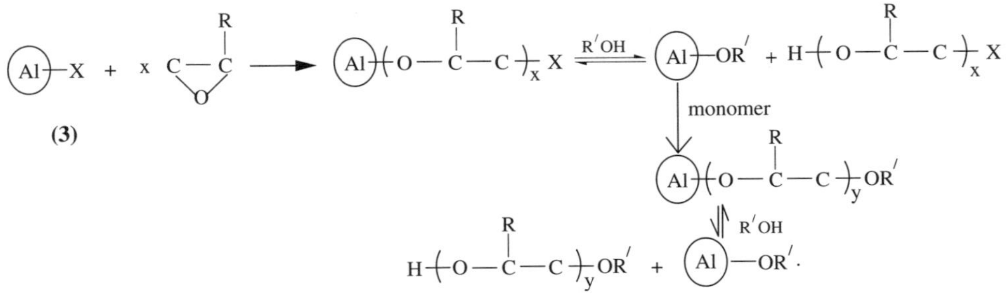

(7.8)

The system is so robust that polymerization is not terminated even by hydrogen chloride since the Al–Cl bond in the resulting chloroaluminum porphyrin is capable of reinitiating polymerization (reactions 7.9 and 7.10). For this ability to continue polymerization even in the presence of such a strong acid as hydrogen chloride, the polymerization was termed 'immortal' by its discoverer.

$$\text{(Al)}\left(\text{O}-\underset{\underset{\text{R}}{|}}{\text{C}}-\text{C}\right)_y \text{OR}' + \text{HCl} \longrightarrow \text{(Al)}-\text{Cl} + \text{H}\left(\text{O}-\underset{\underset{\text{R}}{|}}{\text{C}}-\text{C}\right)_y \text{OR}'. \qquad (7.9)$$

$$\text{(Al)}-\text{Cl} + z\ \underset{\underset{\text{O}}{\diagdown\diagup}}{\text{C}-\underset{\overset{\text{R}}{|}}{\text{C}}} \longrightarrow \text{(Al)}\left(\text{O}-\underset{\underset{\text{R}}{|}}{\text{C}}-\text{C}\right)_z \text{Cl}. \qquad (7.10)$$

Besides oxiranes, the 4-membered cyclic ether oxetane, and various other monomers, such as the cyclic esters (β-, γ-, and ϵ-lactones or lactides), the 6-membered trimethylene carbonate, and the cyclodimethylsiloxane, D_3, also are polymerized livingly by aluminum porphyrins.[23]

The extraordinary living polymerization effected by aluminum porphyrins is attributed to their relatively weak nucleophilicity and moderately high Lewis acidity. This is evident respectively from their low reactivity toward carbon dioxide and high ability to form coordination complexes with nitrogen bases.[23,24] The Lewis acidity enables them to form donor acceptor complexes with the monomers and thereby activate the latter resulting in relatively fast propagation. The weak nucleophilicity helps to suppress side reactions such as monomer transfer.

7.2.1.1 Kinetics

Living anionic polymerization of EO in THF is externally first order in initiator at concentrations below ca., 10^{-4} mol L^{-1} but fractional order at higher concentrations.[26–28] Conductivity studies exclude the presence of ions. Thus, the kinetic results are attributed to propagation solely on ion pairs (reaction 7.12), which exist in equilibrium with aggregated species (Eq. 7.11) that are propagation-inactive.[26–28] Exchange between the active and the inactive species is faster than propagation resulting in polymers with narrow MWD.

$$mP^{\pm} \underset{}{\overset{K_a}{\rightleftharpoons}} (P^{\pm})_m, \qquad K_a = \frac{[(P^{\pm})_m]}{[P^{\pm}]^m}. \qquad (7.11)$$

$$P_n^{\pm} + M \overset{k_p^{\pm}}{\longrightarrow} P_{n+1}^{\pm}. \qquad (7.12)$$

In the exchange equilibrium (7.11), $P^{\pm}$ represents an ion pair of any chain length. Also, the concentration symbols $[P^{\pm}]$ and $[(P^{\pm})_m]$ refer respectively to combined concentrations of non-aggregated and aggregated ion pairs of all chain lengths. The aggregation constant K_a is large such that the equilibrium concentration of ion pairs is negligible in comparison to that of the aggregated species at ionic species concentrations larger than ca., 1 milliequiv./L.[26,27] Irreversible propagation is considered, which is appropriate for polymerization of monomers having relatively large strain such as EO. From Eq. (7.12) the integrated first order rate equation for monomer disappearance follows as

$$\ln \frac{[M]_0}{[M]} = k_p^{\pm}[P^{\pm}]t = k_{app}t. \quad (7.13)$$

Thus, k_{app} is given by the slope of a linear plot of $\ln[M]_0/[M]$ vs. t.
From Eq. (7.11)

$$[P^{\pm}] = K_a^{-1/m}[(P^{\pm})_m]^{1/m}. \quad (7.14)$$

Substituting Eq. (7.14) for $[P^{\pm}]$ in Eq. (7.13) gives[27]

$$k_{app} = k_p^{\pm} K_a^{-1/m}[(P^{\pm})_m]^{1/m}. \quad (7.15)$$

For completed initiation

$$[I]_0 = [P^{\pm}] + m[(P^{\pm})_m] \simeq m[(P^{\pm})_m], \quad (7.16)$$

where $[I_0]$ is the initial initiator concentration in equiv./L.
 Substituting from Eq. (7.16) for $[(P^{\pm})_m]$ in Eq. (7.15) and taking logarithm of both sides gives

$$\log k_{app} = \log[k_p^{\pm}(K_a m)^{-1/m}] + \frac{1}{m}\log[I]_0. \quad (7.17)$$

Thus, the degree of aggregation (m) may be determined from the slope of the linear plot of $\log k_{app}$ vs. $\log[I]_0$. In this way, a value of m = 3 has been derived for ion pair with Cs^+ counterion.[27] From the intercept, $k_p^{\pm} K_a$ may be determined using the value of m. The individual constants, $k_p^{\pm}$ and K_a, have been determined by numerically analyzing the kinetics described by Eqs. (7.11) and (7.12).[26] However, an analytical treatment of the kinetic scheme by Penczek et al. gave the following equation using which the individual constants were obtained.[29]

$$k_{app}^{1-m} = -mK_a(k_p^{\pm})^{1-m} + k_p^{\pm}[I]_0 k_{app}^{-m}.$$

Thus, plotting k_{app}^{1-m} against $[I]_0 k_{app}^{-m}$ the rate constant $k_p^{\pm}$ is directly obtained from the slope and K_a is calculated from the intercept with m and $k_p^{\pm}$ being determined, as discussed above. The procedure gives $K_a = 1.99 \times 10^6$ (L/mol)$^{m-1}$ and $k_p^{\pm} = 3.65$ mol/L.s on PEO$^-$Cs$^+$ ion pair in THF at 70°C. The latter is about an order of magnitude smaller than $k_p^{\pm}$ on (polystyrene)$^-$Cs$^+$ ion pair in styrene polymerization in THF demonstrating lower reactivity of oxide ion pair (Table 7.3, Sec. 7.1.3).[9,30]

7.2.2 Cationic polymerization

Cationic polymerization of cyclic ethers is typified by that of tetrahydrofuran (THF). Initiators used are similar to those used to initiate carbocationic polymerization (Chap. 6). Strong Bronsted acids, such as trifluoroacetic, trifluorosulfonic, and trifluoromethanesulfonic, are efficient initiators.[17] Cationic photoinitiators, which generate strong acids on irradiation with

UV light (Chap. 6, Sec. 6.5), also make efficient initiators. Less strong acids yield only low molecular weight polymers. Initiation occurs by protonation of the ether oxygen producing a *secondary* onium ion, which undergoes nucleophilic attack by monomer forming the corresponding cyclic *tertiary* onium ion, which constitutes the propagating species[12]

However, esters of strong Bronsted acids are better initiators than acids. Due to the similarity in structure between the initiator and the propagating chain, initiation (reaction 7.18) should be almost as fast as propagation.[12,17,18]

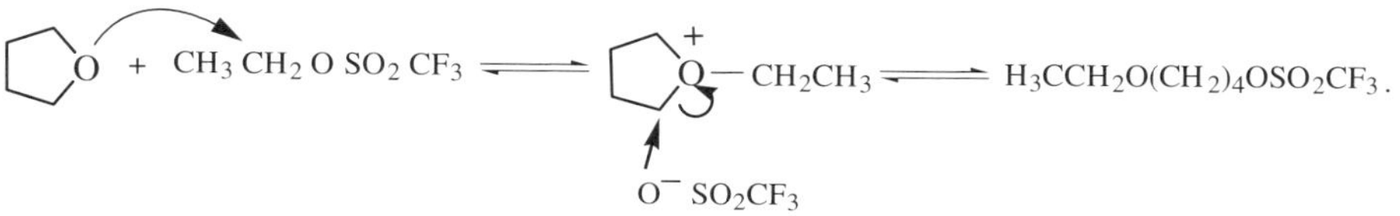

(7.18)

Strong Lewis acids (BF_3, SbF_5, $SbCl_5$, $AlBr_3$, $SnCl_4$, $FeCl_3$, *etc*.,) either used alone or in combination with adventitious protogens or added cationogens are also effective initiators.[17,18,31,32]

Rate of initiation of less reactive monomers by Lewis acids can be increased by addition of a small amount of a more reactive monomer referred to as "promoter" in this role. Thus, epichlorohydrin brings about faster initiation than propagation in the polymerization of THF effected by BF_3.[33] The mechanism of promotion has been suggested to be as follows:[18]

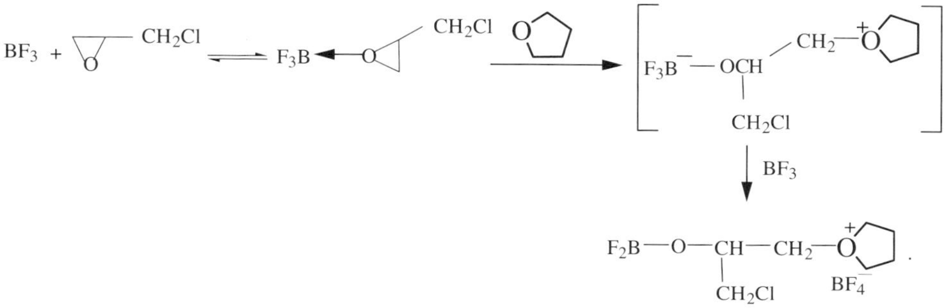

Stable carbenium ion salts, such as trityl and tropylium salts, also initiate polymerization of THF. The carbenium ion of the salt does not directly add to the monomer but abstracts an

α hydride ion from the latter. The resulting carbocation releases a proton producing a strong acid, which is the real initiator[34]

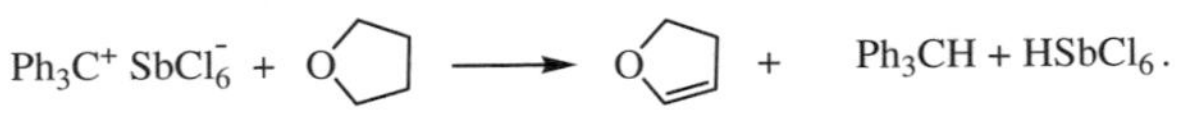

$$Ph_3C^+ \; SbCl_6^- + \; \langle O \rangle \longrightarrow \langle O \rangle + \; Ph_3CH + HSbCl_6.$$

7.2.2.1 Kinetics

The kinetics of cationic ROP depends upon the nature of the counterion, which belongs to one or the other of two types. One type is exemplified by the complex anions, such as BF_4^-, SbF_6^-, $SbCl_6^-$, and $AlCl_4^-$, which can not covalently bind with the oxonium ion in the active center. The other type is exemplified by the noncomplex anions, such as $CF_3SO_3^-$, $CF_3CO_2^-$, and ClO_4^-, which can do so.[12] Thus, even though the polymerization is living with both types of counterions, the active centers are different for the two types. With complex counterions the active centers comprise both ions and ion pairs, whereas with noncomplex counterions not only ions and ion pairs but also covalent esters formed by ion pair collapse constitute the active centers. Inasmuch as propagation is reversible in ROP of common ring monomers, such as THF (Sec. 7.1.1), propagation may be represented as

$$P_n^* + M \; \underset{k_{dp}^{app}}{\overset{k_p^{app}}{\rightleftharpoons}} \; P_{n+1}^*, \qquad (7.19)$$

where k_p^{app} is the apparent rate constant of propagation accommodating the contributions of all kinds of active centers (P^*) present in the system, k_{dp}^{app} is similarly the apparent rate constant of depropagation. The rate of monomer disappearance at any stage of polymerization is given by[30]

$$-\frac{d[M]}{dt} = k_p^{app}[P^*][M] - k_{dp}^{app}[P^*], \qquad (7.20)$$

where [P^*] is the total concentration of free ions and ion pairs irrespective of their chain lengths when the counterion is of the complex ion type, *i.e.*,

$$[P^*] = \sum_{n=1}^{\infty} [P_n^+] + \sum_{n=1}^{\infty} [P_n^\pm].$$

At equilibrium

$$k_p^{app}[P^*][M]_e = k_{dp}^{app}[P^*], \qquad (7.21)$$

where $[M]_e$ is the equilibrium monomer concentration. Equation (7.21) gives

$$k_{dp}^{app} = k_p^{app}[M]_e. \qquad (7.22)$$

Substituting Eq. (7.22) for k_{dp}^{app} in Eq. (7.20) gives

$$-\frac{d[M]}{dt} = k_p^{app}[P^*]([M] - [M]_e). \tag{7.23}$$

With complex counterion

$$k_p^{app} = \alpha k_p^+ + (1-\alpha)k_p^{\pm}, \tag{7.24}$$

where α is the degree of dissociation, k_p^+ and $k_p^{\pm}$ are the rate constants of propagation on ion and ion pair respectively (*vide* Chap. 6).

Since for a living polymerization $[P^*]$ is constant, as is α for a given $[P^*]$, Eq. (7.23) gives on integration between limits $[M] = [M]_0$ at $t = 0$ and $[M] = [M]_e$ at $t = t$[35,36]

$$\ln \frac{[M_0] - [M]_e}{[M] - [M]_e} = k_p^{app}[P^*]t. \tag{7.25}$$

Measurement of $[M]$ at different polymerization times and that of $[M]_e$ at final conversion allows one to make a plot of the left side of Eq. (7.25) *vs*. t. A straight-line plot is indicative of living polymerization. The slope of the line yields the value of $k_p^{app}[P^*]$ from which k_p^{app} can be calculated since $[P^*]$ is known, it being equal to $[In]_0$ for completed initiation. Such linear plots as obtained in the polymerization of THF in nitromethane at different initiator concentrations are shown in Fig. 7.2.

It is found that k_p^{app} does not change with change in the initiator concentration, although the degree of dissociation (α) of the active centers must increase with decrease in the initiator concentration. This can only happen if $k_p^{app} = k_p^+ = k_p^{\pm}$, as would be evident from Eq. 7.24. The same results have been obtained also with noncomplex counterions like $CF_3SO_3^-$ and FSO_3^-. However, the equality between k_p^+ and $k_p^{\pm}$ is not unique with cyclic oxonium ion. It is also observed when the cationic species are carbon-centered, as we have already seen in Chap. 6, which has been attributed to the large size of the counteranion. However, $k_p^+ (= k_p^{\pm})$ on oxonium ion derived from THF and oxepane are 10 to 12 orders of magnitude smaller than that on carbenium ion[35,36] (Table 7.3).

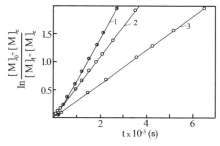

Figure 7.2 First order monomer disappearance plots in the polymerizations of tetrahydorfuran (7 mol L^{-1}) in nitromethane at 25°C using initiator concentrations: (line 1) ethyl trifluoromethanesulfonate, 3.62 × 10^{-2} mol L^{-1}, (line 2) 1,3-dioxolan-2-ylium hexafluoroantimonate (HFA) 2.42 × 10^{-2} mol L^{-1}, (line 3) HFA 1.43 × 10^{-2} mol L^{-1}. "Reprinted with permission from Ref. 35. Copyright © 1985 John Wiley & Sons Inc."

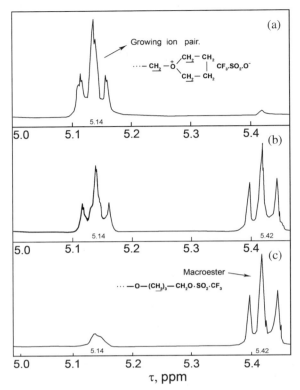

Figure 7.3 300 MHz ^{1}H NMR spectra (the 5.0–5.5 τ region only) of living THF with triflate counter anion at 17° C in various solvents: curves (a) in nitromethane, (b) in methylene chloride, and (c) in carbon tetrachloride. "Reprinted with permission of John Wiley & Sons, Inc. from Ref. 37. Copyright © 1985 John Wiley & Sons, Inc."

With noncomplex counteranions, the ionic species exist in dynamic equilibrium with covalent esters. The proportion of the ionic and covalent species changes with the polarity of the medium. Thus, in the triflate ester initiated living polymerization of THF the proportion of covalent species is small in nitromethane, almost equal to that of ionic species in methylene chloride, and large in carbon tetrachloride, as determined from the 300 MHz ^{1}H NMR spectra of polymerizing THF in the three solvents (Fig. 7.3).[37]

Inasmuch as $k_p^+ \approx k_p^\pm$, the rate of monomer disappearance in these polymerizations can be expressed as

$$-\frac{d[M]}{dt} = (k_p^\pm [P^\pm] + k_p^{cov}[P^{cov}])([M] - [M]_e), \qquad (7.26)$$

where k_p^{cov} and $[P^{cov}]$ refer respectively to the propagation rate constant and total concentrations of the covalent species of all chain lengths, $[P^\pm]$ refers to the total concentration of both ions and ion pairs (no distinction between them being required due to their kinetically equivalent character). $[P^\pm]$ and $[P^{cov}]$ are determined from ^{1}H NMR spectra (Fig. 7.3) and $k_p^\pm$ is assumed to be the same as with the complex counterions.

Thus, it remains to measure the rate of monomer disappearance to calculate k_p^{cov} using Eq. (7.26). The k_p^{cov} thus determined with $CF_3SO_3^-$ as the counteranion in nitrobenzene and carbon tetrachloride is lower than $k_p^{\pm}$ by a factor of 10^2 and 10^3 respectively at 25°C. As would be expected, the covalent propagation is much slower than propagation on ion pair. The former is visualized to take place through nucleophilic attack of monomer on chain end methylene of the macro ester (reaction 7.27)[12,29]

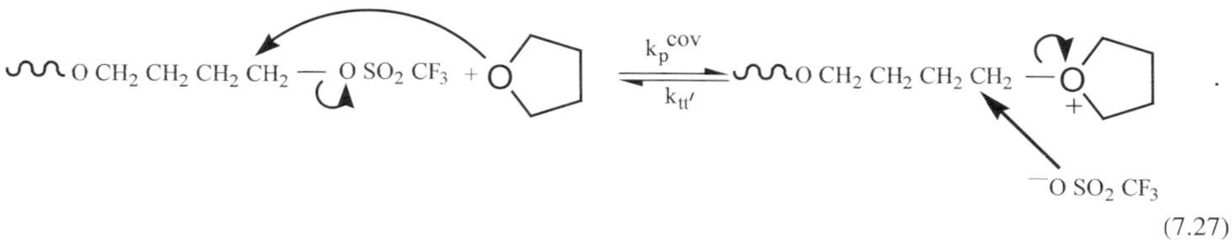

$$\sim\sim O\ CH_2\ CH_2\ CH_2\ CH_2{-}O\ SO_2\ CF_3\ +\ \underset{}{\bigcirc} \underset{k_{tt'}}{\overset{k_p^{cov}}{\rightleftharpoons}} \sim\sim O\ CH_2\ CH_2\ CH_2\ CH_2{-}\overset{+}{O}\!\!\bigcirc \quad {}^-O\ SO_2\ CF_3 \tag{7.27}$$

7.2.2.1.1 Macroion-macroester interconversion

The reversible conversion of macroester to macroion takes place both intermolecularly and intramolecularly. The former path is shown above (reaction 7.27). The intramolecular path is represented by reaction (7.28)[12,38]

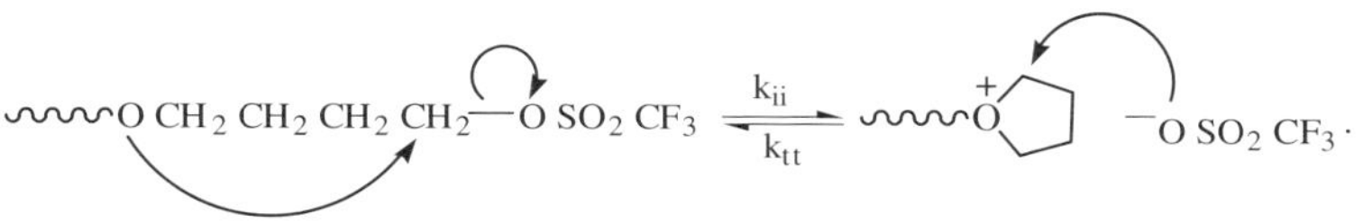

$$\sim\sim O\ CH_2\ CH_2\ CH_2\ CH_2{-}O\ SO_2\ CF_3 \underset{k_{tt}}{\overset{k_{ii}}{\rightleftharpoons}} \sim\sim \overset{+}{O}\!\!\bigcirc \quad {}^-O\ SO_2\ CF_3\ . \tag{7.28}$$

The intramolecular ionization is much faster than the intermolecular one. As a result, the ratio of ion to ester reaches the equilibrium value at early stages of polymerization and does not change with increase in conversion to any significant extent. The reversible switching of the propagating chain between ionic and covalent states is analogous to that of the propagating chain in living cationic polymerization except for the fact that the deactivated state in the latter is propagation-inactive unlike the macroester in the present polymerization.

7.2.2.2 Backbiting and intermolecular interchange

The nucleophilicity of ring oxygen relative to that of in-chain oxygen increases with decrease in ring strain.[39] This makes backbiting insignificant in the cationic polymerization of common ring monomers but competing with propagation in the polymerization of the small ring monomers. For example, 1,4-dioxane (the cyclic dimer) is the major product in the polymerization of ethylene oxide (Eq. 7.29).[12,40]

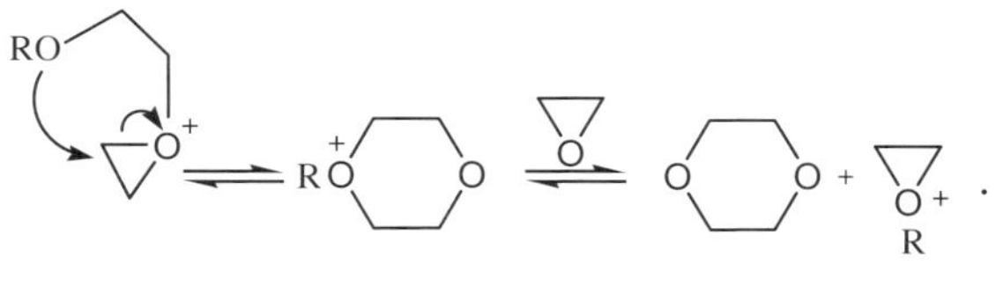

$$\text{RO-structures} \tag{7.29}$$

However, it is possible to direct backbiting to the formation of a desired macrocycle using suitable alkali ion as template for the alkali-ion-specific crown ether.[41] Similarly, in the polymerization of propylene oxide, cyclic oligomers are the major products up to 50% of which being cyclic tetramers.[42]

Due to the presence of the cyclics, the molecular weight of polymer formed is much lower than the theoretical value for a living polymerization

$$\overline{DP}_n(\text{theory}) = \frac{[M]_0}{[I]_0} \cdot f_c, \qquad (7.30)$$

where f_c is the fractional conversion. The cyclics also broaden MWD.

On the other hand, the intermolecular interchange results in MWD broadening without effecting a change in molecular weight particularly if polymerization is continued in monomer-depleted condition (*vide* Sec. 7.1.4).

7.2.3 Activated monomer mechanism of polymerization

The problem of polymer transfer (both backbiting and intermolecular interchange) in the cationic ROP of 3- and 4-membered cyclic ethers is largely eliminated by conducting the polymerizations in the presence of hydroxylic compounds using acid catalysts. In this method, the end of the growing chain remains capped with hydroxyl group, which renders it inactive toward polymer transfer. Monomer is activated by protonation. Propagation involves nucleophilic attack of hydroxylic oxygen present in the chain end on activated monomer (AM).[43]

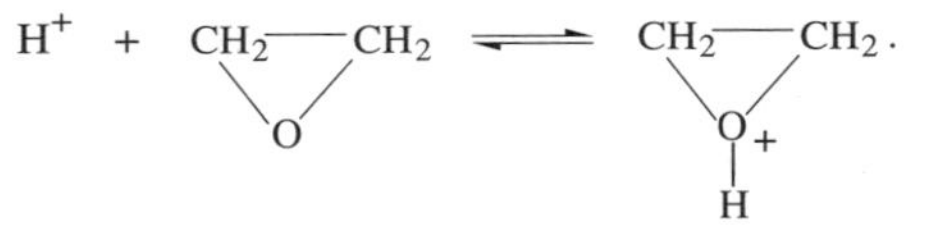

$$H^+ + \underset{O}{CH_2\text{---}CH_2} \rightleftharpoons \underset{\overset{+}{O}\text{---}H}{CH_2\text{---}CH_2}. \qquad (7.31)$$

$$R\text{---}OH + \underset{\overset{+}{O}\text{---}H}{CH_2\text{---}CH_2} \longrightarrow RO\text{---}CH_2\text{---}CH_2\text{---}OH + \underset{\overset{+}{O}\text{---}H}{CH_2\text{---}CH_2}. \qquad (7.32)$$

$$RO\text{---}CH_2\text{---}CH_2\text{---}OH + \underset{\overset{+}{O}\text{---}H}{CH_2\text{---}CH_2} \rightleftharpoons \text{Polymer}. \qquad (7.33)$$

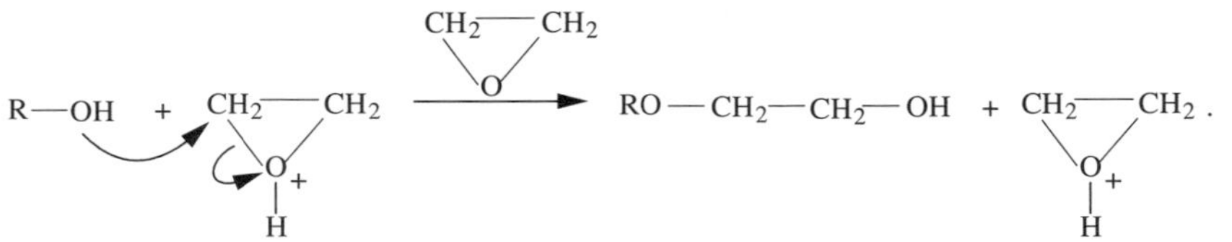

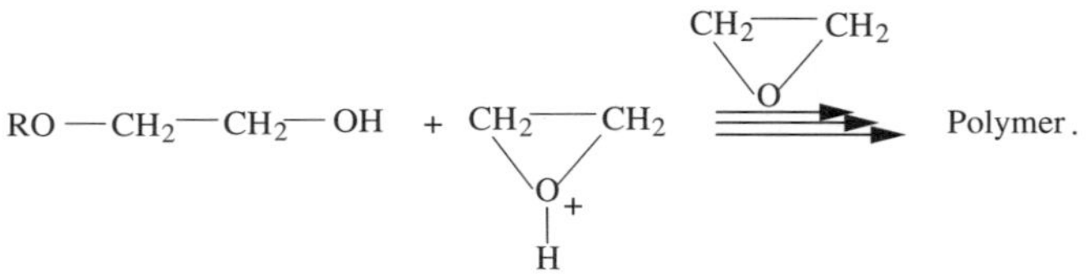

It follows from the mechanism that there exists no termination and the number of alcohol molecules used in the system determines the number of chains.[27] Thus, the degree of polymerization for complete conversion is given by

$$\overline{DP}_n = \frac{[M]_0}{[ROH]_0}. \tag{7.34}$$

However, a parallel propagation mechanism (activated chain end or ACE mechanism) proceeds in the regular course where the active center is located on the chain end as shown in reactions (7.35) and (7.36).

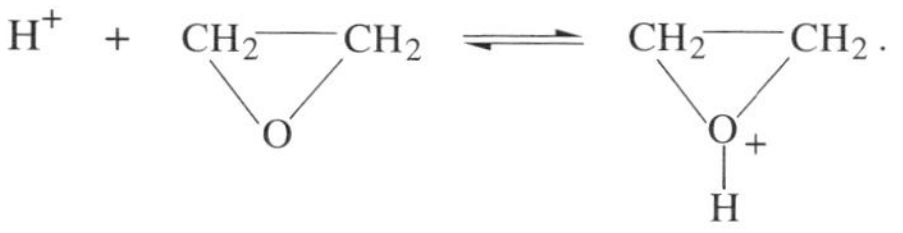

$$H^+ + \underset{O}{CH_2-CH_2} \rightleftharpoons \underset{\underset{H}{\overset{+}{O}}}{CH_2-CH_2}. \tag{7.31}$$

$$\underset{\underset{H}{\overset{+}{O}}}{CH_2-CH_2} + \underset{O}{CH_2-CH_2} \longrightarrow HO-CH_2-CH_2-\overset{+}{O}\underset{CH_2}{\overset{CH_2}{\diagup}}. \tag{7.35}$$

$$HO-CH_2-CH_2-\overset{+}{O}\underset{CH_2}{\overset{CH_2}{\diagup}} + \underset{O}{CH_2-CH_2} \rightleftharpoons \text{Polymer.} \tag{7.36}$$

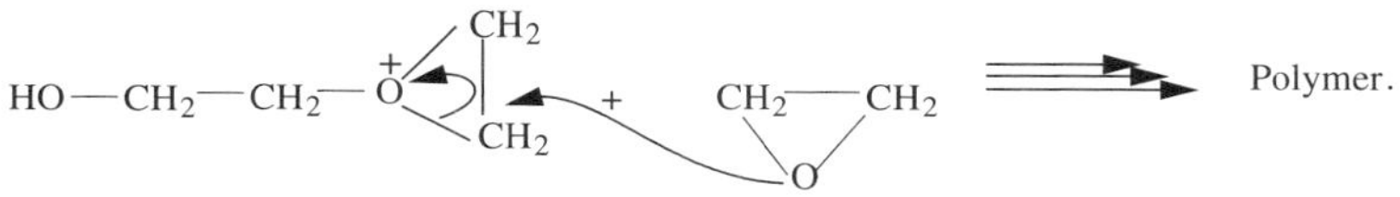

Because of the simultaneous occurrence of chain growth by the AM and ACE mechanisms the polymerization is not free from backbiting. The relative rate of propagation by AM and ACE mechanisms is given by[43]

$$\frac{R_{AM}}{R_{ACE}} = \frac{k_{p(AM)}[M^+H][-OH]}{k_{p(ACE)}[M^+H][M]}, \tag{7.37}$$

where $[M^+H]$ is the activated monomer concentration and $[-OH]$ represents the hydroxylic group concentration. Thus, by maintaining a high $[-OH]/[M]$ ratio during the whole course of polymerization, the ACE mechanism of polymerization and, hence, backbiting can be rendered insignificant. The concentration of hydroxyl functions is limited by the desired molecular weight (Eq. 7.34), whereas the monomer concentration can be reduced at will using a continuous process instead of a batch one keeping the total amount of monomer fed constant.[41,43] To give an example, in a polymerization of propylene oxide, the proportion of the cyclic tetramer is reduced from 50% to 0.95% by changing over from a batch process to a continuous one.[41]

Polymerization of unsymmetrically substituted oxiranes exhibits regioselectivity also, which depends on the interplay of reactivity as well as electronic and steric factors in propagation. Thus, whereas epichlorohydrin yields predominantly the head-to-tail polymer, propylene oxide does not exhibit any such preference.[44]

7.3 Cyclic Acetals

The polymerizability of common ring acetals is shown in Table 7.2. Of the monomers containing only one formal group, both the five-membered 1,3-dioxolane (**4**) and the seven-membered 1,3-dioxepane (**6**) are polymerizable, whereas the six-membered 1,3-dioxane (**5**) is not. On the other hand, the six-membered trioxane (**7**) containing three formal groups is polymerizable. Cationic polymerization occurs with the use of strong Bronsted acids or Lewis acids as initiators[45]

(4)   (5)   (6)   (7)

The propagating species consists of an equilibrium mixture of *tertiary* oxonium and oxocarbenium ionic species (Eq. 7.38). Although the latter is more reactive than the former, it constitutes only a very small fraction of the total ionic species making negligible contribution to propagation, which proceeds by nucleophilic attack of monomer on carbon in the formal group of the oxonium ion followed by ring opening (reaction 7.39).[46,47]

$$\sim\sim\sim\overset{+}{O}\underset{(CH_2)_r}{\overset{CH_2}{<}}O \rightleftharpoons \sim\sim\sim O-(CH_2)_r-O\overset{\pm}{=}CH_2 \cdot \quad (7.38)$$

$$\sim\sim\sim\overset{+}{O}\underset{(CH_2)_r}{\overset{CH_2}{<}}O + O\underset{(CH_2)_r}{\overset{CH_2}{<}}O \longrightarrow \sim\sim\sim O-(CH_2)_r-O-CH_2-\overset{+}{O}\underset{(CH_2)_r}{\overset{CH_2}{<}}O \cdot$$

(7.39)

Backbiting competes with propagation yielding large fraction of cyclics early in the polymerization like that occurs in the polymerization of oxiranes.[12] The cyclics eventually polymerize and may reach equilibrium concentrations at high conversions. Forcible termination of polymerization at an early stage gives largely cyclic products.

The critical monomer concentration for polymerization is relatively high, $\sim$2 mol/L, for 1,3-dioxolane (**4**).[46,52] However, this value based on the equilibrium ring concentration predicted by Jacobson–Stockmayer theory includes the concentration of the monomeric ring, which would be incorrectly predicted due to insufficient number of ring atoms.[50] The polymerization has, however, an $[M]_e = \sim$1 mol/L (*vide* Chap. 1, Table 1.6).

Incidentally, failure to take cognizance of the critical monomer concentration resulted in a bitter controversy among the early workers over the products of polymerization. Thus, whereas the Plesch group obtained exclusively cyclic products,[48,49] the Jaacks group obtained mainly linear polymers.[54] Other workers supported one group or the other.[54–60] The origin of the controversy was ultimately traced to the consistently large difference in monomer concentrations used by the two groups. The Plesch group used monomer concentrations, which are close to or less than the critical (*vide* Sec. 7.1.2), whereas the Jaacks group used much higher concentrations.

Trioxane (**7**) is an important monomer in that it yields polyformaldehyde (acetal resin) when polymerized. However, the polymerization proceeds with an induction period during which equilibrium concentrations of formaldehyde and cationic oxymethylene oligomers are established (Eq. 7.40).[61,62]

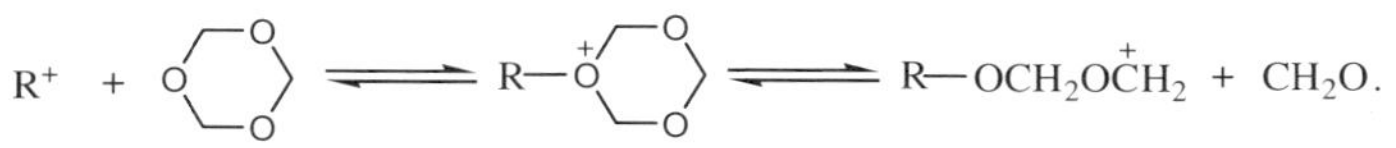

$$R^+ + \underset{(7)}{\text{trioxane}} \rightleftharpoons R\text{–}\overset{+}{O}\text{(ring)} \rightleftharpoons R\text{–}OCH_2O\overset{+}{C}H_2 + CH_2O. \qquad (7.40)$$

At the beginning of the reaction, the rate of depolymerization to formaldehyde is faster than the rate at which trioxane is polymerized. Hence, no polymer forms; instead, formaldehyde accumulates reaching eventually its equilibrium concentration, this being 60 mmol/L at 30°C (*vide* Table 1.6, Chap. 1). Thenceforth, the rate of polymerization of formaldehyde combined with that of trioxane exceeds the rate of depolymerization resulting in polymer accumulation.[61] Polyformaldehyde crystallizes out of the solution, the heat of crystallization making formaldehyde polymerization thermodynamically more favorable.

However, polyformaldehyde produced as above is unstable. It undergoes depolymerization during processing. Depolymerization involves the end hydroxyl groups

$$HO\text{–}(CH_2O)_n\text{–}H \longrightarrow HO\text{–}(CH_2O)_{n-1}\text{–}H + CH_2O.$$

In order to get rid of the problem, trioxane is copolymerized in commerce with 0.1–15 mol % ethylene oxide or 1,3-dioxolane.[63] The comonomer residues in the random copolymer disrupt the continuity of the oxymethylene sequences at various points in the chain. Any progress of depolymerization starting from chain end is arrested at the first disruption. The alternative method of stabilization of the homopolymer uses acetylation of the end hydroxyl groups.

7.4 Cyclic Esters

Cyclic ester monomers include lactones, dilactones, cyclic anhydrides, and cyclic carbonates. Lactones give polylactones, *i.e.*, poly(ω-hydroxyacid)s; dilactones give poly(dilactones), *i.e.*, poly(α-hydroxyacid)s. Of the unsubstituted common ring lactones, the 5-membered γ-butyrolactone is not polymerizable, whereas the 6-membered δ-valerolactone and the 7-membered ε-caprolactone undergo polymerization (Table 7.2). Of the dilactones, the most important monomers are the glycolide (**8a**) and the various isomeric lactides (stereoisomers of **8b**).[64]

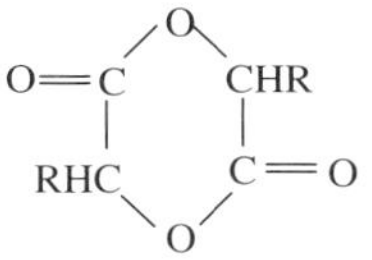

a) R = H   Glycolide

b) R = CH$_3$ Lactide

(**8**)

Living polymerizations of cyclic esters are carried out best using covalent metal alkoxides as initiators.

7.4.1 Anionic polymerization

Weak nucleophiles like alkali metal carboxylates, *viz.*, potassium acetate, and benzoate, are good enough to initiate ROP of highly strained 4-membered β-lactones. The carboxylate ion or the metal carboxylate ion pair attacks the carbon atom bonded with ring oxygen leading to C–O bond scission and ring opening as shown in reaction (7.41).

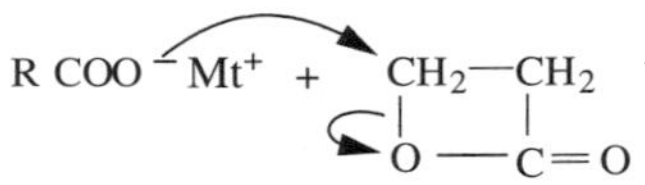

$$R\ COO^-\ Mt^+\ +\ \underset{\underset{O\ -\ C=O}{|\ \ \ \ \ \ \ \ |}}{CH_2-CH_2}\ \longrightarrow\ R\ COOCH_2CH_2COO^-\ Mt^+. \qquad (7.41)$$

Less strained higher lactones require stronger nucleophilic initiators, *e.g.*, metal alkoxides.[66,67,70] However, ring opening occurs in a different way, *viz.*, through the attack of the strong nucleophile on the carbonyl carbon leading to acyl-oxygen scission as occurs in the nucleophilic attack on linear esters

$$RO^-\ Mt^+\ +\ O=C-O\ \longrightarrow\ ROOC\ \ \ O^-\ Mt^+. \qquad (7.42)$$

Propagation proceeds with the alkoxide as chain carrier. Although strong nucleophiles can also effectively initiate polymerization of highly strained β-lactones, initiation occurs following both paths (reactions 7.41 and 7.42) yielding carboxylate as well as alkoxide propagating species. While the former reproduces itself in subsequent propagation steps, the latter converts almost half of itself to carboxylate in each subsequent propagation step. Thus, eventually, all propagating species become carboxylates as the polymerization progresses.[67]

However, polymerization of β-lactones by metal carboxylates proceeds slowly at moderate temperature *ca.*, 70°C.[66,68] At higher temperatures, significant irreversible monomer transfer (reaction 7.43) occurs resulting in ring opening and formation of α,β-unsaturated carboxylate.[69,71] With less strained larger lactones, monomer transfer is much slower due to the reaction not being aided by ring strain release (reaction 7.44). Furthermore, it is reversible.[71] As a result; anionic polymerizations of the larger lactones are living.

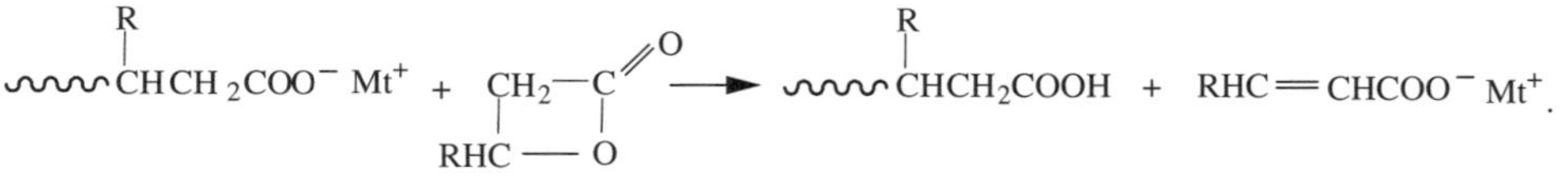

$$\sim\sim\sim\overset{R}{\underset{|}{C}}HCH_2COO^-\ Mt^+\ +\ \underset{\underset{RHC\ -\ O}{|\ \ \ \ \ \ \ \ |}}{CH_2-C}\overset{O}{\diagup}\ \longrightarrow\ \sim\sim\sim\overset{R}{\underset{|}{C}}HCH_2COOH\ +\ RHC=CHCOO^-\ Mt^+.$$

$$(7.43)$$

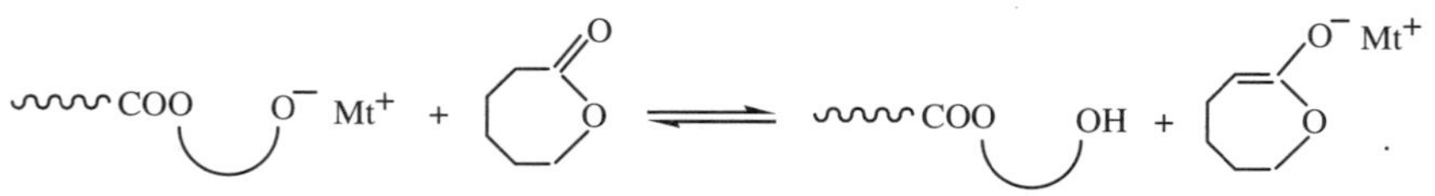

$$\sim\sim\sim COO\ \ \ O^-\ Mt^+\ +\ \bigcirc\!\!=\!\!O\ \rightleftharpoons\ \sim\sim\sim COO\ \ \ OH\ +\ \bigcirc\!\!\!-\!O^-\ Mt^+. \qquad (7.44)$$

However, polymerization of β-propiolactone may also be made living using macrocyclic ligands like crown ethers[66,72] or cryptates[68] to complex the alkali counterion. This makes the polymerization very fast allowing the use of low temperature so that monomer transfer is subdued. The increase in rate is attributed to the conversion of some of the inactive contact ion pairs or their aggregates into the active ligand-separated ion pairs and increased activity as well as increased dissociation due to the greater interionic distance

$$\left(\sim\sim O^- K^+\right)_n \rightleftharpoons n \sim\sim O^- K^+ \rightleftharpoons n \sim\sim O^- (K^+). \qquad (7.45)$$

(◯ = a crown ether or a cryptate).

Although the polymerizations are without termination, side reactions occur like in most ROPs (Sec. 7.1.4). For example, in the polymerization of ε-caprolactone (CL) in THF at 0°C initiated by potassium *tert*-butoxide, the initially formed linear polymer rapidly degrades to give the equilibrated mixture of rings and chains. At monomer concentrations of 0.21 molar or below, the products are all rings.[73,74] Thus, the critical monomer concentration in this case is much lower than that in the polymerization of 1,3-dioxolane discussed earlier (Sec. 7.3).

7.4.2 Coordination polymerization

Initiators in coordination polymerization are covalently bonded metal alkoxides or alkyl alkoxides,[12] *e.g.*, bimetallic μ-oxo-alkoxides,[75,76] aluminum alkoxides,[77–82] dialkylaluminum alkoxides,[83–86] tin(II) alkoxides,[13] tin(II) carboxylates,[87–92] tin(IV) alkylalkoxides,[78,93–96] titanium(IV) isopropoxide,[94–96] ferric alkoxides,[94–96] and so on. Initiation as well as propagation follows a two-step "coordination–insertion" mechanism. In the first step, the ester is coordinated with the metal alkoxide, which is followed by rearrangement of the bonds leading to cleavage of the acyl-oxygen and the metal-oxygen bonds. In the second step, reformation of new metal-oxygen and acyl-oxygen bonds takes place.[64,97–99]

$$O=C-O + Mt-OR \longrightarrow \begin{matrix} O \cdots Mt \\ \| \quad | \\ O-C \cdots OR \end{matrix} \longrightarrow ROC(=O)-O-Mt. \qquad (7.46)$$

A covalent alkoxide is less reactive than an ionic alkoxide so that chain transfer is suppressed leading to living polymerization.[79,81]

Some of the initiators exist in solution in the aggregated form. Thus, aluminum isopropoxide exists as the trimer (A_3) in equilibrium with the tetramer (A_4). The former is very reactive toward initiation of polymerization of ε-caprolactone (CL) but the latter is unreactive. As a result, at ambient temperature the polymerization of CL is initiated exclusively by A_3 resulting in the breakage of the aggregate and the growth of three chains from each Al atom as shown in Eqs. (7.47) and (7.48).[79,80]

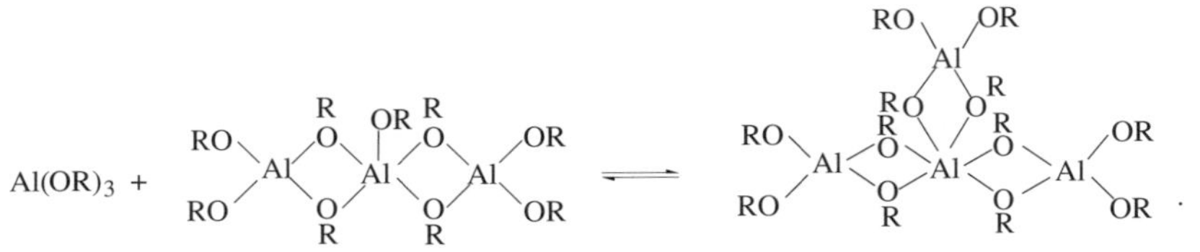

$$Al(OR)_3 + A_3 \rightleftharpoons A_4 \quad (R = CH(CH_3)_2) \tag{7.47}$$

$$A_3 + 9\,nM \longrightarrow 3\,Al\begin{pmatrix}OP_n\\OP_n\\OP_n\end{pmatrix} \tag{7.48}$$

(M = monomer)

Initiation by A_4 as well as its rate of conversion to A_3 or other reactive forms is extremely slow so that it is left unreacted when CL polymerization reaches polymer-monomer equilibrium. If the polymerization is not forcibly terminated, A_4 is incorporated very slowly into the polymer by alkoxide-ester interchange reaction. Eventually, these exchanges give rise to the most probable MWD. However, A_3 and A_4 may be isolated in greater than 90% purity and A_3 separately used in polymerization. The number of polymer molecules formed in this case is three times the number of A_3 molecules (reaction 7.48).

At higher temperatures, the exchange between A_3 and A_4 becomes faster so that A_4 also is used up in initiation. This happens, for example, in the polymerization of the lesser reactive lactide (**8**) at 100°C.

Besides the initiator, the Al anchored polymer chains also undergo aggregation particularly when there exists only one chain attached to Al. More the number of chains per Al atom is more the steric hindrance to aggregation becomes. Thus, unlike in the A_3/CL system the propagating species undergoes aggregation in the $R_2Al\,OR'$/CL system, although at relatively high concentrations. The degree of aggregation of the propagating species can be determined kinetically using the method described in Sec. 7.2.1.1, which is found to be 3 in THF at 25°C.[85] This agrees well with the value determined from the analysis of the ^{27}Al NMR spectrum of the polymerizing mixture, which exhibits separate resonances for the aggregated and the non-aggregated species.[96]

Polymerization of cyclic esters by covalent metal alkoxides is first order in monomer as is evident from the first order monomer-disappearance plots shown in Fig. 7.4 in the polymerization of L,L-lactide in THF with Sn $(OBu)_2$ as the initiator.[13] The rate of monomer consumption obeys Eq. (7.25), which applies to a living equilibrium polymerization. In addition, molecular weight increases linearly with conversion (Fig. 7.5) and is equal to the theoretical value given by

$$\overline{DP}_n = \frac{[M].f_c}{n[I]_0},$$

where f_c is the fractional conversion, n is the number of alkoxy groups per metal atom. The agreement between the experimental and the theoretical $\overline{DP}_n$ proves that all the alkoxy groups in the initiator are used up in initiation and there exists no irreversible chain transfer or termination. Besides, MWD is narrow (Fig. 7.1) unless polymerization is continued in monomer-depleted condition (*vide* Sec. 7.1.4).

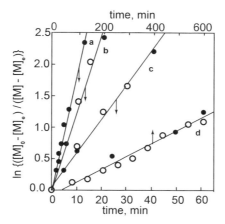

Figure 7.4 First order monomer disappearance plots in the polymerizations of L,L-lactide initiated by Sn(OBu)$_2$ in THF at 80°C. [Initiator]$_0$ (in mmol/L) = 5 (line a), 2 (line b), 1.16 (line c), 0.75 (line d); [M]$_0$ = 1 mol/L. "Reprinted with permission from Ref. 13. Copyright © 2000 American Chemical Society."

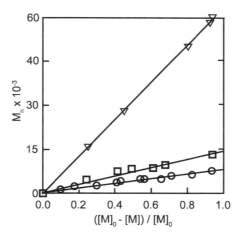

Figure 7.5 Molecular weight evolution with conversion in the living ROP of L,L-lactide in THF using Sn(OBu)$_2$ initiator. [M]$_0$ = 1 mol/L; [Sn(OBu)$_2$]$_0$ (in mmol/L) and polymerization temperature = 1.16 (∇), 5 ($\square$), both at 80°C; 8.9 (O) at 20°C. "Reprinted with permission from Ref. 13. Copyright © 2000 American Chemical Society."

The covalent tin alkoxide initiator may be prepared also *in situ*. Thus, tin(II) bis-2-ethylhexanoate, also called tin octoate, in combination with an alcohol makes efficient initiating system for the living polymerization of cyclic esters.[88–92] The actual initiator is tin alkoxide formed by alkoxide-carboxylate interchange.

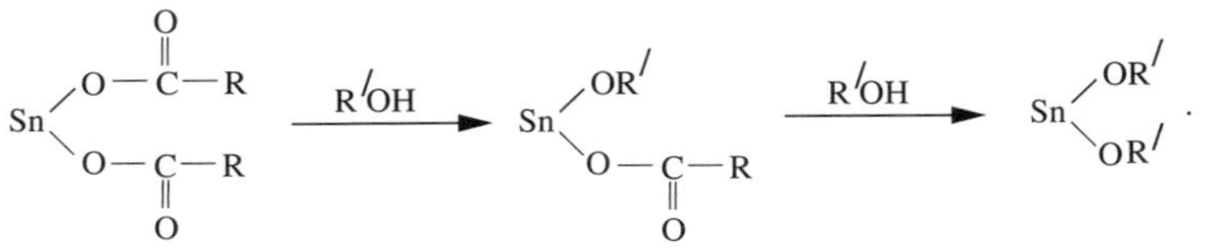

$$\text{Sn}\begin{matrix}\text{O}-\text{C}-\text{R}\\ \| \\ \text{O}\end{matrix} \cdots \begin{matrix}\text{O}\\ \| \\ \text{O}-\text{C}-\text{R}\end{matrix} \xrightarrow{\text{R'OH}} \text{Sn}\begin{matrix}\text{OR'}\\ \\ \text{O}-\text{C}-\text{R}\\ \| \\ \text{O}\end{matrix} \xrightarrow{\text{R'OH}} \text{Sn}\begin{matrix}\text{OR'}\\ \\ \text{OR'}\end{matrix}.$$

The degree of polymerization is given by the ratio of the mols of monomer consumed to the mols of alcohol used and is independent of the tin octoate concentration. The alcohol acts as a reversible chain transfer agent yielding as many polymer molecules as there are alcohol molecules

$$P_nO-Mt + ROH \rightleftharpoons P_nOH + Mt-OR \cdot$$

Some rare earth metal alkoxides (La, Sc, Sm, Y, or Yb) are much more active than Al or Sn alkoxides giving rise to much higher rates of polymerization.[100–105] However, the polymerizations are not free from termination.[101] Besides, backbiting occurs to a significant extent giving rise to cyclic oligomers.[103]

Covalent oxide initiators, *e.g.*, aluminum isopropoxide, also polymerize livingly certain common ring monomers belonging to the cyclic carbonates and cyclic anhydrides families, *e.g.*, 2,2-dimethyltrimethylene carbonate[106] and adipic anhydride[107,108] (**9**) respectively. The active centers are not of the same type in the two cases. In the case of adipic anhydride the active center is aluminum carboxylate (reaction 7.49), whereas in the other case it is aluminum alkoxide.

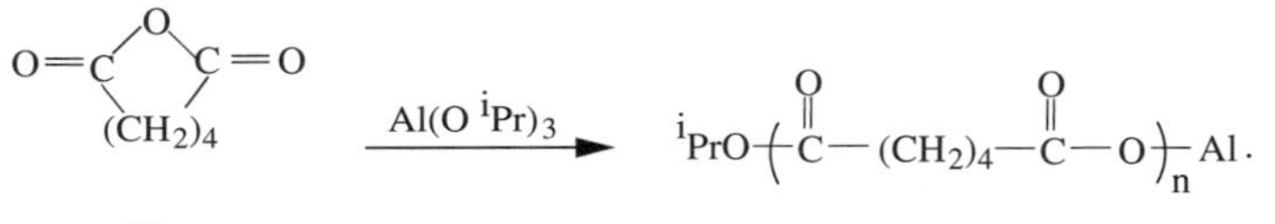

$$O=C\underset{(CH_2)_4}{\overset{O}{\diagdown}}C=O \xrightarrow{Al(O\,^iPr)_3} {}^iPrO\!\left(\!\!\begin{matrix}O\\ \|\\ C\end{matrix}\!-\!(CH_2)_4\!-\!\begin{matrix}O\\ \|\\ C\end{matrix}\!-\!O\!\right)_{\!\!n}\!\!Al. \qquad (7.49)$$

(**9**)

7.4.2.1 Backbiting and intermolecular interchange

As discussed earlier, the rate of polymer transfer reaction depends on the nature of the active center. With ionic (alcoholate) chain ends, backbiting occurs at a fast rate so that a large proportion of cyclics form before all the monomer is consumed (Sec. 7.1.4). In contrast, with covalent alkoxide chain ends (*e.g.*, aluminum alkoxide), cyclics do not appear even at the completion of polymerization but do so after a very long time (*e.g.*, observed in ϵ-caprolactone polymerization after 57 days had elapsed).[83] Nevertheless, both systems would give the same MWD at equilibrium.

Intermolecular interchange also occurs at much slower rate in coordination polymerization than in ionic polymerization. However, MWD broadening occurs when polymerization is continued in monomer-depleted conditions (*vide* Sec. 7.1.4).

7.4.3 Cationic polymerization

The active centers in the cationic ROPs of lactones are cyclic dialkoxycarbenium ionic species (reaction 7.50). Propagation involves alkyl-oxygen bond scission.[109–113] The initiators are the same as those used in the cationic ROP of cyclic ethers.

$$(7.50)$$

Backbiting competes with propagation giving rise to a considerable fraction of cyclic products.

Formation of cyclics can be avoided by conducting the polymerization in the presence of catalytic amounts of water or alcohols as in the case of cyclic ethers (Sec. 7.2.3).[114–117] The polymerization occurs livingly following activated monomer (AM) mechanism yielding polymer of low PDI

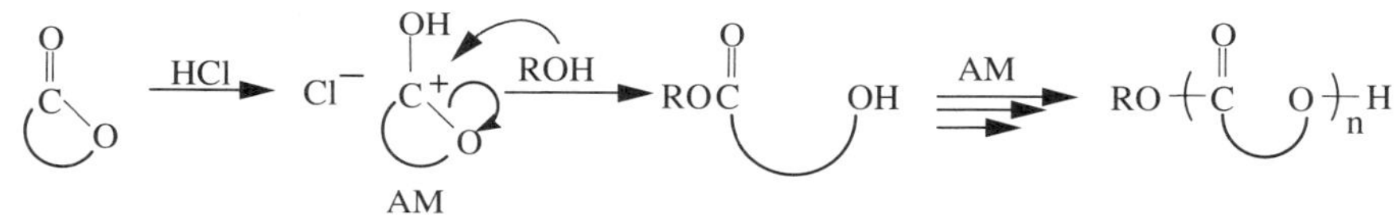

Apart from lactones, cyclic carbonates also undergo living polymerization by this method.[115,117]

7.4.4 Enzymatic polymerization

Many bacteria synthesize and store in the cells aliphatic polyesters. *In vitro* ring-opening polymerization of lactones and cyclic carbonates was, therefore, studied using enzyme catalysts for developing environment–friendly method of synthesis.[118,119] These enzymes are lipases of different origin. Polymerization involves nucleophilic attack of the hydroxylic oxygen of the serine residue in lipase on the carbonyl carbon of the monomer resulting in ring opening with the formation of an acyl-enzyme intermediate. The latter acts essentially as an enzyme-activated monomer (EM)[118]

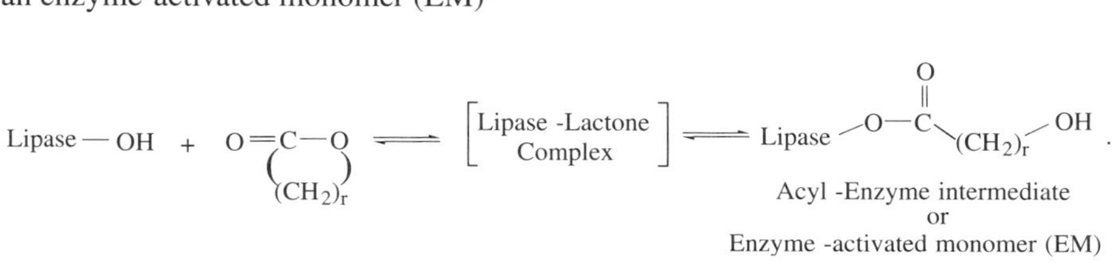

The acyl-enzyme intermediate subsequently undergoes hydrolysis by water present in enzyme or alcoholysis by purposely-added alcohol in the first propagating step.[120]

$$\text{EM} + \text{ROH} \longrightarrow \text{HO}\text{-}(\text{CH}_2)_r\text{-COOR} + \text{lipase}\text{-OH}.$$

(R=H or alkyl)

Thus, the hydroxylic compounds act as the real initiators. Propagation takes place by nucleophilic attack of terminal hydroxyl group in propagating chain on EM.[120]

7.5 Lactams

Initiation of polymerization of lactams can be effected by bases, acids, or water. Polymerization initiated by water, referred to as hydrolytic polymerization, is used in the industry to produce nylon 6. Other nylons are not usually produced by this method.[121] Cationic polymerization initiated by acids gives low yield as well as low molecular weight polymer and hence is not industrially practiced. On the other hand, anionic polymerization initiated by bases proceeds rapidly and gives high molecular weight polymer. Hence, the latter is used when direct polymerization of the monomer in molds is required. All the three kinds of polymerization follow the activated monomer mechanism. Transamidation is a common side reaction in these polymerizations.

7.5.1 Anionic polymerization

Initiators used are lactamate anions (L_0) or their precursors. However, initiation solely by these anions is slow. Hence, an activator, N-acyllactam (P_0), which is similar in structure to the propagating chain end, is used so that initiation becomes nearly as fast as propagation.[122–124] The mechanism of polymerization is shown in reactions (7.51) through (7.53).

The lactamate ion initiates polymerization by nucleophilic attack at the endocyclic carbonyl carbon in the activator resulting in the formation of the amidate ion (A_1). N-acylation of the lactam in the activator increases the electrophilicity of the endocyclic carbonyl carbon facilitating nucleophilic attack by the lactamate ion. Propagation involves activation of the monomer by proton transfer to the amidate ion (A_1) regenerating the lactamate ion and generating an N-acyllactam-ended chain (P_1) (reaction 7.53). The latter in turn is attacked by the lactamate ion (the activated monomer) and the cycle is repeated many times producing the polymer. It thus follows that at each propagation step the activated monomer is needed, and the number of chains in a living system equals that of the activator.

Transamidation, which can be intramolecular (backbiting) as well as intermolecular, occurs with varied facility and consequent effects on molecular weight and molecular weight distribution (*vide* Sec. 7.1.4).

Initiation

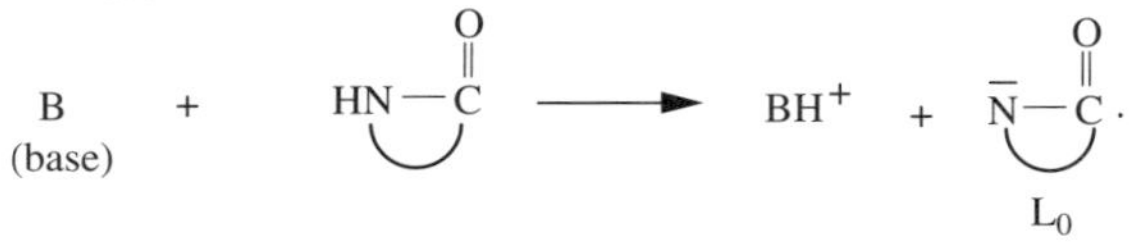

$$\text{B} + \text{HN-C(=O)} \longrightarrow \text{BH}^+ + \underset{L_0}{\overline{\text{N}}\text{-C(=O)}} \quad (7.51)$$

(base)

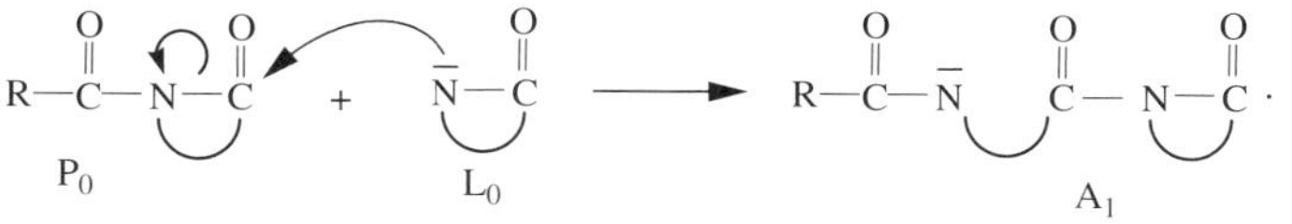

Ring-Opening Polymerization and Ring-Opening Metathesis Polymerization 351

$$R-\underset{\underset{P_0}{\|}}{\overset{O}{C}}-\bar{N}-\underset{\|}{\overset{O}{C}} + \underset{L_0}{\bar{N}-\overset{O}{\overset{\|}{C}}} \longrightarrow R-\underset{\|}{\overset{O}{C}}-\bar{N}\underset{A_1}{\overset{O}{\overset{\|}{C}}-N-\overset{O}{\overset{\|}{C}}} \cdot \quad (7.52)$$

Propagation

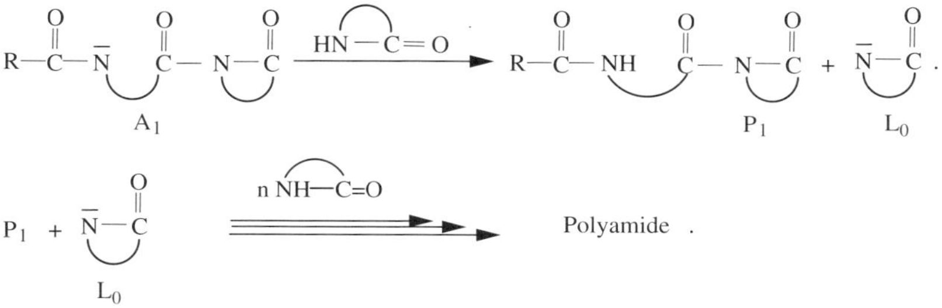

$$R-\underset{\|}{\overset{O}{C}}-\bar{N}\underset{A_1}{\overset{O}{\overset{\|}{C}}-N-\overset{O}{\overset{\|}{C}}} \xrightarrow{HN-C=O} R-\underset{\|}{\overset{O}{C}}-NH\underset{P_1}{\overset{O}{\overset{\|}{C}}-N-\overset{O}{\overset{\|}{C}}} + \underset{L_0}{\bar{N}-\overset{O}{\overset{\|}{C}}} \cdot \quad (7.53)$$

$$P_1 + \underset{L_0}{\bar{N}-\overset{O}{\overset{\|}{C}}} \xrightarrow{n\ NH-C=O} \text{Polyamide} \ .$$

In the unsubstituted lactams of high ring-strain, the protons attached to α-carbon atoms neighboring the carbonyl are acidic enough to be abstracted by amidate and/or lactamate ion causing termination[121]

$$\underset{HN\ -\ C=O}{\overset{CH_2-CH_2}{|\quad\quad|}} + A_i^- \longrightarrow \underset{HN\ -\ C=O}{\overset{CH_2-\bar{CH}}{|\quad\quad|}} + A_i\ H,$$

where A_i is the lactamate or the growing amidate ion.

7.5.2 Cationic polymerization

In cationic polymerization, the monomer is activated by protonation

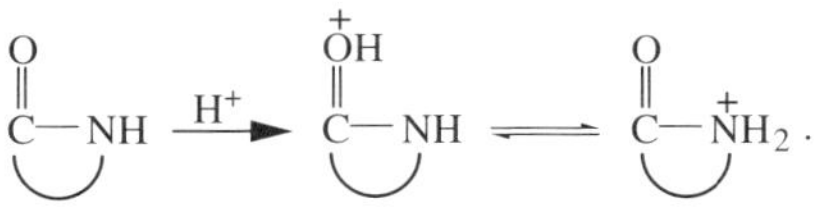

$$\underset{}{\overset{O}{\overset{\|}{C}}-NH} \xrightarrow{H^+} \underset{}{\overset{\overset{+}{OH}}{\overset{\|}{C}}-NH} \rightleftharpoons \underset{}{\overset{O}{\overset{\|}{C}}-\overset{+}{N}H_2} \cdot$$

Initiation takes place by the nucleophilic attack of an unactivated monomer on the carbonyl carbon of an activated monomer.[12,125] This is followed by proton transfer from the resulting adduct to a monomer liberating the conjugate base of the former and regenerating a protonated monomer. Thus, only a catalytic amount of protonic acid is needed.

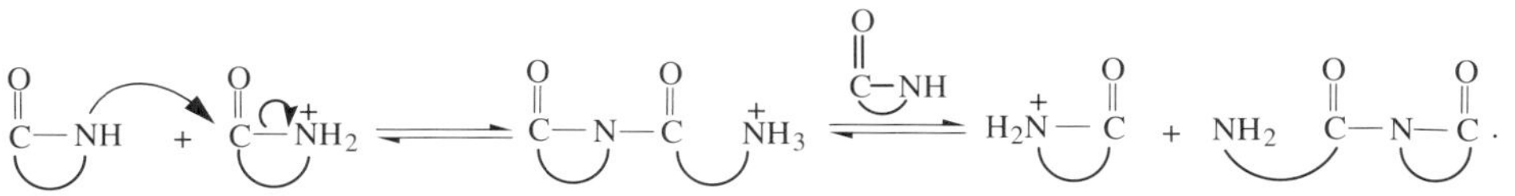

$$\overset{O}{\overset{\|}{C}}-NH + \overset{O}{\overset{\|}{C}}-\overset{+}{N}H_2 \rightleftharpoons \overset{O}{\overset{\|}{C}}-N-\overset{O}{\overset{\|}{C}} \overset{+}{N}H_3 \xrightarrow{\overset{O}{\overset{\|}{C}}-NH} H_2\overset{+}{N}-\overset{O}{\overset{\|}{C}} + NH_2\overset{O}{\overset{\|}{C}}-N-\overset{O}{\overset{\|}{C}} \cdot$$

Propagation proceeds by the nucleophilic attack of the amino nitrogen at the chain end on the carbonyl carbon of the activated monomer, which is restored in each propagation step

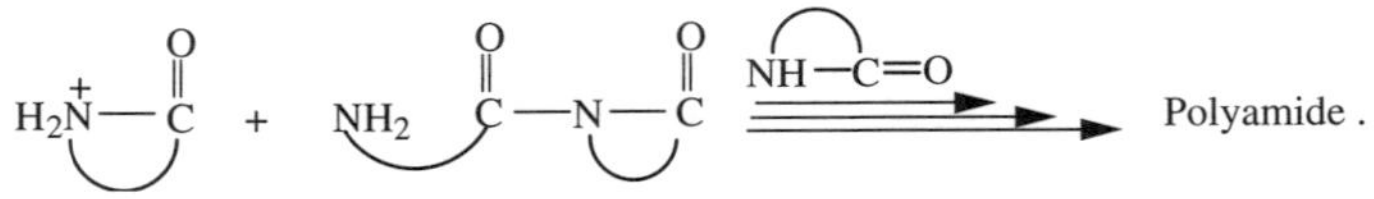

$$\overset{+}{H_2N}-\overset{O}{\overset{\|}{C}} \quad + \quad NH_2 \overset{O}{\overset{\|}{C}}-N-\overset{O}{\overset{\|}{C}} \overset{NH-C=O}{\rightleftharpoons} \quad \text{Polyamide}.$$

Step polymerization may also proceed simultaneously. This involves activation of the lactam end of a chain by protonation followed by nucleophilic attack on it by the terminal amine of another chain

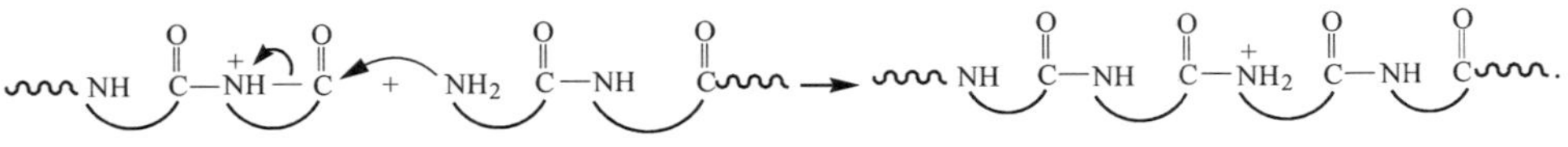

$$\sim NH \overset{O}{\overset{\|}{C}}-\overset{+}{NH}-\overset{O}{\overset{\|}{C}} \quad + \quad NH_2 \overset{O}{\overset{\|}{C}}-NH \overset{O}{\overset{\|}{C}}\sim \rightarrow \sim NH \overset{O}{\overset{\|}{C}}-NH \overset{O}{\overset{\|}{C}}-\overset{+}{NH_2} \overset{O}{\overset{\|}{C}}-NH \overset{O}{\overset{\|}{C}}\sim.$$

Even though the in-chain carbonyls are much less basic than the ring carbonyls, there is always a finite possibility of protonation of the in-chain carbonyls. This may then be followed by amidation. The possibility of this side reaction increases with increase in chain length.

7.5.3 *Hydrolytic polymerization*

As has been already introduced, hydrolytic polymerization refers to polymerization initiated by water. Nylon 6 is produced industrially by hydrolytic polymerization of caprolactam. An induction period occurs during which the true initiating species ω-aminocaproic acid is formed by the hydrolysis of the monomer

$$\underset{(CH_2)_5}{HN-CO} + H_2O \rightleftharpoons H_2N(CH_2)_5 COOH.$$

Indeed, the induction period can be eliminated by initially adding in the polymerization charge ω-aminocaproic acid or hexamethylenediamine salt of adipic acid.[123,126–128] Industrially, the polymerization is carried out at 250–280°C in the presence of 5–10% water for 12 to 24 h.[127] A monofunctional reactant, viz., acetic acid, is usually used to limit the degree of polymerization. Propagation occurs by both ring opening and condensation.

Ring-opening polymerization
$$H_2N(CH_2)_5 COOH + (x-1) \underset{(CH_2)_5}{HN-CO} \rightleftharpoons H\text{+}HN(CH_2)_5 CO\text{+}_x OH.$$

Condensation polymerization
$$H\text{+}HN(CH_2)_5CO\text{+}_x OH + H\text{+}HN(CH_2)_5CO\text{+}_y OH \rightleftharpoons H\text{+}HN(CH_2)_5CO\text{+}_{x+y} OH.$$

However, the ring-opening polymerization proceeds much faster than condensation so that the latter becomes important only after about 95% monomer is consumed.[128] The ring-opening propagation occurs through the nucleophilic attack of the chain end NH_2

on the carbonyl carbon of the monomer, which is activated by protonation with the carboxylic proton.[129] Thus, an activated monomer mechanism of propagation operates

$$\sim\sim CO(CH_2)_5 NH_2 + \overset{+}{NH_2}-\underset{(CH_2)_5}{CO} \longrightarrow \sim\sim CO(CH_2)_5 NHCO(CH_2)_5 NH_2 .$$

7.6 N-Carboxy-α-aminoacid Anhydrides

N-carboxyanhydrides of α-amino acids (NCA, **10**) undergo polymerization by bases to yield polypeptides.[130–137] The method finds favor for the large scale preparation of polypeptides, which finds increasing applications in biotechnology, medicine, and medical diagnostics. Polymerization can occur following two different mechanisms, their relative contribution being determined by the relative nucleophilicity and the basicity of the bases.

1. Amine (nucleophile) mechanism

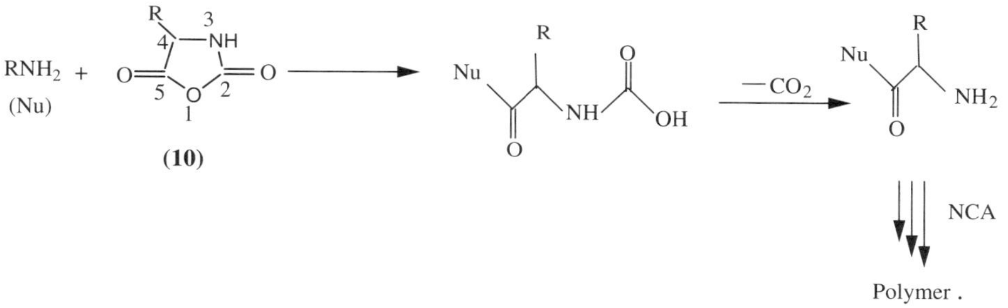

(7.54)

In the amine mechanism, a nucleophile attacks the monomer at the C5 position resulting in ring opening through acyl–oxygen bond scission. The resultant carbamate eliminates CO_2 following protonation generating a primary amine, which subsequently acts as a nucleophile, and the process is repeated in subsequent propagation steps. The polymer formed retains the initiator nucleophile covalently bonded at one of its ends.

2. Activated monomer mechanism (with nonnucleophilic bases)

The activated monomer mechanism (reaction sequence 7.55) is similar to that described for the anionic polymerization of lactams.[9] The base abstracts an N–H proton from the monomer generating the NCA anion (M^-), which acts as a nucleophile to attack the unactivated monomer (M) at the C5 position resulting in ring opening through acyl–oxygen bond scission in the latter. The carbamate product (P_1^-) eliminates CO_2 following protonation generating N-aminoacyl NCA (P_1) and regenerating the base. The latter repeats the proton abstraction from the monomer generating another NCA anion (M^-), which in turn attacks the P_1 on C5 of the NCA ring. Propagation occurs with the activated monomer attacking the NCA ring at the propagating chain end.

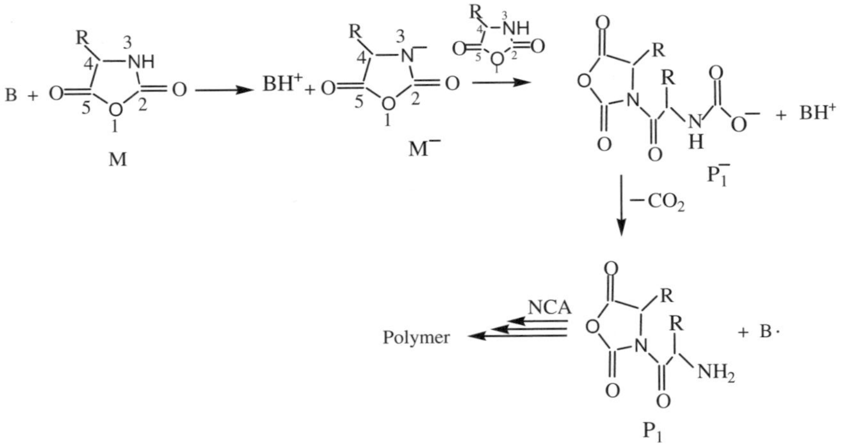

$$B + M \longrightarrow BH^+ + M^- \longrightarrow P_1^- + BH^+ \quad (-CO_2) \quad (7.55)$$

$$P_1^- + NCA \rightleftarrows Polymer + B\cdot$$

Hydroxides, alkoxides, and even water, among others, effect polymerization by this mechanism. However, since no N-acylated monomer is used as an activator unlike in lactam polymerization (Sec. 7.5.1), initiation is slower than propagation. As a result, polymers are polydisperse although they are mostly living. Besides, the polymerization switches back and forth between the two mechanisms (activated monomer and amine) due to the amino end of the propagating chain being able to propagate through the amine mechanism. In addition, the N-aminoacyl intermediates can self-condense giving rise to high molecular weight polypeptides as monomer conversion approaches 100%. This contributes to further broadening of MWD.

Polymerization initiated by nucleophilies also suffers from side reactions. The most important one is due to the polymerization switching back and forth between the two mechanisms with the amine acting as both nucleophile and base.

To make NCA polymerization living, it is required to circumvent the activated monomer mechanism. Two methods are available for achieving this; one of these uses zero valent nickel and cobalt initiators,[133–135] e.g., bpyNi(COD) (COD = 1,5-cyclooctadiene and bpy = 2,2'-bipyridine) and (PMe$_3$)$_4$Co, which react with the monomer forming metal amine complex capable of initiating polymerization. The presence of free amine is avoided by its complexation with the metal ion. Polypeptides of molecular weight ranging between 500 and 50 000 with PDI < 1.2 may be prepared by this method. The other method uses dry primary amine hydrochloride instead of free amine as the initiator.[138–141] The amine hydrochloride dissociates into free *primary* amine and hydrogen chloride. The latter converts any NCA$^-$ ion into NCA, thus preventing polymerization by the activated monomer mechanism.[139,141] Monodisperse (PDI < 1.03) homopolymers as well as block copolymers may be prepared by this method. An added attraction of the method is the ease of synthesis of hybrid block copolymers, e.g., polystyrene-b-polypeptide, which may be done using dry polystyrene–NH$_2$ hydrochloride as macroinitiator of NCA polymerization.[141]

7.7 Oxazolines (Cyclic Imino Ethers)

Polymerization of oxazolines is initiated by cationic initiators such as strong protonic acids or their salts or esters, alkyl halides, acid anhydrides, and Lewis acids.[142,143] The mechanism of polymerization of 2-alkyl-2-oxazoline (**11**) follows the reaction sequence (7.56).[144]

The active centers exist in equilibrium between ionic and covalent species. In this respect, the situation is analogous to that observed in the cationic polymerization of tetrahydrofuran.[12]

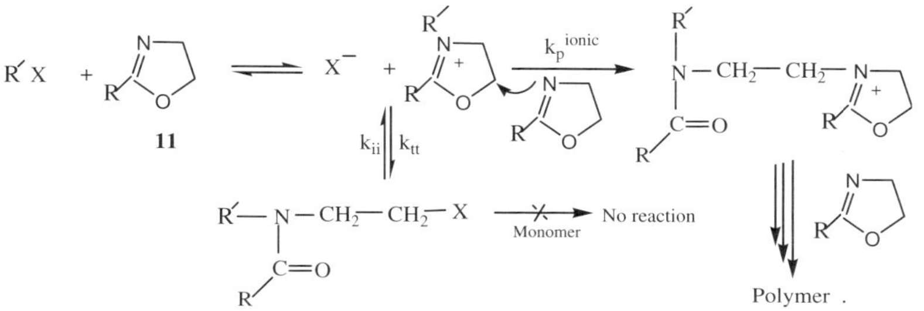

$$(7.56)$$

The equilibrium constant of reversible ionization of the covalent species, $K = k_{ii}/k_{tt}$ (where the subscripts ii and tt stand for intramolecular ionization and temporary termination respectively) depends upon the nucleophilicity of the counter ion, the dielectric constant of the medium, the nature of the substituents, and the temperature of polymerization. With iodide, triflate, or tosylate as the counterion, the active centers are predominantly ionic in acetonitrile.[145,146] With bromide as the counterion, the proportion of ionic species is ca., 60% in nitrobenzene at 25°C but less than 3% in CCl_4, whereas, with chloride as the counterion, the ionic species is not detectable by NMR even in nitrobenzene. Following a procedure similar to that used to determine the various rate constants in the strong Bronsted acids initiated polymerizations of THF or oxepane described in Sec. 7.2.2, k_p^{ionic} in the polymerization of 2-methyl-2-oxazoline was determined to be 2.8×10^{-5} L mol^{-1}s^{-1} at 25°C in nitrobenzene (Table 7.3).

The living polymerization of oxazolines has been successfully used in the preparation of telechelic polymers, macromonomers, block copolymers, and (co)polymers of various architectures.[142,143,145]

7.8 Cyclic Amines

Five- and six-membered cyclic amines are not polymerizable (Table 7.2). Others can be polymerized cationically by initiators such as protonic acids, Lewis acids, organic halides, esters of strong acids, onium salts, *etc*.[147,148] Polymerization by protonic acids follows both activated monomer (AM) and activated chain end (ACE) mechanisms (*vide* Sec. 7.2.3), as discussed below with aziridine (ethylene imine) as the monomer.[125]

Initiation

$$\text{azetidine} + H^+ \longrightarrow \text{azetidinium}^+ \cdot$$

$$\text{azetidinium}^+ + \text{azetidine-H} \longrightarrow \overset{+}{N}H_3-CH_2-CH_2-N\triangleleft \xrightarrow{\text{azetidine}} NH_2-CH_2-CH_2-N\triangleleft + \text{azetidinium}^+ \cdot$$

Propagation

$$\begin{array}{c}
\text{azetidinium} \xleftarrow{a} H_2N-CH_2-CH_2-N\triangleleft \\
\xrightarrow{b}
\end{array}
\begin{cases}
\xrightarrow{a} NH_2-CH_2-CH_2-\overset{+}{N}H_2-CH_2-CH_2-N\triangleleft \\
\xrightarrow{b} H_2N-CH_2\,CH_2-\overset{+}{N}\triangle-CH_2-CH_2-NH_2
\end{cases}$$

(7.57)

Propagation proceeds with nucleophilic attack of either of the two nucleophilic end groups of the propagating chain on the activated monomer. Product of reaction 7.57(a) transfers a proton to the monomer and nucleophilic attack of the primary amine on the protonated monomer follows (AM mechanism). In contrast, reaction 7.57(b) introduces iminium rings in the chain. These rings become subject to nucleophilic attack by the monomer or the amino end of the propagating chain (ACE mechanism) leading to branched polymers. Determination of polymer microstructure by ^{13}C NMR proved the presence of all three types of amino groups: *primary*, *secondary*, and *tertiary*.[149]

Involvement of cyclic iminium species in the polymerization of cyclic amines was established from the studies of 300 MHz ^1H NMR spectra of polymerizing 1,3,3-trimethyl-azetidine initiated by triethyloxonium tetrafluoroborate[150]

$$\text{azetidine} + Et_3OBF_4 \longrightarrow \text{iminium}^+ BF_4^- + Et_2O .$$

7.9 Cyclic Sulfides

Five- and six-membered cyclic sulfides are not polymerizable (Table 7.2). Three- and four-membered ones undergo both anionic and cationic polymerizations.[151] Cationic polymerization of thiiranes unlike that of oxiranes gives rise to dormant polymer, which undergoes chain extension on being fed with additional monomer.[152,153] The chain end of the dormant polymer formed through backbiting is a 12-membered cyclic tertiary sulfonium ion. It reinitiates polymerization releasing the cyclic tetramer and reforming the active chain.

$$R^+ + \triangle_S \longrightarrow \overset{+S-R}{\triangle} \longrightarrow RSCH_2CH_2-\overset{+}{S}\triangleleft \rightleftharpoons R\mathord{-}(SCH_2CH_2)_n\mathord{-}\overset{+}{S}\triangleleft .$$

(7.58)

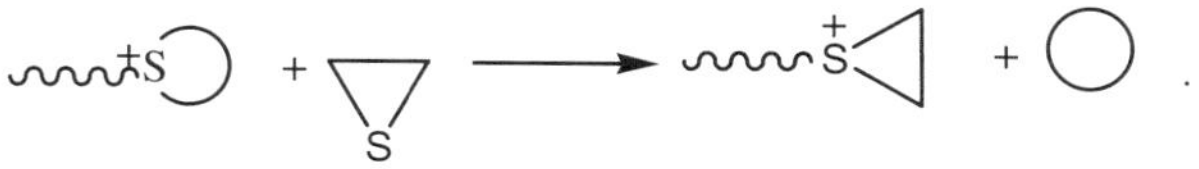

$$\text{~~~S}^+ + \triangle_S \longrightarrow \text{~~~S}^+ + \bigcirc (7.59)$$

12-membered ring ended polymer   12-membered ring oligomer

In contrast, polymerization of the substituted four-membered cyclic sulfide, 3,3-dimethylthietane stops short of completion due to irreversible intermolecular polymer transfer resulting in the formation of the linear *tertiary* sulfonium ion, which is incapable of reinitiating chains[154,155]

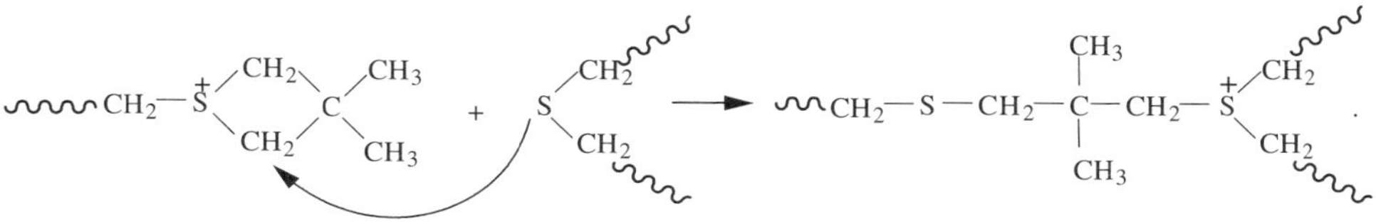

(7.60)

7.10 Cyclosiloxanes

Hexamethylcyclotrisiloxane (D_3) and octamethylcyclotetrasiloxane (D_4) are the two commonly studied monomers of this class. However, D_3 has moderate ring strain, whereas D_4 is strain-free (*vide* Sec. 7.1.1). Hence, the critical monomer concentration in polymerization is very low for D_3 but considerably high for D_4 (*ca.*, 18 vol % in THF at room temperature).[156] Polymerization can be effected both by anionic and cationic initiators.

7.10.1 *Anionic polymerization*

Strong bases such as quaternary ammonium or phosphonium hydroxide, alkali hydroxides, alkoxides or trimethylsilanoates, alkali napthalenides and alkyls can initiate the anionic polymerization. Initiation as well as propagation involves nucleophilic attack of the anion on Si atom in the monomer resulting in Si–O bond scission[157–159]

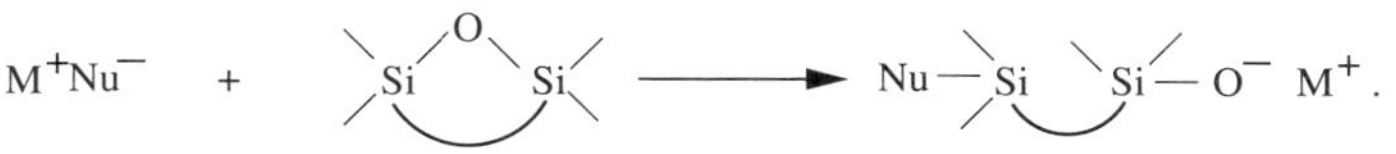

$$M^+Nu^- + \overset{O}{\underset{Si Si}{\frown}} \longrightarrow Nu-Si Si-O^- M^+ . (7.61)$$

Propagation involves similar nucleophilic attack of the silanoate chain end on the Si atom in the monomer

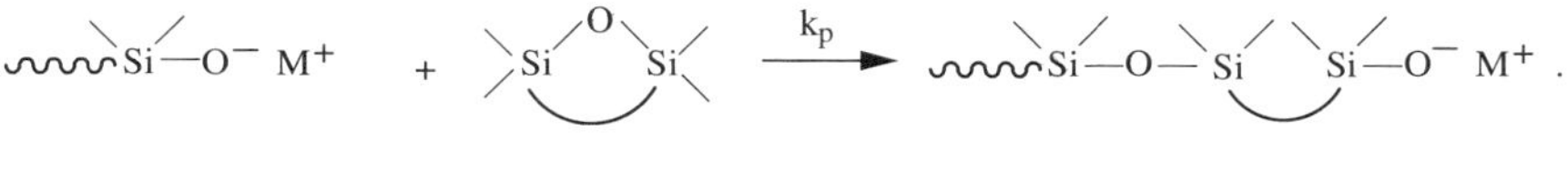

$$\text{~~~Si}-O^- M^+ + \overset{O}{\underset{Si Si}{\frown}} \xrightarrow{k_p} \text{~~~Si}-O-Si Si-O^- M^+ . (7.62)$$

Polymerization is usually slow. Free ion is not present. The external order in the active center is fractional indicating that the reactive ion pair species is in equilibrium with the unreactive aggregated species. However, the rate is sharply increased in the presence of nucleophiles like hexamethylphosphoramide, dimethylsulfoxide, dimethylformamide, or macroheterocyclic complexing agents of alkali metal ions, *e.g.*, crown ethers and cryptates.[158,159] These substances form complexes with the metal ions and thereby reduce the cation anion interaction resulting in increased reactivity of ion pairs.

Although termination and irreversible transfer are absent, polymerization is not free from side reactions of which backbiting is the most important. However, initiation as well as propagation of D_3 is fast enough in the presence of nucleophilic promoters to render backbiting unimportant. Nevertheless, the importance of backbiting increases as monomer concentration is depleted (*vide* Sec. 7.1.4). Higher cyclics D_4, D_5, and D_6 appear in the product if the reaction mixture is allowed to stand for 1–2 h after polymerization.[157] It is therefore desirable to terminate the polymerization before complete conversion, which can be done, for example, by adding chlorotrimethylsilane, which converts the chain end to the unreactive siloxane group.

7.10.2 *Cationic polymerization*

Strong acids such as $HClO_4$, H_2SO_4, CH_3SO_3H, and CF_3SO_3H commonly initiate the cationic polymerization of cyclosiloxanes. However, Lewis acids also effect polymerization acting as coinitiators with adventitious protogenic initiators.[158–161]

The mechanism of polymerization is very complex and not well understood.[12] Polymerization by triflic acid occurs rapidly in methylene chloride at ambient temperature. Backbiting competes with propagation so that cyclics appear from the very beginning of polymerization. However, the distribution of cyclics differs for D_3 from that for any of the other D_ns. With D_3, the predominant cyclic product is D_6; the other cyclic products are multiples of D_3 (D_9, D_{12}, *etc.*), macrocycles ($M_n \sim 10^4$), and minor amounts of D_4 and D_5 ($D_4 > D_5$). In contrast, with the other D_ns cyclics are formed in amounts, which follow the same order as obtained in an equilibrium distribution, *i.e.*, $D_4 > D_5 > D_6 \ldots$, *etc*.[162] Additionally, these cyclics reach their equilibrium concentrations before the high polymer appears.[163,164] The different cyclics distribution with D_3 presumably arises from the differences in ring strain and basicity between D_3 and other monomers (D_n, $n > 3$).[165]

The rate of backbiting relative to that of propagation depends on the nature of the initiator. Thus, with $HCl/SbCl_5$ as the initiating system the polymerization of D_3 gives high polymer with controlled molecular weight and only small amounts of cyclics ($\sim 10\%$) in contrast to the high concentration of cyclics obtained with the triflic acid initiator, as discussed above.[166]

Initiation in the exclusive absence of protonic acid may be effected by using the $R_3SiH/Ph_3C^+(C_6F_5)_4^-$ initiating system,[167] which forms the trialkylsilicium ion by Corey hydride transfer from R_3SiH to Ph_3C^{+168}

$$R_3SiH + Ph_3C^+ \rightarrow R_3Si^+ + Ph_3CH.$$

The silicium ion reacts with D_n to generate the trisilyl oxonium ion chain carrier, which has been identified in the ^{29}Si NMR spectra of mixtures of D_3, Me_3SiH, and $Ph_3C^+B(C_6F_5)_4^-$ at low temperature

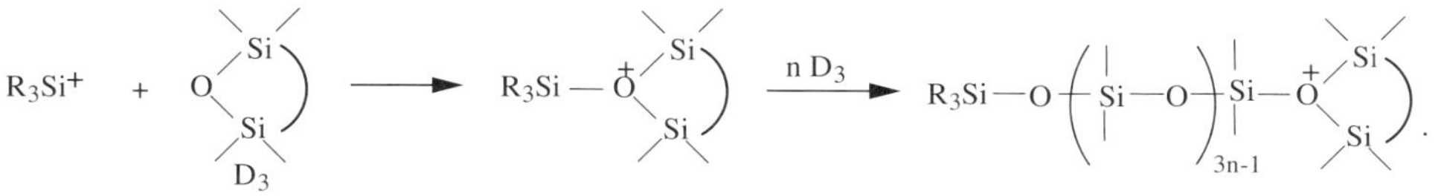

$$R_3Si^+ + D_3 \longrightarrow R_3Si-O\begin{pmatrix}Si\\Si\end{pmatrix}^+ \xrightarrow{nD_3} R_3Si-O\left(Si-O\right)_{3n-1}Si-O\begin{pmatrix}Si\\Si\end{pmatrix}^+.$$

Polymerization of D_3 as well as D_4 occurs livingly in CH_2Cl_2 at room temperature giving 70–90% yield of high polymers; cyclic oligomers are formed concurrently with the high polymers resulting in relatively high polydispersity (PDI $=$ 1.8–2.8). Nevertheless, the living polymers can be used to form block or graft copolymers. One distinguishing feature of this initiating system is that unlike the protonic acid or the Lewis acid/protogen initiator it gives rise to active center at only one end of the growing chain. This makes possible to prepare triblock copolymer of A–B–C type. In contrast, protonic acid initiation leads to propagating chains with active centers at both ends, which can produce triblock copolymer of A–B–A kind only.[167]

7.11 Cyclotriphosphazenes

Structurally, polyphosphazene chain is constituted of alternating phosphorous and nitrogen atoms with each of the former having two side groups (**12**).

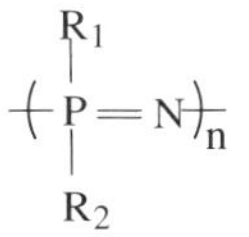

$$\left(\begin{array}{c}R_1\\|\\P=N\\|\\R_2\end{array}\right)_n$$

(**12**)

The nature of R groups on phosphorous atoms determines the properties of the polymers. The polymers have great potential for specialty applications such as high performance elastomers, matrices for solid polymer electrolytes, fuel cell membranes, membranes for gas and liquid separation, biomaterials, and nonlinear optical materials.[169]

The polymer is prepared by the ROP of hexachlorocyclotriphosphazene (**13**) followed by the nucleophilic substitution of chlorine atoms in the formed poly(dichlorophosphazene) (**14**) with a desired nucleophile or a combination of nucleophiles (reaction 7.63). The cyclic monomer is prepared by reacting phosphorous pentachloride with ammonium chloride.[170]

Polymerization can be conducted either in melt or in solution[171–175]

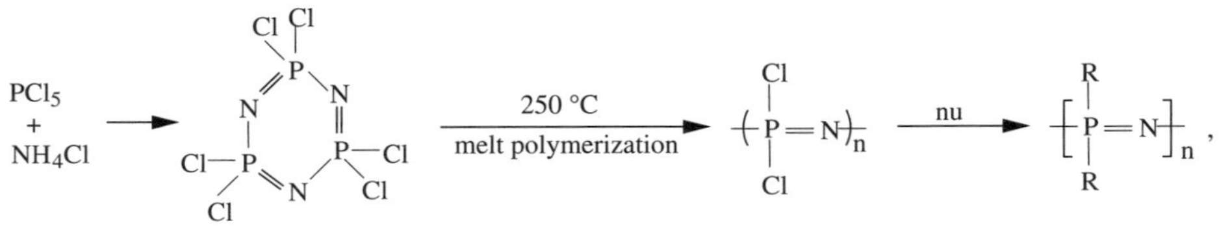

$$PCl_5 + NH_4Cl \longrightarrow (13) \xrightarrow[\text{melt polymerization}]{250\ °C} \left(\underset{Cl}{\overset{Cl}{P}} = N \right)_n \xrightarrow{nu} \left[\underset{R}{\overset{R}{P}} = N \right]_n ,$$

(13)   (14)

(7.63)

where nu = a nucleophile, which may be RNH_2, ROM, RM (M is an alkali metal), *etc.*

7.12 Cyclic Olefins

Under the influence of suitable transition metal catalysts cyclic olefins undergo ring-opening metathesis polymerization (ROMP), which involves cleavage of carbon-carbon double bond and intermolecular reassembly of the opened fragments forming a linear polymer with the unsaturation now linking the opened rings[176–178]

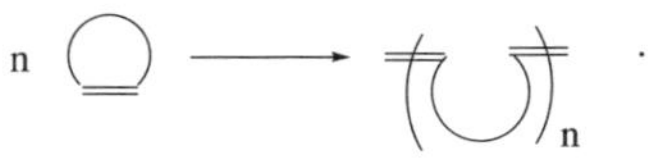

$$n\ \bigcirc \longrightarrow \left(\right)_n$$

(15)

(7.64)

Polymerization of a cycloalkene, *e.g.*, cyclobutene, cyclopentene, cycloheptene, or cyclooctene produces the corresponding polyalkenamer (15). Cyclohexene does not polymerize for unfavorable thermodynamics of polymerization as discussed in Sec. 7.1.1.

Early catalysts constituted of transition metal compounds in their high oxidation states such as $MoCl_5$ and WCl_6 used in combination with strong Lewis acids cocatalysts, aluminum alkyls/alkyl halides.[179–184] However, these catalysts are ill-defined since the true catalysts formed, *in situ*, the alkylidene complexes (*vide infra*), decompose in the course of polymerization under the conditions used. Nevertheless, some of these catalysts found applications in the production of some commercially important polymers, notably polynorbornene (16) and polyoctenamer (17).

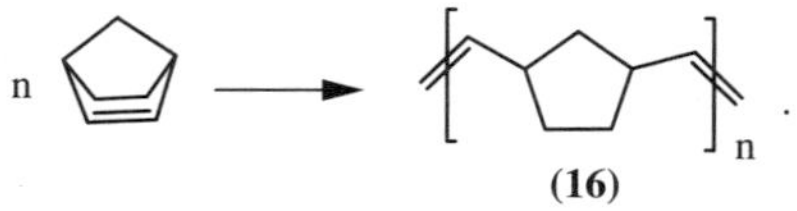

$$n\ \bigcirc\!\!\!= \longrightarrow \left(\right)_n .$$

(16)

(7.65)

$$n\ \bigcirc \longrightarrow =\!\!\left(HC-(CH_2)_6-CH \right)_n\!\!= .$$

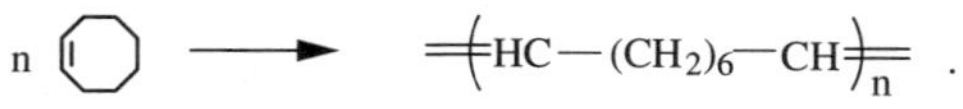

(17)

(7.66)

The well-defined soluble initiators that give rise to living ROMP are mostly the Grubbs initiators, which are metallacycles and alkylidene complexes of Ti and Ru respectively, and the Schrock initiators, which are alkylidene complexes of Ta, Mo, or W.

Initiation

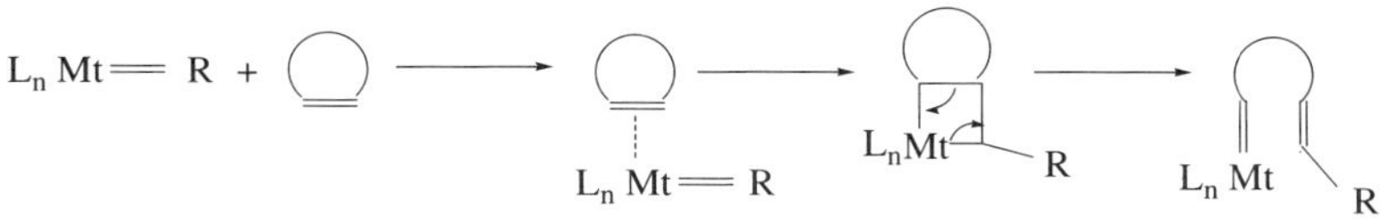

$L_n Mt = R +$ ⬡ ⟶ ⬡ ⟶ [L$_n$Mt─R cyclobutane] ⟶ [L$_n$Mt=...R]

                      L$_n$Mt=R

(18)   **(19)**   **(20)**

Propagation

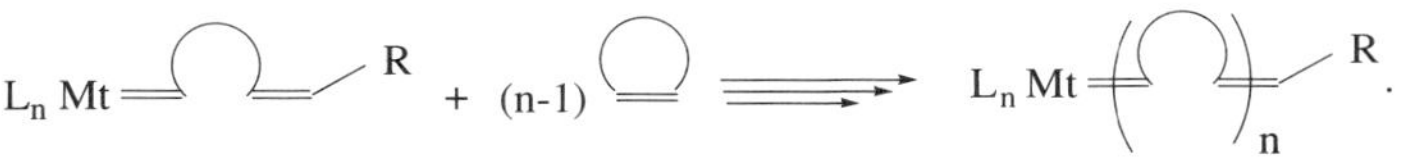

$L_n Mt = \!\!\!\diagup\!\!\!\diagdown\!\!\!R \; + \; (n-1) \; \bigcirc \; \rightleftharpoons \; L_n Mt = (\!\!\!\diagup\!\!\!\diagdown\!\!)_n R$

Scheme 7.1 Mechanisms of initiation and propagation in ROMP. "Reprinted with permission from Ref. 189, Living ring-opening metathesis polymerization. Copyright © 2006 Elsevier Ltd."

The mechanism of polymerization has been established to be as shown in Scheme 7.1.[185–189]

Initiation involves coordination of the cyclic olefin monomer to the transition metal alkylidene (**18**) (directly supplied or formed *in situ* from the initiator) and subsequent formation of a four-membered metallacyclobutane intermediate (**19**) by [2+2]-cycloaddition. The latter undergoes retro [2+2] cleavage forming a new metal alkylidene linked to one end of the ring-opened monomer (**20**). Propagation involves repetition of the above steps on the metal alkylidene chain end.

7.12.1 *Initiators*

The titanium-based initiators are substituted titanacyclobutanes. The substituents should have either relatively large bulk (**21**) or strained rings (**22**) in order that the retro [2+2] cleavage of the initiator becomes reasonably fast at temperatures *ca.*, ≥60°C with the expelling of bulky olefin or releasing ring strain through opening of the strained ring respectively. The titanium alkylidene complex thus formed is the real initiator. These initiators are low in activity so that they can only polymerize monomers, which have relatively high ring strain. Thus, norbornene is polymerized, whereas cyclopentene is not.

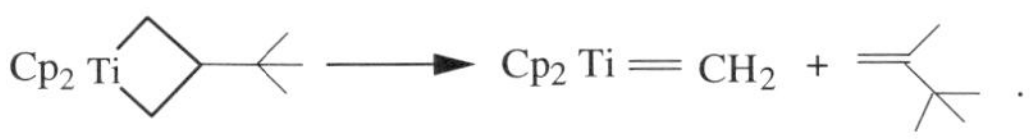

$$Cp_2 Ti\diagup\!\!\!\diagdown\!\!\!\diagup \!\!\!\!\! \diagdown \longrightarrow Cp_2 Ti = CH_2 \; + \; \diagup\!\!\!\diagdown\!\!\!\! \diagdown \qquad (7.67)$$
                 (21)

$$Cp_2 Ti\diagup\!\!\!\diagdown\!\!\!\bigtriangleup \longrightarrow Cp_2 Ti \diagdown\!\!\!\diagup\!\!\!\diagdown \qquad (7.68)$$
               (22)

Typical examples of the more active metal-alkylidene complexes initiators are given in Fig. 7.6. There exists great scope of tuning the initiators by the appropriate choice of ligands to effect living ROMP. Thus, the activity of the highly active tantalum-based initiator (**23**) is decreased by replacing phenoxide ligands with bulky and electron rich diisopropylphenoxide ligands.[190] Similarly, the activity of the W and Mo based initiators (**24** and **25**) can be

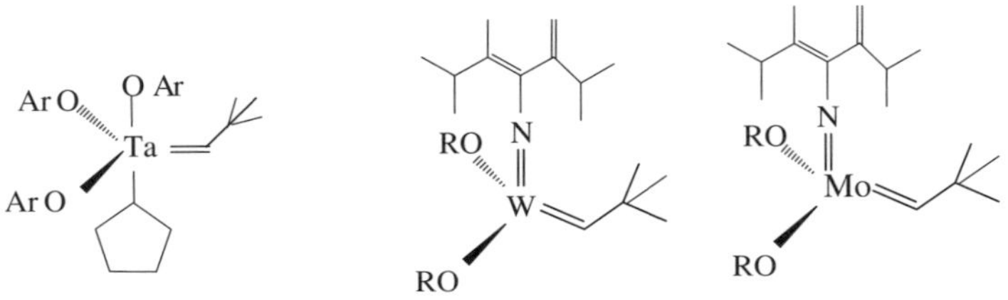

Figure 7.6 Examples of some transition metal alkylidenes initiators.[187,189]

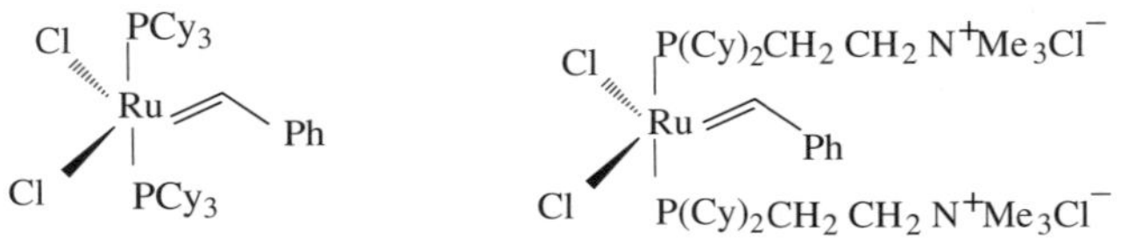

modulated by modifying the alkoxide ligands some of which are shown in the figure.[192–194] Electron-withdrawing ligands increase the activity. However, the Ru-based initiators exhibit the opposite behavior. For instance, activity of **26** is decreased by replacing bulky and electron rich PCy_3 ligands with PPh_3.[189,191]

7.12.2 *Livimg character*

Ring-opening metathesis polymerizations effected by suitable Grubbs or Schrock initiators exhibit living character. They yield polymer of predicted molecular weight and low polydispersity (PDI < 2). The molecular weight increases linearly with conversion. The polymer formed may be chain extended with other monomers of higher ring strain to yield block copolymers. The polymer can be "killed", for example, by Wittig-type reaction using aldehydes or ketones.[188,189] Ruthenium alkylidene ended polymer is not attacked. It may, however, be "killed" by alkyl vinyl ethers (*vide infra*).

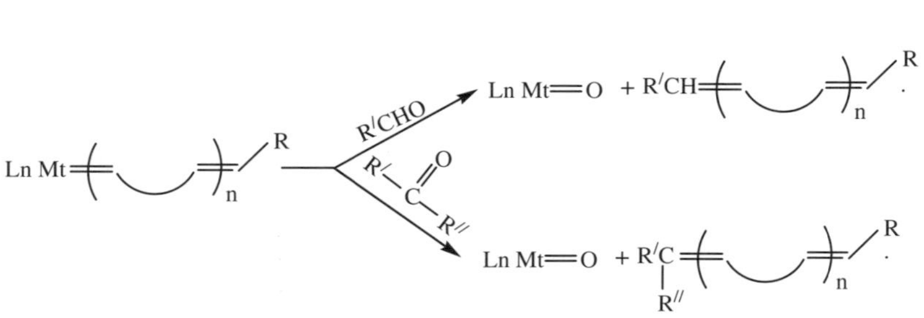

$$Ln\,Mt=O + R'CH= \quad (7.69)$$

$$Ln\,Mt=O + R'\underset{R''}{\overset{|}{C}}= \quad (7.70)$$

7.12.3 Initiators with high functional group tolerance

In general, functional group tolerance of Ti- and Ta-based initiators is low, whereas that of W-, Mo-, or Ru-based initiators is high.[188,189] The Ru-based ones are the best and the Mo-based initiators are better than the W-based ones.[189,195] Both the Mo- and W-based initiators are, however, not tolerant to aldehydes and ketones. As a result, the living polymers prepared by using them may be "killed", for example, by benzaldehyde.

The low oxophilicity of Ru makes the Ru-based initiators extraordinarily tolerant toward functional groups and stable in protic solvents. Thus, **26** is capable of polymerizing livingly a wide variety of functionalized norbornenes and cyclobutenes with such functional groups as hydroxyl, amino, amido, ester, and keto. Also, importantly, the polymerizations can be carried out in protic media.[196,197] Even, water soluble Ru based initiators have been developed by incorporating charged groups in the phosphine ligands (**27**).[190] These initiators are capable of polymerizing water soluble norbornene derivatives at moderate temperatures $ca.$, 45°C in the presence of a protonic acid.

In view of the high tolerance to a wide range of functional groups, which include carbonyl, the ruthenium alkylidene ended living polymers are not "killed" by aldehydes or ketones unlike the other metal alkylidene ended living polymers discussed above. However, suitable "killing" agents are vinyl alkyl ethers or their derivatives[189]

$$\sim\!\!\sim\!\!\sim =RuL_n + CH_2\!\!=\!\!CHOCH_2X \longrightarrow \sim\!\!\sim\!\!\sim = + L_nRu=CHOCH_2X\ .$$
(metathesis-inactive) (7.71)

7.12.4 Synthesis of polyacetylene

ROMP may be used to prepare soluble and processable precursors to polyacetylene. Thus, the method developed at the University of Durham (called the Durham route) prepares the soluble precursor polymer, poly(7,8-bis(trifluoromethyl)-tricyclo[4.2.2.0]deca(3,7,9-triene) (**28**) using W-based catalysts.[198,199]

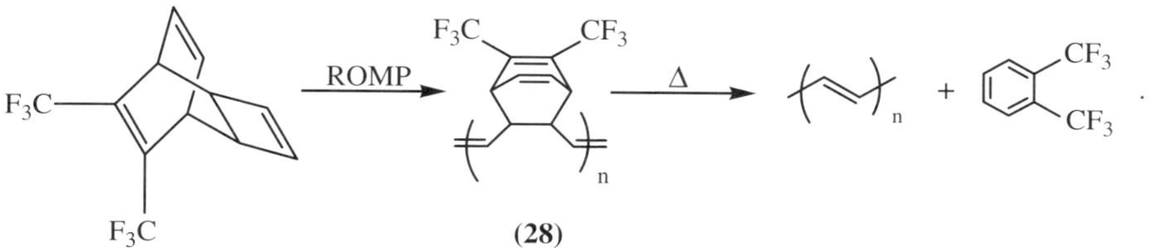

(7.71)

Upon heating, **28** eliminates the volatile bistrifluoromethylbenzene, which escapes leaving behind a polyacetylene film.

ROMP may be used also to prepare polyacetylene from cyclooctatetraene using titanium or tungsten alkylidene complexes as catalysts. However, the polymer is intractable. The polymerization is also nonliving due to the occurrence of backbiting.[200,201]

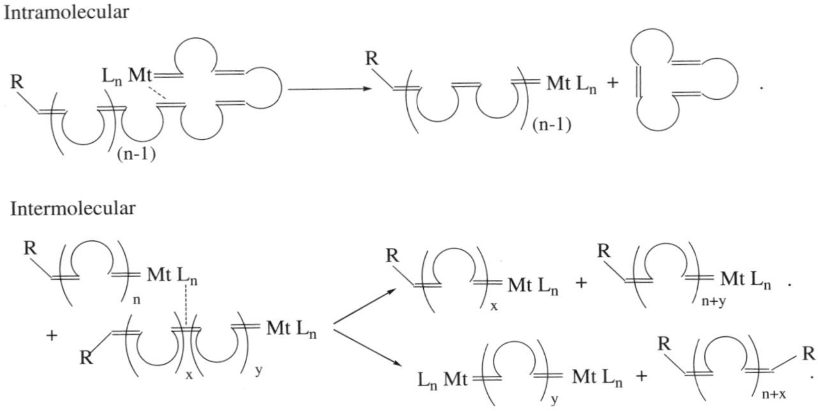

Scheme 7.2 Backbiting and intermolecular exchange in ROMP of cyclic olefins. "Reprinted with permission from Ref. 189, Living ring-opening metathesis polymerization. Copyright © 2006 Elsevier Ltd."

7.12.5 Backbiting and intermolecular interchange

Like ROP of heterocyclic monomers, ROMP of cyclic olefins also suffers from the polymer transfer reactions, both intramolecular (backbiting) and intermolecular (Scheme 7.2).[187–189]

These side reactions become more important with a more active initiator, i.e., a metal alkylidene complex, which is not dependent on ring strain for effecting metathesis. A highly active initiator can polymerize cyclopentene, which is low in strain. However, such initiators are also capable of effecting metathesis of linear olefins. Hence, they impair the living character of ROMP. The side reactions may be suppressed but not completely eliminated by working at low temperatures, e.g., −40°C, with the W-based initiators.[202,203]

7.12.6 Alicyclic diene metathesis (ADMET) polymerization

Although not an ROMP, ADMET polymerization has been included in this chapter since it uses some of the Schrock and Grubbs initiators discussed above. For example, 1,9-decadiene is polymerized by Schrock initiators to polyoctenamer (reaction 7.75). Ethylene is eliminated as a by-product. Step-growth nature of the polymerization has been established from the studies of kinetics, molecular weight, and molecular weight distribution.[204] With dienes containing less than 10 chain atoms, ring-closing metathesis is favored.

$$n\,CH_2{=}CH{-}(CH_2)_6{-}CH{=}CH_2 \longrightarrow {-}(CH{=}CH{-}(CH_2)_6)_n{-} + n\,CH_2{=}CH_2 \,. \quad (7.75)$$

The mechanism of polymerization may be represented as in Scheme 7.3.[205] The diene can be a monomer or an oligomer. Thus, B may also be a metal alkylidene complex of an oligomer instead of that of the monomer shown in the scheme. It may react with either an oligomer or a monomer splitting off the metal methylidene complex and producing oligomers of longer chain lengths. At any stage of polymerization, molecules of all sizes starting from monomer would be present, which is a characteristic feature of step polymerization. The

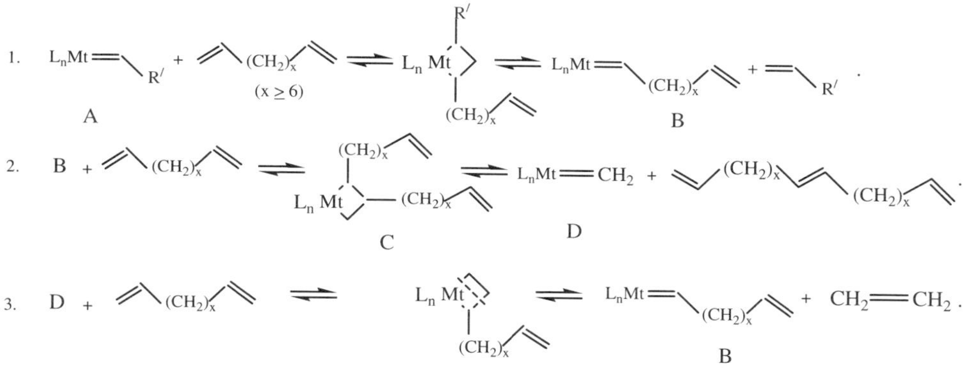

Scheme 7.3 Mechanism of ADMET polymerization.[205]

true catalyst is the methylidene complex D, which reacts with a diene (monomer or oliogmer as shown in step 3 with a monomer) producing an alkylidene complex B and the byproduct ethylene. Elimination of the small molecule ethylene also indicates the polymerization to be belonging to the condensation class. Removal of ethylene under reduced pressure pushes the polymerization toward completion. Polymerization progresses through the repetitions of the cycle involving steps 2 and 3.

References

1. F.S. Dainton, K.J. Ivin, *Quart. Rev.* (1958), 12, 61.
2. F.S. Dainton, T.R.E. Devlin, P.A. Small, *Trans. Faraday Soc.* (1955), 51, 1710.
3. K.J. Ivin, W.K. Busfield, *Polymerization Thermodynamics*, in, *Ency. Polym. Sci. and Eng.*, H. Mark, N.M. Bikales, C.G. Overberger, G. Menges Eds., Wiley-Interscience, New York, Vol. 12, p. 555 (1988).
4. E.L. Eliel, *Stereochemistry of Carbon Compounds*, Tata McGraw-Hill, New Delhi, Chap. 7, p. 198 (1975).
5. (a) H.A. Liebhafsky, *Silicones under the Monogram*, John Wiley & Sons., Inc. New York (1978), A.J. Barry, H.N. Beck in F.G.A. Stone, W.A.G. Graham Eds., *Inorganic Polymers*, Academic Press, Inc., New York (1962).
(b) R.H. Baney, C.E. Voight, J.W. Mentele in H. Seymour Ed., Structure-Solubility Relationship in Polymers, Academic Press, Inc., New York (1977).
6. J.B. Carmichael, R. Winger, *J. Polym. Sci.* (1965), A3, 971.
7. B. Hardman, A. Torkelson, *Silicone*, in, *Ency. Polym. Sci. & Eng.*, H. Mark, N.M. Bikales, C.G. Overberger, G. Menges Eds., Wiley-Interscience, New York, Vol. 15, p. 204 (1989).
8. Y. Yamashita, *Adv. Polym. Sci.* (1978), 28, 1.
9. M. Szwarc, *Adv. Polym. Sci.* (1965), 4, 1.
10. H. Jacobson, W.H. Stockmayer, *J. Chem. Phys.* (1950), 18, 1600.
11. J.F. Brown, G.M.J. Slusarczuk, *J. Am. Chem. Soc.* (1965), 87, 931.
12. S. Penczek, M. Cyprik, A. Duda, P. Kubisa, S. Slomkowski, *Prog. Polym. Sci.* (2007), 32, 247.
13. A. Kowalski, J. Libiszowski, A. Duda, S. Penczek, *Macromolecules* (2000), 33, 1964.
14. K.J. Ivin, J.C. Mol, *Olefin Metathesis and Metathesis Polymerization*, Academic Press, San Diego (1997).

15. F. Andruzzi, G. Pilcher, Y. Virmani, P.H. Plesch, *Makromol. Chem.* (1977), 178, 2367.
16. S. Slomkowski, A. Duda, *Anionic Ring-Opening Polymerization*, in, D.J. Brunelle Ed., Ring-opening polymerization. Mechanisms, Catalysis, Structure, Utility. Munich: Hanser, p. 87 (1993).
17. S. Penczek, P. Kubisa, *Cationic Ring-Opening Polymerization*, in, D.J. Brunelle Ed., Ring-opening polymerization. Mechanisms, Catalysis, Structure, Utility, Munich, Hanser, p. 13 (1993).
18. K.J. Ivin, T. Saegusa Eds., *Ring-Opening Polymerization*, Elsevier Applied Science, London (1984).
19. P.J. Flory, *Principles of Polymer Chemistry*, Cornell University Press, Ithaca, New York (1953).
20. P.J. Flory, *J. Am. Chem. Soc.* (1942), 64, 2205.
21. G.-E. Yu, F. Heatley, C. Booth, T.G. Blease, *J. Polym. Sci. Polym. Chem.* (1994), 32, 1131.
22. R.P. Quirk, G.M. Lizarraga, *Macromol. Chem. Phys.* (2000), 201, 1395.
23. T. Aida, S. Inoue, *Acc. Chem. Res.* (1996), 29, 39.
24. T. Aida, *Prog. Polym. Sci.* (1994), 18, 469.
25. T. Aida, Y. Maekawa, S. Asano, S. Inoue, *Macromolecules* (1988), 21, 1195.
26. K.S. Kazanskii, A.A. Solovyanov, S.G. Entelis, *Eur. Polym. J.* (1971), 7, 1421.
27. A. Duda, S. Penczek, *Macromolecules* (1994), 27, 4867.
28. A. Duda, S. Penczek, *Macromol. Rapid Commun.* (1994), 15, 559.
29. A. Duda, S. Penczek, *Macromol. Symp.* (1991), 47, 127.
30. M. Szwarc, *Ionic Polymerization Fundamentals*, Munich: Hanser (1996).
31. J. Meerwin et al., *J. Prakt. Chem.* (1939), 154, 83.
32. T. Saegusa, S. Matsumoto, *J. Macromol. Sci. Chem.* (1970), A4, 873.
33. T. Saegusa, S. Matsumoto, *Macromolecules* (1968), 1, 442.
34. M.P. Dreyfuss, J.C. Westfall, P. Dreyfuss, *Macromolecules* (1968), 1, 437.
35. K. Matyjaszewski, S. Slomkowski, S. Penczek, *J. Polym. Sci. Polym Chem.* (Ed.) (1979), 17(1), 69.
36. K. Brzezinska, K. Matyjaszewski, S. Penczek, *Makromol. Chem.* (1978), 179, 2387.
37. K. Matyjaszewski, S. Penczek, *J. Polym. Sci. Polym. Chem Ed.* (1974), 12(9), 1905.
38. S. Penczek, K. Matyjaszewski, *J. Polym. Sci. Polym. Symp.* (1976), 56, 255.
39. K. Matyjaszewski, *J. Macromol. Sci. C, Rev. Macromol. Chem.* (1986), 26, 1.
40. J. Libiszowski, R. Szymanski, S. Penczek, *Makromol. Chem.* (1989), 190, 1225.
41. J. Dale, G. Borgen, K. Daasvatn, *Ger. Pat.* (1974), 2, 401, 126.
42. M. Wojtania, P. Kubisa, S. Penczek, *Macromol. Symp.* (1986), 6, 201.
43. P. Kubisa, S. Penczek, *Prog. Polym. Sci.* (1999), 24, 1409.
44. T. Biedron, R. Szymanski, P. Kubisa, S. Penczek, *Macromol. Symp.* (1990), 32, 155.
45. P. Kubisa, *Cationic Polymerization of Heterocyclics*, in, *Cationic Polymerization: Mechanisms, Synthesis and Applications*, K. Matyjaszewski Ed., Marcel Dekker, New York, Chap. 6 (1996).
46. R. Szymanski, P. Kubisa, S. Penczek, *Macromolecules* (1983), 16, 1000.
47. S. Penczek, R. Szymanski, *Polym. J.* (1980), 12, 617.
48. P.H. Plesch, P.H. Westermann, *J. Polym. Sci.* (1968), C16, 3837.
49. Y. Firat, F.R. Jones, P.H. Plesch, P.H. Westermann, *Makromol. Chem.* (1975), Supp. 1, 203.
50. H. Jacobson, W.H. Stockmayer, *J. Chem. Phys.* (1950), 18, 1600.
51. J.A. Semlyen, *Adv. Polym. Sci.* (1976), 21, 41.
52. J.M. Andrews, J.A. Semlyen, *Polymer* (1972), 13, 142.
53. K. Matyjaszewski et al., *Makromol. Chem.* (1980), 181, 1469.
54. V. Jaacks, K. Boehlke, E. Eberius, *Makromol. Chem.* (1968), 118, 354.
55. R.C. Schulz et al., *ACS Sym. Ser.* (1977), 59, 77.
56. R.C. Schulz et al., *Pure App. Chem.* (1981), 53, 1763.
57. Y. Yamashita, M. Okada, H. Kasahara, *Makromol. Chem.* (1968), 117, 256.

58. Y. Yamashita, Y. Kawakami, *ACS Symp. Ser.* (1977), No. 59, 99.
59. S. Penczek, *J. Polym. Sci. Polym. Chem.* (2000), 38, 1919.
60. Z.N. Nysenko et al., *Vys. Soed.* (1976), 18, 1696.
61. W. Kern, V. Jaacks, *J. Polym. Sci.* (1960), 48, 399.
62. L. Leese, M.W. Bauber, *Polymer* (1965), 6, 269.
63. T.J. Dolce, J.A. Grates, *Acetal Resins*, in, *Ency. Polym. Sci. & Eng.* 2nd ed., H.F. Mark, N.M. Bikales, C.G. Overberger, G. Menges Eds., Wiley-Interscience, New York, Vol. 1, p. 42 (1985).
64. D. Mecerryes, R. Jerome, P. Dubois, *Adv. Polym. Sci.* (1999), 147, 1.
65. D.J. Brunelle, In *Ring-Opening Polymerization*, D.J. Brunelle Ed., Hanser Verlag, Munich, p. 309 (1993).
66. S. Slomkowski, S. Penczek, *Macromolecules* (1976), 9, 367.
67. A. Hofman, S. Slomkowski, S. Penczek, *Macromol. Chem.* (1984), 185, 91.
68. A. Deffieux, S. Boileau, *Macromolecules* (1976), 9, 369.
69. A. Duda, *J. Polym. Sci. Polym. Chem.* (1992), 30, 21.
70. S. Sosnowski, S. Slomkowski, S. Penczek, *J. Macromol. Sci. Chem.* (1983), A20, 979.
71. J. Dale, J.E. Schwarz, *Acta. Chem. Scand.* (1986), B40, 559.
72. S. Slomkowski, S. Penczek, *Macromolecules* (1980), 13, 229.
73. K. Ito, Y. Hashizuka, Y. Yamashita, *Macromolecules* (1977), 10, 821.
74. K. Ito, Y. Yamashita, *Macromolecules* (1978), 11, 68.
75. T. Ouhadi, C. Stevens, R. Jerome, P. Teyssie, *Macromolecules* (1976), 9, 927.
76. A. Hamitou, T. Ouhadi, R. Jerome, P. Teyssie, *J. Polym. Sci. Polym. Chem.* (1977), 15, 865.
77. T. Ouhadi, C. Stevens, P. Teyssie, *Makromol. Chem. Suppl.* (1975), 1, 191.
78. H.R. Kricheldorf, M. Berl, N. Scharnagl, *Macromolecules* (1988), 21, 286.
79. A. Duda, S. Penczek, *Macromol. Rapid Commun.* (1995), 16, 67.
80. A. Duda, S. Penczek, *Macromolecules* (1994), 27, 4867.
81. A. Duda, S. Penczek, *Macromolecules* (1995), 28, 5981.
82. C. Jacobs, P. Dubois, R. Jerome, P. Teyssie, *Macromolecules* (1991), 24, 3027.
83. A. Hofman, S. Slomkowski, S. Penczek, *Macromol. Chem. Rapid Commun.* (1987), 8, 387.
84. S. Penczek, A. Duda, *Macromol. Symp.* (1991), 47, 127.
85. A. Duda, S. Penczek, *Macromol. Rapid Commun.* (1994), 15, 559.
86. T. Biela, A. Duda, *J. Polym. Sci. Polym. Chem.* (1996), 34, 1807.
87. A. Kowalski, A. Duda, S. Penczek, *Macromol. Rapid Commun.* (1998), 19, 567.
88. A. Kowalski, A. Duda, S. Penczek, *Macromolecules* (2000), 33, 689.
89. A. Kowalski, A. Duda, S. Penczek, *Macromolecules* (2000), 33, 7359.
90. H.R. Kricheldorf, J.K.-Saunders, A. Stricker, *Macromolecules* (2000), 33, 702.
91. R.F. Storey, J.W. Sherman, *Macromolecules* (2002), 35, 1504.
92. J. Baran et al., *Macromol. Symp.* (1997), 123, 93.
93. H.R. Kircheldorf, S.R. Lee, S. Bush, *Macromolecules* (1996), 29, 1375.
94. A. Duda, S. Penczek, *Am. Chem. Symp. Ser.* (2000), 764, 160.
95. A. Duda, S. Penczek, *Macromolecules* (1990), 23, 1636.
96. S. Penczek, A. Duda, R. Szymanski, T. Biela, *Macromol. Symp.* (2000), 153, 1.
97. A. Lofgren, A.C. Albertsson, P. Dubois, R. Jerome, *J. Macromol. Sci. Rev. Macromol. Chem. Phys.* (1995), 35, 379.
98. P. Leconte, R. Jerome, *Ency. Polym. Sci. & Tech.*, 3rd ed., H. Mark Ed., John Wiley & Sons, Hoboken, New Jersey, Vol. 11, p. 547 (2004).
99. A.J. Nijenhuis, D.W. Griypma, A.J. Pennings, *Macromolecules* (1992), 25, 6419.
100. S.J. Mclain, N.E. Drysdale, *Polym. Prepr. (Am. Chem. Soc. Div. Polym. Chem.)* 33(2), 463 (1992).
101. S.J. McLain et al., *Polym. Prepr. (Am. Chem. Soc. Div. Polym. Chem.)* 35(2), 534 (1994).
102. H. Yasuda, E. Ihara, *Macromol. Chem. Phys.* (1995), 196, 2417.

103. N. Spassky, V. Simic, M.S. Montaudo, L.G. Hubert-Pfalzgraf, *Macromol. Chem. Phys.* (2000), 201, 2432.
104. M. Save, A. Soum, *Macromol. Chem. Phys.* (2002), 203, 2591.
105. W.M. Stevels, M.J.K. Ankone, P.J. Dijkstra, J. Feijen, *Macromolecules* (1996) 29, 6132.
106. S. Kuhling, H. Keul, H. Hocker, *Makromol. Chem.* (1992), 193, 1207.
107. N. Ropson, P. Dubois, R. Jerome, P. Teyssie, *Macromolecules* (1992), 25, 3820.
108. N. Ropson, P. Dubois, R. Jerome, P. Teyssie, *J. Polym. Sci. Polym. Chem.* (1997), 35, 183.
109. A. Hofman, R. Szymanski, S. Slomkowski, S. Penczek, *Makromol. Chem.* (1984), 185, 655.
110. A. Hofman, S. Slomkowski, S. Penczek, *Makromol. Chem.* (1987), 188, 2027.
111. H.R. Kricheldorf, J.M. Jonte, R. Dunsing, *Makromol. Chem.* (1986), 187, 771.
112. H.R. Kircheldorf, M.V. Sumbel, *Makromol. Chem.* (1988), 189, 317.
113. H.R. Kircheldorf, R. Dunsing, A. Serra, *Macromolecules* (1987), 20, 2050.
114. Y. Okamoto, *Macromol. Symp.* (1991), 42/43, 117.
115. T. Endo, Y. Shibasaki, F. Sanada, *J. Polym. Sci. Polym. Chem.* (2002), 40, 2190.
116. Y. Shibasaki et al., *Macromolecules* (2000), 33, 4316.
117. X. Lou, C. Detrembleur, R. Jerone, *Macromolecules* (2002), 35, 1190.
118. S. Kobayashi, *J. Polym. Sci. Polym. Chem.* (1999), 37, 3041.
119. R.A. Gross, A. Kumar, B. Kalra, *Chem. Rev.* (2001), 101, 2097.
120. A. Duda et al., *Macromolecules* (2002), 35, 4266.
121. K. Hashimoto, *Prog. Polym. Sci.* (2000), 25, 1411.
122. J. Sebenda in G. Allen, J.C. Bevington Eds. *Comprehensive Polymer Science*, Vol. 3, Oxford, U.K.; Pergamon Press, p. 511 (1989), (G.C. Eastmond, A. Ledwith, S. Russo, P. Sigwalt, Vol. Eds.).
123. H. Sekiguchi in: K.J. Ivin, T. Saegusa Eds., *Ring-Opening Polymerization*, Vol. 2, London, Elsevier, p. 809 (1984).
124. J. Sebenda, *J. Macromol. Sci. Chem.* (1972), 6, 1145.
125. P. Kubisa, S. Penczek, *Prog. Polym. Sci.* (1999), 24, 1409.
126. W. Sweeny, J. Zimmerman, *Polyamides*, in, *Ency. Polym. Sci. & Tech.* 1st ed., H.F. Mark, N.G. Gaylord, N.M. Bikales Eds., Interscience, New York, Vol. 10, p. 483 (1969).
127. A. Anton, B.P. Baird, *Polyamides, Fibers*, in, *Ency. Polym. Sci. & Tech.* 3rd ed., H.F. Mark Ed., Wiley-Interscience, Hoboken, New Jersey, Vol. 3, p. 584 (2003).
128. H.K. Reimschuessel, *J. Polym. Sci. Macromol. Rev.* (1977), 12, 65.
129. O.E. Snider, J. Richardson, Polyamide Fibers, in, *Ency. Polym. Sci. & Tech.* 1st ed., H.F. Mark, N.G. Gaylord, N.M. Bikales, Eds., Interscience, New York, Vol. 10, p. 347 (1969).
130. R.R. Blout, R.H. Karlson, P. Doty, B. Hargitay, *J. Am. Chem. Soc.* (1945), 76, 4492.
131. R.R. Blout, R.H. Karlson, *J. Am. Chem. Soc.* (1956), 78, 941.
132. C.H. Bamford, H. Block, *J. Chem. Soc.* (1961), 4989, 4992.
133. T.J. Deming, *J. Am. Chem. Soc.* (1998), 120, 4240.
134. T.J. Deming, *J. Polym. Sci. Polym. Chem.* (2000), 38, 3011.
135. T.J. Deming, *Nature* (1997), 390, 386.
136. H.R. Kricheldorf, *α-Aminoacid-N-Carboxyanhydrides and Related Materials*: Springer-Verlag, New York (1987).
137. H.R. Kricheldorf, in *Models of Biopolymers by Ring-Opening Polymerization*, S. Penczek, Ed., CRC, Boca Baton, Florida, p. 160 (1990).
138. Y. Knobler, S. Bittner, D. Virov, M. Frankel, *J. Chem. Soc.* (1969), 1821.
139. Y. Knobler, S. Bittner, M. Frankel, *J. Chem. Soc.* (1964), 3941.
140. I. Dimitrov, H. Schlaad, *Chem. Commun.* (2003), 2944.
141. I. Dimitrov, H. Kukula, H. Colfen, H. Schlaad, *Macromol. Symp.* (2004), 215, 383.
142. S. Kobayashi, H. Uyama, *J. Polym. Sci. Polym. Chem.* (2002), 40, 192.

143. B.M. Culbertson, *Prog. Polym. Sci.* (2002), 27, 579.
144. A. Dworak, *Macromol. Chem. Phys.* (1998), 199, 1843.
145. T. Saegusa, S. Kobayashi, A. Yamada, *Makromol. Chem.* (1976), 177, 2271.
146. S. Kobayashi, H. Ikeda, T. Saegusa, *Macromolecules* (1973), 6, 808.
147. D.A. Tomalia, G.R. Killat, *Alkyleneimine Polymer*, in, *Ency. Polym. Sci. & Eng.* 2nd, ed., H.E. Mark, N.M. Bikales, C.G. Overberger, G. Menges, Eds., Wiley-Interscience, New York, Vol. 1, p. 680 (1985).
148. E.J. Goethals, *Cationic Polymerization: Amines and N-containing Heterocycles (Part 1)*, in, G. Allen, J.C. Bevington, Eds., Comprehensive Polymer Science, Vol. 3, Pergamon Press, New York, p. 837 (1988).
149. G.M. Lukovkin, V.S. Pshezetsky, G.A. Murtazaeva, *Eur. Polym. J.* (1973), 9, 559.
150. E.J. Goethals, E.W. Schacht, *J. Polym. Sci. Polym. Letters* (1973), 11, 497.
151. P. Sigwalt, N. Spassky, *Cyclic Compounds Containing Sulfer in the Ring*, in, *Ring-Opening Polymerization* Vol. 3, K.J. Ivin, T. Saegusa, Eds., Elsevier, Amsterdam, p. 489 (1984).
152. D. Van Ooteghem, E.J. Goethals, *Makromol. Chem.* (1974), 175, 1513.
153. D. Van Ooteghem, E.J. Goethals, *Makromol. Chem.* (1976), 177, 3389.
154. E.J. Goethals, W. Drijvers, *Makromol. Chem.* (1970), 136, 73.
155. E.J. Goethals, W. Drijvers, *Makromol. Chem.* (1973), 165, 329.
156. M. Morton, E.E. Bostick, *J. Polym. Sci.* (1964), A2, 523.
157. S. Bywater, *Anionic Polymerization,* in, *Ency. Polym. Sci. & Eng.* (2nd) ed., H.F. Mark, N.M. Bikales, C.G. Overberger, G. Menges Eds., Wiley-Interscience, New York, Vol. 2, p. 1 (1985).
158. T.C. Kendrick, B.M. Parbhoo, J.W. White, *Polymerization of Cyclosiloxanes*, in, Comprehensive Polymer Science, The synthesis, characterization, reactions and applications of polymers, G.C. Eastmond, A. Ledwith, S. Russo, P. Sigwalt, G. Allen, J.C. Bevington, Eds., Chain Polymerization Part II, Vol. 4, Oxford, Pergamon Press, p. 459 (1989).
159. J. Chojnowski, M. Cyprek, *Synthesis of Linear Polysiloxanes,* in, R.G. Jones, W. Ando, J. Chojnowski, Eds., Silicone Containing Polymers. The science and technology of their synthesis and applications, Dodrecht: Kluwer Academic Publishers, p. 1 (2000).
160. G. Sauvet, J.-J. Lebrun, P. Sigwalt, in *Cationic Polymerization and Related Processes*, E.J. Goethals Ed., Academic Press, New York, p. 237 (1984).
161. S. Penczek, P. Kubisa, K. Matyjaszewski, *Adv. Polym. Sci.* (1985), 68–69, 216.
162. C. Gobin, M. Masure, G. Sauvet, P. Sigwalt, *Macromol. Symp.* (1988), 6, 237.
163. P. Nicol, M. Masure, P. Sigwalt, *Macromol. Chem. Phys.* (1994), 195, 2327.
164. P. Sigwalt, P. Nicol, M. Masure, *Makromol. Chem. Suppl.* (1989), 15, 15.
165. L. Wilezek, S. Rubinsztajn, L. Chojnowski, *Makromol. Chem.* (1986), 187, 39.
166. G. Toskas, M. Moreau, M. Masure, P. Sigwalt, *Macromolecules* (2001), 34, 4730.
167. O. Wang *et al.*, *Macromolecules* (1996), 29, 6691.
168. J.Y. Corey, *J. Am. Chem. Soc.* (1975), 97, 3237.
169. R. Wycisk, P.P. Pintauro, *Ency. Polym. Sci. & Tech.* 3rd ed., H. Mark. Ed., Wiley-Interscience, Vol. 7, p. 603 (2003).
170. R. Schenck, G. Romer, *Chem. Ber.* (1924), 57B, 1343.
171. H.R. Allcock, R.L. Kugel, *J. Am. Chem. Soc.* (1965), 87, 4216.
172. H.R. Allcock, R.L. Kugel, K.J. Valan, *Inorg. Chem.* (1966), 5, 1709.
173. H.R. Allcock, R.L. Kugel, *Inorg. Chem.* (1966), 5, 1716.
174. A.N. Majumdar, S.G. Young, R.L. Merker, J.H. Magill, *Makromol. Chem.* (1989), 190, 2293.
175. A.N. Majumdar, S.G. Young, R.L. Merker, J.H. Magill, *Macromolecules* (1990), 23, 14.
176. R.H. Grubbs, *Handbook of Metathesis, Vol. 3*, Wiley-VCH, Weinheim (2003).
177. N. Calderon, *Rev. Macromol. Chem.* (1972), 7, 105.

178. G. Natta, G. Dall'Asta, Elastomers from Cyclic Olefins, in, *Polymer Chemistry of Synthetic Elastomers*, Part II, J.P. Kennedy, E.G.M. Tornquist Eds., High Polymer Ser., Interscience, Vol. 23, p. 703 (1969).
179. W.L. Truett, D.R. Solmson, I.M. Robinson, B.A. Montague, *J. Am. Chem. Soc.* (1960), 82, 2337.
180. G. Natta, G. Dall'Asta, G. Mazzanti, G. Mortoni, *Makromol. Chem.* (1963), 69, 163.
181. G. Natta, G. Dall'Asta, G. Mazzanti, *Angew. Chem. Intl. Ed.* (1964), 3, 723.
182. G. Natta, G. Dall'Asta, I.W. Bassi, G. Carella, *Makromol. Chem.* (1966), 91, 87.
183. N. Calderon, *Acc. Chem. Res.* (1972), 5, 127.
184. N. Calderon et al., *J. Am. Chem. Soc.* (1968), 90, 4133.
185. J.L. Herisson, Y. Chauvin, *Makromol. Chem.* (1970), 141, 161.
186. Y. Chauvin, *Angew. Chem. Intl. Ed. Eng.* (2006), 45, 3740.
187. R.P. Schrock, *Acc. Chem. Res.* (1990), 23, 158.
188. R.H. Grubbs, W. Tumas, *Science* (1989), 243, 907.
189. C.W. Bielawski, R.H. Grubbs, *Prog. Polym. Sci.* (2007), 32, 1.
190. K.C. Wallace, A.H. Liu, J.C. Dewan, A.R. Schrock, *J. Am. Chem. Soc.* (1988), 110, 4964.
191. P. Schwab, R.H. Grubbs, J.W. Ziller, *J. Am. Chem. Soc.* (1996), 118, 100.
192. R.R. Schrock et al., *J. Am. Chem. Soc.* (1988), 110, 1423.
193. E. Khosravi, A.A. Al-Hajaji, *Polymer* (1998), 39, 5619.
194. E. Khosravi et al., *J. Mol. Cat. A: Chem.* (2000), 160, 1.
195. R.R. Schrock et al., *J. Am. Chem. Soc.* (1990), 112, 3875.
196. G.C. Bazan et al., *Polymer Commun.* (1989), 30, 258.
197. G.C. Bazan et al., *J. Am. Chem. Soc.* (1990), 112, 8378.
198. J.H. Edwards, W.J. Feast, *Polymer* (1980), 21, 595.
199. J.H. Edwards, W.J. Feast, D.C. Bott, *Polymer* (1984), 25, 395.
200. Y.V. Korshak, V.V. Korshak, G. Kanischka, H.A. Hocker, *Macromol. Chem. Rapid Commun.* (1985), 6, 685.
201. F.L. Klavetter, R.H. Grubbs, *J. Am. Chem. Soc.* (1988), 110, 7807.
202. R.R. Schrock et al., *Macromolecules* (1989), 22, 3191.
203. P. Dounis, W.J. Feast, A.M. Kenwright, *Polymer* (1995), 36, 2787.
204. K.B. Wagner et al., *Macromolecules* (1997), 30, 7363.
205. K.B. Wagner, J.M. Boncella, J.G. Nel, *Macromolecules* (1991), 24, 2649.

Problems

7.1 In what step of the mechanism of a conventional chain reaction does ROP exhibit deviation and what are the consequences?

7.2 Although cationic and anionic charge carriers in ring-opening polymerizations are much less reactive than carbocations and carbanions respectively, they undergo significant intermolecular as well as intramolecular polymer transfer reactions unlike their carbon-centered counterparts. What could be the reason?

7.3 The impact of intermolecular polymer transfer on polydispersity index of polyesters prepared by ROP of lactones using covalent metal alkoxides initiators increases as polymerization is continued under monomer–starved conditions. Explain why this should be so.

7.4 What is the theoretically maximum PDI of polymer produced in ROP or ROMP accompanied by intermolecular polymer transfer? What kind of molecular weight distribution gives this PDI?

7.5 Backbiting proceeds at comparable rates with propagation in the cationic ROP of the small ring cyclic ethers but not of the common ring ones. Explain why this should be so.

7.6 Both backbiting and intermolecular polymer transfer are more prevalent in the ROMP of cyclopentene than in that of norbornene. What could be the reason?

7.7 In the cationic ROP of ethylene oxide, the proportion of the dimeric cyclic product is relatively large. Can you use the Jacobson Stockmayer theory to justify the result?

7.8 Cyclic monomers do not yield high polymers unless the monomer concentration exceeds certain critical value. Explain why this should be so.

7.9 Linear polyester may be prepared by either step polymerization or ring-opening polymerization. Which method should be used if (i) the polymer is required to have a narrow MWD and (ii) high molecular weight polymer (DP $\geq$ 500) is required. Explain your answer.

7.10 In the polymerization of D_4 by protonic acid catalyst, the initial proportion of cyclics differs from that predicted by the Jacobson–Stockmayer theory of ring-chain equilibria. Does the proportion remain unchanged with time?

Chapter 8

Chain Copolymerization

Chain copolymerization refers to simultaneous chain polymerization of two or more monomers producing polymer with monomer units of each kind incorporated in the chain. When two monomers are used, the product is a binary copolymer (ordinarily referred to as a copolymer); when three are used, the ternary copolymer formed is called a terpolymer. Binary copolymerization is the most commonly practiced and thoroughly studied. In this chapter, we shall confine our discussion to it.

The instantaneous copolymer composition is determined by the relative rate of incorporation of the monomers in the polymer chain at an instant. In a batch copolymerization, it ordinarily changes with progress of copolymerization. The property of a copolymer not only depends upon its overall composition but also its microstructure, *i.e.*, the monomer distribution in the chain. When the monomer distribution obeys known statistical laws, *e.g.*, Markovian statistics of zeroth, first, second, or a higher order, the copolymer is called a statistical copolymer. Truly random copolymer is obtained when the monomer distribution is zeroth order Markovian (Bernoullian distribution) or first order Markovian (in the special case of ideal copolymerization). In living copolymerization, the copolymer composition gradually changes from one end of the chain to the other. This is because all chains grow either continuously or intermittently throughout the duration of polymerization. Such a copolymer is called a gradient copolymer.

However, when one monomer is polymerized separately in the presence of a different polymer, the newly grown polymer may become chemically bonded to the other polymer under suitable circumstances. When the polymer is bonded to the end(s) of a different one a block copolymer results, whereas when it is bonded to the chain backbone instead of to the ends, a graft copolymer results.

Nomenclature

The nomenclature uses the term "poly" followed by the names of the two monomers in a parenthesis, a short name in italics for the copolymer type being placed in between the names of the two monomers. Examples are given below:

Random copolymer: poly(isobutylene– *ran* –isoprene).
Statistical copolymer: poly(styrene – *stat* – methyl methacrylate).

Alternating copolymer: poly(styrene – *alt* –maleic anhydride).
Gradient copolymer: poly(styrene – *grad* – methyl methacrylate).

When the type of monomer distribution is not specified, the copolymer is represented by placing -*co*- (short form of copolymer) in between the names of the two monomers, *e.g.*, poly(styrene-*co*-methyl methacrylate). Block and graft copolymers are named with these words in italics placed in between the names of the homopolymers constituting the respective copolymers as shown in the examples below.

Block copolymers:

> *Diblock*: polystyrene-*block*-poly(methyl methacrylate).
> *Triblock*: polystyrene-*block*-poly(butadiene)-*block*-polystyrene.

Graft copolymer: polystyrene-*graft*-poly(methyl methacrylate).

8.1 Terminal Model of Copolymerization

The composition of copolymer formed at any instant during binary copolymerization depends upon the relative rate of homo- and cross-propagations at that instant. However, in order to relate the instantaneous copolymer composition with the corresponding monomer feed composition a copolymerization model is required to be used. This is true also for predicting monomer sequence distribution and propagation rate coefficient. Many models have been used. Some of these are designed for systems in which side reactions like the complex formation between the monomers themselves or between monomers and solvents or active centers, *etc.*, are important. However, there are two models, which focus only on propagation and ignore the side reactions. Obviously, these are applicable to copolymerization in which side reactions are unimportant. The simpler but the inferior of the two is the terminal model, which assumes that the reactivity of an active center depends only on the nature of the terminal monomer unit of the growing chain. The preceding units have no influence.[1-4] The other, the penultimate model, assumes the reactivity to be dependent on not only the terminal monomer unit but also the penultimate unit.[5] The monomer distribution in copolymer is governed by the first order Markov statistics in the terminal model; whereas it is governed by the second order Markov statistics in the penultimate model. However, the terminal model has been used extensively producing the bulk of the copolymerization literature. We shall deal with it first.

8.1.1 *Copolymer composition equation*

The terminal model considers just four propagation reactions as shown here for radical copolymerization[2]

$$M_1^\bullet + M_1 \xrightarrow{k_{11}} M_1^\bullet, \tag{8.1}$$

$$M_1^\bullet + M_2 \xrightarrow{k_{12}} M_2^\bullet, \tag{8.2}$$

$$M_2^\bullet + M_2 \xrightarrow{k_{22}} M_2^\bullet, \tag{8.3}$$

$$M_2^\bullet + M_1 \xrightarrow{k_{21}} M_1^\bullet, \tag{8.4}$$

where $M_1^\bullet$ and $M_2^\bullet$ represent chain radicals respectively with terminal monomer units M_1 and M_2 bearing the radicals. It is to be noted that this representation of chain radicals is different from that followed with chain radicals in homo polymerization (*vide* Chap. 3) where a subscript in a radical symbol refers to the degree of polymerization.

Reactions (8.1) and (8.3) represent homo propagations, whereas (8.2) and (8.4) represent cross-propagations. The rate constant k_{ij} refers to the addition of chain radical $M_i^\bullet$ to monomer M_j (i, j = 1 or 2).

Assuming stationary state polymerization it follows

$$\frac{d[M_1^\bullet]}{dt} = R_{in,1} + k_{21}[M_2^\bullet][M_1] - k_{12}[M_1^\bullet][M_2] - R_{t,1} = 0, \tag{8.5}$$

$$\frac{d[M_2^\bullet]}{dt} = R_{in,2} + k_{12}[M_1^\bullet][M_2] - k_{21}[M_2^\bullet][M_1] - R_{t,2} = 0, \tag{8.6}$$

where $R_{in,i}$ and $R_{t,i}$ represent respectively the rate of initiation and that of termination concerning the radical i. For long chains, $R_{in,i}$ and $R_{t,i}$ terms are negligible compared to the propagation terms. This approximation gives Eq. (8.7) from either of the two equations.

$$k_{12}[M_1^\bullet][M_2] = k_{21}[M_2^\bullet][M_1]. \tag{8.7}$$

Thus, in the stationary state, the rates of interconversion of the two radicals equal each other.

The rates of consumption of monomers M_1 and M_2 are given by

$$-\frac{d[M_1]}{dt} = k_{11}[M_1^\bullet][M_1] + k_{21}[M_2^\bullet][M_1]. \tag{8.8}$$

and

$$-\frac{d[M_2]}{dt} = k_{22}[M_2^\bullet][M_2] + k_{12}[M_1^\bullet][M_2]. \tag{8.9}$$

Dividing Eq. (8.8) by (8.9) and eliminating the radical concentrations using Eq. (8.7) yields the instantaneous copolymer composition equation[2]

$$\frac{d[M_1]}{d[M_2]} = \frac{[M_1]}{[M_2]} \left(\frac{r_1[M_1]/[M_2] + 1}{[M_1]/[M_2] + r_2} \right), \tag{8.10}$$

where r_1 and r_2 are the monomer reactivity ratios of monomer 1 and 2 respectively and defined as

$$\left. \begin{array}{l} r_1 = k_{11}/k_{12} \\ r_2 = k_{22}/k_{21} \end{array} \right\}. \tag{8.11}$$

The parameter r_i is the ratio of the rate constant of addition of monomer M_i to its own radical to that of the addition of the other monomer M_j to the same radical. A value of unity for r_i

($i = 1$ or 2) means that the radical i has equal preference for both the monomers; a larger value indicates preference of the radical for its own monomer over the other.

The copolymer composition Eq. (8.10) may be expressed in terms of mol fractions instead of molar ratios. Thus, representing mol fractions of monomers 1 and 2 in the instantaneous copolymer by F_1 and F_2 respectively and those in the monomer feed by f_1 and f_2, we have

$$F_1 = d[M_1]/(d[M_1] + d[M_2]) = 1 - F_2, \tag{8.12}$$

and

$$f_1 = [M_1]/([M_1] + [M_2]) = 1 - f_2. \tag{8.13}$$

Substituting into Eq. (8.10) from Eqs. (8.12) and (8.13) one gets

$$F_1 = (r_1 f_1^2 + f_1 f_2)/(r_1 f_1^2 + 2 f_1 f_2 + r_2 f_2^2). \tag{8.14}$$

It is evident from Eq. (8.10) or (8.14) that except for the case of $r_1 = r_2 = 1$, the instantaneous copolymer composition ($d[M_1]/d[M_2]$) ordinarily differs from the composition of the monomer mixture and, therefore, both change with the progress of polymerization.

Besides the copolymer composition, the other aspects of copolymerization, *viz.*, monomer sequence distribution in copolymer and mean propagation rate constant, may also be related to monomer reactivity ratios and monomer feed composition (*vide infra*). A valid model should be able to describe all these aspects of copolymerization with the same set of monomer reactivity ratios. We shall see later that the terminal model fails in this respect except for copolymerizations involving monomers of very similar reactivity. Thus, the monomer reactivity ratios in the terminal model are primarily adjustment parameters of copolymer composition equation and have limited physical significance.[6] Nevertheless, this model is very much in vogue due to the existing vast literature on the monomer reactivity ratios and on both the reactivity and the polarity parameters of monomers and radicals derived using monomer reactivity ratios.

Integral copolymer composition

Mayo and Lewis integrated Eq. (8.10) and obtained Eq. (8.15). The equation has been used to determine graphically the monomer reactivity ratios from the monomer conversion and copolymer composition data and *vice versa*.[2]

$$r_2 = \frac{\log \frac{[M_2]_0}{[M_2]} - \frac{1}{p} \log \frac{1 - p \frac{[M_1]}{[M_2]}}{1 - p \frac{[M_1]_0}{[M_2]_0}}}{\log \frac{[M_1]_0}{[M_1]} + \log \frac{1 - p \frac{[M_1]}{[M_2]}}{1 - p \frac{[M_1]_0}{[M_2]_0}}}, \tag{8.15}$$

where $r_1 = p(r_2 - 1) + 1$ and the subscript 0 refers to the initial condition.

Subsequently, Skeist provided a convenient graphical integration method for the determination of the conversion-composition relationship and, from this, the distribution of compositions in the copolymer. In addition, the method is also applicable to copolymerization of more than two monomers. The outline of the method for the binary copolymerization is given below.[7]

Consider a binary copolymerization in which $M(=M_1+M_2)$ represents the combined mols of the two monomers at an instant and the mol fraction of M_1 is f_1. Polymerization of an infinitesimally small amount dM of total monomers will result in the incorporation of $F_1 dM$ amount of monomer 1 in the copolymer, F_1 being the mol fraction of monomer 1 in the increment of copolymer formed. Correspondingly, the monomer composition in the feed changes from f_1 to $f_1 + df_1$, df_1 being positive when $f_1 > F_1$, and negative in the opposite case.

Using material balance for M_1

$$F_1 dM = f_1 M - (M - dM)(f_1 + df_1). \tag{8.16}$$

Rearrangement of the equation neglecting $dMdf_1$ gives

$$\frac{dM}{M} = \frac{df_1}{f_1 - F_1}. \tag{8.17}$$

Integration between limits f_1 and f_1^0 corresponding to M and M_0 respectively gives

$$\ln \frac{M}{M_0} = \int_{f_1^0}^{f_1} \frac{df_1}{f_1 - F_1}. \tag{8.18}$$

The integration is performed graphically for which F_1, f_1 data are needed, which may be calculated from Eq. (8.14) with values of r_1 and r_2 known or determined experimentally. From the value of the integral, for change in monomer composition from f_1^0 to f_1, the degree of conversion is computed. Similar procedure with other f_1 values makes possible to draw a graph relating conversion with f_1 or F_1. The average composition of the copolymer formed up to a given conversion is now calculable from the change in F_1 that occurs in the conversion interval.[7,8]

Integration of the right side of Eq. (8.18) gives for binary copolymerization[9]

$$\frac{M}{M_0} = \left(\frac{f_1}{f_1^0}\right)^\alpha \left(\frac{f_2}{f_2^0}\right)^\beta \left(\frac{f_1^0 - \delta}{f_1 - \delta}\right)^\gamma, \tag{8.19}$$

where $\alpha = r_2/(1-r_2)$, $\beta = r_1/(1-r_1)$, $\gamma = (1-r_1 r_2)/(1-r_1)(1-r_2)$ and $\delta = (1-r_2)/(2-r_1-r_2)$. The monomer reactivity may be calculated using this equation and conversion–composition data.[10]

8.1.1.1 Types of Copolymerization[4]

Equation (8.14) may be used to construct F_1 vs. f_1 curves for chosen values of r_1 and r_2. The results are discussed below for two broad divisions of copolymerization.

Case 1: Ideal copolymerization ($r_1 r_2 = 1$)

When $r_1 r_2 = 1$, i.e., $k_{11}/k_{12} = k_{21}/k_{22}$, both the active centers show the same preference for one of the monomers to the other. In this case, the copolymer composition

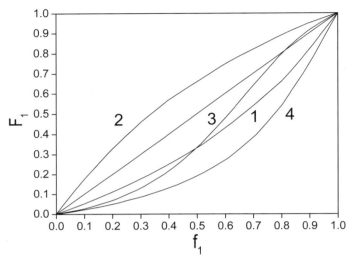

Figure 8.1 Calculated F_1 vs. f_1 curves: curve 1, $r_1 = 0.5$, $r_2 = 2$; curve 2, $r_1 = 2$, $r_2 = 0.5$; curve 3, $r_1 = 2$, $r_2 = 5$; curve 4, $r_1 = 0.4$, $r_2 = 5$. "Reprinted in part with permission from Ref. 4. Copyright © 1950 American Chemical Society."

equation (Eq. 8.14) reduces to

$$F_1 = r_1 f_1 / (r_1 f_1 + f_2). \tag{8.20}$$

The F_1 vs. f_1 plot yields a straight line with a slope of unity represented by the diagonal in Fig. 8.1 for the special case $r_1 = r_2 = 1$. The copolymer formed has always the same composition as the monomer mixture; no change in composition of either occurs with the progress of copolymerization. In other cases, curved plots are obtained. For instance, curves 1 and 2 represent ideal copolymerization respectively with $r_1 = 1/r_2 = 0.5$ and 2.[4] In these systems, the composition of the instantaneously formed copolymer is different from that of the monomer feed. As a result, during the course of copolymerization the compositions of both the instantaneous copolymer and the monomer mixture change with conversion. When r_1 is greater than unity, the instantaneous copolymer is richer in M_1 than the monomer feed is. Consequently, the compositions of both move along their respective axes toward the origin as the copolymerization progresses. Similarly, when r_1 is less than unity, the reverse is true, i.e., the monomer feed is richer in M_1 than the instantaneous copolymer is. Therefore, the monomer feed composition moves along the f_1 axis toward $f_1 = 1$ and the copolymer composition moves along the F_1 axis toward $F_1 = 1$ as the copolymerization progresses. However, there are only few radical copolymerizations, which are close to but not perfectly ideal. On the other hand, many ionic copolymerizations are more or less ideal in nature (vide infra).

Curves 1 and 2 are closely analogous to vapor vs. liquid composition curves at equilibrium in ideal binary liquid mixtures.[3] Hence, a copolymerization with $r_1 r_2 = 1$ is referred to as ideal copolymerization. The r_1 in Eq. (8.20) is analogous to the ratio of the vapor pressure (P_1^0/P_2^0) of the pure components (1 and 2) of the ideal mixture. The copolymer corresponds to the "vapor" and the monomer feed to the liquid of the ideal mixture.

The ideality condition makes the probability of addition of a monomer to a radical of its own kind equal to that of its addition to a radical of the other kind. Accordingly, the monomer distribution in the copolymer is random in character.

Case 2: $r_1 r_2 \neq 1$, i.e., $r_1 \neq 1/r_2$

In this case, unlike in the ideal one, the two active centers show different selectivity for the monomers, i.e.,

$$\frac{k_{11}}{k_{12}} \neq \frac{k_{21}}{k_{22}}.$$

Two situations may arise.

(a) $r_1 r_2 > 1$.

This situation obtains either when both the reactivity ratios are individually greater than unity or one of the reactivity ratios is less than unity and the other is greater than the reciprocal of the former. The copolymer would differ from the random copolymer of the same composition in having sequences of like monomer units in greater abundance. This is encountered in some ionic copolymerizations but not in radical ones. Representative F_1 vs. f_1 plots are shown in Fig. 8.1 as curves 3 and 4 respectively.

b) $r_1 r_2 < 1$.
i) $r_1 < 1$ and $r_2 < 1$.

In these copolymerizations, both active centers prefer the other monomer to their own. The more the r_is are closer to zero more is the tendency of alternate placement of the two monomers in the copolymer chain ($F_i \rightarrow 0.5$). This is evident from the calculated curves 1 and 2 shown in Fig. 8.2. Curve 2 refers to the radical copolymerization of maleic anhydride ($r_1 = 0.005$) and styrene ($r_2 = 0.05$).[11] In this system, nearly alternating copolymer is formed over the monomer feed composition (f_1) ranging from 0.2 to 0.5. Perfectly alternating copolymer is predicted only when $r_1 = 0$ and $r_2 = 0$.

The alternating tendency is observed in the radical copolymerization of an electron-donor monomer and an electron-acceptor one. It is immensely increased in the presence of a Lewis acid such as i-butylaluminum dichloride, ethylaluminum sesquichloride, and diethylaluminum chloride.[12–14] The Lewis acid forms complex with the functional group in the acceptor monomer, like acrylonitrile, methyl acrylate, or methyl methacrylate, thereby making the monomer a stronger electron acceptor. Copolymerization of the complexed monomer and an electron-donor monomer such as ethylene, propylene, styrene, or isobutylene gives rise to an alternating copolymer.

n CH$_2$=CH + n CH$_2$=C(Y)(Z) → (CH$_2$—CH—CH$_2$—C(Y)(Z))$_n$
    | |
    X → L.A.   X → L.A.

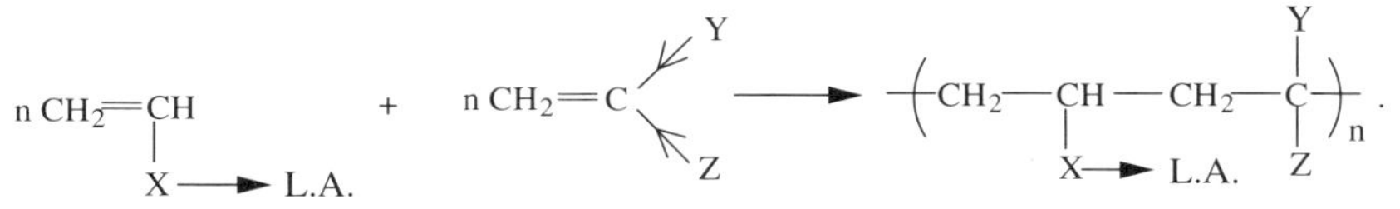

(L.A. = Lewis Acid)

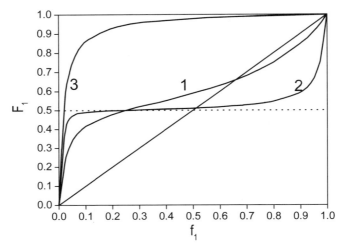

Figure 8.2 Calculated F_1 vs. f_1 curves for $r_1 r_2 < 1$: curve 1, $r_1 = 0.5$, $r_2 = 0.05$; curve 2, $r_1 = 0.05$, $r_2 = 0.005$; curve 3, $r_1 = 50$, $r_2 = 0.01$. The dotted line represents $F_1 = 0.5$.

An alternative view considers the alternating copolymerization to be the result of the polymerization of the 1:1 complex formed between an electron-donor monomer and an electron-acceptor one, which is facilitated by the complexation of the latter with a Lewis acid.15 − 17

(ii) $r_1 \gg 1$, $r_2 \ll 1$, and $r_1 r_2 < 1$.

Many radical copolymerizations exist in which one of the reactivity ratios is very small, whereas the other is very large but $r_1 r_2$ is still less than unity. A well-known example is that of styrene ($r_1 = 55$) and vinyl acetate ($r_2 = 0.01$) at 60°C.[18] Styrene is much more reactive than vinyl acetate so that styrene radical prefers to add to styrene, whereas vinyl acetate radical almost exclusively adds to styrene. As a result, the copolymer formed is highly rich in styrene until most of the styrene is consumed. Vinyl acetate principally polymerizes only after nearly all the styrene is consumed. This will be evident from curve 3 in Fig. 8.2, which is drawn for a closely analogous system with $r_1 = 50$ and $r_2 = 0.01$.

An examination of Figs 8.1 and 8.2 shows that some of the curves cross the $F_1 = f_1$ line. At the monomer feed composition corresponding to the point of intersection, the copolymer formed has the same composition as that of the monomer feed, which remains unchanged through the completion of copolymerization. In analogy with the distillation of binary liquid mixtures such copolymerization is referred to as *azeotropic*, and so is the copolymer.

The copolymer composition curves intersect the $F_1 = f_1$ line when both r_1 and r_2 are either less than unity or greater than unity. At the azeotropic composition, the copolymer composition equation reduces to

$$\frac{d[M_1]}{d[M_2]} = \frac{[M_1]}{[M_2]} = \frac{f_{1,az}}{1 - f_{1,az}}. \tag{8.21}$$

Substituting Eq. (8.21) for $\frac{d[M_1]}{d[M_2]}$ in Eq. (8.10) and solving for f_1 one gets

$$f_{1,az} = \frac{r_2 - 1}{r_1 + r_2 - 2}.$$

8.1.1.2 Determination of monomer reactivity ratios

It is convenient to use the instantaneous copolymer composition equation (Eq. 8.14) rather than the integrated one (Eq. 8.15) for the determination of monomer reactivity ratios. The instantaneous copolymer composition required for the purpose can be approximated best by restricting the conversion to as low a level as practicable (*ca.*, <5% conversion). Figures 8.1 and 8.2 reveal that the copolymer composition is more sensitive to changes in f_1 and f_2 near both ends in the monomer feed composition range. Accordingly, several copolymerization are performed covering the entire range of monomer feed composition with more emphasis on the extremes in composition. The copolymer formed in each experiment is then analyzed for its composition. Alternatively, the change in feed composition due to the polymerization is measured. Thus, a set of (F_1, f_1) data are obtained, which can be fitted to the copolymer composition equation (8.10) or (8.14) with the reactivity ratios as adjustable parameters.[19]

However, there are several linearized methods available. In the Mayo-Lewis method, the f_1, F_1 values in a single copolymerization are used in Eq. (8.14) or the corresponding concentration ratios are used in Eq. (8.10) and r_2 is plotted against arbitrarily chosen values of r_1 to get a straight line.[2] The same procedure is repeated for each of several copolymerizations. All the lines should intersect at a single point should the experimental data be accurate and the terminal model be applicable. The coordinate of the point of intersection gives the values of r_2 and r_1. Usually, however, the lines intersect at more than one point, which are close to each other. An average of the values of the coordinates of the points of intersection is chosen in such cases.

In the method of Fineman and Ross, Eq. (8.14) is rearranged in the linear form

$$f_1(1 - 2F_1)/(1 - f_1)F_1 = r_2 + [f_1^2(F_1 - 1)/(1 - f_1)^2 F_1] r_1. \tag{8.22}$$

Using the experimental f_1, F_1 data the left side of the equation is plotted against the coefficient of r_1 in the right side. The slope and intercept of the line give r_1 and r_2 respectively.[20]

Equation (8.22) is reduced to the simple form

$$G = r_1 F - r_2, \tag{8.23}$$

where $F = x^2/y$ and $G = x(y - 1)/y$ with $x = f_1/f_2$ and $y = F_1/F_2$. Using experimental f_1 and F_1 data G is plotted against F, which yields a straight line to give r_1 from the slope and r_2 from the intercept.

The problems with the Mayo-Lewis and Fineman-Ross plots are that the data points are not evenly distributed resulting in incorrect values of r for some systems. Kelen and Tudos

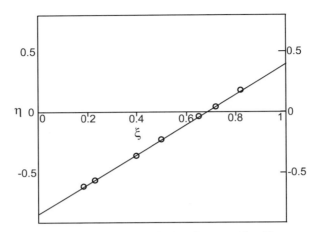

Figure 8.3 Kelen-Tudos plot in the cationic copolymerization of styrene and *p*-chlorostyrene. Ref. 22. Copyright ©1985 John Wiley & Sons, Inc., "Reprinted with permission of John Wiley & Sons, Inc."

solved this problem by introducing a parameter α defined as

$$\alpha = (F_m \cdot F_M)^{1/2}, \tag{8.24}$$

where F_m and F_M are the lowest and the highest F values respectively.[21] The variables F and G in Eq. (8.23) are then replaced by

$$\xi = \frac{F}{\alpha + F}, \tag{8.25}$$

and

$$\eta = \frac{G}{\alpha + F}. \tag{8.26}$$

Their equation relates η and ξ as[21,22]

$$\eta = \left(r_1 + \frac{r_2}{\alpha}\right)\xi - \frac{r_2}{\alpha}. \tag{8.27}$$

Equation (8.27) is equivalent to Eq. (8.23), which may be verified by substituting in it Eq. (8.25) for ξ and Eq. (8.26) for η and performing the necessary simplification.

A plot of η vs. ξ gives a straight line with the data points evenly distributed between zero and one for ξ, as shown in Fig. 8.3 for a cationic copolymerization. The intercepts of the line at the two ordinate axes $\xi = 0$ and $\xi = 1$ respectively give r_2/α and r_1. The procedure has been used to reevaluate originally published monomer reactivity ratios.[23,24]

Aside from the above methods, the reactivity ratios may be determined also from the measurement of monomer sequence distributions in copolymers, *e.g.*, fractions of various triad sequences[25–27] (*vide infra*).

8.1.1.3 Substituent effect on monomer reactivity

The reciporal monomer reactivity ratio ($1/r = k_{ij}/k_{ii}$) provides a measure of the relative reactivity of a different *vs.* own monomer with a given radical (i).[4] Values of this parameter for several copolymerizations are reproduced in Table 8.1. In the table, the monomers are arranged in an approximate order of decreasing average reactivity from top downwards in the first column. The order was arrived at by examining the 1/r values of 34 monomers with each of several radicals.[4]

However, the 1/r values do not maintain the same order with different radicals, *i.e.*, the order of monomer reactivity is not the same with each radical. This is indicative of some specificity in the reaction, which is attributed to polar effect in radical reaction, as was discussed in Chapter 3, and will be dealt with further in a later section.

8.1.1.4 Substituent effect on radical reactivity

The rate constants of addition (k_{ij}) of five radicals to each of their monomers are presented in Table 8.2. These are obtained by multiplying the corresponding 1/r values given in Table 8.1 with homopropagation rate constants (k_{ii}) for which the new k_p values determined by the PLP-SEC method (Table 3.9, Chap. 3) have been used. The k_{ij} values reveal that the radical reactivity decreases in the order of radicals as

$$VA > MA > MMA > S > BD.$$

Vinyl acetate radical is the most reactive and butadiene is the least. The more stable is the radical (*vide* Table 3.1, Chap. 3) the less is its reactivity. However, the relative reactivity of monomers is not the same with each radical, as discussed above.

The effect of a substituent on radical reactivity is much larger than the effect on monomer reactivity. Thus, while styrene is only about 50 times more reactive than vinyl acetate toward a given radical, the vinyl acetate radical is about 1000 times more reactive than styrene radical toward a given monomer.

Table 8.1 Relative reactivity ($1/r = k_{ij}/k_{ii}$) of selected monomers with various polymer radicals (*ca.* 60°C).

Monomer \ Radical:	Butadiene[a]	Styrene	Methyl methacrylate	Acrylonitrile	Methyl acrylate	Vinyl chloride	Vinyl acetate	Allyl chloride	Maleic anhydride
Chloroprene	16	19	12	22	12	—	—	—	—
Butadiene[a]	1.0	1.3	4	20	20	—	—	—	—
α–Methylstyrene	—	—	2	17	—	—	—	—	—
Styrene	0.7	1.0	2.2	20	5.5	30	>50	>30	>100
Methyl methacrylate	1.3	1.9	1.0	5.5	—	—	70	—	50
Methacrylonitrile	2.8	4	1.5	—	—	—	>50	—	—
Acrylonitrile	3	2.4	0.74	1.0	—	>15	18	20	—
Methyl acrylate	1.3	1.3	—	—	1.0	12	>5	—	50
Vinyl chloride	—	0.05	0.07	0.30	0.11	1.0	3.5	—	120
Allyl chloride	—	0.03	0.02	0.33	—	—	1.4	1.0	—
Vinyl acetate	—	0.02	0.05	0.2	0.11	0.5	1.0	1.5	300
Allyl acetate	—	0.01	0.04	—	0.2	0.9	1.7	—	—
Ethyl vinyl ether	—	0.01	—	0.2	0.3	0.5[b]	0.3	—	—
Maleic anhydride	—	>20	0.15	0.16	0.36	3.5	18	—	—

[a] Values indistinguishable from isoprene; [b] *iso*butyl vinyl ether. "Reprinted in part with permission from Ref. 4. Copyright ©1950 American Chemical Society."

Table 8.2 Rate constants (k_{ij}) for addition of polymer radicals to monomers at 60°C.[a]

Monomer	Radical				
	Butadiene (BD)	Styrene (S)	Methyl methacrylate (MMA)	Methyl acrylate (MA)	Vinyl acetate (VA)
Butadiene	200	450	3320	555600	—
Styrene	140	340	1830	152900	>397000
Methyl methacrylate	260	650	830	—	556000
Methyl acrylate	260	450	—	27800	>39700
Vinyl acetate	—	7	40	3060	7940

[a]The rate constants are recalculated using the k_p values determined by the PLP-SP method (Table 3.9, Chap. 3). "Adapted with permission from Ref. 4. Copyright © 1950 American Chemical Society."

8.1.1.5 Resonance and polar effects in radical addition

In Chap. 3 (Secs. 3.1.2 and 3.8.2), the effects of resonance and polar factors in radical reaction have been discussed with reference to two approaches, which deal with the subject from a quantitative standpoint. The applications of those approaches in copolymerization is discussed below.[28-36]

8.1.1.5.1 Q, e Scheme

In analogy with the Hammett equation, Alfrey and Price proposed Eq. (8.28) for the rate constant (k_{ij}) of the addition of radical i to monomer j[30]

$$k_{ij} = P_i Q_j e^{-e_i e_j}, \qquad (8.28)$$

where P_i and Q_j are parameters representing the general reactivity of the radical i and monomer j respectively, while e_i and e_j are the corresponding polarity parameters. The same e value was assigned to a monomer and its radical. The monomer reactivity ratios (r_1 and r_2) therefore follow as

$$r_1 = \frac{k_{11}}{k_{12}} = \frac{P_1 Q_1 e^{-e_1^2}}{P_1 Q_2 e^{-e_1 e_2}} = \frac{Q_1}{Q_2} e^{-e_1(e_1-e_2)}, \qquad (8.29)$$

and

$$r_2 = \frac{k_{22}}{k_{21}} = \frac{P_2 Q_2 e^{-e_2^2}}{P_2 Q_1 e^{-e_1 e_2}} = \frac{Q_2}{Q_1} e^{-e_2(e_2-e_1)}. \qquad (8.30)$$

From the experimentally measured values of r_1 and r_2, the Q and e values of a monomer can be determined if the corresponding values of the other monomer are known. Price chose styrene as the reference monomer for which $Q = 1$ and $e = -0.8$ were arbitrarily assigned.[30] Using the monomer reactivity ratio values of several monomer pairs with styrene as the common comonomer the Q and e values of all such monomers may be determined. These monomers are referred to as the primary monomers.

Using the monomer reactivity ratios of other monomer pairs with any or more of these primary monomers as comonomers the Q and e values of these other monomers may be

Table 8.3 Q and e values of primary monomers.[a]

Monomer	Q	e
Butadiene	1.70	−0.5
Vinyl acetate	0.026	−0.88
Styrene	1	−0.8
Methyl methacrylate	0.78	0.40
Methyl acrylate	0.45	0.64
Methacrylonitrile	0.86	0.68
Acrylonitrile	0.48	1.23
Acrylic acid	0.83	0.88

[a]Reference 31. Copyright © 1999 John Wiley & Sons Inc., Reprinted with permission of John Wiley & Sons, Inc.

determined, and so on. The values of a large number of monomers so determined have been tabulated; those of the primary monomers are given in Table 8.3.[31] With Q and e values of a monomer pair known, their monomer reactivity ratios may be determined, if necessary, when the experimentally determined values are not available.

As discussed in Chap. 3 (Sec. 3.8.2), the Q, e scheme has also been used to radical transfer reactions. However, as pointed out earlier, the scheme is at best semi-empirical. The assignment of the same polarity parameter for both monomer and its corresponding radical has no theoretical basis.[32]

However, the radical polarity parameter determined by applying the Q, e Scheme to the radical reactivity ratios (R_1, R_2) is found to be close to the monomer polarity parameter.[33] For example, Theil considered two radical reactivity ratios (R_1 and R_2) analogous to monomer reactivity ratios

$$R_1 = \frac{k_{11}}{k_{21}}, \quad R_2 = \frac{k_{22}}{k_{12}}. \tag{8.31}$$

R_1 and R_2 can be calculated using the known monomer reactivity ratios (r_1 and r_2) and known k_ps (k_{11} and k_{22}) and expressed according to the Q, e Scheme (Eq. (8.28)) as

$$R_1 = \frac{P_1}{P_2} e^{-e_1(e_1 - e_2)}, \tag{8.32}$$

and

$$R_2 = \frac{P_2}{P_1} e^{-e_2(e_2 - e_1)}. \tag{8.33}$$

Using styrene radical as the calibrant for which $P_R = 1$ and $e_R = -0.8$, P_R and e_R values of several radicals were calculated giving the above-mentioned result. However, e_r ($= -0.22$) is more negative than e_R ($= -0.027$) in case of vinyl acetate[33] (for a detailed disucssion of the vinyl acetate case, see Chap. 3, Sec. 3.8.2).

8.1.1.5.2 The revised pattern scheme

The radical reactivity pattern scheme developed by Bamford, Jenkins, and Johnston for separating and quantifying the resonance and polar effects in radical transfer or addition reactions has already been discussed in Chap. 3 (Sec. 3.8.2.1). According to the scheme, the rate constant k_{12} for the addition of monomer 2 to radical 1 may be written as[34]

$$\log k_{12} = \log k_{tr,t} + \alpha_2 \sigma_1 + \beta_2, \tag{8.34}$$

where $k_{tr,t}$ is the rate constant of chain transfer of radical 1 to toluene (t), σ_1 is the Hammett *para* σ parameter of the α substituent in radical (1), α_2 is analogous to ρ in Hammett equation, and β_2 is the general reactivity parameter of monomer (2). The difficulty with Eq. (8.34) is that the $k_{tr,t}$ data for many radicals are not readily available. In order to overcome this problem, Jenkins replaced $k_{tr,t}$ by k_{1s} (rate constant of addition of radical 1 to styrene) as a measure of general reactivity of radical.[35,36] This step transforms Eq. (8.34) to

$$\log k_{12} = \log k_{1s} + \alpha_2 \sigma_1 + \beta_2 + \gamma, \tag{8.35}$$

where γ is the difference between $\log k_{tr,t}$ and $\log k_{1s}$.

The number of parameters is reduced by writing the equation as

$$\log k_{12} = \log k_{1s} + \alpha_2 \sigma_1 + \nu_2. \tag{8.36}$$

Subtraction of $\log k_{11}$ from both sides of Eq. (8.36) gives

$$\log r_{12} = \log r_{1s} - \alpha_2 \sigma_1 - \nu_2. \tag{8.37}$$

Another difficulty with the pattern scheme is the limited availability of reliable Hammett σ parameters of substituents. This problem was overcome by replacing σ with a polarity parameter determined from polymerization studies.[35,36] The procedure specifically requires the knowledge of reactivity ratio of monomer 1 with acrylonitrile as comonomer (monomer 2).

8.1.2 Statistical treatment of copolymerization

The copolymerization equation may also be derived by a statistical method.[37]

Consider the set of propagation reactions (8.1) through (8.4). The probability of monomer M_1 adding to radical $M_1^\bullet$ is given by

$$P_{11} = \frac{k_{11}[M_1^\bullet][M_1]}{k_{11}[M_1^\bullet][M_1] + k_{12}[M_1^\bullet][M_2]} = 1 - P_{12} \tag{8.38}$$

$$= \frac{r_1[M_1]}{r_1[M_1] + [M_2]}. \tag{8.39}$$

Similarly, three other probabilities can be defined as

$$P_{12} = \frac{[M_2]}{r_1[M_1]+[M_2]} = 1 - P_{11} \tag{8.40}$$

$$P_{22} = \frac{r_2[M_2]}{[M_1]+r_2[M_2]} = 1 - P_{21} \tag{8.41}$$

$$P_{21} = \frac{[M_1]}{[M_1]+r_2[M_2]} = 1 - P_{22}. \tag{8.42}$$

P_{ij} is referred to as transition probability of terminal radical M_i to M_j (i, j = 1 or 2).[38] The probability of finding a sequence of exactly n units of M_1 monomer in a copolymer chain is given by

$$P_n = F_n = P_{11}^{n-1} P_{12}, \tag{8.43}$$

where F_n is the number or mol fraction of all M_1 sequences comprising n units.

The number average sequence length of M_1 monomer units is given by

$$\bar{l}_1 = \frac{\text{Total number of } M_1 \text{ units}}{\text{Total number of } M_1 \text{ sequences}}$$

or

$$\bar{l}_1 = \sum_{n=1}^{\infty} n F_n. \tag{8.44}$$

Substituting Eq. (8.43) for F_n in Eq. (8.44) gives[37,39]

$$= P_{12} + 2P_{11}P_{12} + 3P_{11}^2 P_{12} + \cdots \tag{8.45}$$

$$= \frac{P_{12}}{P_{11}} \sum_{n=1}^{\infty} n P_{11}^n \tag{8.46}$$

$$= \frac{P_{12}}{P_{11}} \frac{P_{11}}{(1-P_{11})^2}. \tag{8.47}$$

$$= \frac{1}{P_{12}} \quad \text{(since } 1 - P_{11} = P_{12}\text{)}. \tag{8.48}$$

Substituting Eq. (8.40) for P_{12} in Eq. (8.48) gives

$$\bar{l}_1 = r_1 \frac{f_1}{f_2} + 1. \tag{8.49}$$

Correspondingly, the number average sequence length of M_2 monomer units is given by

$$\bar{l}_2 = \frac{1}{P_{21}}. \tag{8.50}$$

Substituting Eq. (8.42) for P_{21} in Eq. (8.50) gives

$$\bar{l}_2 = r_2 \frac{f_2}{f_1} + 1. \tag{8.51}$$

The instantaneous copolymer composition equation follows as

$$\frac{d[M_1]}{d[M_2]} = \frac{\bar{l}_1}{\bar{l}_2} = \frac{P_{21}}{P_{12}} = \frac{[M_1]}{[M_2]}\left(\frac{r_1[M_1]+[M_2]}{[M_1]+r_2[M_2]}\right), \qquad (8.52)$$

which is identical with Eq. (8.10) derived by the kinetic method.

However, strictly speaking, the copolymer composition should be given by Eq. (8.53) rather than by Eq. (8.52)[40]

$$\frac{d[M_1]}{d[M_2]} = \frac{s_1\bar{l}_1}{s_2\bar{l}_2}, \qquad (8.53)$$

where s_i refers to the total number of sequences (run number) of monomer i of all lengths: monads, diads, triads, *etc.*

$$s_i = n(M_i) + n(M_iM_i) + n(M_iM_iM_i) + \cdots. \qquad (8.54)$$

Each run is bounded at each end by an unlike monomer unit. In linear statistical copolymer of infinite molecular weight, the run numbers of the two monomers are equal. So are the numbers of 12 and 21 sequences.[41,42] This may be verified by examining a chain with the two ends connected together as to approximate a chain with no end effects (an infinite chain).[42] In a finite chain the numbers differ at most by one unit. Thus, for high molecular weight copolymer we may write

$$s_1 = s_2, \qquad (8.55)$$

and

$$n_{12} = n_{21}. \qquad (8.56)$$

For the equality condition Eq. (8.56) to be attained, the rate of formation of 12 diads (v_{12}) should equal that of 21 diads (v_{21})[40]

$$v_{12} = v_{21}. \qquad (8.57)$$

This is the condition for the stationary state of the radicals, which is assumed in the kinetic method (Sec. 8.1.1) and implicit in the statistical method under discussion.

8.1.2.1 *Monomer sequence distribution*

The property of a copolymer depends not only on the copolymer composition but also on the monomer distribution (microstructure) in the copolymer chain. The latter may be expressed in terms of the abundance of various monomer sequence lengths: diads, triads, and higher order n-ads. The number fractions of homogeneous sequences of n units of M_1 may be calculated using Eq. (8.43) and analogously those of n units of M_2, provided monomer reactivity ratios are known. Complete distribution of homogeneous sequences of all possible lengths is not, in general, experimentally measurable. However, abundances of triads and pentads in many copolymers have been measured using NMR spectroscopy, ^{13}C NMR in particular.[42-45] However, a higher order sequence may be considered to be comprised of several lower order ones. For example, an M_1 pentad bounded by two M_2 units, one on each end, comprises four homogeneous M_1 diads and two heterogeneous mirror image diads M_2M_1 and M_1M_2 (Fig. 8.4) or three M_1 centered homogeneous $(M_1M_1M_1)$ and two heterogeneous mirror image triads $(M_2M_1M_1 = M_1M_1M_2)$ (Fig. 8.5). The various diads and triads are indicated in Figs. 8.4 and 8.5 respectively by curved lines encompassing their monomer units. The mirror image diads or triads are indistinguishable since the polymer chain direction cannot be specified usually.

Figure 8.4   Diad components of a pentad identified as connected with curved lines.

$M_2 \quad M_1 \quad M_1 \quad M_1 \quad M_1 \quad M_1 \quad M_2$

Figure 8.5   Triad components of a pentad identified with curved lines linking the end units of the triads.

8.1.2.1.1 Triad fractions and monomer reactivity ratios

The monomer reactivity ratios may be determined from the number average sequence length of the monomer units, which in turn may be determined from the various diad and triad fractions.[25–27] Thus, focusing our attention to M_1 sequences, all of them are bounded by M_2 units on both ends. Such sequences up to the length of four M_1 units and the number of various diads and triads present in them are shown in Table 8.4. It may be verified that in each sequence the number of monomer units may be equated with the number of diads or M_1 centered triads as

$$n_1 = n_{11} + \tfrac{1}{2} n_{12} = n_{212} + n_{211} + n_{111}.$$

Similar equation may be written for monomer 2.

$$n_2 = n_{22} + \tfrac{1}{2} n_{12} = n_{121} + n_{122} + n_{222}.$$

Thus, the mol fraction of M_1 units in the copolymer may be equated with the diad or triad fractions as

$$F_1 = F_{11} + \tfrac{1}{2} F_{12} = F_{212} + F_{211} + F_{111}. \tag{8.58}$$

The corresponding normalization relations are

$$F_1 + F_2 = F_{11} + F_{12} + F_{22} = F_{212} + F_{211} + F_{111} + F_{121} + F_{122} + F_{222} = 1.$$

Since each sequence of M_1 units starts with a 2,1 linkage and ends with a 1,2 linkage the number of such sequences is given by half the number of 12 diads (remembering that this number includes the 21 diads also). Thus, the number average sequence length of M_1 monomers is given by

$$\bar{l}_1 = \frac{\text{Total number of } M_1 \text{ units}}{\text{Total number of } M_1 \text{ runs}} = \frac{n_1}{n_{12}/2} = \frac{F_1}{F_{12}/2}. \tag{8.59}$$

Table 8.4 M_1 sequences and the numbers of the diads and the triads.

Sequence[a]	No. of diads			No. of triads	
	11	12[b]	212	211[c]	111
212	0	2	1	0	0
2112	1	2	0	2	0
21112	2	2	0	2	1
211112	3	2	0	2	2

[a]M_1 sequences bounded by monomer 2 units at both ends;
[b]Value includes the number of the mirror image diads 21;
[c]Value includes the number of the mirror image triads 112.

The fraction of 12 diads may be equated with that of the M_1 centered triads (*vide* Table 8.4)

$$F_{12} = 2F_{212} + F_{112}. \tag{8.60}$$

Substituting Eq. (8.58) for F_1 and Eq. (8.60) for F_{12} in Eq. (8.59) gives[25]

$$\bar{l}_1 = \frac{F_{11} + F_{12}/2}{F_{12}/2} = \frac{F_{212} + F_{211} + F_{111}}{F_{212} + F_{112}/2}. \tag{8.61}$$

Proceeding similarly for M_2 sequences

$$\bar{l}_2 = \frac{F_{22} + F_{12}/2}{F_{12}/2} = \frac{F_{121} + F_{221} + F_{222}}{F_{212} + F_{112}/2}. \tag{8.62}$$

When only the M_1-centered triad fractions are measurable, the normalization relation becomes

$$F_{111} + F_{212} + F_{112} = 1. \tag{8.63}$$

Substituting Eq. (8.63) for the numerator and Eq. (8.49) for $\bar{l}_1$ in Eq. (8.61) gives[26]

$$\bar{l}_1 = \left(F_{212} + \frac{F_{112}}{2} \right)^{-1} = r_1 \frac{f_1}{f_2} + 1. \tag{8.64}$$

Relating $\bar{l}_2$ with $\bar{l}_1$ and writing Eq. (8.51) for $\bar{l}_2$ gives[26]

$$\bar{l}_2 = \left(\frac{F_2}{F_1} \right) \bar{l}_1 = r_2 \frac{f_2}{f_1} + 1. \tag{8.65}$$

Thus, from Eq. (8.64), the slope of the linear plot of $\bar{l}_1$ vs. f_1/f_2 gives r_1, and analogously from Eq. (8.65), the slope of the linear plot of $\bar{l}_2$ vs. f_2/f_1 gives r_2. Obviously, the determination of r_2 requires the additional knowledge of the copolymer composition F_2/F_1.

An alternative method uses the probability relations (8.39) through (8.42).[27] Thus, from Eq. (8.39)

$$\frac{1}{P_{11}} = 1 + \frac{1}{r_1} \frac{f_2}{f_1}. \tag{8.66}$$

P_{11} may be related to the triad fraction F_{111} with the normalization involving only the M_1 centered triads (Eq. (8.63)) as

$$F_{111} = P_{11}^2. \tag{8.67}$$

Substituting P_{11} from Eq. (8.67) in Eq. (8.66) gives

$$\left(\frac{1}{F_{111}}\right)^{1/2} = \frac{1}{P_{11}} = 1 + \frac{1}{r_1}\frac{f_2}{f_1}. \tag{8.68}$$

Similarly, from Eq. (8.40)

$$\frac{1}{P_{12}} = 1 + r_1 \frac{f_1}{f_2}. \tag{8.69}$$

P_{12} may be related to the triad fraction F_{212} as

$$F_{212} = P_{21} \cdot P_{12} = P_{12}^2. \tag{8.70}$$

Substituting P_{12} from Eq. (8.70) in Eq. (8.69) gives

$$\left(\frac{1}{F_{212}}\right)^{1/2} = \frac{1}{P_{12}} = 1 + r_1 \frac{f_1}{f_2}. \tag{8.71}$$

The remaining M_1 centered triad fraction F_{211} is related to the transition probabilities as

$$F_{112} = 2P_{11}P_{12} \quad \text{(the factor 2 accounting for the mirror image triad 211).} \tag{8.72}$$

or

$$F_{112} = 2 P_{12}(1 - P_{12}). \tag{8.73}$$

Equation (8.73) gives on expansion and transportation

$$P_{12}^2 - P_{12} + \frac{F_{112}}{2} = 0. \tag{8.74}$$

Solving the quadratic for P_{12} and substituting Eq. (8.40) for P_{12} gives

$$\alpha = \frac{2}{1 \pm (1 - 2F_{112})^{1/2}} = \frac{1}{P_{12}} = 1 + r_1 \frac{f_1}{f_2}. \tag{8.75}$$

Thus, according to Eq. (8.68) by plotting $(1/F_{111})^{1/2}$ vs. f_2/f_1, $1/r_1$ is evaluated from the slope of the line. Similarly, according to Eq. (8.71), by plotting $(1/F_{212})^{1/2}$ vs. f_1/f_2, or according to Eq. (8.75), by plotting α vs. f_1/f_2 one gets r_1 from the slopes of the respective lines. Although, the use of any of the three different M_1 centered triad concentrations allows determination of r_1, use of the data of the other two triad fractions allows one to crosscheck the results.[27]

Analogously, the M_2 centered triad fractions (F_{222}, F_{221}, and F_{121}) may be used to obtain equations relating them with r_2 and the latter determined using a similar procedure as is used for r_1.

8.1.3 *Kinetics of radical copolymerization based on terminal model*

The terminal model of copolymerization involves four propagation reactions (8.1) through (8.4) and three termination reactions

$$\left. \begin{array}{l} M_1^\bullet + M_1^\bullet \xrightarrow{k_{t11}} \\ M_1^\bullet + M_2^\bullet \xrightarrow{k_{t12}} \\ M_2^\bullet + M_2^\bullet \xrightarrow{k_{t22}} \end{array} \right\} \text{dead copolymer.} \tag{8.76}$$

The rate of copolymerizations may be expressed by an equation analogous to that of radical homo polymerization (Eq. 3.20, Chap. 3) as

$$R_p = \bar{k}_p R_i^{1/2}[M]/(2\bar{k}_t)^{1/2}, \tag{8.77}$$

where [M] is the combined concentration of both monomers M_1 and M_2, $\bar{k}_p$ and $\bar{k}_t$ being the mean rate constant of propagation and termination respectively. Alternatively, R_p may be expressed in terms of $\bar{k}_p$ and concentrations of propagating radical and monomers as

$$R_p = \bar{k}_p[M^\bullet][M], \tag{8.78}$$

where

$$[M^\bullet] = [M_1^\bullet] + [M_2^\bullet] \quad \text{and} \quad [M] = [M_1] + [M_2].$$

It may be expressed also in terms of the individual propagation reactions

$$R_p = \left(k_{11}[M_1^\bullet] + k_{21}[M_2^\bullet]\right)[M_1] + \left(k_{22}[M_2^\bullet] + k_{12}[M_1^\bullet]\right)[M_2]. \tag{8.79}$$

Equating (8.78) with (8.79) gives

$$\bar{k}_p = \frac{\left(k_{11}[M_1^\bullet] + k_{21}[M_2^\bullet]\right)[M_1] + \left(k_{22}[M_2^\bullet] + k_{12}[M_1^\bullet]\right)[M_2]}{([M_1^\bullet] + [M_2^\bullet])([M_1] + [M_2])}. \tag{8.80}$$

Eliminating the radical concentrations using Eq. (8.7), as described earlier (Sec. 8.1.1), one gets from Eq. (8.80)

$$\bar{k}_p = \frac{\left(k_{11} + k_{12}\frac{f_2}{f_1}\right)f_1 + \left(k_{12} + k_{12}r_2\frac{f_2}{f_1}\right)f_2}{1 + \left(\frac{k_{12}}{k_{21}}\right)\left(\frac{f_2}{f_1}\right)}, \tag{8.81}$$

where f_1, f_2, and r_2 are the same as defined earlier.

Dividing both numerator and denominator of the equation by k_{12} gives[46]

$$\bar{k}_p = \frac{r_1 f_1^2 + 2f_2 f_1 + r_2 f_2^2}{\dfrac{f_1}{k_{12}} + \dfrac{f_2}{k_{21}}} \tag{8.82}$$

$$= \frac{r_1 f_1^2 + 2f_1 f_2 + r_2 f_2^2}{\dfrac{r_1 f_1}{k_{11}} + \dfrac{r_2 f_2}{k_{22}}}. \tag{8.83}$$

As regards termination, the mean rate constant is expressed in two different forms depending on whether termination is reaction-controlled or diffusion-controlled.

For reaction-controlled termination[47,48]

$$\bar{k}_t = k_{t11}p_1^2 + 2k_{t12}p_1p_2 + k_{t22}p_2^2, \qquad (8.84)$$

where $p_i = [M_i^{\bullet}]/[M^{\bullet}]$ and k_{tij} refers to the rate constant of termination between i and j radicals. A cross-termination factor ϕ has been defined as the ratio of cross-termination constant to that of the geometric mean of the homo termination constants

$$\phi = k_{t12}/(k_{t11}k_{t22})^{1/2}.$$

Preference for cross-termination is indicated with values of $\phi > 1$.

For diffusion-controlled termination the mean rate constant has been defined using the "ideal" diffusion model by Eq. (8.85), which assumes that $\bar{k}_t$ depends on the average composition of the low conversion copolymer represented by mol fractions of monomer units (F_i) in the copolymer.

$$\bar{k}_t = F_1 k_{t1} + F_2 k_{t2}, \qquad (8.85)$$

where the k_{ti} s refer to the homopolymerizations.[49]

8.1.4 Ionic copolymerization

In general, ionic polymerization exhibits much higher selectivity toward monomers than radical polymerization does. The higher is the nucleophilicity of a monomer the higher is its rate of addition to a given cation. Conversely, higher is the electronegativity of a monomer higher is its rate of addition to a given anion. Hence, both propagating species M_1^* and M_2^* in a given ionic copolymerization prefer the same monomer to the other. As a result, ionic copolymerization frequently exhibits ideal or semi ideal behavior.[39]

The subject may be discussed from a quantitative standpoint for cationic copolymerization. As introduced in Chapter 6, the rate constants of homo propagation and cross propagation in cationic copolymerization may be expressed following Mayr as[50,51]

$$\log k_{ii}(20°C) = s_i(N_i + E_i), \qquad (8.86a)$$

$$\log k_{ij}(20°C) = s_j(N_j + E_i), \quad (i, j = 1 \text{ or } 2), \qquad (8.86b)$$

where k is the rate constant, E is the electrophilicity parameter of the propagating ion, N is the nucleophilicity parameter of the monomer, and s is the nucleophile-dependent slope parameter, which is close to unity. Subtraction of one equation from the other gives[52]

$$\log r_i = -\log r_j \approx (N_i - N_j). \qquad (8.87)$$

Thus, the copolymerization would be ideal. However, larger the nucleophilicity difference ($|\Delta N|$) between the two monomers is richer the copolymer is in the monomer with the higher

Table 8.5 Monomer reactivity ratios in cationic polymerization.[a]

Monomer 1	Monomer 2	N_1	N_2	r_1	r_2	$r_1 r_2$	Ref.[b]
Styrene	Isoprene	0.78	1.10	0.47	0.53	0.25	53 (22)
Styrene	Isobutylene	0.78	1.11	0.70	2.70	1.89	54 (22)
Styrene	p-Cl-styrene	0.78	0.21	1.5–2.1	0.22–0.54	0.37–1.06	55,56 (22)
Styrene	p-Me-Styrene	0.78	1.70	0.35–0.65	2.5–3.1	1.08–1.91	(57)
Styrene	Chloroprene	0.78	−1.59	6.93–36.5	0.12–0.60	0.42–4.38	58 (22)
Isobutylene	Isoprene	1.11	1.10	~1	~1	~1	59a,b
Isobutylene	Butadiene	1.11	−0.87	43–130	0–0.02	0-2.6	59b,60
2-Cl-EVE[c]	p-OMe-St	–	3.30	1.16–2.56	3.40–5.02	3–17	62 (22)
2-Cl-EVE[c]	p-Me-St	–	1.70	3.29	1.11	3.62	62 (22)
Isobutylene	Chloroprene	1.11	−1.59	No copolymerization			61
Isobutylene	Alkyl vinyl ether	1.11	~4	Reported copolymerization questionable[d]			(22)

[a]The range of r values, when given, refers to copolymerization using different reaction variables, or conditions such as different initiator, coinitiator, solvent, and temperature. [b]Reference in parenthesis gives the recalculated reactivity ratios by Kelen–Tudos method which are given in the table, when available. [c]2-chloroethyl vinyl ether. [d]Meaningless Kelen-Tudos plots.[22]

nucleophilicity. When the nucleophilicity difference is too large, the copolymer contains so little monomer of the lower nucleophilicity that virtually no copolymerization occurs.

Table 8.5 gives the values of the monomer reactivity ratios in some cationic copolymerizations. A range of values reflects the variation of r with experimental variables, *viz.*, initiator, coinitiator, solvent, and temperature. The absence of a range does not mean that the variation does not occur. For these copolymerizations the data are either not available or the review of the published data recommends only one set.[22] In copolymerizations with relatively small $|\Delta N|$ (say, $<\sim 1$), one of the reactivity ratios is close to or less than unity, while the other is close to or greater than unity. Besides, the product of the two reactivity ratios under some experimental conditions is close to unity. These represent ideal or semi ideal copolymerizations. When $|\Delta N|$ is relatively large, say, between 2 and 2.5, one of the reactivity ratios is very large, while the other is less than unity being close to zero in some case (isobutylene and butadiene being an example). As $|\Delta N|$ becomes larger still, the copolymerization does not take place.

Disagreement with the $|\Delta N|$ theory, of course, exists. For example, in the copolymerization of styrene and isoprene both the reactivity ratios are less than unity indicating alternating tendency in the monomer placement in the copolymer chain, although ideal polymerization was expected. Conversely, in the copolymerization of 2-chloroethyl vinyl ether and p-methoxystyrene or p-methylstyrene both the r values are appreciably greater than unity indicating blockiness in copolymer composition.[22,62] Obviously, these results are not explicable solely by the difference in the nucleophilicity of monomers.

In the anionic copolymerization, the principal property that determines copolymerizability is the electronegativity difference between the monomers involved. Monomer pairs with only small difference in electronegativity are successfully copolymerized.[2]

As mentioned above, change in initiator, coinitiator, solvent, and temperature may bring about change in the reactivity ratios. This is true also in anionic or coordinated anionic copolymerization.[63,64] The effect has been attributed to various factors such as solvation of ionic species by solvent or monomer, ion pair dissociation or inapplicability of the terminal model.[65]

8.1.5 *Validity of the terminal model*

Copolymer composition and feed composition data of most copolymerization could be fitted in the copolymer composition equation (Eq. 8.10 or 8.14). As a result, the terminal model gained the preeminent status of a basis model based on which the vast literature on monomer reactivity ratios was developed. The small number of nonconforming copolymerization was considered exceptions rather than evidence of the invalidity of the model. Penultimate model and/or various side reaction models were used to describe their copolymer composition. However, the validity of the terminal model could be checked if other aspects of copolymerization such as the mean propagation rate constant and the monomer sequence distribution in copolymer, could be predicted by monomer reactivity ratios derived from copolymer composition.[46,66,67] Unfortunately, these checks could not be performed until 1980s due to the lack of accurate methods of determining k_p and sequence distribution. The advent of pulsed laser polymerization-size exclusion chromatographic method for determining k_p (Chap. 3 Sec. 3.10)[46,68–70] and development in ^{13}C NMR spectroscopy for determining sequence distribution made performing these tests possible.

The tests revealed wide discrepancies between the experimentally measured and the predicted k_ps in radical copolymerization[6,65] excepting those involving monomers of similar reactivity such as methyl methacrylate and dodecyl methacrylate,[71] methyl methacrylate and n-butyl methacrylate,[71] styrene and *p*-methoxystyrene,[72] *p*-chlorostyrene and *p*-methoxystyrene,[73] among others. For instance, Figure 8.6 shows how greatly the measured $\bar{k}_p$s differ from the predicted values (Eq. 8.83) in the copolymerization of styrene and methyl methacrylate.[46]

Besides, the predicted monomer sequence distribution does not agree with the experimentally measured distribution in many copolymerizations. This includes that of styrene and methyl methacrylate.[6] These results led to the conclusion that the terminal model reactivity ratios should be regarded merely as adjustment parameters in copolymer composition equation, which have limited physical meaning.[6,65]

8.2 Penultimate Model of Copolymerization

In quest of a better model, several models were proposed to replace the terminal model. Many of these models explained the deviations of the experimental results from the terminal model as arising from side reactions[65] such as monomer partitioning, depropagation, and complex formation between radicals and solvents, radicals and monomers, monomers and monomers, *etc.*[74–78]

Monomer partitioning may occur from different causes. For instance, one of the monomers may be preferentially absorbed by the copolymer if it is a good solvent, while

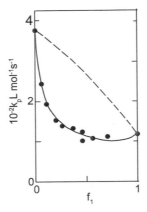

Figure 8.6 Test of the validity of the terminal model in the radical copolymerization of styrene and methyl methacrylate. The dashed curve is calculated based on the terminal model using the monomer reactivity ratios based on the same model (Eq. 8.83). The points are the measured k_p values. The solid curve is obtained by least-squares curve-fitting of k_p equation based on the implicit penultimate model (Eq. 8.119) to the experimental points (see text below and note that in the text k_p is appropriately written as $\bar{k}_p$). "Reprinted with permission from Ref. 46. Copyright ©1985 American Chemical Society."

the other is a poor solvent or a nonsolvent for the copolymer. As a result, the ratio of the two monomers at the active center becomes different from that in the bulk.

Effect of depropagation is usually encountered in binary copolymerization where one of the comonomers, e.g., CO, SO_2, or α-methylstyrene, has a ceiling temperature near to that of the polymerization temperature.

Side reactions are, however, not a general phenomenon. A general model is therefore called for. The penultimate model turns out to be the one, which adequately describes various aspects of copolymerization in the absence of significant side reactions. For this reason, the penultimate model has replaced the terminal model as the basis model.[6,65] It assumes that the radical reactivity depends on not only the terminal monomer unit bearing the radical but also the penultimate monomer unit.[5,6] Accordingly, eight different propagation reactions are involved, (reactions 8.88 through 8.95) as against four in the terminal model (Sec. 8.1). In addition, four monomer reactivity ratios follow as against two in the terminal model. There are two radical reactivity ratio parameters as well (Eqs. 8.100 and 8.101). The model is referred to as explicit penultimate model when it is applied to systems in which the terminal model fails to describe even the copolymer composition.

However, several systems exist in which the copolymer compositions can be described adequately by the terminal model but not the $\bar{k}_p$ values, which exhibit wide deviation from the measured values. For such systems, the penultimate model is used to describe the rate of propagation, whereas the terminal model is used to describe the copolymer composition (Fig. 8.6).[46] This model is referred to as the implicit penultimate model. Copolymerization of styrene and methyl methacrylate is a typical example. In effect, the model assumes that the radical reactivity depends on both its terminal and penultimate units, whereas the monomer selectivity of the radical depends only on its terminal unit. It reduces to the terminal model in describing the copolymer composition, but differs from the terminal model in describing the propagation rate constant. However, it is simpler than the explicit model due to the number of monomer reactivity ratio parameters being reduced to two from four, while the radical reactivity ratio parameters remained unchanged (*vide infra*). However, several experimental studies on small radical reactions as well as theoretical studies reveal that the assumptions underlying the implicit model are untenable.[6,65]

8.2.1 *Mean rate constant of propagation*

The propagation reactions considered in the penultimate model are[5]

$$M_1M_1^\bullet + M_1 \xrightarrow{k_{111}} M_1M_1M_1^\bullet, \tag{8.88}$$

$$M_1M_1^\bullet + M_2 \xrightarrow{k_{112}} M_1M_1M_2^\bullet, \tag{8.89}$$

$$M_2M_1^\bullet + M_1 \xrightarrow{k_{211}} M_2M_1M_1^\bullet, \tag{8.90}$$

$$M_2M_1^\bullet + M_2 \xrightarrow{k_{212}} M_2M_1M_2^\bullet, \tag{8.91}$$

$$M_2M_2^\bullet + M_2 \xrightarrow{k_{222}} M_2M_2M_2^\bullet, \tag{8.92}$$

$$M_2M_2^\bullet + M_1 \xrightarrow{k_{221}} M_2M_2M_1^\bullet, \tag{8.93}$$

$$M_1M_2^\bullet + M_1 \xrightarrow{k_{121}} M_1M_2M_1^\bullet, \tag{8.94}$$

$$M_1M_2^\bullet + M_2 \xrightarrow{k_{122}} M_1M_2M_2^\bullet. \tag{8.95}$$

The four monomer reactivity ratios defined are

$$r_1 = r_{11} = \frac{k_{111}}{k_{112}}, \tag{8.96}$$

$$r_1' = r_{21} = \frac{k_{211}}{k_{212}}, \tag{8.97}$$

$$r_2 = r_{22} = \frac{k_{222}}{k_{221}}, \tag{8.98}$$

$$r_2' = r_{12} = \frac{k_{122}}{k_{121}}. \tag{8.99}$$

Thus, r_1 and r_1' refer to the reactivity of monomer 1 relative to that of monomer 2 toward propagating radicals centered on M_1 terminal unit with M_1 and M_2 as penultimate units respectively. Similarly, r_2 and r_2' refer to the reactivity of M_2 relative to that of M_1 toward propagating radicals centered on M_2 terminal unit with M_2 and M_1 as penultimate units respectively.

The two radical reactivity ratios defined are[6,46]

$$s_1 = \frac{k_{211}}{k_{111}}, \tag{8.100}$$

and
$$s_2 = \frac{k_{122}}{k_{222}}. \tag{8.101}$$

The radical reactivity ratio refers to the relative reactivity of a propagating radical with a particular terminal unit bearing the radical toward its own monomer when the penultimate unit is different from it to that when the penultimate unit is same as it.

Considering the stationary state for the copolymerization one obtains for long chains

$$\frac{d[M_1 M_1^\bullet]}{dt} = k_{211}[M_2 M_1^\bullet][M_1] - k_{112}[M_1 M_1^\bullet][M_2] = 0, \tag{8.102}$$

$$\frac{d[M_2 M_2^\bullet]}{dt} = k_{122}[M_1 M_2^\bullet][M_2] - k_{221}[M_2 M_2^\bullet][M_1] = 0. \tag{8.103}$$

Representing total concentrations of all radicals centered on M_1 terminal unit as $[M_1^\bullet]$ we have

$$[M_1^\bullet] = [M_1 M_1^\bullet] + [M_2 M_1^\bullet]. \tag{8.104}$$

Now, two mean rate constants may be defined as

$$\overline{k}_{11}[M_1^\bullet] = k_{111}[M_1 M_1^\bullet] + k_{211}[M_2 M_1^\bullet], \tag{8.105}$$

and

$$\overline{k}_{12}[M_1^\bullet] = k_{112}[M_1 M_1^\bullet] + k_{212}[M_2 M_1^\bullet], \tag{8.106}$$

where $\overline{k}_{11}$ and $\overline{k}_{12}$ are the mean rate constants of the addition of $M_1^\bullet$ radicals to their own and the different monomer respectively.

Substitution of Eq. (8.104) for $[M_1^\bullet]$ in Eq. (8.105) gives

$$\overline{k}_{11} = \frac{k_{111}[M_1 M_1^\bullet] + k_{211}[M_2 M_1^\bullet]}{[M_1 M_1^\bullet] + [M_2 M_1^\bullet]}. \tag{8.107}$$

Eliminating radical concentrations using Eq. (8.102) one obtains

$$\overline{k}_{11} = \frac{k_{111}[M_1] + k_{112}[M_2]}{[M_1]\left(1 + \frac{k_{112}[M_2]}{k_{211}[M_1]}\right)} \tag{8.108}$$

$$= \frac{k_{111}(r_{11} f_1 + f_2)}{r_{11} f_1 + \left(\frac{f_2}{s_1}\right)}. \tag{8.109}$$

Proceeding in the same way as above one gets for $\overline{k}_{12}$

$$\overline{k}_{12} = \frac{k_{111}(r_{21} f_1 + f_2)}{r_{21}\left(r_{11} f_1 + \frac{f_2}{s_1}\right)}. \tag{8.110}$$

Dividing Eq. (8.109) by Eq. (8.110)

$$\overline{r}_1 = \frac{\overline{k}_{11}}{\overline{k}_{12}} = r_{21} \frac{f_1 r_{11} + f_2}{f_1 r_{21} + f_2}. \tag{8.111}$$

Similarly, it may be derived that

$$\bar{k}_{22} = k_{222} \frac{r_{22}f_2 + f_1}{r_{22}f_2 + \left(\frac{f_1}{s_2}\right)}, \tag{8.112}$$

$$\bar{k}_{21} = \frac{k_{222}(r_{12}f_2 + f_1)}{r_{12}\left(r_{22}f_2 + \frac{f_1}{s_2}\right)}, \tag{8.113}$$

$$\bar{r}_2 = \frac{\bar{k}_{22}}{\bar{k}_{21}} = r_{12} \frac{f_2 r_{22} + f_1}{f_2 r_{12} + f_1}. \tag{8.114}$$

With the mean rate constants $\bar{k}_{ij}$ defined as above, the following four propagation reactions suffice to describe the penultimate model binary copolymerization. In these reactions, $M_i^\bullet$ represents all radicals centered on the terminal unit M_i irrespective of the identity of the penultimate unit.

$$M_1^\bullet + M_1 \xrightarrow{\bar{k}_{11}} M_1^\bullet, \tag{8.115}$$

$$M_1^\bullet + M_2 \xrightarrow{\bar{k}_{12}} M_2^\bullet, \tag{8.116}$$

$$M_2^\bullet + M_2 \xrightarrow{\bar{k}_{22}} M_2^\bullet, \tag{8.117}$$

$$M_2^\bullet + M_1 \xrightarrow{\bar{k}_{21}} M_1^\bullet. \tag{8.118}$$

Proceeding as in the case of terminal model kinetics of radical copolymerization Eq. (8.119) analogous to Eq. (8.83) is easily derived for the mean propagation rate constant

$$\bar{k}_p = \frac{\bar{r}_1 f_1^2 + 2f_1 f_2 + \bar{r}_2 f_2^2}{(\bar{r}_1 f_1 / \bar{k}_{11}) + (\bar{r}_2 f_2 / \bar{k}_{22})}, \tag{8.119}$$

where $\bar{r}_1 = \bar{k}_{11}/\bar{k}_{12}$ and $\bar{r}_2 = \bar{k}_{22}/\bar{k}_{21}$ (as defined above).

In the implicit penultimate model, $r_{21} = r_{11} = r_1$ and $r_{12} = r_{22} = r_2$. With these assumptions made, the only unknowns in Eq. (8.119) are s_1 and s_2, which define $\bar{k}_{11}$ and $\bar{k}_{22}$ respectively [*vide* Eqs. (8.109) and (8.112)]. Equation (8.119) may therefore be least-squares curve fitted to experimental $\bar{k}_p$ to yield s_1 and s_2 values. Such a procedure was used to draw the solid curve in Fig. 8.6.[46]

8.2.2 The copolymer composition equation

Again, proceeding as with deriving Eq. 8.14 based on the terminal model, the copolymer composition equation is easily derived based on the propagation reactions (Eqs. 8.115 through 8.118)

$$F_1 = \frac{\bar{r}_1 f_1^2 + f_1 f_2}{\bar{r}_1 f_1^2 + 2f_1 f_2 + \bar{r}_2 f_2^2}. \tag{8.120}$$

8.3 Living Radical Copolymerization

It was stated in the introduction of this chapter that living copolymerization ordinarily gives rise to gradient copolymer in contrast to nonliving copolymerization, which ordinarily forms statistical copolymer. However, monomer reactivity ratios applicable to the two polymerizations should be the same. Experimental verification requires the use of the integrated copolymer composition data of high conversion living copolymers.[79] The commonly used method based on the composition data of low conversion copolymer may show some deviation. This is partly due to the short chain polymers formed at low conversions in living polymerization to which the copolymer composition equation is not strictly applicable. A more important reason applicable to the living copolymerization working on the principle of reversible deactivation is the slow establishment of the cross-propagation equilibrium. As a result, the stationary state assumption becomes invalid at low conversions.[80,81] In fact, a simulation study of atom transfer radical copolymerization of styrene and butyl acrylate reveals that the stationary state is not established before the overall monomer conversion reaches as much as 15%.[80]

The living and nonliving copolymerizations also differ in another respect. This concerns the rate of copolymerization. Even when the ratio of the active center concentrations, $[M_1^\bullet]:[M_2^\bullet]$, is the same in both the systems, the total concentrations of the active centers may not necessarily be the same in both. Thus, the rate of copolymerization may be different in the two systems. $[M_1^\bullet]$ and $[M_2^\bullet]$ are governed by the equilibrium constants of the respective reversible terminations as well as by the ratio of the concentrations of the dormant polymer and the deactivator in living radical copolymerization.

The problem was quantitatively treated by Charleux et al. in nitroxide mediated copolymerization, which can be suitably adapted to other living copolymerizations based on reversible deactivation.[79] In a nitroxide mediated living copolymerization, the two deactivation equilibria may be represented as

$$M_1 - Y \underset{}{\overset{K_1}{\rightleftharpoons}} M_1^\bullet + Y^\bullet \ ; \quad K_1 = \frac{[M_1^\bullet][Y^\bullet]}{[M_1 - Y]}, \qquad (8.121)$$

and

$$M_2 - Y \underset{}{\overset{K_2}{\rightleftharpoons}} M_2^\bullet + Y^\bullet \ ; \quad K_2 = \frac{[M_2^\bullet][Y^\bullet]}{[M_2 - Y]}, \qquad (8.122)$$

where $Y^\bullet$ represents a nitroxide radical and $M_i - Y$ ($i = 1$ or 2) represents the polymer-nitroxide adduct with the terminal monomer unit i in the polymer. The average equilibrium constant may be expressed as

$$\overline{K} = \frac{[M^\bullet][Y^\bullet]}{[M - Y]}, \qquad (8.123)$$

where

$$[M^\bullet] = [M_1^\bullet] + [M_2^\bullet], \qquad (8.124)$$

and

$$[M-Y] = [M_1-Y] + [M_2-Y]. \tag{8.125}$$

Substituting from Eqs. (8.121) and (8.122) respectably for $[M_1^\bullet]$ and $[M_2^\bullet]$ in Eq. (8.124) one gets

$$[M^\bullet] = K_1 \frac{[M_1-Y]}{[Y^\bullet]} + K_2 \frac{[M_2-Y]}{[Y^\bullet]}. \tag{8.126}$$

The mean equilibrium constant follows by substituting Eq. (8.126) for $[M^\bullet]$ in Eq. (8.123)

$$\overline{K} = \frac{K_1[M_1-Y] + K_2[M_2-Y]}{[M-Y]}. \tag{8.127}$$

Now, the rate of copolymerization is given by

$$-\frac{d[M]}{dt} = R_p = \overline{k}_p[M^\bullet][M], \tag{8.128}$$

where $\overline{k}_p$ is the mean rate constant of propagation.
Substituting from Eq. (8.123) for $[M^\bullet]$ in Eq. (8.128) gives

$$-\frac{d[M]}{dt} = \overline{k}_p \overline{K}[M-Y][M]/[Y^\bullet] \tag{8.129}$$

or

$$\frac{d \ln[M]}{dt} = -\overline{k}_p \overline{K}[M-Y]/[Y^\bullet]. \tag{8.130}$$

Equations relating $\overline{k}_p \overline{K}$ with reactivity ratios and other parameters have also been derived.[79]
Thus, using the terminal model stationary state condition

$$k_{12}[M_1^\bullet][M_2] = k_{21}[M_2^\bullet][M_1] \tag{8.131}$$

one obtains

$$\frac{[M_1^\bullet]}{[M_2^\bullet]} = \frac{r_1 k_{22}}{r_2 k_{11}} \frac{f_1}{f_2}. \tag{8.132}$$

Alternatively, dividing Eq. (8.121) with Eq. (8.122) one gets

$$\frac{[M_1^\bullet]}{[M_2^\bullet]} = \frac{K_1[M_1-Y]}{K_2[M_2-Y]}. \tag{8.133}$$

Equating. (8.132) with (8.133) and transposing one gets

$$\frac{[M_1-Y]}{[M_2-Y]} = \frac{K_2}{K_1} \frac{r_1}{r_2} \frac{k_{22}}{k_{11}} \frac{f_1}{f_2}. \tag{8.134}$$

Now, since $[M-Y] = [M_1-Y] + [M_2-Y]$, one may easily obtain equations for $[M_1-Y]/[M-Y]$ and $[M_2-Y]/[M-Y]$ from Eq. (8.134). Substituting these equations for the corresponding $[M_i - Y]/[M - Y]$ in Eq. (8.127) gives

$$\overline{K} = \frac{K_1 K_2 r_1 k_{22} f_1 + K_2 K_1 r_2 k_{11} f_2}{K_2 r_1 k_{22} f_1 + K_1 r_2 k_{11} f_2}. \tag{8.135}$$

Dividing both the numerator and the denominator in Eq. (8.135) by $k_{11} k_{22} K_1 K_2$ one gets

$$\overline{K} = \frac{\frac{r_1 f_1}{k_{11}} + \frac{r_2 f_2}{k_{22}}}{\frac{r_1 f_1}{k_{11} K_1} + \frac{r_2 f_2}{k_{22} K_2}}. \tag{8.136}$$

Multiplying Eq. (8.82) with Eq. (8.136) one gets

$$\overline{k_p}\overline{K} = \frac{r_1 f_1^2 + 2 f_1 f_2 + r_2 f_2^2}{\frac{r_1 f_1}{k_{11} K_1} + \frac{r_2 f_2}{k_{22} K_2}}. \tag{8.137}$$

The parameters k_{11}/r_1 and k_{22}/r_2 refer to the cross-propagation constants k_{12} and k_{21} respectively.

Referring to Eq. (8.129) the rate of copolymerization is determined by the value of the $\overline{k_p}\overline{K}$ parameter at given ratios of $[M - Y]/[Y^\bullet]$. Equation (8.137) reveals that if one of the two K_is (say K_1) is much smaller than the other one and also the cross-propagation rate constants k_{12} and k_{21} are such that $K_1 k_{12}$ is much smaller than $K_2 k_{21}$, $\overline{k_p}\overline{K}$ is essentially controlled by it and so is the rate of copolymerization.

Similarly, the $\overline{k_p}\overline{K}$ relation applicable to the implicit penultimate model may be deduced to give

$$\overline{k_p}\overline{K} = \frac{r_1 f_1^2 + 2 f_1 f_2 + r_2 f_2^2}{\frac{r_1 f_1}{\overline{k}_{11} K_1} + \frac{r_2 f_2}{\overline{k}_{22} K_2}}, \tag{8.138}$$

where $\overline{k}_{11}$ and $\overline{k}_{22}$ are given by Eqs. (8.109) and (8.112) respectively.

References

1. T. Alfrey, Jr., G. Goldfinger, *J. Chem. Phys.* (1944), 12, 115, 205, 332.
2. F.R. Mayo, F.M. Lewis, *J. Am. Chem. Soc.* (1944), 66, 1594.
3. F.T. Wall, *J. Am. Chem. Soc.* (1944), 66, 2050.
4. F.R. Mayo, C. Walling, *Chem. Rev.* (1950), 46, 191.
5. E. Merz, T. Alfrey, G. Goldfinger, *J. Polym. Sci.* (1946), 75, 1.
6. M.L. Coote, T.P. Davis, *Prog. Polym. Sci.* (1999), 24, 1217.
7. I. Skeist, *J. Am. Chem. Soc.* (1946), 68, 1781.
8. P.J. Flory, Principles of Polymer Chemistry, Cornell University Press, Ithaca, New York, Chap. 5-1, p. 178, (1953).
9. V.E. Meyer, G.G. Lowry, *J. Polym. Sci. A* (1965), 3, 2843.
10. V.E. Meyer, *J. Polym. Sci.* A-1, (1966), 4, 2819 .
11. T. Alfrey, E. Lavin, *J. Am. Chem. Soc.* (1945), 67, 2044.
12. J. Furukawa, *Alternating Copolymers* in *Ency. Polym. Sci. & Eng.*, H.F. Mark, N.M. Bikales, C.G. Overberger, G. Menges Eds., Wiley-Interscience, New York, Vol. 4, p. 233 (1986).
13. J.M.G. Cowie, *Alternating Copolymerization* in Comprehensive Polymer Science, G.C. Eastmond, A. Ledwith, S. Russo, P. Sigwalt Eds., Pergamon Press, Oxford, Vol. 4, p. 377 (1989).
14. H.T. Ban *et al.*, Macromolecules (2000), 33, 6907.
15. H. Hirai, *J. Polym. Sci. Macromol. Rev.* (1976), 11, 47.

16. G.B. Butler, C.H. Do, Makromol. *Chem. Suppl.* (1989), 15, 93.
17. K.G. Olson, G.B. Butler, *Macromolecules* (1984), 17, 2486.
18. F.R. Mayo, C. Walling, F.M. Lewis, W.F. Hulse, *J. Am. Chem. Soc.* (1948), 70, 1537.
19. A.M. van Herk, *J. Chem. Edn.* (1995), 72, 138.
20. F. Fineman, S.D. Ross, *J. Polym. Sci.* (1950), 5, 259.
21. T. Kelen, F. Tudos, *J. Macromol. Sci. Chem.* (1975), A9, 1.
22. J.P. Kennedy, T. Kelen, F. Tudos, *J. Polym. Sci. Polym. Chem. Ed.* (1975), 13, 2277.
23. R.Z. Greenley, *J. Macromol. Sci.* (1980), A14, 445.
24. R.Z. Greenley, *Free Radical Copolymerization Reactivity Ratios*, in, Polymerization Handbook 4th ed., J. Brandrup, E.H. Immergut, E.I. Grulke Eds., Wiley Interscience, New York, Section II, p. 181 (1999).
25. A. Rudin, K.F. O'Driscoll, M.S. Rumack, *Polymer* (1981), 22, 740.
26. K.F. O'Driscoll, *J. Polym. Sci. Polym. Chem. Ed.* (1980), 18, 2747.
27. J.J. Uebel, F.J. Dinan, *J. Polym. Sci. Polym. Chem. Ed.* (1983), 21, 917.
28. C.C. Price, *J. Polym. Sci.* (1946), 1, 83.
29. C. Walling, E.R. Briggs, K.B. Wolfstirn, F.R. Mayo, *J. Am. Chem. Soc.* (1948), 70, 1537.
30. (a) T. Alfrey, C.C. Price, *J. Polym. Sci.* (1947), 2, 101.
    (b) C.C. Price, *J. Polym. Sci.* (1948), 3, 772.
31. R.Z. Greenley, *Q and e Values for Free Radical Copolymers of Vinyl Monomers and Telogens*, in, Polymerization Handbook 4th ed., J. Brandrup, E.H. Immergut, E.I. Grulke eds., Wiley Interscience, New York, Section II, p. 309 (1999).
32. P.J. Flory, Principles of Polymer Chemistry, Cornell University Press, Ithaca, New York, Chap. 5-1, p. 178 (1953).
33. M.H. Theil, *J. Macromol. Sci. Chem.* (1983), A20, 377.
34. C.H. Bamford, A.D. Jenkins, R. Johnston, *Trans Faraday Soc.* (1959), 55, 418.
35. A.D. Jenkins, *J. Polym. Sci. Polym. Chem. Ed.* (1999), 37, 113.
36. A.D. Jenkins, *Eur. Polym. J.* (1989), 25, 721.
37. G. Goldfinger, T. Kane, *J. Polym. Sci.* (1948), 3, 462.
38. T. Fukuda, K. Kubo, Y-D. Ma, *Prog. Polym. Sci.* (1992), 17, 875.
39. G.E. Ham, in, Copolymerization, High Polymer Series, no. 18, G.E. Ham, Ed., Interscience, New York, Chap. 1, p. 1 (1964).
40. M. Farina, Makromol. *Chem.* (1990), 191, 2795.
41. M.H. Theil, *J. Polym. Sci., Polym. Chem. Ed.* (1983), 21, 633.
42. J.C. Randall, Polymer Sequence Determination, Academic Press, New York (1977).
43. F.A. Bovey, P.A. Mirau, NMR of Polymers, Acad. Press, New York (1996).
44. J.L. Koenig, Chemical Microstructure of Polymer Chains, John Wiley & Sons, Inc., New York (1980).
45. A.E. Tonelli, NMR Spectroscopy and Polymer Microstructure, VCH, New York (1989).
46. T. Fukuda, Y-D. Ma, H. Inagaki, Macromolecules (1985), 18, 17.
47. H.W. Melville, B. Noble, W.F. Watson, *J. Polym. Sci.* (1947), 2, 229.
48. C.J. Walling, *J. Polym. Sci.* (1949), 71, 1930.
49. J.N. Atherton, A.M. North, *Trans. Faraday Soc.* (1962), 58, 2049.
50. H. Mayr et al., *J. Am. Chem., Soc.* (2001), 123, 950.
51. H. Mayr et al., *Acc. Chem. Res.* (2003), 36, 66.
52. H. Mayr, M. Patz, *Macromol. Symp.* (1996), 107, 99.
53. N.T. Lipscomb, W.K. Matthews, *J. Polym. Sci.* A-1 (1971), 9, 563.
54. Y. Imanishi, T. Higashimura, S. Okamura, *J. Polym. Sci.*, A (1965), 3, 2455.
55. C.G. Overberger, L.H. Arond, J.J. Taylor, *J. Am. Chem. Soc.* (1951), 73, 5541.
56. C.G. Overberger, R.J. Ehring, D. Tanner, *J. Am. Chem. Soc.* (1954), 76, 772.
57. J.P. Kennedy, E. Marechal, Carbocationic Polymerization, Wiley, New York, p. 297 (1982).

58. C.G. Overberger, V.G. Kamath, *J. Am. Chem. Soc.* (1963), 85, 446.
59. (a) F. Williams, A. Shinkawa, J.P. Kennedy, *J. Polym. Sci. Polym. Symp.* (1976), 56, 421.
    (b) R.M. Thomas, W.J. Sparks, U. S. Pat. (1944), 2, 356, 128.
60. J.P. Kennedy, N.H.J. Canter, *J. Polym. Sci.* A1 (1967), 5, 2455.
61. V.A. Anosov, A.A. Korotkov, Vys. Soed. (1960), 2, 354.
62. (a) T. Masuda, T. Higashimura, *Polymer J.* (1971), 2, 29.
    (b) T. Masuda, T. Higashimura, S. Okamura, *Polymer J.* (1970), 1, 19.
63. M. Morton, in, Copolymerization, High Polymer Series no. 18, G.E. Ham Ed., interscience, New York, Chap. 7, p. 421 (1964).
64. G. Crespi, A. Valvassori, G. Sartori, in, *Copolymerization, High Polymer* Series no. 18, G.E. Ham Ed., Interscience, New York, Chap. 4c, p. 231 (1964).
65. C.B. Kowollik *et al.*, *Copolymerization*, in, *Ency. Polym. Sci. & Tech.*, 3rd ed., H. Mark Ed., Wiley-Interscience, Hoboken, New Jersey, Vol. 9, p. 394 (2004).
66. M. Berger, I. Kuntz, *J. Polym. Sci.*, A (1964), 2, 1687.
67. D.J.T. Hill, J.H. O'Donnell, P.W. O'Sullivan, Macromolecules (1982), 15, 960.
68. O.F. Olaj, I. Bitai, F. Hinkelmann, *Makromol. Chem.* (1987), 188, 1689.
69. Y. Kuwae, M. Kamachi, S. Nozakura, *Macromolecules* (1986), 19, 2912.
70. M. Kamachi, Adv. Polym. Sci. (1987), 82, 207.
71. K. Ito, K.F. O'Driscoll, *J. Polym. Sci. Polym. Chem. Ed.* (1979), 17, 3913.
72. M.C. Piton, M.A. Winnik, T.P. Davis, K.F. O'Driscoll, *J. Polym. Sci. Polym. Chem. Ed.* (1990), 28, 2097.
73. M.L. Coote, T.P. Davis, *Macromolecules* (1999), 32, 3626.
74. Y.D. Semchikov, Macromol. Symp. (1996), 111, 317.
75. G.A. Egorochkin *et al.*, *Eur. Polym. J.* (1992), 28, 681.
76. D.J.T. Hill, J.J. O'Donnell, P.W. O'Sullivan, *Prog. Polym. Sci.* (1982), 8, 215.
77. J.A. Seiner, M. Litt, Macromolecules (1971), 4, 308.
78. M. Kamachi, Adv. Polym. Sci. (1981), 38, 56.
79. B. Charleux, J. Nicolas, O. Guerret, *Macromolecules* (2005), 38, 5485.
80. B.K. Klumperman, G. Champard, R.H.G. Brinkhuis, ACS *Symp. Ser.* (2003), 854, 180.
81. K. Matyjaszewski, *Macromolecules* (2002), 35, 6773.

Problems

8.1 What polymerization process should be adopted to obtain copolymers of uniform composition when the instantaneous copolymer differs in composition from the monomer feed?

8.2 In radical copolymerization of styrene and vinyl acetate, hardly a copolymer is obtained. Explain why it is so.

8.3 Arrange the following monomers and the corresponding radicals in order of increasing average reactivity: ethylene, vinyl acetate, styrene, butadiene, methyl acrylate, acrylonitrile, and methyl methacrylate. What is the major determining factor in the general reactivity? Under what situation does the reactivity order differ from the average reactivity order?

8.4 What kind of copolymer will be formed if styrene and methyl methacrylate are anionically copolymerized?

8.5 Under what conditions would there be no change in copolymer composition with progress of copolymerization in a batch process?

8.6 In the application of the instantaneous copolymer composition equation to determine the monomer reactivity ratio what restriction to monomer conversion should be imposed and why?

8.7 What kind of copolymer will be obtained when styrene and acrylonitrile are radically copolymerized in the presence of $ZnCl_2$?

8.8 Arrange the following monomers in two groups, one with positive e values in the Alfrey-Price Q, e scheme, and the other with negative e values: butadiene, methyl methacrylate, styrene, methyl acrylate, vinyl acetate, α-methylstyrene, and acrylonitrile. Explain the grouping.

8.9 Arrange the same monomers as given in Problem 8.8 in the order of decreasing Q values. Explain the arrangement.

8.10 Does the Q, e scheme apply to ionic copolymerization as well? Explain.

8.11 What factors determine the copolymerizability of monomers in radical, cationic, and anionic copolymerization?

Chapter 9

Heterophase Polymerization

In the previous chapters we have dealt with homogeneous phase polymerization. However, some polymerization may be conducted also in the heterogeneous phase with advantage. In this case, the polymer growth occurs to varying degree in isolated polymer particles, either suspended in or precipitated from the polymerization medium, depending on the process used. Radical polymerization is the most widely practiced because of its tolerance to environment-friendly liquids like water and lower alcohols, which are but natural choices as promoters of system heterogeneity.

The simplest heterophase polymerization is, of course, the precipitation polymerization conducted in the absence of surface active agents, which occurs when the polymer formed is insoluble in either its own monomer[1] or in diluents (when one such is used) or in both. Some examples are bulk polymerization of vinyl chloride,[2] vinylidene chloride,[3] or acrylonitrile,[1] and polymerization of styrene in lower alcohols,[4,5] of methyl methacrylate in cyclohexane,[5] or of vinyl acetate in n-hexane.[6] The propagating chains initiated in the liquid phase grow partly there and partly in the precipitated particles. Inside the particles, bimolecular termination is hindered due either to the isolation of the radicals or to the gel effect leading to the acceleration of polymerization and the increase of molecular weight.

However, various heterogeneous polymerization processes (Table 9.1) have been developed in which a variety of surface active agents, additives, and agitation methods has been used to produce dispersions of polymer particles covering a wide range of sizes from *ca.*, 50 nm to several microns in diameter.

9.1 Particle Stabilization Mechanisms

Particles in a dispersion medium experience attractive force toward each other, which is van der Waals in origin. The large number of atoms or molecules that constitutes a particle contributes to the attraction. However, only the dispersion force component of the van der Waals forces contributes.[7,8] This is because the other two components, *viz.*, the dipole-dipole and the dipole-induced dipole, are basically binary in nature due to the proper alignment of the dipoles required for attractive interaction. In contrast, the dispersion force component can operate between all neighboring molecules at the same time.[7]

Table 9.1 Approximate size range of polymer particles prepared by various heterophase polymerization processes. Particle diameters are in nanometers.

Suspension	Dispersion	Direct	Inverse	Emulsion Surfactant-free	Mini	Micro
$5000-2 \times 10^6$	$100-20\,000$	$100-300$	$50-300$	$100-2000$	$50-500$	$10-50$

The potential energy of attraction (V_A) due to the London dispersion force between two equal-size spherical particles is given by the Hamaker equation (9.1)[9]

$$V_A \approx -\frac{Ar}{12h} \quad \text{(for } r \gg h\text{)}, \tag{9.1}$$

where A is the Hamaker constant, r is the particle radius, and h is the distance between surfaces of the two particles. The potential is of long range, it being proportional to the inverse first power of the surface-to-surface distance between the particles. When the particles are suspended in a medium, the attraction is reduced to some extent. Thus, the potential energy of attraction between latex particles is given by the approximate equation[10]

$$V_A \approx -\frac{(A_{11}^{1/2} - A_{22}^{1/2})^2 r}{12h} = -\frac{A_c r}{12h}, \tag{9.2}$$

where A_{11} and A_{22} refer to Hamaker constant for suspended particles and particles of dispersion medium respectively and A_c is the composite Hamaker constant having a value of the order of kT.[11]

For latex coagulation to occur, the particles must come closer such that the potential energy of attraction becomes numerically greater than the thermal energy (3/2 kT) experienced by the particles, which works against particle aggregation.[8] Thus, in order to prevent latex coagulation, a repulsive interaction between particles must exist to keep the particles apart such that

$$\frac{A_c r}{12h} < \sim \frac{3}{2} kT \tag{9.3}$$

or

$$h > \sim r/18 \quad \text{(since } A_c \text{ is } \sim 1kT\text{)}. \tag{9.4}$$

It follows from Eq. (9.4) that the critical interparticle surface-to-surface distance, below which particle aggregation occurs, varies linearly as the particle diameter. To give some concrete values, the critical distance is $\sim$3 nm for 100 nm particles and $\sim$0.3 nm for 10 nm particles.

There exist usually three kinds of repulsive interaction between latex particles, *viz.*, electrostatic, steric, and electrosteric (*i.e.*, combination of electrostatic and steric).[8] The electrostatic repulsion occurs between particles, which have surface charge of the same kind. This they acquire by adsorbing ionic surfactant used in polymerization and/or due to ionic end groups (derived from an ionic initiator) or mid chain ionic groups (derived from ionic comonomers). These charges form electrical double layers around the particles. When two particles come close to each other due to the Brownian motion such that an overlap of the electrical double layers occurs, they are repelled. Addition of electrolytes to the dispersion medium increases the ionic strength of the medium leading to a decrease in the double layer thickness. Adequate amount of electrolytes can reduce the double layer thickness to the extent that particles can come closer than the critical distance promoting aggregation.

Steric stabilization mechanism operates when the particles have nonionic stabilizers adsorbed on their surfaces. Close approach of particles causing overlap of stabilizer layers would be prevented by two effects: (i) volume restriction and (ii) osmotic.[8,12] Volume restriction is a consequence of the stabilizer molecules facing an impenetrable barrier as two particles come close enough such that the surface-to-surface distance between the bare particles becomes less than the thickness of the stabilizer layer. The consequent loss of configurational entropy of the stabilizer would prevent the occurrence of such an event. The osmotic effect is a consequence of overlapping of stabilizer layers that occurs following close approach of particles. This would effect an increase in the concentration of the stabilizer in the space between the two particles resulting in an increase in the local osmotic pressure, which would be prevented by liquid (of dispersion medium) rushing in and separating the particles.

Electrosteric stabilization occurs when the particles have both ionic and nonionic surfactants adsorbed on their surfaces or when a polymeric surfactant carries charges in its molecules, *e.g.*, polyelectrolytes, proteins, and some polysaccharides.

9.2 Suspension Polymerization

In this process, the monomer droplets are suspended in water by agitation in the presence of suspending agents. The latter may be non-micelle forming polymeric surfactants, the so-called "protective colloids", such as starch, poly(vinyl alcohol), and poly(vinylpyrrolidone), or water insoluble inorganic compounds, which include talc, magnesium or aluminum hydroxide, and barium, calcium, or magnesium sulfate, carbonate, silicate, and phosphate.

The water : monomer ratio varies from 1:1 to 4:1 (w/w). Polymerization takes place inside the monomer droplets with the help of an oil soluble initiator. At the end of the polymerization, the droplets become hard particle beads, which look like pearls and the polymerization, therefore, is referred to as 'bead' or 'pearl' polymerization. Polymerization kinetics is the same as that of bulk polymerization.

The most serious problem of suspension polymerization occurs when the polymerization has proceeded to ~20% conversion. At that point, the monomer droplets containing dissolved polymer become sticky and vulnerable to agglomeration forming a lump. Prevention of this requires proper selection of surfactant and type of agitation. Particle size depends on the type and concentration of the surfactant, speed and type of agitation, water: monomer ratio, and the viscosities of both the suspended and the continuous phase.[13]

Suspension polymerization offers certain operational advantages over bulk or solution polymerization. These relate to the use of water as the suspension medium. The bulk viscosity of the suspension remains near that of water during most of the time of polymerization. This makes energy cost of agitation moderate. In addition, due to the relatively high heat capacity and thermal conductivity of water the heat of polymerization is easily removed.

9.3 Emulsion Polymerization

Emulsion polymerization was developed originally for oil soluble monomers with water as the continuous medium using an oil-in-water emulsifier. Subsequently, polymerization of water soluble monomers with oil as continuous medium was developed as well. These are referred to as inverse emulsion polymerization.

Due to the use of water as the polymerization medium and relatively low viscosity of the latex, emulsion polymerization offers the operational advantages that are discussed under suspension polymerization. These include ease of removal of heat of polymerization, lower energy cost of agitation, and ease of handling of the dispersions. Being water-borne, it is environment-friendly as is suspension polymerization. In addition, it has attractive kinetic features (*vide infra*). Accordingly, a large fraction of vinyl and diene polymers are produced industrially by this technique.[14]

An emulsifier is an amphiphilic substance with the molecule containing a hydrophilic part and a lipophilic part. Depending on the proportion of these parts, it may be water soluble or oil soluble and used as oil-in-water or water-in-oil emulsifier respectively. Some examples of these two types of emulsifiers are given in Table 9.2.

An empirically developed index ranging from 1 to 20 referred to as hydrophile-lipophile balance (HLB) has been used to categorize amphiphiles for various applications.[15] The index is given by the equation

$$\text{HLB} = 7 + \text{(hydrophilic group number)} - 0.475\, n_c,$$

Table 9.2 Some examples of emulsifiers.

Anionic	Cationic	Nonionic
	Oil-in-water emulsifiers	
Soaps		
Sodium dodecyl sulfate (SDS)	Cetyltrimethylammonium bromide (CTAB)	Polyoxyethylene(40) nonylphenyl ether (branched) (Igepal® CO-890)
Sodium dodecylbenzene sulfonate	Cetylpyridinum bromide	
Bis-1, 3-dimethylbutyl sodium–sulfosuccinate (Aerosol®MA)		Polyoxyethylenesorbitan monooleate (Tween®80)
		Polyoxyethylenesorbitan monostearate (Tween®60)
		Polyoxyethylene(20) oleyl ether (Brij®98)
		Polyoxyethylene(20) stearyl ether (Brij®18)
		Polyoxyethylene (7-8) *tert*-octylphenyl ether (Triton®X-114)
	Water-in-oil emulsifiers	
Di-2-ethylhexyl sodium sulfosuccinate (AOT®)		Sorbitan monooleate (Span®80)
		Sorbitan monostearate (Span®60)
		Polyoxyethylene-polyoxypropylene-ethylenediamine adduct (Tetronic®1102)

where n_c is the number of CH_2 groups in the lipophilic part and the group number has been assigned by Davis and Rideal.[16] Water-in-oil emulsifiers have HLB index in the range of 4–6, while oil-in-water types have it in the range of 8–18. The HLB ranges of emulsifiers yielding optimally stable lattices have been worked out for several lattices. Some of these are 11.8–12.4 for poly(ethyl acrylate), 12.1–13.7 for poly(methyl methacrylate), 13.3–13.7 for polyacrylonitrile, 13–16 for polystyrene, and 14.5–17.5 for poly(vinyl acetate).[14]

The emulsifiers do not remain molecularly dispersed in solutions even at very low concentrations. At concentrations above certain critical values, they form aggregates, which exist in equilibrium with the monomeric emulsifier and are known as micelles. The critical micelle concentration (CMC) depends on many factors, *e.g.*, HLB, temperature, and ionic strength of the medium (for ionic emulsifiers).

In spherical micelles dispersed in water, the lipophilic part of the oil-in-water emulsifier orients away from water toward the interior of the micelle forming a lipophilic core, while the hydrophilic part orients toward the micelle exterior forming a hydrophilic shell. In contrast, in micelles formed in oil, the water-in-oil emulsifer takes opposite orientation forming a hydrophilic core and a lipophilic shell (Fig. 9.1). These micelles are referred to, accordingly, as invert or reverse micelles. Their formation requires the presence of traces of water, at the least, which may be adventitiously present in the system, if not purposefully added.

A unique characteristic of emulsion polymerization is the simultaneous attainment of high rate of polymerization and high molecular weight polymer. This is an incompatible combination in homogeneous radical polymerization in which the degree of polymerization is inversely related with the rate of polymerization (Eq. 3.39, Chap. 3).[17] Accordingly, monomers with low value of the k_p^2/k_t parameter may be polymerized in the homogeneous phase to high molecular weight polymers only at very low rates of polymerization. In contrast, they may be emulsion polymerized to give simultaneously high rates of polymerization and high molecular weight polymers. For instance, solution polymerization of isoprene, which has a relatively low k_p^{18}, *ca.*, 100 L mol^{-1} s^{-1}, gives limited yield (*vide* dead end polymerization, Fig. 3.11, Chap. 3)[19]. In contrast, emulsion polymerization gives high yield (Fig. 9.2) and high molecular weight polymer.[20] As we shall soon see, these

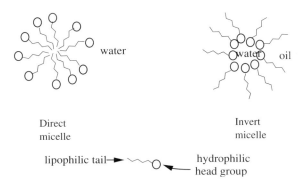

Figure 9.1   Two-dimensional representations of the cross-sections of direct and invert micelles.

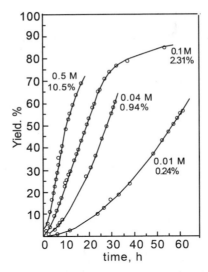

Figure 9.2 Percent yield *vs.* time curves in the emulsion polymerization of isoprene at 50°C effected by 0.3 g potassium persulfate per 100 g isoprene suspended in 180 g water. The amount of potassium laurate soap used is indicated on the curves in both weight percent and molal concentrations with respect to monomer. The concentration is below CMC for the curve in the extreme right. "Reprinted with permission from Ref. 20. Copyright ©1947 American Chemical Society."

characteristic features of emulsion polymerization originate from polymerization in isolated loci with radicals supplied from outside.

Emulsion polymerization usually uses an amount of monomer, which constitutes 30 to 50% of the total volume (monomer + water). A water soluble thermal initiator, usually potassium persulfate, is used at 50°C or above, whereas at low temperatures, a redox initiator, *e.g.*, cumene hydroperoxide/Fe(II), is used (Sec. 3.4.3, Chap. 3).[21] As regards emulsifier, the anionic variety is mostly used.

A typical recipe is the "standard" or "mutual" recipe that was adopted in USA in the large scale production of GR-S (abbreviation of Government Rubber Program from Styrene) rubber from butadiene and styrene during World War II.[21] It comprises butadiene 75, styrene 25, water 180, soap 5, potassium persulfate 0.3, and lauryl mercaptan 0.5, all in parts by weight. The soap used was a mixture of sodium palmitate, stearate, and oleate.

At the outset, the emulsion contains emulsifier micelles, emulsified monomer droplets, and water with molecularly dissolved initiator, emulsifier, and a small amount of monomer.[20] The spherical soap micelles are 5–10 nm in diameter, each containing 50 to 100 or more soap molecules. Some amount of monomer (1–2%) is solubilized in the micelles, which have the number density in the range of 10^{17}–10^{18} per mL water. Higher is the soap concentration higher is the number density. In comparison, the emulsified monomer droplets are orders of magnitude larger, 1–10 μm in diameter, and smaller in number density, <10^{12} per mL water. In the end, the dispersed polymer particles have the average size in the range of 100–300 nm diameters and the number density in the range of 10^{13}–10^{15} per mL water.

9.3.1 *The three periods of polymerization*

The evolution of rate of polymerization with time exhibits some characteristic features as would be evident from the typical yield *vs.* time curve obtained in the emulsion polymerization of isoprene using 2.3% soap (the standard amount) shown in Fig. 9.2 along with the curves at other concentrations of the soap.[20] The rate of polymerization accelerates from an initial low value to reach a practically constant one as the conversion increases to about 5%. This period is known as 'interval I'. The constant rate period continues until the conversion reaches about 65%. Then the rate begins to decelerate. The constant rate period is referred to as 'interval II' and the deceleratory rate period ending in the completion of polymerization is referred to as 'interval III'. With the use of much higher concentration of soap (10.5%) the polymerization proceeds much faster but the interval III begins early at lower conversion. With the soap concentration reduced to two fifths the standard amount, the interval I extends to about 20% conversion. With the soap concentration reduced still further to below CMC, the interval I extends to still higher conversion (the curve in the extreme right). However, in the last two systems, the polymerization was not continued long enough to exhibit the beginning of interval III.

9.3.2 *Harkins qualitative theory*

Harkins proposed two kinds of reaction loci, one located in the micelles during the nucleation of polymer particles, and the other exclusively in the polymer particles after the particle nucleation is completed.[20]

Polymerization starts mainly in the monomer swollen micelles. A radical generated in the aqueous phase diffuses into one of the micelles in almost total exclusion of the monomer droplets. This assumption is justified since the total surface area of the former is enormously larger than that of the latter, as can be easily calculated from their respective number densities and sizes given above. The radical hardly adds a monomer molecule before entering into the micelle since the dissolved monomer concentration in the aqueous phase is very small, *e.g.*, 4.3 mmol/L for styrene at 50°C,[40] and it exists in the aqueous phase for too short a time. However, it initiates a polymer chain in the micelle, which is swollen with monomer. The chain grows beyond the size of the micellar core, the monomer being continually supplied from the monomer droplets through the intervening aqueous medium. Soap adsorbs onto the polymer radical-water interface causing disintegration of the micelle. More soap is required to stabilize the polymer particle as the polymer chain grows. This is supplied by the micelles that have not yet received a radical. The radical is terminated when another radical enters the polymer particle from the aqueous phase. As more and more micelles are converted to polymer particles, a stage arises when micelles disappear. Major polymerization then proceeds exclusively inside the polymer particles.

Harkins also postulated the existence of a minor locus of the initiation of particle nuclei, which is the aqueous phase containing the dissolved monomer, the initiator, and the nonmicellar soap. Apparently, this should be the sole locus of particle nucleation when no soap is present or when the soap concentration is below the CMC. However, even in these cases, micellar nucleation is plausible with monomers of low water solubility since the micelles could be formed in situ (*vide infra*).

Given the numbers of micelles and the polymer particles, out of every 10^3 to 10^4 micelles one becomes a locus of polymerization. The rest of the micelles become disbanded supplying soap for the stabilization of the particles, which is evidenced by the abrupt rise in surface tension of the latex early in the polymerization.[20] Particle nucleation ceases with the disappearance of micelles. The interval from the start of polymerization to this stage corresponds to interval 1. The rate accelerates in this interval due to the number density of particles increasing until becoming practically constant at the end of the interval.

Apart from the micelles, which have not received a radical, some soap molecules from the emulsified monomer droplets also are used up to stabilize the particle nuclei. This is evident from the observation that if the stirring is stopped even when the conversion is low, the monomer droplets coalesce together forming a layer. The polymer content in the monomer layer is insignificant even when polymerization has sufficiently advanced. This observation rules out monomer droplets having acted as polymerization loci at any stage of polymerization.[20]

After the disappearance of micelles, the polymer particles become the sole loci of polymerization. They remain saturated with the monomer, which diffuses into them through the aqueous phase from the monomer droplets. The monomer concentration in the particle remains constant due to the monomer in the particle and in the aqueous phase being in thermodynamic equilibrium as long as the monomer droplets are present. The number of polymer particles also remains substantially constant. Thus, due to the combined effects of these factors the rate of polymerization remains practically constant during interval II, which continues up to 60 to 70% conversion with the use of the standard amount of soap. With the disappearance of monomer droplets at the end of interval II, the monomer concentration in the particles decreases with further progress of polymerization resulting in the deceleration of the rate of polymerization (interval III).

The possibility of particle nucleation by polymerization of dissolved monomer in the aqueous phase was also examined in view of the fact that polymerization occurs even in the absence of soap with monomer fed from gas phase into water containing potassium persulfate. This particular way of introducing monomer ensures that no monomer droplets form in water. However, polymerization is very slow yielding much larger-sized polymer particles even at low conversions.[20] Consequently, the number of particles is orders of magnitude lower than that formed in the presence of micelles.

9.3.3 *Quantitative theory of Smith and Ewart*

Soon after Harkins proposed his theory on the polymerization loci Smith and Ewart developed a quantitative theory deriving equations for rate, molecular weight, and number of polymer particles.[22] Three different cases were considered.

Case 1 is that in which the average number of radicals per polymer particle is much less than unity. This may occur if the monomer transfer constant is comparatively high, *e.g.*, in vinyl chloride and vinyl acetate polymerizations (*vide* Chap. 3, Table 3.6).[23,24] This is due to the faster rate of diffusion of the transferred small radical out of the particle. This case may occur also if the polymer radicals have some water affinity so that termination occurs in the water phase.[25–27]

Case 2 is that in which the average number of radicals per polymer particle is approximately 0.5. This means that about one-half of the particles are active at any instant, while the other half are inactive.[22] This situation arises when a particle becomes active following the entry of a radical from the aqueous phase and remains so until another radical enters into it causing instantaneous termination of both radicals. The particle becomes active again receiving another radical after an average time interval determined by the frequency of radical entry into the particles. For example, consider a polymerization system typically containing 10^{14} polymer particles per mL aqueous phase with 10^{13} radicals per mL being generated in the aqueous phase per second. According to Harkins theory, nearly all the radicals enter into the polymer particles. Thus, on average, a particle receives one radical every ten seconds. It remains active for an average period of ten seconds following the entry of the radical and inactive for an equal period, on average, following the entry of a second radical. The cycle continues repeating itself during the whole time of growth of the particle. However, the frequency of radical entry may change due to the decay of the initiator and any change in particle concentration that may occur with time.

Case 3 is that in which the particle is large so that more than one radical can coexist inside it.

Case 2 is typical of emulsion polymerization. It gives rise to the characteristic emulsion polymerization kinetics, *viz.*, combination of high rate of polymerization and high molecular weight.[23] Since one-half of the particles are active at any instant, the rate of polymerization per mL aqueous phase is given by

$$R_p = k_p \left(\frac{N}{2}\right) [M], \qquad (9.5)$$

where N is the number of polymer particles per mL and [M] is the monomer concentration in the polymer particles. Thus, the rate should be proportional to the concentration of polymer particles. The monomer concentration in the particles remains constant in the constant rate regime (interval II), as discussed above. The monomer is supplied to the polymer particles by diffusion through the aqueous phase from the monomer droplets. Thus, it follows from Eq. (9.5) that the rate of diffusional entry of monomer into a particle is k_p [M]/2. Using $k_p = 100$ L mol^{-1}s^{-1} for isoprene and [M] = 5 mol L^{-1} (typical value in interval II), the diffusion current (I) works out to be 250 monomer molecules per second.[17] Flory used the Debye equation for diffusion current to calculate the concentration gradient required for the monomer to maintain this level of I. The equation, relates I with the difference in concentration (ΔC) of the diffusing species between the surface of a spherical particle of radius r and away from it as

$$I = 4\pi r D \Delta C, \qquad (9.6)$$

where D is the diffusion constant, which is of the order of 10^{-5} cm^2 s^{-1} for small molecules.[20] Using a typical value of r = 50 nm it may be calculated that the required value of ΔC is about 10^{-9} mol/L.[17] Thus, the monomer must have a minimum solubility of 10^{-9} mol/L in the aqueous phase to maintain the diffusion current discussed above. The solubility of monomer in the aqueous phase far exceeds this value so that the required concentration gradient is easily maintained.

When the monomer droplets disappear, a gradual fall in the rate of polymerization occurs with the progress of polymerization due to the decrease in the monomer concentration in the particle (interval III).

Experimental proof in support of Harkins theory and Eq. (9.5) was provided also by Smith.[23] He showed that the number density of particles does not change by more than 30% when polystyrene latex particles prepared in a separate polymerization (95% conversion) were used as seed in a new polymerization with the addition of fresh styrene and initiator but not soap. The polymer yield was as much as seventy fold the weight of the seed polymer. This result proves that no new particles form in the absence of micellar soap; the initially used stable latex particles (seed particles) acted as the sole loci of polymerization. From other series of seeded polymerizations with no added soap it was shown that the rate of polymerization per particle remains substantially constant and independent of particle size over a hundred-fold range in particle number density ca., 10^{12} to 10^{14} per mL water or of potassium persulfate concentration over a sixteen-fold range, 0.044 to 0.7%. However, the rate of polymerization increased at increased conversion with the initiator concentration at the upper end of the range, which could be suggestive of new particle formation at high initiator concentrations.

As regards the degree of polymerization, it should be equal to the average number of monomer molecules polymerized in an active phase of a particle. Since a radical adds k_p [M] number of monomer molecules per second, $\overline{DP}_n$ is given by[22]

$$\overline{DP}_n = \frac{k_p[M]N}{\rho}, \tag{9.7}$$

where N/ρ is the average frequency of radical entry into a particle, ρ being the rate of radical generation in numbers per mL per second in the aqueous phase. Equations (9.5) and (9.7) explain the unique characteristics of emulsion polymerization, *i.e.*, both rate and degree of polymerization are simultaneously increased by increasing the concentration of polymer particles. This is easily done by increasing the soap concentration at a fixed concentration of the initiator.

Smith showed that, in accordance with Eq. (9.7), the average degree of polymerization at a fixed persulfate concentration increases linearly with increase in the number density of particles provided the latter is not too high (not above ca., 10^{14} per mL in styrene polymerization at 50°C with 0.172% potassium persulfate). In the indicated high particle concentration region, the molecular weight is very high so that it tends to approach a limiting value due to chain transfer to monomer.[23]

9.3.4 *Applicability of Smith-Ewart theory with oil soluble initiators*

An oil soluble initiator such as benzoyl peroxide or azobisisobutyronitrile dissolves in monomer droplets, monomer-swollen micelles, and monomer-polymer particles. It is to be expected that the dissolved initiator in the micelles and in the polymer particles is responsible for initiation effecting nucleation and growth respectively of the polymer particles. However, initiator efficiency should be low due to the enhanced cage effect engendered by the high viscosity in the interior of the micelles as well as in the polymer particles. Thus, only a very small fraction of the generated radicals should be available for initiation leading to case 1 kinetics and relatively slow polymerization. Surprisingly, similar kinetics as observed with persulfate initiator follows also with benzoyl peroxide in styrene polymerization.[28] The similarity is ascribed to initiation being predominantly effected by the part of the benzoyl peroxide initiator, which dissolves in the aqueous phase due to the solubilization by the dissolved emulsifier and the monomer. In effect, the oil soluble initiator behaves like a water soluble one. Some polymerization, of course, takes place also in the monomer droplets, which is akin to that of suspension polymerization obeying bulk polymerization kinetics yielding lower molecular weight polymer. However, the polymer formed in the monomer droplets can account for several percent of the total polymer when the conversion is low $ca.$, <20%.[29-31]

9.3.5 Number density of particles

Smith and Ewart derived also Eq. (9.8) relating N with ρ and soap concentration (c_s) in g per mL water

$$N = k(c_s a_s)^{0.6} (\rho/\mu)^{0.4}, \qquad (9.8)$$

where k is a numerical constant with a value between 0.37 and 0.53, a_s is the interfacial area occupied by a gram of soap during the particle formation stage and μ is the average rate of increase in the volume of a particle due to polymerization. Smith provided the first experimental proof of the proportionality of N with 0.6 power of soap concentration and 0.4 power of initiator concentration in emulsion polymerization of styrene.[23] The number density of particles was determined from the measurement of particle size by electron microscopy and the polymer content in latex. Subsequently, several other investigators found agreement with either or both power exponents (Eq. 9.8) in the emulsion polymerization of not only styrene[23,24,32,33] but also methyl methacrylate,[34] chloroprene,[33] butadiene,[35] vinyl acetate,[36] and vinylidene chloride,[37,38] among others. However, strong deviation from the equation occurs particularly with monomers having relatively high water solubility for which particle nucleation occurs in the aqueous phase or with strongly polar monomers for which soap does not adsorb efficiently on the particle.[39]

9.3.6 Particle nucleation in the aqueous phase

Harkins' assumption of micellar nucleation of particles by almost every radical generated in the aqueous phase is theoretically justifiable for diffusional entry of radicals into micelles.[17] However, radical entry from polar aqueous environment into non polar environment inside micelle faces an energy barrier. In order to overcome this barrier the radical must add sufficient monomer molecules in the aqueous phase and become surface active. From the dependence of the experimentally measured rate coefficient of the entry of SO_4^- ion radical into the micelles of sodium dodecyl sulfate on the concentrations of various ingredients in the emulsion polymerizations of styrene, the critical size for entry of oligomer radical has been deduced to be dimeric.[40] However, this result does not rule out micellar nucleation. It only suggests that the rate-controlling step in micellar nucleation is the formation of the critical sized oligomer radical for irreversible entry into micelles.

However, the micellar nucleation theory faced another challenge when experimental measurements showed no discontinuity in particle number density as the emulsifier concentration was increased from low to CMC and above.[41,42] Besides, the Smith-Ewart equation of particle number density (Eq. 9.8) was also derived assuming nucleation in the aqueous phase in sharp distinction from nucleation inside the micelles.[43]

Aqueous phase nucleation occurs as the growing polymer radical separates out from the aqueous solution as it becomes insoluble with increase in chain length. The radical so precipitated becomes swollen with monomer and stabilized against flocculation by the adsorption of emulsifier. It continues to grow until being terminated with the entry of a second radical into the particle from the aqueous phase or by the capture of an inadequately stabilized particle containing a growing radical. A radical may also enter into a micelle when the latter is present and nucleate a particle. The particle number density for a given initiator concentration is determined by the concentration of emulsifier irrespective of whether micelles are present or not. Roe considered this mechanism applicable even to the polymerization of styrene, which has low water solubility. However, using the solubility data of styrene and solution polymerization kinetics Vanderhoff calculated that polystyrene of degree of polymerization of only 7.3 forms at 60°C at potassium persulfate concentration of 0.1%.[44] The oligostyrene formed is thus essentially an emulsifier. This argument goes in favor of micellar nucleation.

It is generally accepted that, at one extreme, micellar nucleation applies to polymerizations of monomers with low water solubility $ca.$, <20 mmol/L, styrene and the diene monomers being typical examples. At the other extreme, aqueous phase nucleation is prevalent, in the polymerizations of monomers with significantly high water solubility $ca.$, >200 mmol/L, vinyl acetate, methyl acrylate, ethylene, and acrylonitrile being examples.[14,45–47] In between the two extremes, both mechanisms of nucleation apply. Some monomers belonging to this last class are vinylidene chloride, vinyl chloride, ethyl acrylate, and methyl methacrylate, which have intermediate water solubility $ca.$, 65 to 150 mmol/L. However, many investigators favor aqueous phase nucleation in the polymerizations of these monomers as well.[14]

The aqueous phase nucleation theory was quantitatively developed by Hansen, Ugelsted, Fitch, and Tsai for soapless emulsion polymerization. It is referred to as HUFT theory after the names of these investigators.[48–51] According to the theory, a propagating radical on attaining a critical chain length for insolubility separates from the aqueous phase and nucleates a polymer particle, which is stabilized by the surface charge contributed by initiator-derived ionic end groups.[52–55]

The critical chain length, however, may be too long to make the surface charge density on the particle sufficient for stability. Some of the nucleated particles, therefore, may coalesce together forming stable particles with large enough surface charge density. The stable particles grow by capturing oligo radicals from the aqueous phase. The oligomeric chains continue to propagate inside the particles until being mutually terminated.[44]

9.4 Inverse Emulsion Polymerization

Some examples of water-in-oil emulsifiers suitable for emulsifying water soluble monomers in inverse emulsion polymerization are given in Table 9.2. These emulsifiers have low HLB values and form invert micelles (Sec. 9.3). The primary difference in parameters between emulsion polymerization and inverse emulsion polymerization lies in the size of the emulsified monomer droplets, 20–500 nm in inverse emulsion *vs.* 1000–10 000 nm in direct emulsion.[56,57] However, the size range of polymer particles produced in both processes is similar (Table 4.1). The small size enables the monomer droplets compete effectively with the invert micelles for the capture of radicals generated in the continuous oil phase from an oil-soluble initiator. Thus, particle nucleation occurs in both the monomer-swollen micelles and monomer droplets.

However, both oil-soluble and water-soluble initiators are used. Surprisingly, studies of the inverse emulsion polymerization of sodium 4-styrenesulfonate revealed that Smith-Ewart Case 1 kinetics is obeyed when benzoyl peroxide is the initiator, whereas Case 2 kinetics is obeyed when potassium persulfate is the initiator.[44] The opposite should have been observed. Radicals generated in the continuous oil phase from benzoyl peroxide would enter into dispersed polymer particles favoring Case 2 kinetics. In contrast, radicals from persulfate initiator would be generated in pairs inside the dispersed polymer particles, and the cage effect would leave only a very small fraction of particles with single radicals favoring Case 1 kinetics. However, the anomaly may be resolved considering that persulfate solubility in the oil phase is increased due to the presence of the oil soluble emulsifier, and benzoyl peroxide solubility in the aqueous phase is increased due to the presence of the monomer, as already discussed in Sec. 9.3.4.[44]

Inverse emulsion polymerization of acrylamide is also effected using both oil soluble and water soluble initiators.[58,59] However, polymerization kinetics using water soluble persulfate initiator resembles that of a solution polymerization.[60–62] This suggests that polymerization is to be treated as a microsuspension one where each monomer droplet contains a number of radicals. The highly viscous medium inside the monomer polymer droplet makes the termination translational diffusion-limited allowing the presence of multiple radicals.[62]

9.5 Miniemulsion Polymerization

Miniemulsion polymerization is a variant of emulsion polymerization with the difference that the emulsified monomer droplets are much smaller in size.[14,63] The size range is, in fact, similar to that in inverse emulsion polymerization (Sec. 9.4) and, like in the latter, the particle nucleation and growth occur in the monomer droplets. However, in many systems not all the droplets become polymer particles; only about 20 percent become so, the rest act as reservoirs supplying monomer to the polymer particles. Miniemulsions require the use of a combination of a surfactant and a costabilizer, which is usually a fatty alcohol[64,65] or a higher alkane, *e.g.* hexadecane.[66] The fatty alcohol is also referred to as a cosurfactant.[14,63] The surfactant is used below its CMC so that no micellar phase is present to act as particle nucleation locus. Both ionic and nonionic surfactants are used, as are oil soluble and water soluble initiators.[63]

Very high shearing force is required to break the bulk monomer into the miniemulsion droplets. Sonication often suffices. However, high-pressure agitators are required to make extremely stable miniemulsion specially when a large alkane is used as a costabilizer.[63]

The costabilization action of fatty alcohol is attributed to the presence of a crystalline complex in the interface between monomer/particle and water. The complex is formed between the alcohol and the surfactant.[67] The existence of the complex has been proved by electron microscopy and electron diffraction studies.[68,69] Presumably, the collision energy between monomer droplets is insufficient to dislodge the crystalline complex from the interface imparting stability to miniemulsions. Besides coalescence, another mechanism operates in destabilizing miniemulsion. This involves diffusion of monomer through water from smaller droplets to larger ones. The driving force for this diffusion is the higher activity of monomer in smaller droplets due to the latter's greater curvature. This process of diffusive degradation is known as "Ostwald ripening". Presence of a water insoluble compound such as hexadecane in the monomer droplets prevents the diffusion.[70] Diffusion of monomer would raise the chemical potential of hexadecane in the smaller particles and lower that in the larger ones, which is not permitted thermodynamically.

Miniemulsion polymerization provides the only means of preparing polymer latexes from highly water insoluble monomers, *e.g.*, dodecyl methacrylate, stearyl methacrylate *etc.*, using modest amount of stabilizers. However, it is also used in the polymerizations of lesser hydrophobic monomers, *e.g.*, vinyl chloride, vinyl acetate, methyl methacrylate, and styrene, *etc.*[63] Another important use is found in its adoption in living radical polymerization.

9.6 Microemulsion Polymerization

The monomer droplets in these polymerizations are essentially monomer-laden micelles, which are very small in size *ca.*, 5–10 nm with a narrow size distribution, and thermodynamically stable. Due to the very small size of the droplets microemulsions are transparent.[71,72]

Particle nucleation and growth occur in the monomer droplets yielding larger particles (although, still very small *ca.*, < 50 nm). This suggests that not all the droplets are nucleated. Some of them act as monomer reservoirs supplying monomers to the nucleated droplets where polymerization takes place. Both direct (oil-in-water)[73–75] and inverse (water-in-oil)[76,77] microemulsion polymerizations have been developed. The emulsifier used is usually a mixture of an ionic surfactant and a cosurfactant (mostly an alcohol). However, much larger proportion of surfactant to monomer is needed than in any of the other types of emulsion polymerization. For example, a typical recipe for the microemulsion polymerization of styrene uses 2% each of cetyltrimethylammonium bromide and styrene, as well as 1% hexanol (cosurfactant) in water.[74] Initiation may be effected either thermally or photochemically using an oil soluble initiator, *e.g.*, azobisisobutyronitrile.

Inverse microemulsion requires still larger quantity of surfactant. For example, a recipe for the inverse microemulsion polymerization of acrylamide uses typically 60–75% toluene, 15–25% sodium-bis-2-ethylhexyl sulfosuccinate (AOT), 8–10% water, 2–5% acrylamide, and an oil soluble initiator. The monomer itself acts also as a cosurfactant, some of which being located at the water-oil interface.[77] As polymerization proceeds in the nucleated monomer droplets, monomer supply is maintained either by diffusion of aqueous monomer solution from the nonnucleated droplets through toluene or by rapid exchange of monomer (located at the interface) during collision between the nucleated and nonnucleated droplets.[77] The final dispersion contains not only polymer particles but also some AOT micelles. Each particle contains very few (2 to 10) polymer molecules.

9.7 Dispersion Polymerization

Dispersion polymerization is a form of precipitation polymerization in which the precipitating polymer is kept dispersed in the polymerization medium using a polymeric steric stabilizer. It starts with a homogeneous solution of monomer, initiator, and stabilizer. The polymer formed being insoluble in the polymerization medium separates out from the latter forming a dispersion.[78] In order to obtain coagulum-free dispersions, the polymeric stabilizer should adsorb strongly at the interface between dispersed polymer particle and polymerization medium. One way to achieve this is to use as stabilizer an amphipathic diblock copolymer, a graft or a comb copolymer in each of which one of the component polymer is soluble in the dispersion medium, whereas the other is insoluble.[78-81] The latter adheres strongly to the dispersed polymer particle anchoring the stabilizer onto the particle. Anchoring also occurs when the insoluble part is miscible with the dispersed polymer. The part of the stabilizer soluble in the dispersion medium, referred to as the stabilizing moiety, extends out of the particle into the medium and prevents particle coagulation by steric stabilization mechanism (Sec. 9.1). Thus, Barrett used a comb polymer with a poly(methyl methacrylate) trunk and poly(12-hydroxy stearic acid) teeth ($\overline{DP_n} = \sim 5$) as stabilizer in the early reported dispersion polymerization of methyl methacrylate in an aliphatic hydrocarbon medium.[78,79] The trunk of the comb polymer acts as the anchor for the soluble teeth, which act as the stabilizing moieties.

Although the amphipathic stabilizers are expected to be very efficient, they need to be used at high concentrations, *ca.*, 2 to 5 percent. However, these polymerizations conducted in hydrocarbon media yield only submicronic particles. In contrast, near monodisperse micron-sized particles (up to 15 μm) of nonpolar polymers may be prepared using homopolymeric stabilizers and alcoholic or aqueous alcoholic dispersion media.[82-89] Dispersions of water soluble polymers, *e.g.*, polyacrylamide may also be prepared in aqueous alcoholic media by this method.[90]

Monodisperse polymer partciles have found a number of important applications such as standards for instrument calibration, pore size determination in membranes, column packing materials in chromatography, medical assays, *etc.*

Some of the homopolymers used as stabilizers are hydroxypropylcellulose (HPC), poly(acrylic acid) (PAA), poly(vinylpyrrolidone) (PVP), and poly(vinyl methyl ether) (PVME). The method is also used to prepare processable dispersions of intractable conducting polymers, *e.g.*, polypyrrole and polyaniline.[91-98] However, in homopolymer stabilized systems the real stabilizer is the graft copolymer formed *in situ* following polymer transfer.[84,86–89] The grafted chains, being constituted of the same polymer as that constitutes the polymer particle, anchor the stabilizer to the particle surface. Presence of 10–20 nm thick stabilizer layers on polystyrene particle surface was demonstrated using transmission electron micrography with HPC, PVP, or PAA stabilizers.[89] Although the graft copolymer has not been isolated, there are indirect evidences of its formation. Thus, both HPC- and PVP-stabilized polystyrene particles freed of unadsorbed stabilizers are converted to new particles with similar surface stabilized morphology following dissolution in dioxane and precipitation by regulated addition of methanol indicating that the stabilizer is not physically adsorbed.[88]

However, Shen et al. observed that a part of PVP adsorbed on PMMA particles prepared in methanol by using AIBN initiator and PVP stabilizer can be removed by serum replacement with methanol. The remaining part is not removed even after two weeks of treatment. This suggests that both grafted and ungrafted PVP act as stabilizers.[99]

Another approach for making stable dispersions makes use of reactive stabilizers.[99–103] Thus, a macromonomeric stabilizer is copolymerized with large molar excess (*ca.*, 50 to 500 times) of a monomer, which is to be dispersion polymerized. The stabilizer becomes chemically bound to polymer particles giving stable dispersions. A typical macromonomer suitable for use as a reactive stabilizer for dispersion polymerization of vinyl monomers in aqueous alcoholic media is poly(ethylene glycol) methyl ether methacrylate,

$$CH_2 = \underset{CH_3}{C} - \underset{O}{C} + OCH_2CH_2 \xrightarrow{}_n OCH_3 .$$

A significant advancement in dispersion polymerization refers to the use of the environment-friendly (recyclable) supercritical carbon dioxide ($scCO_2$) as dispersion medium. For example, dispersion polymerization of methyl methacrylate is effected in this medium using $[-CH_2 - CH(COOCH_2(CF_2)_6CF_3)-]_n$ as the stabilizer. The lipophilic acrylic main chain of the stabilizer anchors it on the polymer particle and the CO_2–philic fluorinated pendant short chains provide steric stabilization.[104]

9.8 Heterophase Living Radical Polymerization

Water-borne living radical polymerization (LRP) in emulsion systems is attractive for large-scale production of controlled polymers due to the virtues of emulsion polymerization mentioned earlier. However, combining LRP with conventional emulsion polymerization is difficult to accomplish without loss of control on polymerization due to partitioning of monomer, initiator, LRP mediator, *etc.*, between the multiple phases that exist: micelles, polymer particles, monomer droplets, and aqueous dispersion medium. Nevertheless, successful ATR emulsion polymerizations of styrene, butyl acrylate, and butyl methacrylate were achieved using hydrophobic copper complexes such as CuBr/dNbpy as activator. Also, a nonionic surfactant, *e.g.*, Brij 98, is required to be used since the transferable halide ion in copper complexes is replaced with the nontransferable surfactant anion of the usually used anionic surfactants. As a result, deactivation of the polymer radicals by halogen atom transfer fails to occur resulting in uncontrolled polymerization.[105]

The problem of partition of reactants and transport through the aqueous phase is avoided in miniemulsion,[106–111] microemulsion,[112] and dispersion polymerization[113–118] systems where a separate monomer phase does not exist. Hence, these have been the preferred processes.

References

1. C.H. Bamford, W.G. Barb, A.D. Jenkins, P.F. Onyon, The Kinetics of Vinyl Polymerization by Radical Mechanism, Butterworths, London, Chap. 4 (1958).
2. W.I. Bengough, R.G.W. Norrish, *Proc. Roy Soc.* (1950), A200, 301.
3. J.D. Burnett, H.W. Melville, *Trans. Faraday Soc.* (1950), 46, 976.
4. J. Abere, G. Goldfinger, H. Naidus, H.F. Mark, *J. Phys. Chem.* (1945), 49, 211.
5. R.G.W. Norrish, R.R. Smith, *Nature* (1942), 150, 336.
6. G.M. Burnett, H.W. Melville, *Proc. Roy. Soc.* (1947), A189, 494.
7. K. Tauer, Heterophase Polymerization in *Ency. Polym. Sci. & Tech.* 3rd ed., H. Mark Ed., Wiley-Interscience Vol. 6, p. 410 (2003).
8. D.H. Napper, Polymeric Stabilization of Colloidal Dispersions, Academic Press, London, Chap. 1 (1983).
9. H.C. Hamaker, *Physica* (1937), 4, 1058.
10. R.H. Ottewill in Scientific Methods for the Study of Polymer Colloids and Their Applications, F. Candau, R.H. Ottewill Eds., NATO ASI Series C: Mathematical and Physical Sciences, Vol. 303, Kluwer Academic Pub., Dordrecht, p. 129 (1990).
11. D.B. Hough, L. White, *Adv. Colloid Interf. Sci.* (1980), 14, 3.
12. Dispersion Polymerization in Organic Media, K.E.J. Barett Ed., Wiley, New York (1975).
13. (a) E. Ferber, in, *Ency. Polym. Sci. & Tech.* 1st ed., H. Mark, N.G. Gaylord, N.M. Bikales Eds., Wiley-Interscience, New York, Vol. 13, p. 552 (1970).
    (b) E.A. Grulke, in, *Ency. Polym. Sci. & Eng.*, 2nd ed., H. Mark, N.M. Bikales, C.G. Overberger, G. Menges, Eds., Wiley-Interscience, New York, Vol. 16, p. 443 (1989).
14. M.S. El-Aasser, in, Scientific Methods for the Study of Polymer Colloids and Their Applications, F. Candau, R.H. Otlewill Eds., Nato ASI Series C, Kluwer Academic Publishers, Dordreclit, Vol. 303, Chap. 1, p. 1 (1990).
15. W.C. Griffin, *J. Soc. Cosmet. Chem.* (1949), 1(5), 311.
16. J.T. Davies, E.K. Rideal, Interfacial Phenomenon, *Acad.* Press (1963).
17. P.J. Flory, Principles of Polymer Chemistry, Cornell University Press, Ithaca, New York, Chap. 5-3, p. 203 (1953).
18. M. Morton, P.P. Salatiello, H. Landfield, *J. Polym. Sci.* (1952), 8, 215, 279.
19. R.H. Gobran, M.B. Berenbaum, A.V. Tobolsky, *J. Polym. Sci.* (1960), 46, 431.
20. W.D. Harkins, *J. Am. Chem. Soc.* (1947), 69, 1428.
21. F.A. Bovey, I.M. Kolthoff, A.J. Medalia, E.J. Meehan, *Emulsion Polymerization*, in, High Polymers Ser. H. Mark, H.W. Melville, C.S. Marvel, G.S. Whitby Eds., Interscience, New York (1955).
22. W.V. Smith, R.H. Ewart, *J. Chem. Phys.* (1948), 16, 592.
23. W.V. Smith, *J. Am. Chem. Soc.* (1948), 70, 3695; (1949), 71, 4077.
24. R.G. Gilbert, D.H. Napper, *J. Chem. Soc. Faraday* 1 (1974), 71, 391.
25. M. Litt, R. Pastiga, V. Stannet, *J. Polymer Sci.*, A-1 (1970), 3607.
26. M. Nomura, M. Harada, *J. App. Polym. Sci.* (1981), 26, 17.
27. H. Sakai, Y. Kihara, K. Fujita, T. Kodami, M. Nomura, *J. Polym. Sci. Polym. Chem.* (2001), 39, 1005.
28. J.W. Vanderhoff, E.B. Bradford, Tappi (1956), 39, 650.
29. M. Nomura, J. Ikoma, K. Fujita, *Polym. Mater. Sci. Eng.* (1991), 64, 310.

30. M. Nomura et al., *J. Polym. Sci. Polym. Chem. Ed.* (1991), 29, 987.
31. M. Nomura, J. Ikoma, K. Fujita, in, Polymer Latexes, *Am. Chem. Soc.*, p. 55 (1992).
32. E. Barthalome, H. Gerrens, R. Harbeck, H.M. Weitz, *Z. Elektrochem.* (1956), 60, 334.
33. Z. Manyasek, A. Rezabek, *J. Polym. Sci.* (1962), 56, 47.
34. H. Gerrens, Ber. Bunsenges. *Physik. Chem.* (1963), 67, 741.
35. M. Morton, P.P. Salatiello, H. Landfield, *J. Polym. Sci.* (1952), 8, 111.
36. A.S. Dunn, P.A. Taylor, *Makromol. Chem.* (1965), 83, 207.
37. H. Wiener, *J. Polym. Sci.* (1951), 7, 1.
38. P.M. Hay et al., *J. App. Polym. Sci.* (1961), 5, 23.
39. J.L. Gardon, *J. Polym. Sci.* A-1 (1968), 6, 643.
40. I.A. Maxwell, B.R. Morrison, D.H. Napper, R.G. Gilbert, *Macromolecules* (1991), 24, 1629.
41. B.M.E. van der Hoff, *J. Polym. Sci.* (1958), 33, 487; (1960), 48, 175.
42. S. Okamura, T. Motoyama, *J. Polym. Sci.* (1962), 58, 221.
43. C.P. Roe, *Ind. Eng. Chem.* (1968), 60 (9), 20.
44. J.W. Vanderhoff, *J. Polym. Sci.* C (1985), 72, 161.
45. W.J. Priest, *J. Phys. Chem.* (1952), 56, 1077.
46. B. Jacobi, *Angew. Chem.* (1952), 64, 539.
47. R. Pastiga, M. Litt, V. Stannett, *J. Phys. Chem.* (1960), 64, 801.
48. R.M. Fitch, C.H. Tsai, in, *Polymer Colloids*, R.M. Fitch Ed., Plenum Press, New York, p. 73 (1971).
49. F.K. Hansen, J. Ugelstad, *J. Polym. Sci.*, A-1 (1978), 16, 1953; (1979), 17, 3033, 3047, 3069.
50. F.K. Hansen, J. Ugelstad, *Makromol. Chem.* (1979), 180, 2423.
51. J. Ugelstad, F.K. Hansen, in, *Emulsion Polymerization*, I. Pirma Ed., Acad. Press, New York (1982).
52. G.S. Whitby, M.D. Gross, J.R. Miller, A.J. Costanza, *J. Polym. Sci.* (1955), 16, 549.
53. V. Stannett, *J. Polym. Sci.* (1956), 21, 343.
54. D.H. Napper, A.G. Parts, *J. Polym. Sci* (1962), 61, 113.
55. J.W. Goodwin, J. Hearn, C.C. Ho, R.H. Ottewill, *Colloid Polymer Sci.* (1974), 252, 464.
56. F. Candau, Y.S. Leong, G. Pouyet, S.J. Candau, in, *Physics of Amphiphiles: Micelles, Vesicles, and Microemulsions* 1985, XC Corso, Soc. Italiana di Fisica, Bologna, Italy.
57. J.W. Vanderhoff et al., *Adv. Chem. Ser.* (1962), 34, 32.
58. C. Graillat, C. Pichot, M.S. El-Aasser, *J. Polym. Sci. Polym. Chem.* (1986), 24, 427.
59. W. Baade, K.H. Reichert, *Eur. Polym. J.* (1984), 20, 505.
60. D. Hunkeler, A.E. Hamielec, W. Baade, *Polymer* (1989), 30, 127.
61. S.K. Ghosh, B.M. Mandal, *Polymer* (1993), 34, 4287.
62. V.E. Kurenkov, V.A. Myagchenkov, *Polym.-Plast. Technol. Eng.* (1991), 30, 367.
63. J.M. Asna, *Prog. Polym. Sci.* (2002), 27, 1283.
64. J. Ugelstad, M.S. El-Aasser, J.W. Vanderhoff, *Polym. Lett.* (1973), 11, 505.
65. J. Ugelstad, F.K. Hansen, S. Lange, *Makromol. Chem.* (1974), 175, 507.
66. J. Ugelstad, *Makromol. Chem.* (1978), 179, 815.
67. Y.J. Chou, M.S. El-Aasser, J.W. Vanderhoff, in, *Polymer Colloids* II, R.M. Fitch Ed., Plenum, New York, p. 619 (1981); *J. Dispersion Sci. & Tech.* (1980), 1, 129.
68. W.L. Grimm, M.S. Thesis, Lehigh University (1982), cited in Ref. 69.
69. J.W. Vanderhoff, *J. Polym. Sci.* C (1985), 72, 161.
70. W.I. Higuchi, J. Misra, *J. Pharm. Sci.* (1962), 51, 459.
71. T.P. Hoar, H.H. Schulman, *Nature* (1943), 152, 102.
72. K.S, Shinoda, S. Friberg, *Adv. Colloid Interface Sci.* (1975), 4, 281.
73. P-L. Kuo et al., *Macromolecules* (1987), 20, 1216.
74. S.S. Atik, J.K. Thomas, *J. Am. Chem. Soc.* (1981), 103, 4279.
75. M.R. Ferrick, J. Murtagh, J.K. Thomas, *Macromolecules* (1989), 22, 1515.

76. Y.S. Leong, F. Candau, *J. Phys. Chem.* (1982), 86, 2269.
77. F. Candau, Y.S. Leong, G. Pouyet, S. Candau, *J. Colloid Interface Sci.* (1984), 101, 167.
78. K.E.J. Barrett, Ed. Dispersion Polymerization in Organic media, Wiley, London (1975).
79. K.E.J. Barrett, H.R. Thomas, *J. Polym. Sci.*, A-1 (1969), 7, 2621.
80. M.D. Croucher, M.A. Winnik, in, *Scientific Methods for the Study of Polymer Colloids and Their Applications*, F. Candau, R.H. Ottewill Eds., Kluwer Acad. Publishers, Dordrecht, p. 35 (1990).
81. J.V. Dawkins, D.J. Neep, P.L. Shaw, *Polymer* (1994), 35, 5366.
82. Y. Almog, S. Reich, M. Levy, *Brit. Polym. J.* (1982), 14, 131.
83. C.K. Ober, K.P. Lok, *Macromolecules* (1987), 20, 268.
84. K.P. Lok, C.K. Ober, *Can. J. Chem.* (1985), 63, 209.
85. S. Shen, E.D. Sudol, M.S. El-Aasser, *J. Polym. Sci. Polym. Chem. Ed.* (1993), 31, 1393.
86. A.J. Paine, W. Luymes, J. McNulty, *Macromolecules* (1990), 23, 3104.
87. A.J. Paine, *Macromolecules* (1990), 23, 3109.
88. A.J. Paine, *J. Colloid Interface Sci.* (1990), 138, 157.
89. A.J. Paine, Y. Deslandes, P. Gerroir, B. Henrissat, *J. Coll. Interface Sci.* (1990), 138, 171.
90. B. Roy, B.M. Mandal, *Langmuir* (1997), 13, 2191.
91. S.P. Armes, J.F. Miller, B. Vincent, *J. Colloid Int. Sci.* (1987), 118, 410.
92. B.M. Mandal, P. Banerjee, S.N. Bhattacharyya, in, *Polymeric Materials Encyclopedia: Synthesis, Properties and Applications*: J.C. Salamone Ed., CRC Press, Boca Raton, Fl., Vol. 9, p. 6670 (1996).
93. S.P. Armes, M. Aldissi, *Chem. Commun.* (1989), 88.
94. M.L. Digar, S.N. Bhattacharyya, B.M. Mandal, *Chem. Commun.* (1992), 18.
95. P. Banerjee, S.N. Bhattacharyya, B. M. Mandal, *Langmuir* (1995), 11, 2414.
96. N. Gospodinova, P. Mokrova, L. Terlemezyan, *Chem. Commun.* (1992), 923.
97. J. Stejscal, P. Kratochvil, M. Helmstedt, *Langmuir* (1996), 12, 3389.
98. D. Chattopadhyay, B.M. Mandal, *Langmuir* (1996), 12, 1585.
99. K. Ito, S. Yokoyama, F. Arakawa, *Polym. Bull.* (1986), 16, 345.
100. S. Kawaguchi, M.A. Winnik, K. Ito, *Macromolecules* (1995), 28, 1159.
101. S. Kawaguchi, K. Ito, *Adv. Polym. Sci.* (2005), 175, 299.
102. A. Guyot, K. Tauer, *Adv. Polym. Sci.* (1994), 111, 43.
103. K. Ito, S. Kawaguchi, *Adv. Polym. Sci.* (1999), 142, 129.
104. J.M. DeSimone *et al.* *Science* (1994), 265, 356.
105. J. Qiu, S.G. Gaynor, K. Matyjaszewski, *Macromolecules* (1999), 32, 2872.
106. K. Matyjaszewski, J. Qiu, N.V. Tsarevsky, B. Charleux, *J. Polym. Sci. Polym. Chem.* (2000), 38, 4724.
107. H. de Brouwer, J.G. Tsavalas, F.J. Schork, M.J. Monteiro, *Macromolecules* (2000), 33, 9239.
108. K. Min, H. Gao, K. Matyjaszewski, *J. Am. Chem. Soc.* (2005), 127, 3825.
109. A. Butte, G. Storti, M. Morbidelli, *Macromolecules* (2000), 33, 3485.
110. J.J. Vosloo *et al.*, *Macromolecules* (2002), 35, 4894.
111. M. Lansalot, T.P. Davis, J.P.A. Heuts, *Macromolecules* (2002), 35, 7582.
112. K. Min, K. Matyjaszewski, *Macromolecules* (2005), 38, 8131.
113. J.-S. Song, F. Tronc, M.A. Winnik, *J. Am. Chem. Soc.* (2004), 126, 6562.
114. M. Hoelderle, M. Baumert, R. Muelhaupt, *Macromolecules* (1997), 30, 3420.
115. L.I. Gabaston, R.A. Jackson, S.P. Armes, *Macromolecules* (1998), 31, 2883.
116. S.E. Shim *et al.*, *J. Polym. Sci. Polym. Chem.* (2007), 45, 348.
117. J.-S. Song, M.A. Winnik, *Macromolecules* (2005), 38, 8300; (2006), 39, 8318.
118. K. Min, K. Matyjaszewski, *Macromolecules* (2007), 40, 7217.

Problems

9.1 A polymer emulsion prepared by using an oil-in-water ionic emulsifier can be coagulated by brine. How does this happen? Would the same thing happen if an emulsion stabilized by a nonionic emulsifier is used? Explain your answer.

9.2 What are the unique features of emulsion polymerization kinetics *vis-à-vis* homogeneous radical polymerization kinetics?

9.3 If the initiator concentration is increased in an emulsion polymerization keeping the soap concentration unchanged, what will be the effect on the rate of polymerization?

9.4 What would happen if the above polymerization (problem 9.3) were a seeded one and carried out in the absence of added soap with initiator concentrations, which are not too high?

9.5 Discuss the mechanism of charge stabilization and steric stabilization of colloidal particles. How may the destabilization of the particles be effected?

9.6 What is Ostwald ripening? What is the thermodynamic reason? How can it be prevented?

9.7 What do you understand by solubilization? How do surfactants solubilize styrene and to what extent?

9.8 Nanoparticles of gold (< 5 nm) may be stabilized in hydrocarbon dispersion medium using a small molecular alkane thiol as stabilizer, whereas a polymeric stabilizer is needed to stabilize micron size particles. Explain why this should be so.

Index

α,ω-dichloropolyisobutylene, synthesis of, 308–310
β-agostic interaction, 278
β-hydride elimination, 273
[2+2]-cycloaddition, 362
2,6-di-*t*-butylpyridine as proton trap in LCP, 302
4-membered ring transition state, 269
4th generation Z–N catalysts, 265

acetal resin, 344
acetylene polymerization, 272
activated chain end or ACE mechanism, 341, 355, 356
activated magnesium chloride, 262, 264
activated $MgCl_2$–supported Z–N catalysts, 262, 271
activated monomer (AM) mechanism, 328, 329, 342, 350, 351, 353, 354
activation energy:
    in radical polymerization, 140, 142
    of MW in cationic polymerization, 296
    of living anionic polymerization of styrene in THF, 232, 235
activation-deactivation cycles, frequency of, effect on PDI, 157, 299
activator regenerated by electron transfer, 181
active center, 7, 8, 9, 113, 268, 337, 356, 374, 379, 396
active center concentration, 7
active site in Z–N catalyst, 265, 266
active site isomerization, 268, 275
acyclic diene metathesis (ADMET) polymerization, 88, 365
acyl-oxygen scission, 345
addition polymers, definition of, 5, 6
addition-fragmentation chain transfer, 120

ADMET polymerization, 88, 365
adventitious initiation, 299
adventitious protogenic initiators, 291, 359
adventitious protogens, 290, 302, 335
adventitious termination in living polymerization, 215
after-effect, 135
AGET ATRP, 180
aggregated ion pairs, 238, 333
aggregation constant, 333
AIBN, decomposition products of, 101
aliphatic diacyl peroxides, 106
aliphatic polyamides, 22, 74
aliphatic-aromatic polyesters, 22, 77
aliphatic polysulfides, 72
alkali-ion-specific crown ethers, 341
alkoxide as chain carrier, 331, 344
alkoxide-carboxylate interchange, 348
alkoxide-ester interchange, 346
alkyd resins, 67
alkyl migration, 269
alkyl transfer in Z–N polymerization, 273
alkyl-oxygen bond scission, 350
alkylacrylamides, living anionic polymerization of, 242
alkylated $TiCl_3$ crystals, 260
alkylation by cocatalyst in Z–N catalyst systems, 265
alkyllithium aggregates, 223
alkyllithiums, 223
allophanate crosslinks, 68
allyltrimethylsilane, 311
alternating copolymer, 374, 379
amidate ion, 351
amidoenolate, 242
amine (nucleophile) mechanism in ROP of NCA, 354

amine hydrochloride as initiator of NCA, 355
amino end-capped oligoimide, 86
amino resins, 9, 65
aminoalkyl radical, 108, 109
amorphous polymers, 14, 20
amorphous regions, in semicrystalline polymer, 22
amphipathic diblock copolymer, 332, 421
amphiphilic substance, 410
angle strain, 42, 325
anhydride end-capped oligoimide, 86
anionic active center, 328, 329
ansa metallocenes, 275
apparent rate constant of:
    depropagation, 337
    polymerization, 177
    propagation, 227, 337
aqueous phase nucleation, of particles in emulsion polymerization, 418
ARGET ATRP, 181, 183
aromatic polyamides, 75
aromatic polyesters, 78
asymmetric star polymers, 245
atactic blocks, 263
atactic fraction, 265
atactic polymer, regiospecificity in, 266
ate complexes, 226
atom transfer radical addition, 167
atom transfer radical polymerization, 167–185
ATRP of lowly reactive monomers, 178
ATRP with activator generated in situ, 179–182
attractive force between polymer particles, 407
attractive interaction, between dispersed particles, 407
autoaccelerated nature of gelation, 57
autoacceleration in radical polymerization, 143
autoinhibition, 118
autosolvation of carbocations in vinyl ether polymerization, 299
average degree of functionality, 47
average reactivity of monomers, 383
azeotropic copolymer, 380
azo initiators, 106

B-strain effect, 169, 192, 299
backbiting, 43, 122, 248, 326, 328, 332, 340, 342, 343, 349, 350, 357, 359, 365

backward fragmentation in RAFT polymerization, 190
Bakelite, 2, 37
base metal alkyl, 259
basis model of copolymerization, 395
batch copolymerization, instantaneous composition in, 373
'bead' polymerization, 409
bicyclic monomers, ROP products of, 330
bidentate donors as components of 5th generation Z–N catalysts, 262
bifunctional monomer, introduction to, 9
bifunctional polycondensation, 48–51
bimetallic active center in Z–N polymerization, 268
bimolecular chain termination, 100–102
binary copolymer, 373
binary copolymerization, introduction to, 374
biomedical applications, 323, 332
bisphenol A polyarylate, 79
bisphenol A polycarbonate, 79
bisphenol A polysulfone, 84
biuret crosslinks, 68
block copolymerization, see table of contents for various chapters on synthesis of block copolymers
block copolymers, definition and nomenclature, 373, 374
blockiness in copolymer composition, 394
bond dissociation energy, 106, 124, 170
branched polymers, 12
branching coefficient, 59
branching in polyethylene, 122, 273, 277
bridged cyclic monomers, ROP products of, 330
broadening of MWD in living polymerization, causes of, 156, 216
Bronsted acids as initiators, 286
Brownian motion, 408
bulk polymerization of methyl methacrylate, gel effect in, 144
butyl rubber, 289

C–N bond dissociation in alkyl nitroxides, 160
C-alkylation vs. O-alkylation, 64, 84
cage effect, 101
cage recombination, 101
cage return, 101
carbon fibers, wetting with polyimide resins, 86
Carothers classification, 6

Carothers equation, 47–49, 56, 62, 63, 65
  departure from, 51
catalytic chain transfer (CCT), 123
cationic active center, 285
cationic copolymerization, 382, 394
cationic photoinitiators, 310, 334
ceiling temperature of polymerization, 29
cellulose, a condensation polymer, 5
cesium as counterion in living anionic polymerization, 233, 235
chain carrier, 149
chain equilibration, in RAFT polymerization, 187
chain extension of DGEBA, 70
chain flexibility, 14, 18, 19, 23
chain molecular nature of polymers, 2
chain packability, 14, 18
chain-growth polymer, mechanism of growth of, 7
chain polymerization, characteristic of, 7–9
chain reactions, 8
chain skeleton, 19
chain termination, a component of chain polymerization, 8, 9
chain transfer, 9, 113–129, 266–268, 291–296, 306
chain transfer coefficient, 189, 195
chain transfer constant, 114, 273
  determination by Mayo method, 115
  determination by CLD method, 123
chain transfer constants of solvents in radical polymerization, 115
chain transfer to initiator, 118, 150, 309
chain transfer to monomer, 116, 286, 291–297, 332, 416
chain transfer to polymer, 121, 328–330, 339, 348, 357, 364
chain-length dependence of k_t, 142
chain-length dependence of k_p, 142
chain-transfer agent, 114, 149
"chain walking" mechanism, in late transition-metal-catalyzed polymerization of propylene, 277
chiral metallocene, 275
chirality of transition metal atoms in Z–N catalysts, 268
chlorine atom vacancies in $TiCl_3$ crystals, 265, 268
chlorine-terminated polymer, 184, 296

chlorohydrocarbon solvents, as medium in cationic polymerization, 287
chlorotelechelics, 309
chlorotrimethylsilane, 359
cis-1,4-polybutadiene, 18, 264
cis-polyacetylene, 25, 272
CLD method, chain transfer constant determined by, 123, 124
coagulum-free dispersions, 421
cobalt mediated living radical polymerization, 149, 167
cocatalyst, 259, 264, 290
coinitiator, 290
collision frequency, 40
colloidal molecular hypothesis, 1
colloidal molecules, 1
comb polymers, 246, 421
combination of radicals, 100, 101, 113
commodity polymers, 2
common ion effect, 230, 303
common rings, 42, 325
common ring monomers, polymerizability of, 325
complex counterions, 337
complexed (with Lewis acid) monomers, copolymerization of, with electron-donor monomers, 379
composite Hamaker constant, 408
concerted four center reaction, 269
condensation polymer, 2, 3, 5, 6, 323
condensed amorphous polymer, 324
conducting polyheterocycles, 26, 27
conducting polymers, 25–27
conformational entropy, 324, 326
conjugated polymers, 25, 28, 88
constant rate period, in emulsion polymerization, 413, 415
contact ion pair, experimental proof for existence, 232
controlled initiation, in cationic polymerization, 290
controlled radical polymerization, 149
controlled termination, in cationic copolymerization, 291
conventional initiator, 162
conventional radical polymerization (CRP), 192
conversion-composition relationship in copolymerization, 376
coordinated anionic initiator, 332

coordinated anionic mechanism in Z–N polymerization, 266
coordination sites of magnesium, vacant sites in magnesium chloride lattice, 262, 270
coordination–insertion mechanism, in Z–N polymerization, 266, 346
copolymer composition equation, 375, 376, 381, 399
copolymerization models, 374
copolymerization of ethylene and:
    cycloolefins, 274
    alpha olefins, 274
    polar monomers, 178, 276
costabilizer in miniemulsion polymerization, 419, 420
cosurfactant, 419, 420
covalent metal alkoxide chain ends, 349
covalent esters, propagation on, 338, 339
covalent metal alkoxides, initiators in living ROP of cyclic esters, 329, 347, 349
covalent propagation, 338, 339
critical distance in particle coagulation, 408, 409
critical extent of reaction in gelation, 10
critical micelle concentration (CMC), 411
critical monomer concentration, 45, 327, 342, 345, 357
cross-coupling, 152
cross-linking, 70
cross-propagation, 375, 402
cross-termination, 101, 393
crosslinked network, 65
crosslinked polymers, 12, 18, 121
crosslinked polystyrene, 11
crown ethers, 346
cryptates, 346
crystal modifications of $TiCl_3$, 260
crystalline polymer, 14
crystalline regions in semicrystalline polymer, 22
crystalline $TiCl_3$, 260
crystallinity, measurement of, 265
crystallites, 16, 19
crystallizable polymers, 20
Cu(I) complexes, disproportionation of, 175
cumyl chloride, 299, 301, 309
curing, 70, 71, 87
curing agents, 70, 71, 87
cyclic acetals, 343
cyclic amines, 356

cyclic anhydrides, 344, 349
cyclic carbonates, 344, 349
cyclic esters, 329, 344
cyclic ethers, 332
cyclic iminium species, 357
cyclic olefins, 361
cyclic oxonium ion, 335
cyclic siloxanes, 71, 357
cyclic sulfides, 357
cycloalkenes, 42, 323
cyclodimethylsiloxanes, 326, 328
cyclosiloxanes, polymerization of, 357, 358

$\overline{DP}_n$ evolution, typical patterns of, 7, 48, 155, 156
D_n (cyclosiloxanes), 71, 326
DC (diffusion clock) method, 314
deactivation equilibrium, 149, 299, 400
deactivation frequency, 157, 198, 303
dead end polymerization, 144, 411
deashing, 262
Debye equation for diffusion current, 415
deceleratory rate period, in emulsion polymerization, 413
degenerative transfer, 149, 186, 248, 332, 333
degenerative transfer agent, 192
degenerative transfer mechanism of exchange in GTP, 248
degradative chain transfer, 118
degree of dissociation of ion pairs, 229, 293, 338
degree of functionality, 47
degree of polymerization, 47, 49, 53, 216, 332, 342, 416
degree of stereoregularity, determination of, 265
degree of stereospecificity of initiation, 270
dendrimers, 14
dendrimers with chlorosilane groups, 245
dendrons, 14
departure from Carothers equation, 51
depolymerization, 344
depropagation, 30, 396
determination of k_d in NMP, 165
determination of monomer reactivity ratios, 381
determination of polymer end groups, 110
determination of rate of initiation in radical polymerization, 131
DGEBA, 69
diacid monomers, 9

diad and triad fractions in copolymers, 389
diamine monomers, 9
diaryliodonium salts, 310
dichain step polymer, 147
dicumyl methyl ether, 308
diffusion clock (DC) method, 311
diffusion current, 415
diffusion limited entry of radicals into dispersed polymer particles, 417
diffusion of molecules in simple liquids, 41
diffusion of segments in polymer coil, 41
diffusion of monomers through the aqueous phase, 415
diffusion-controlled termination, 141, 393
diffusion-limited reaction, rate constant of, 311
difunctional initiator oligo (α-methylstyrene) disodium, 222, 238
difunctional organolithium initiators, 219
diglycidyl ether of bisphenol A, 69
dihydroxy-poly (oxytetramethylene) polyether, 78
dilactones, 344
dimeric halide initiator, 155, 300
dimerization of radical ion, 221
direct esterification, 77
disappearance of micelles, 414
dispersion force, 18, 407
dispersion medium, 407, 409, 422
dispersion polymerization, 421
disproportionation of radicals, 100–102, 113, 164
disproportionation of ATRP catalysts, 175
dissociation mechanism of initiation in GTP, 247
dissociation-combination (DC) equilibrium, 149, 151
dissociative mechanism of GTP, 248
distribution of polymer molecular weights:
    in step polymerization, 52
    in conventional radical polymerization, 145
    in living polymerization, 156, 216, 306, 330
disulfide linkages, 72
dithiocarbamates, 188
dithioesters, 187
divinyl monomers, 11
divinylbenzene, 11
donor acceptor complex between nucleophiles and carbocation, 301

dormant polymers, 148, 357
double layer thickness, role in electrostatic stabilization, 409
DP equation:
    in living polymerization, 156, 216, 306, 330
    in step polymerization, 47–51
    in conventional radical polymerization, 114
duration of collision between molecules, 41
Durham route, acetylene polymerization by, 364
dye partition method of end group determination, 109, 110
dynamic equilibrium, 55, 150, 303, 339

ease of ring formation, 42
eco-friendly processing of polyamic acid, 85
elastomeric property, structural requirement for, 18–20
electrical double layers, 408, 409
electrochemical polymerization, 222
electrochromic polymer, polyaniline, 27
electroluminescent polymers, 28
electron affinity, monomers with high, 221
electron donors:
    effect of, on PDI in cationic polymerization, 306
    as component of higher generations Z–N catalysts, 262
electron-rich ligands, 362
electron transfer initiation, 220–222
electron-acceptor monomers, behavior in radical reactions, 98, 379, 380
electron-acceptor substrates, chain transfer with, 126
electron-donor monomers, behavior in radical reactions, 98, 379, 380
electron-donor substrates, chain transfer with, 126
electron-rich double bond, 285
electron-withdrawing ligands, role in activity of ROMP catalysts, 363
electronegativity difference between monomers,
    role in anionic block copolymerization, 244
electrophiles, 285
electrophilic catalysts for GTP, 98, 249
electrophilic monomers, 219

electrophilic radicals, 125, 130
electrophilicity, 328
electrophilicity parameter, 286, 393
electrosteric repulsion, 408
electrosteric stabilization, 409
emulsified monomer droplets, 412, 414
emulsifiers, 410
emulsion polymerization;
    unique characteristics of, 410, 411
    qualitative theory, 413
    quantitative theory, 414
end group, determination of, 110
end-capped oligoimides, 86, 87
end-functionalized polymers, 214, 242, 308
energy barrier to rotation, 18
engineering plastics, 73, 76
enolic character of chain end in PMMA-alkali, 238, 240
enthalpy change of polymer with temperature, 15
enthalpy of polymerization, 326
entropy change in ring-chain equilibrium, 44
entropy of activation in cyclization, 42
entropy of polymerization, 325
enzymatic polymerization, 350
enzyme-activated monomer, 350
EPDM rubbers, 264
epoxy resins, 69
equilibrated interchange, 47, 328
equilibrium between ionic and covalent species in ROP, 339, 356
equilibrium concentrations of cyclics, 328, 343, 344
equilibrium distribution of cyclics, 358
equilibrium monomer concentration, 30, 336
equilibrium polymerization, 325, 326
equilibrium polymerization temperature, 30
erythro and *threo* isomers, 270
ethylene copolymers, 264, 274, 278
ethylene copolymerization by ATRP, 178
ethylene-propylene copolymers, 20, 264, 278
evolution of the transient and persistent radicals in LRP, 152
exchange between active and dormant states, 149. See also individual living polymerizations.
exciplexes, 108
exoentropic aggregation processes, 29
exothermic aggregation, 29
extended chain conformation, 19

external donors in higher generations Z–N catalysts, 262, 270
fatty alcohol, cosurfactant in miniemulsion polymerization, 420
fiber shaped β-TiCl$_3$ lattice, 261
fibers, molecular property of, 20
fifth generation Z–N catalysts, 262
Fineman-Ross plots, 381
first generation Z–N catalysts, 261
first monomer (propylene) insertion into Mt-C bond, stereochemistry of, 270
first order kinetics, as test of livingness, 303
flexible bonds, 19, 73
flexible foams, 68
floor temperature, 30
fluorides initiators in ATRP, 171
forcible termination ("killing") of living ends, 242
forward fragmentation in RAFT polymerization, 190
fourth generation telechelics from living cationic polymerization, 309
free energy of polymerization, 28, 324
frequency of deactivation, role on PDI of dormant polymers, 157, 198, 217, 307
frequency of exchange, 157, 198, 217, 307
Friedel Crafts halides, 289, 290
functional end groups, 242, 291, 308
functional group tolerance:
    of late transition metal catalysts, 276
    of ROMP catalysts, 363
functionality, definition of, 9
functionalized conjugated polymers, synthesis of, 89

gel effect in conventional radical polymerization, 143
gel effect in living radical polymerization, 154
gel fraction, 11
gel permeation chromatography, 33
gel point, 11, 56, 59
gelation, 11, 56, 59
general reactivity, of monomers and radicals, 384
general reactivity order of radicals, 125, 383
glass filled PET, 82
glass transition temperature, 14–17
glycolide, 344

gradient copolymer, 373, 374, 400
graft copolymer, definition and nomenclature, 373, 374
grafting methods, 246
group transfer polymerization, 247–250
Grubbs initiators, 361

Haber-Weiss mechanism, 109
halide exchange effect in ATRP, 171
halogenophilicity of ATRP catalysts, 174
Hamaker constant, 408
Hamaker equation, 408
Hammett *para* σ parameter, 127, 386
Hammett equation, 128, 384, 386
Harkins qualitative theory of emulsion polymerization, 413
HDPE, 22, 264, 274
head-to-head, tail-to-tail, addition, 24
head-to-head, tail-to-tail linkages in polypropylene, 267
head-to-tail addition, 24
heat capacity, variation pattern at transition temperatures, 16
heat of crystallization, favorable effect of, in formaldehyde polymerization, 343
Heck coupling, 89
heterocyclic monomers, polymerizability of, 325
heterotactic triads, 265
high density polyethylene, 22, 264
high performance polymers, introduction to, 73
high pressure polymerization of ethylene, oxygen as initiator in, 133
high unsaturation rubbers, molecular properties of, 19
higher generations Z–N catalysts, 262, 263
highly branched polyethylene, 277, 278
hindered Lewis base, agent for improving PDI in living cationic polymerization, 302
hindrance to rotation, effect of substituents on, 23
HLB index, 411
Hofmann elimination, as side reaction in LAP of electrophilic monomers with quarternary ammonium salts of bases as initiators, 250
homolytic dissociation of C–X bonds, relative dissociation constants, 170
homopolymeric stabilizers, for dispersion polymerization, 421
HUFT theory, 418

hybrid block copolymers, PS-b-polypeptide, 355
hydrogen as chain transfer agent, in coordination polymerization, 273
hydrolytic polymerization of nylon, 6, 352
hydrophile-lipophile balance (HLB), 410
hydroxy telechelics, 242, 309
hydroxyl-ester interchange reaction, in synthesis of polyester and polycarbonate, 76–79
hyperbranched polymers, general scheme for synthesis, 13

ICAR ATRP, 182
ideal copolymerization, 378, 394
imidation, 85
immobile gel, 10
immortal polymerization, 333
implicit penultimate model in copolymerization, 396, 399, 402
indanyl end group, 287, 303
induced decomposition of peroxides, 104
induction period, 129
industrial photocuring, 133, 310
infinite network, 10, 11, 57
inhibition and retardation of polymerization, 129
inhibitors of polymerization, 129, 131
inifers, 119, 309
iniferters, 150
initiation by electron transfer, 219
initiator efficiency, 101, 106, 111, 237, 241, 294, 303, 417
initiator transfer, 118, 309
inner-sphere electron transfer, 132
insertion of monomer into metal-carbon bond, mechanisms of, 266, 271, 276, 277
instantaneous copolymer, 378
instantaneous copolymer composition, 373, 374, 376, 381
integrated copolymer composition, 376, 400
interfacial polycondensation, 51, 75, 79
interionic distance in ion pair, 314
intermolecular attraction, 14, 20
intermolecular interchange, 46, 55, 329, 339, 340, 348
intermolecular polymer transfer, 121, 329
intermolecular transetherification, 332
internal donor, as component in higher generations Z–N catalyst, 262, 270

internal olefins, polymerizability of, 30
interunit linkage, 43, 46, 52, 55
intramolecular chain transfer, 122, 326, 328
intramolecular condensation, 53, 61
intramolecular dimerization of living ends, 238
intramolecular interchange, 43, 46, 328
intramolecular ionization:
    in THF polymerization, 339
    in oxazoline polymerization, 355
intramolecular polymer transfer, 122, 238
intramolecular solvation of ions, 239, 240, 299
intrinsic flexibility, 18
intrinsic viscosity, relation with viscosity molecular weight, 32
inverse emulsion polymerization, 410, 419
inverse microemulsion, 420
invert micelles, 411, 419
ion association beyond ion-pair, 235
ion pair, contact and solvent-separated, 232–235
ion pair collapse, 287, 301, 336
ionic alkoxide initiators, 219, 346
ionic copolymerization, 378, 379, 393
isomerization of catalytic sites in Z–N catalysts, 270
isospecific catalyst, 260, 275
isotacticity index, 265
isotactic polymer, 23, 265, 266
isotactic propagation, 267, 275
isotactic triads, 265
isotactoid polymer, 266

Jacobson–Stockmayer equation, 44, 46, 75, 329

Kelen-Tudos plot, 382
Kevlar, 81
"killing" agents for living polymers, 242, 309, 362, 363
kinetic chain length, 111, 113, 306
kinetic parameter of radical polymerization, 103, 117
kinetic scheme of conventional radical polymerization, 100
kinetics of living radical polymerization with reversible termination, 153
kinetics of equilibrium polymerization, 30, 336
kinetics of conventional radical polymerization, 99
kinetics of polyesterification, 38–41

kinetics of radical copolymerization, 392, 397–402
kinetics of RAFT polymerization, 192, 193
Kumuda coupling, 89

lactamate anions, 351
lactides, 344, 347
lactones, 344
large rings, 42, 326
late transition metal catalysts, 276
late transition state, 99
latex coagulation, 408
lattices of α-, γ-, and δ- $TiCl_3$, 260, 261
liquid crystalline polymers, 81, 82
LDPE, 22, 122
Lewis acids as coinitiators of cationic polymerization, 289, 291, 299, 335, 342, 349, 355, 356, 358
ligated living anionic polymerization of polar monomers, 240
linear low density polyethylene, 22, 122
linear polyesters, 76
lipases as ROP catalysts of cyclic esters, 350
liquid crystalline polyamides, 75
liquid rubber, 72
living anionic polymerization at elevated temperatures, 226
living anionic polymerization, 213–250
living cationic polymerization, 298–308
living character of ROMP, impairment of, 364
living polymerizations:
    characteristics of, 7, 9,148, 213, 215
    living anionic polymerization, 214–250, see table of contents for Chap. 4
    living coordination polymerization, 279
    living cationic polymerization, 298–308, see table of contents for Chap. 6
    living radical polymerization, 148–199, see table of contents for Chap. 3
        nitroxide mediated polymerization (NMP), 158–166
        metalloradical mediated polymerization, 167
        atom transfer radical polymerization (ATRP), 167–185
        reversible-addition fragmentation transfer polymerization (RAFT), 186–198
    living ring-opening polymerization , see below for individual monomers

living ring-opening metathesis
    polymerization, 362–364
  molecular weight evolution in, 155, 216
  molecular weight distribution and
    polydispersity index in, 156, 216
living ring-opening polymerization:
  of cyclic esters, 345, 348
  of cyclosiloxanes, 357, 359
  of ethylene oxide, 331
  of oxazolines, 355
  of propylene oxide, 332
  of THF, 336
  of N-carboxy anhydrides of alpha amino
    acids, 354
living polymers, anionic, long term stability
  of, 218
living radical copolymerization, 400
living Z–N polymerization, 279
LLDPE, 22, 264, 274
local osmotic pressure, relevance in steric
  stabilization of dispersed particles, 409
loci of emulsion polymerization, 413, 414
locus of particle nucleation, 413
long chain branches, origin of, 121
low HLB values, amphiphiles with, 419
low oxophilicity, catalysts with, 276, 363
low unsaturation hydrocarbon elastomers, 20
lyotropic LCPs, 81

M_1-centered triad fractions, 390
M_2-centered triad fractions, 391
macrocycles, formation mechanism in ROP of,
  328, 340, 358
macrocyclic ligands, 345
macroinitiators, 150, 166, 184, 308, 354, 355
macroion-macroester interconversion, 340
macromolecular design by interchange of
  xanthates, 188
macromolecular hypothesis, 1, 2
macromonomeric stabilizer in dispersion
  polymerization, 422
macromonomers, synthesis of, 243, 309, 356
MADIX, 188
magnesium chloride, 262
magnesium ions, 269
main equilibrium in RAFT polymerization, 197
maleimide end-capped oligomer, 87
MAO, 274
Mark-Houwink-Sakurada equation, 32

Markovian statistics, applicability of, to
  copolymerization, 373, 374
Mayo equation, 103, 115, 120, 124, 195, 292
Mayo plots, 116, 195, 293, 294
Mayr equation, 285, 296, 393
mean propagation rate constant in
  copolymerization, 395, 398, 399, 401
measurement of crystallinity, 265
mechanical properties, 17
mechanism of activation of metallocene
  catalysts, 275
mechanism of ATRP, 168
mechanism of Z–N polymerization, 269
mechanism of ROMP, 361
medium rings, 42, 326
melamine formaldehyde resins, 66
melt blending, 47
melt polymerization, 43
melt processable liquid crystalline
  polyesters, 82
melt processing, 47
melt transesterification, 79
melting temperature, 15, 16, 43
mesogens, 81, 82
metal alkoxide, covalent, as catalysts of living
  ROP of cyclic esters, 345, 346
metal alkylidene ended living polymers, 364
metal ions impurities in Z–N catalysts, 274
metal methylidene complex, the true catalyst in
  ADMET polymerization, 364
metal–free living anionic polymerization, 250
metallacycles of titanium, Grubbs initiators of
  relatively low activity, 360
metallacyclobutane intermediates, 361
metallocene catalysts, 274
metathesis of linear olefins, as a side reaction in
  living ROMP, 364
methacrylate-telechelic polymers, 243, 309
methallyl end group in polyisobutylene, 309
methallyltrimethylsilane terminating agent in
  living cationic polymerization, 309
methylaluminoxane, chemical structure of, 274
micellar nucleation in emulsion
  polymerization, 414, 418
micelle exterior, 411
micelles, 411
Michael type addition, 87, 248
microemulsion polymerization, 420
microstructure, 22, 264, 265, 373, 388
miktoarm star polymers, 245

miniemulsion droplets, 420
miniemulsion polymerization, 419
mobility of charge carriers, in conducting
   polymers, 25
mode of insertion of isobutylene into Mt-C
   bond, 266
mode of radical termination, 102
model alkoxyamines, suitability of, as initiators
   in NMP, 158
model halide initiators, suitability of, as
   initiators in ATRP or living cationic
   polymerization, 169, 299
model RAFT agents, suitability of, as
   mediators in RAFT polymerization, 192
modified MAO, 277
moisture-cure formulations of
   polyurethanes, 68
molecular weight, various averages of, 31, 32
molecular weight distribution (MWD):
   in step polymerization, 52
   in conventional radical polymerization,
      145
   in living polymerization,156, 216,
      306, 330
molecular weight evolution with conversion, 8,
   155, 156, 347
molecularly oriented fibers, 81
monofunctional reactant, use of, in producing
   stable step-growth polymers, 49, 50, 71
monomer concentration in particles in emulsion
   polymerization, 415
monomer-depleted condition, deleterious effect
   on MWD in ROP continued under, 329, 331,
   341, 348, 349
monomer distribution in copolymer, 373, 379
monomer droplets, particle nucleation in,
   419, 420
monomer-polymer particles in emulsion
   polymerization, 416
monomer reactivity, in radical polymerization,
   98, 139, 383
monomer reactivity ratios:
   definition of, 375, 376, 397
   determination of, 381
monomer selectivity of radical, 396
monomer sequence distribution in copolymer,
   374, 376, 382, 395
monomer sequence lengths, in copolymer, 388
monomer swollen micelles, 413, 416, 419, 420
monomer transfer, 116, 273, 287, 291, 333

monomer-transfer dominated polymerization,
   117, 118, 296
monomeric alkyllithium, 223
monomeric living ends, in animonic
   polymerization, 241
monometallic active centers, in Z–N
   catalysts, 268
monomethylated metallocene, 275
most probable distribution of molecular
   weights, 55, 146, 217, 330, 347
multiblock copolymers, 69, 78, 185
MWD broadening, causes of, 156, 216,
   307, 330

N-vinylcarbazole, cationic polymerization of,
   285–287, 300
N and s parameters of monomers in cationic
   polymerization, 286
N-acyllactams, activators in anionic
   polymerization of lactams, 351
n-doping, of conjugated polymers, 25
nanocrystalline magnesium chloride, 262
neoprene, 19
net activation energy of living anionic
   polymerization, 235
network polymer, 11
nickel impurity centres, in Z–N catalysts, 274
nitrile rubber, 19
nitroxide radicals, 129, 158
nitroxide mediated polymerization, 158–166
nomenclature of polymers, 2, 373
nonbonded interaction, 270
noncomplex counterions, 337–339
nonconventional step polymerization, 88
nonionic stabilizers in heterophase
   polymerization, 409
nonlinear polymers, 10
nonnucleophilic bases, 302, 328, 353
nonnucleophilic polar solvents, 299
nonstationary state kinetics, 134
nonstoichiometric mixtures of monomers, use
   of, for DP control in step polymerization, 50
noryl, 80
novolac, 63
nucleated monomer droplets in microemulsion
   polymerization, 421
nucleation in the aqueous phase in emulsion
   polymerization, 417
nucleophiles, parameters of, 285
nucleophilic aromatic substitution, 83, 85

nucleophilic catalysts for GTP, 247–249
nucleophilic counteranion, 287, 302
nucleophilic promoters in anionic ROP of cyclosiloxanes, 359
nucleophilic radical, 98
nucleophilic substitution mechanism of ROP, 327
nucleophilicity and electrophilicity scales, 285
nucleophilicity difference of comonomers, role in cationic copolymerization, 393
nucleophilicity of ring oxygen vs. in-chain oxygen, 339
nucleophilicity parameters of vinyl monomers, 286, 393
number average molecular weight, definition of, 31
number average sequence length of monomer sequences in copolymer, 387, 389
number density of particles in emulsion polymerization, 414, 416, 417
number distribution of degree of polymerization, 53, 145
nylon-6, 20, 22, 66, 74, 353

O-alkylation vs. C-alkylation, 64, 84
oil-in-water emulsifiers, 410, 411
oligo (α-methylstyrene) disodium, as difunctional initiator for living anionic polymerization, 222
oligo (butylene terephthalate), 78
onium salts, as cationic photoinitiators, 310
order of monomer reactivity in radical polymerization, 383
organoboron, 89
organocopper, 90
organolithium, 219, 222, 223
organotin, 80, 89
organozinc, 89
osmotic effect factor in steric stabilization of dispersed particles, 409
Ostwald ripening, as destabilization mechanism in miniemulsion polymerization, 420
outer-sphere electron transfer, 132
overall activation energy of polymerization, 142
overlapping of stabilizer layers, prevention of, 409
oxidative and reductive termination of polymer radicals, 131
oxidative polymerization, 80

oxidative termination, 131, 132
oxocarbenium ionic species, 343
oxophilicity of ROMP catalysts, 363
oxygen as an initiator, 133
oxygen as an inhibitor, 133

p-acetoxybenzoic acid, 78
p-dicumyl chloride, as difunctional initiator in living cationic polymerization, 310
p-doping of conjugated polymers, 25
packability into crystal lattice, 18
particle nucleation, 414, 417, 419
particle stabilization mechanisms, 407
Pd based catalysts, 89, 90
PDI equation, 146, 148, 217, 156–158, 306, 307
PEDOT, 27, 28
PEEK, 83, 84
penultimate model of copolymerization, 374, 395, 396, 399
penultimate monomer unit, 238, 374, 396
perchlorate ester end group, 288
perchloric acid initiated polymerization, 288
peripherally solvated ion pair, 238
peroxide and hydroperoxide initiators, 104–106
persistent radical effect, 151, 152, 168
PF resins, 62–65
phenolic resins, 62–65
Phillips catalysts for polyethylene synthesis, 264
phosphor ylide dormant species, 250
photo fragmentation, 107
photochemical after-effect, 135
photocuring, 133, 310
photoinitiated cationic polymerization, 310
photoinitiation efficiency, 108
photoinitiators, 107, 108, 134, 136, 310
photoreducible organic dyes, 108
photosensitizers, 108
physical crosslinks, 69, 144
physical state, 14
plastics, molecular properties of, 17
PLED, 28
PLP-SEC method, 136
PMMA-alkali ion pairs, 238–241
PMR (polymerization of monomer reactants), 86
Poisson distribution, 156
polar additives, effect on living anionic polymerization of nonpolar monomers in hydrocarbon solvents, 225

polar aprotic solvents as medium in synthesis of high performance polymers, 76, 83, 86
polar effect in radical reactions, 98, 99, 124–128, 383–386
polarity parameters of monomers, 126–128, 160, 384–386
poly (1,4-benzamide), 73, 81
poly (butylene terephthalate), 78
poly (trimethylene terephthalate), 78
poly (vinyl alcohol), crystallinity in, 23
poly (vinylidine chloride), cystallinity in, 24
poly (vinylidine fluoride), crystallinity in, 24
poly(α-hydroxyacid)s, 343
poly(ω-hydroxyacid)s, 343
poly(p-phenylene terephthalamide), 81
poly(p-phenylene terephthalate), 82
poly(p-phenylene vinylene), 26, 28, 90
poly(p-phenylene)s, 26, 73, 89
poly(p-hydroxybenzoic acid), 78
poly(alkyl thiophene)s, regioregular, 89
poly(2,6-dimethyl-1,4-phenylene ether), 80
poly(4-hydroxybenzoic acid), 82
poly(alkylene sulfide), 72
poly(arylene ether)s, 83, 84
poly(arylene ethynylene)s, 28, 90
poly(arylene vinylene)s, 28, 89
poly(dichlorophosphazene), 360
poly(dilactones), 343
poly(ether ether ketone ether ketone)(PEEKEK), 84
poly(ether ether ketone)(PEEK), 84
poly(ether ketone ether ketone ketone)(PEKEKK), 84
poly(ether ketone)(PEK), 84
poly(ether sulfone)(PES), 84
poly(ethylene oxide), 331
poly(ethylene terephthalate), 20, 21, 77
poly(ethylenedioxythiophene), 27
poly(hexamethylene adipamide), 3, 55
poly(methyl methacrylate), plastics property, origin of, 20, 23
poly(N,N-diethylacrylamide), 242
poly(oxadiazole)s, 81
poly(p-phenylene sulfide), 83
poly(phenylene oxide), 80
poly(phenyleneethynylene)s, 89, 90
poly(phenylenevinylene)s, 89
poly(phenylquinoxaline), 87
poly(propylene oxide), flexible bonds in, 20
poly(vinyl acetate), head-to-head linkage in, 24

poly(vinyl chloride), 20
polyacetylene, 25, 26, 272
polyacrylonitrile, crystallinity of, 23
polyamic acid, 85
polyamidation, 9, 85
polyamide, 5, 20–22, 43, 46, 74
polyamide, nylon 6 and nylon 66, effect of structural regularity on properties of, 20, 21
polyamide imide, 86
polyanhydrides, 46
polyaniline, 25, 26
polyarylenes, 28, 89
polybenzimidazoles, 73, 81, 87
polybenzothiazoles, 73, 81, 88
polybenzoxazoles, 73, 81, 88
polybutadiene polyols, 68
polycarbonate polyols, 68
polycarbonates, 46, 79
polycondensation, 52, 56
polydiene rubbers, crosslinking of, 11
polydienyllithium, 225, 243
polydimethylsiloxane, 20
polydisperse polymer, 355
polydispersity index (PDI), 32, 56, 156, 157, 178
polyesters, melting points of, 21, 22
polyester diols, 68
polyester LCPs, 82
polyester polyols, 68
polyesterification, kinetics of, 38–41
polyether diols, 68
polyether polyols, 68
polyetherimides, 85
polyethers, flexibility of, 20
polyethylenes, microstructures of, 22
polyfluorenes, 28
polyformaldehyde, 344
polyfunctional monomers, gel forming compositions, 57
polyfunctional polycondensation, 56
polyfunctional polymers, 11
polyfunctional vinyl ethers, used in photocuring, 311
polyheterocyclization, 85, 87
polyimides, 73, 84
polyisobutylene, molecular property of, 20
polymer particles, 414
polymer standards, for GPC calibration, 33
polymer-nitroxide adduct, 158–167, 400
polymeric LEDs, 28

polymeric RAFT agent, 186
polymeric stabilizer, 421
polymeric surfactant, 409
polymerizability of ring monomers, 324
polymerizability of internal olefins, 30
polymerization in isolated loci, 412
polymerization in the solid state, 75, 78, 88
polymerization of small ring monomers, 340
polymerization of undiluted methyl methacrylate, gel effect in, 144
polymers with designed functional end groups, 242, 308
polynorbornene, 330, 360, 363
polyoctenamer, 360, 364
polypeptides, large scale preparation of, 354
polyphenylene, 87
polyphosphazene, 360
polypropylene, stereochemical structures of, 53, 259
polypyrrole, 26
polyquinolines, 87
polysiloxanes, 70, 71
polystyrene, plastics property, origin of, 20
polystyrene-*b*-polyisobutylene-*b*-polystyrene, 308
polystyrene-b-polypeptide, 354
polystyrene-*b*-polybutadiene-*b*-polystyrene (SBS), 244
polystyrene-*b*-polyisoprene-*b*-polystyrene, 244
polystyrenesulfonic acid, 27
polysulfides, elastomeric property, origin of, 20
polysulfone, 84
polytetrahydrofuran, elastomeric property, origin of, 20
polythiophene, 26, 28, 89
polyureas, 67
polyurethanes, 6, 67
polyurethanes/ureas, 67
popcorn polymerization, 144
potential energy of attraction, 408
power law kinetics in LRP, 164
PPS, 83
PPVE, 28
pre-equilibrium in RAFT polymerization, 186
pre-equilibrium regime in LRP, 152, 155, 186, 195
precatalyst, 259, 266, 277
precipitation polymerization, 407, 421
predicted monomer sequence distribution in copolymer, 395

predicted range of gel forming compositions, 57–59, 61
prediction of gel point, 56, 59
primary decomposition of peroxides, 104
primary radicals, 101
principle of equal reactivity of functional groups, 37, 38, 52, 55
probability of ring conformation, importance of, in ring formation, 42
processable precursors to polyacetylene, 364
propagation rate constant, values of, 139, 140, 232, 239, 312, 327, 339, 376
propagation-active enolate, 248
propagation-active monomeric ion pair, 240
propagation-inactive dimeric ion pair, 238, 240
proportion of cyclics in ROP, general discussion of, 329
protective colloids, 409
protogens, 290
proton traps, 301, 302
protonic acid initiators, 286, 302
pseudocationic polymerization, 288
pulsed laser polymerization-size exclusion chromatographic method, 136, 395
purposeful termination, 214, 242, 309

Q, e Scheme, 126, 384
quantifying the resonance and polar effects, 126
quantum yield of photoinitiation, 108
quasi-equilibrium in LRP, 152
quaternary ammonium or phosphonium halide, influence on living cationic polymerization, 303

racemic α-olefin, 268
racemic polymer, 268
radiation induced polymerization, 296
radical addition to monomer, effect of resonance, polar and steric factors on, 98, 99
radical anion as initiator in anionic polymerization, 220
radical reactivity, 98, 99, 139, 383
radical reactivity pattern scheme, 127, 386
radical reactivity ratios, 385, 397
radical stabilization energy, 98, 139
radical transfer reactions, 113–129,
radiochemical tracer technique, application to the determination of polymer end group, 110
RAFT agents, 189
RAFT polymerization, 186–198

random copolyesters as processable LCPs, 82, 373
random copolymers, 373, 379
randomly branched polymers, 12
rare earth metal alkoxides, initiators in the living polymerization of cyclic esters, 348
rate constant of activation in LRP, determination of, 165, 166, 197
rate constants of termination, values of, 140
rate constants of dissociation of model- and polymer-nitroxides, 159
reactable functional groups, relevance to Carothers equation, 49, 50
reaction diffusion, mechanism of radical termination at high conversion by, 141
reaction-controlled propagation, 143
reaction-controlled termination, 393
reactive oligomers, 86
reactive stabilizers, use in dispersion polymerization, 422
reactivity and polarity parameters, values of, 385
real catalysts, *in situ* generation of, in coordination polymerization, 260, 274
real initiator, *in situ* generated, 302, 361
redox initiators, 67, 108, 128, 144, 412
redox transfer, 122
reducing nucleophilicity, of polystyrene anion, need and means of, 243
reduction potential, dependence of activity of ATRP catalysts on, 174
reductive termination of polymer radicals, 132
regio- and stereoregularity, induction in polyisobutylene by vanadium catalysts, 263
regioblock, 266
regiochemical modes of monomer insertion on Mt-C bond, 266
regiochemical regularity, 24, 266
regiochemistry, 267
regiochemistry of addition, of monomer to active center, 24, 99, 266
regioregular poly(alkylthiophene)s, 89
regioregularity, 266, 267
regiospecificity, 266, 267
regulators, 120
relative rate of polymerization, of monomers in CRP and ATRP, 104, 177
relative reactivity of monomers, 383
repeating unit, 3
repulsive interaction between particles, 408

resole, 64
resonance effect, role in radical reaction, 98, 125, 384
resonance-stabilization energies of radicals, 98
retardation of polymerization, 193
retarded anionic polymerization, 226
retarders, 129, 131
retro [2+2] cleavage, 362
reverse ATRP, 179
reversible (degenerative) chain transfer, 148
reversible activation in GTP, LRP, LCP, RMP and ROMP, see respective chapter sections
reversible addition–fragmentation chain transfer, 150, 186
reversible deactivation, 148, 149, 186, 296–298, 400
reversible fragmentation in RAFT polymerization, 150, 187
reversible termination, 148, 149, 186, 296–298, 400
reversible (degenerative) transfer, 148–150
rigid rings in chain skeleton, influence on mechanical property, 73
ring conformation, role in ring formation, 41–44
ring formation, ease of, 42, 43
ring strain, 41, 42, 324
ring vs. chain formation, 41
ring-chain equilibrium, 44, 326
ring-opening metathesis polymerization (ROMP), 360–364
rod shaped polymer, 81
room temperature vulcanizing silicone elastomers, 71
rotating sector method, 133–136
rotational entropy, 324
rubbers, molecular properties of, 18
run number, of monomer sequences in copolymer, 388

SBR, molecular properties of, 19
second generation Z–N catalysts, 262
second order transition temperature, 16
secondary insertion of propylene into V–C bond, 271
secondary onium ion, 336
seeded emulsion polymerization, 416
segmental diffusion in polymer, 41, 42
segmented polyurethanes, 68
self-catalyzed polyesterification, 38

self-condensation of functional groups, 41, 46, 64
self-reinforced thermoplastics, 82
self-termination of chain radicals, 101, 152
semiconductivity, 28
semicrystalline polymers, 14, 20
sequential block copolymerization, 184, 243
short branches, 273
short chain branching in polyethylene, mechanisms of, 122, 273, 278
Shrock initiators, 362
side reactions, 183, 236, 323, 332, 346, 351, 353, 355, 359, 365, 374, 395, 396
silicium ion, 359
silicone rubber, flexibility of chain bonds in, 20
silicones, 46, 70, 71, 357–359
silyl ketene acetal ended polymer, 247
silyl ketene acetal initiator, 247
single electron transfer-ATRP (SET-ATRP), 175
single pulse-pulsed laser polymerization (SP-PLP) method, determination of termination rate constant by, 138
single-site catalysts, 274
size exclusion chromatography, 33
size of counterions, 314, 338
size range of emulsified monomer droplets, 419, 420
size range of polymer particles, 408
slow initiation in living polymerization effect of, on PDI, 157
small ring monomers, polymerizability of, 29
Smith–Ewart theory of emulsion polymerization, 414
soapless emulsion polymerization, 418
sodium naphthalene, 214, 221
soft phase in segmented polyurethane, 69
sol fraction, 11
solid-state post polymerization, 75, 78, 88
soluble Z–N catalyst systems, 263
solvent-separated ion pair, 232–235
source-based nomenclature, 2
SP-PLP method, determination of termination rate constant by, 138
specific volume, application in glass transition temperature measurement, 16
specificity in radical polymerization, 98, 383
spherical micelles, 411
spontaneous chain transfer in carbontionic polymerization, 287, 292, 302

SR and NI ATRP, 180
stabilized step polymer, 49, 53
stable carbenium ion salts as initiators, 300, 330, 336
stable free radical polymerization (SFRP), 149
stable radicals, 111, 129, 149, 151
"standard" or "mutual" recipe, of emulsion polymerization, 412
star polymers, 12, 186, 198, 244, 309
stationary state, 102, 197, 292, 375, 388, 400
stationary state radical polymerization, kinetics of, 99–103, 164, 192
statistical copolymer, 373, 400
statistical treatment of copolymerization, 386
statistical treatment of gelation, Flory's method, 59, 63
stereoblocks, 265, 266, 270
stereochemical regularity, 24, 267
stereochemical structures, 23
stereochemistry of monomer insertion, 267
stereoisomers of polybutadiene, 264
stereoregularity of first monomer insertion, 271
steric and polar factors, role in radical polymerization, 24, 98, 99, 124–128, 382–386
steric stabilization mechanism, of polymer particles, 409, 421
Stille coupling, 89
stoichiometric imbalance, effect of, on DP in step polymerization, 50, 53, 323
stress-crystallizable elastomers, 20
strictly linear polyethylene, synthesis of, 279
structural composites, 86
structural irregularity, 14, 20
structural unit, 3
structure-less MWD, 137
structure-property relationship, 17, 22
structured molecular weight distribution, 137
substituent effect on radical reactivity, 383
substituent effect on radical stabilization energy, 98, 160
supported transition metal oxide catalysts, 264
supported Ziegler catalysts, 262, 264
surface active agents, 407
surface charge on polymer particles, 408, 418
suspension polymerization, 409
suture-less surgery, 219
Suzuki coupling, 89
syndiotactic polymer, 23

syndiotactic blocks in stereoblock polypropylene, 263, 265
syndiotactic placement, 272
syndiotactic polymerization, of propylene, 263, 264, 275
syndiotactic triads, 265
syndiotactoid configuration, 266
syndiotactic diads, 265

tail-to-head linkage, 24, 266
tail-to-tail linkage, head-to-head linkage, 24, 266
telechelic polyisobutylenes, various generations of, 308–310
telechelic polymers, 242, 308, 356
telogens, 128
telomers, 128
telomerization, 128
template for alkali ions, use in directed synthesis of crown ethers, 340
TEMPO mediated polymerization, 158–164
temporary termination, 339, 356
terminal model of copolymerization, 374, 392, 395
terminal unsaturation, 303
termination of polymer radicals, diffusion control in, 141
ternary copolymer, 373
terpolymer, 373
tetrafunctional monomer, stoichiometric requirement for gelation in reaction with bifunctional monomer, 62, 65
tertiary oxonium ion, chain-carrier in cationic polymerization of cyclic ethers, 343
thermal curing, 87
thermal initiators, 104, 106, 115, 145
thermal polymerization of styrene, 112
thermal transitions, 14
thermodynamic feasibility, in ROP, 324–326
thermodynamics of polymerization, 28, 324–326
thermoplastic elastomers, synthesis of, 243, 244, 308, 331
thermoplastic polyimides, synthesis of, 85, 86
thermosets, 66
thermosetting polyimides, 86, 87
thermosetting resins, 62
thermostable adhesives, 86
thermotropic LCPs, 81
thiocarbonylthio compounds, 150, 187

third generation Z–N catalysts, 263
third order initiation, in thermal polymerization of styrene, 112
three periods of emulsion polymerization, 413
$TiCl_3$ macromolecule, 264
timethylsiloxy groups, 71
topologically asymmetric star-block copolymers, 245
trans polyacetylene, 25
trans-1,4-polybutadiene, molecular property of, 19
trans-1,4-polychloroprene, molecular property of, 19
trans-1,4-polyisoprene, molecular property of, 41, 19
transamidation, 351
transfer agents, 114, 272
transient radical, 152
transition metal alkylidene, 362
transition metal coupling polymerization, 28, 88–90
transition probabilities in copolymerization, 387, 391
transition state, early or late, 99
transition temperatures, 16
translational diffusion, 40, 141
translational diffusion rate, 41
translational diffusion-control, in polymer radical termination, 41, 140, 143, 144
translational entropy, 324
tri- chloro telechelics, 310
triad fractions, 265, 389
triarylsulfonium salts, as initiator in cationic photopolymerization, 310
tricumyl chloride, as trifunctional initiator in living cationic polymerization of isobutylene, 310
trifunctional monomer, 10, 63
triple ion formation in oligostyrene dicesium in THF, 235, 236
triplet excited state, 107
trithiocarbonates, as RAFT agents, 187
trityl salts as cationic polymerization initiators, 300, 335
true catalysts:
    from substituted titanacyclobutane, 360, 361
    in ADMET polymerization, 364
tuning the initiators, in ROMP, 362
tuning the RAFT agent, 188

two-ended living polymers:
    initiators for the synthesis of, 219–222
    intramolecular association of living ends, 235, 238

unimer halide, as initiator in ATRP or living cationic polymerization, 169, 299
unimolecular chain termination, effect on MWD, 146
universal mediator for NMP, 164
unreactive aggregated species, 238, 358
unsaturated polyesters, 66
urea-formaldehyde (UF) resins, 65
urethane linkage, 67

validity of the terminal model, 395
van der Waals forces, 407
variation of k_t with conversion, in radical polymerization, 141
vibrational entropy, 324
vinyl acetate, polarity of, 385
vinyl esters, 187
vinyl ethers, living cationic polymerization of, 300–302
vinyl monomers, functionality of, 10
vinyl polymers, introduction to, 5, 22
vinyl pyridines, living anionic polymerization of, 242

visible light curing, of dental and orthopedic filling materials, 108
volume expansion coefficient, application of, in determination of transition temperatures, 16
volume restriction, role in steric stabilization, 409
vulcanization, 12, 20

water soluble thermal initiator of radical polymerization, 106, 412
water-borne living radical polymerization, 422
water-in-oil emulsifer, 411
weight average degree of polymerization, for most probable distribution, 56, 148
weight distribution, 53
weight fraction of rings, in ring-chain equilibrated systems, 45

xanthates as RAFT agents, 187–189
xylene index, as measure of polypropylene crystallinity, 265

Ziegler–Natta catalysts, various generations of, 259, 263
zinc diethyl, as transfer agent in Z–N polymerization, 273
zirconocene, 274
Z- and Z+1-average molecular weights, 32